# Deregulation
## and
## Environmental
## Quality

*New Titles from*
*QUORUM BOOKS*

The Politics of Taxation
THOMAS J. REESE

Modern Products Liability Law
RICHARD A. EPSTEIN

U.S. Multinationals and Worker Participation in Management:
   The American Experience in the European Community
TON DEVOS

Our Stagflation Malaise: Ending Inflation and Unemployment
SIDNEY WEINTRAUB

Employee Stock Ownership and Related Plans: Analysis and Practice
TIMOTHY C. JOCHIM

Psychology and Professional Practice:
   The Interface of Psychology and the Law
FRANCIS R. J. FIELDS AND RUDY J. HORWITZ, EDITORS

American Court Management: Theories and Practices
DAVID J. SAARI

Budget Reform for Government:
   A Comprehensive Allocation and Management System (CAMS)
MICHAEL BABUNAKIS

Railroads: The Free Enterprise Alternative
DANIEL L. OVERBEY

Managing Terrorism: Strategies for the Corporate Executive
PATRICK J. MONTANA AND GEORGE S. ROUKIS, EDITORS

The Export-Import Bank at Work: Promotional Financing in the Public Sector
JORDAN JAY HILLMAN

Supply-Side Economics in the 1980s: Conference Proceedings
FEDERAL RESERVE BANK OF ATLANTA AND EMORY UNIVERSITY
LAW AND ECONOMICS CENTER, SPONSORS

Craig E. Reese

# Deregulation and Environmental Quality

## The Use of Tax Policy to Control Pollution in North America and Western Europe

Q

QUORUM BOOKS
WESTPORT, CONNECTICUT • LONDON, ENGLAND

**Library of Congress Cataloging in Publication Data**

Reese, Craig E.
  Deregulation and environmental quality.

  Bibliography: p.
  Includes index.
  1. Environmental policy.     2. Fiscal policy.
3. Taxation.     I. Title.
HC79.E5R433     1983     363.7'36     82-11266
ISBN 0-89930-018-9     (lib. bdg.)

Library of Congress Catalog Card Number: 82-11266
ISBN: 0-89930-018-9

First published in 1983 by Quorum Books

Greenwood Press
A division of Congressional Information Service, Inc.
88 Post Road West, Westport, Connecticut 06881

Printed in the United States of America

10 9 8 7 6 5 4 3 2 1

# CONTENTS

Figures                                                                ix
Tables                                                                 xi
Preface                                                               xiii
Acknowledgments                                                        xv
Abbreviations                                                        xvii

1.  Introduction                                                        3
2.  Conceptual Framework for Analyzing Environmental
    Policy                                                             13
3.  Canada                                                             49
4.  France                                                             87
5.  West Germany                                                      139
6.  Sweden                                                            193
7.  The United Kingdom                                                235
8.  The United States                                                 281
9.  U.S. Taxation of Pollution Control Investments                    329

10.   Environmental Excise Taxes on U.S. Crude Oil,
      Chemicals, and Hazardous Wastes                              373

11.   A Comparative Analysis of Pollution Control Policy           385

12.   Summary and Conclusions                                      409

## APPENDICES

   I.   Waste Water Charges Act                                    422

  II.   Section 169 Regulations                                    430

 III.   EPA Regulations                                            444

  IV.   EPA Guidelines                                             449

   V.   EPA Regional Offices                                       455

  VI.   State Water Quality Agencies                               457

 VII.   State Air Quality Agencies                                 462

VIII.   State Solid Waste Agencies                                 467

  IX.   Environmental Excise Taxes                                 472


        Selected Bibliography                                      479
        Index                                                      485

# FIGURES

2-1  The Flow of Materials in an Economy                              28

2-2  Optimal Level of Pollution:
     Macroeconomic Perspective                                        36

2-3  Optimal Level of Pollution:
     Microeconomic Perspective                                        37

6-1  Environmental Administration in Sweden                          207

# TABLES

2-1  Estimated Emissions of Air Pollutants
     by Weight Nationwide, 1971                                    16

6-1  1969 Swedish Public Opinion Poll
     on Environmental Issues                                      195

6-2  Government Subsidies for Emission Control Measures
     Fiscal Years 1968–1969 to 1975–1976                          218

7-1  1980 Surface Water Quality                                   266

8-1  State and Local Tax Incentives of Air and
     Water Pollution Control Investments                          296

9-1  Three-Year Cost Recovery Percentages                         330

9-2  Five-Year Cost Recovery Percentages                          331

9-3  Ten-Year Cost Recovery Percentages                           331

9-4  Fifteen-Year Public Utility Cost Recovery Percentages        332

9-5  Fifteen-Year Real Property Cost Recovery Percentages         333

9-6  Optional Straight-line Recovery Periods                      333

9-7  Section 179 Current Expense Election                          334

9-8  ITC Recapture Percentage                                      342

10-1  Tax on Petrochemicals                                        375

10-2  Tax on Inorganic Chemicals and Heavy Metals                  376

11-1  Comparison of Vital Statistics, 1977–1978                    386

11-2  Comparison of Fiscal Systems                                 390

11-3  Comparison of Environmental Planning Systems                 393

11-4  Comparison of Water Pollution Control Policy                 398

11-5  Comparison of Air Pollution Control Policy                   401

11-6  Comparison of Solid and Hazardous Waste
      Control Policy                                               404

12-1  Comparison of the Use of Tax Incentives for Investment
      in Pollution Control                                         411

12-2  Comparison of Effluent Charge/Tax Systems                    413

# PREFACE

Although a significant portion of the text was completed by early 1982, it has been updated through June 1982. Most of the updated material, describing the environmental policy for each of six countries, appears at the end of Chapters 3 to 8. The monetary references in Chapters 3 to 7 have been translated to U.S. dollars at the New York exchange rates for June 30, 1982. Notwithstanding the continual evolution of environmental policy in the six countries, public policymakers in each country rely almost exclusively on regulation to control pollution. Therefore, much of the updated text reflects further development of pollution control regulation.

Let me state my preference for the use of "pollution taxes" to complement minimum regulatory standards wherever practicable as a means of water resource management. The French and German experience with effluent charges demonstrates that effluent charges can be used to effect compliance and to finance necessary wastewater treatment facilities. In fact, my conviction is the impetus for a current research project to design an effluent charge system for Texas, which has a water problem of growing seriousness.

# ACKNOWLEDGMENTS

Since 1971, when the idea for this study germinated, many individuals have supported my efforts. I would like to thank Ray M. Sommerfeld (University of Texas) for his early encouragement and guidance in the development of my idea and Michael L. Moore (University of Southern California) for taking Ray's place as committee chairman and persevering until the end. I am also indebted to Jack Kramer (University of Florida) for his assistance with the methodology and objectives of my study, to Jared Hazleton (Texas Research League) who guided the final reorganization and "cleaning up" of the manuscript, and to Pat Blair (University of Texas) for his helpful comments on the next-to-final draft. Needless to say, after ten years of development and five relocations, many typists contributed their efforts, including Linda Stewart, Valerie Sinclair, Patti Blystone, Marti Walsh, Jill Hebda, Phyllis Odom, Catherine Fang, Lori Wrotenbery, Marcia Woosley, Barbara Phillips, Mollie Guion, and Gail Giebink. I owe Howell Lynch, a doctoral student at the University of Texas, a special word of thanks for his research assistance. I am also indebted to Kathryn Stevens, Lori Cheek, Teresa Byrd, John Bernhagen, and—last, but not least of all—Lynn Taylor and the Quorum Books staff for editorial assistance. Finally, I dedicate this volume to my son Dave and my mother who both inspired me to complete this project.

# ABBREVIATIONS

ACRS        Accelerated Cost Recovery System
BOD         Biological Oxygen Demand
CMHC        Central Mortgage and Housing Corporation
CFC         Chlorofluorcarbous
CLADR       Class Life—Asset Depreciation Range
CAA         Clean Air Act
CERCLA      Comprehensive Environmental Response Compensation
            and Liability Act
CPSWR       Control of Pollution Special Waste Regulations
CEEQ        Council of Experts for Environmental Questions
CEQ         Council on Environmental Quality
DATAR       Delegation à l'Amonagement du Territoire et à l' Action
            Regionale
DOE         Department of Environment
DREE        Department of Regional Economic Expansion
ECE         Economic Commission of Europe (United Nations)
ERTA        Economic Recovery Tax Act
EARP        Environmental Assessment and Review Process
EAC         Environmental Assimilation Capacity
ECA         Environmental Contaminants Act

| | |
|---|---|
| EIS | Environmental Impact Statement |
| EIS | Environmental Information System |
| EPA | Environmental Protection Act (Sweden) |
| EPA | Environmental Protection Agency |
| EPO | Environmental Protection Ordinance |
| EPS | Environmental Protection Service |
| EEC | European Economic Community |
| ERP | European Recovery Program |
| FCL | Federal Chemicals Law |
| ECA | Federal Emissions Control Act |
| FEM | Federal Environment Ministry |
| BUA | Federal Environmental Office |
| BGA | Federal Health Office |
| BAU | Federal Office for Occupational and Safety Policy |
| WDA | Federal Waste Disposal Act |
| WWCA | Federal Waste Water Charges Act |
| FWPCA | Federal Water Pollution Control Act |
| WRMA | Federal Water Resources Management Act |
| FIANE | Fund for Intervention and Action for Nature and the Environment |
| GAO | General Accounting Office |
| GLC | Greater London Council |
| HPA | Hazardous Products Act |
| HSC | Health and Safety Commission |
| IDB | Industrial Development Revenue Bonds |
| IEE | Initial Environmental Evaluation |
| IMCO | Intergovernmental Maritime Consultation Organization |
| CIANE | International Action Committee for Nature and the Environment |
| IJC | International Joint Commission |
| ITC | Investment Tax Credit |
| MARPOL | Marine Pollution Convention |
| MAP | Mediterranean Action Plan/Program |
| MSA | Municipal Sanitation Act |
| NAREW | National Agency for the Recycling and Elimination of Waste |

| | |
|---|---|
| NBUP | National Board of Urban Planning |
| NEPB | National Environment Protection Board |
| NEPA | National Environmental Policy Act (United States) |
| NEPA | National Environmental Protection Act |
| NFB | National Franchise Board |
| NPDES | National Pollutent Discharge Elimination System |
| NWC | National Water Council |
| OSHA | Occupational Safety and Health Administration |
| OMB | Office of Management and Budget |
| OSM | Office of Surface Mining |
| OEM | Ontario Environmental Ministry |
| OECD | Organization of Economic Cooperation and Development |
| OPEC | Organization of Petroleum Exporting Countries |
| PUK | Pechiney—Ugine—Kuhlmann |
| PBB | Ploybrominated Biphenyls |
| PSI | Pollutants Standards Index |
| PCA | Pollution Control Act |
| PCB | Polychlorinated Biphenyls |
| PVC | Polyvinylchloride |
| PSD | Prevention of Significant Deterioration |
| PHC | Public Health Committee |
| RCRA | Resource Conservation and Recovery Act |
| RCEP | Royal Commission on Environmental Pollution |
| SDD | Scottish Development Department |
| SCL | Societe Crusot-Loire |
| SMCRA | Surface Mining Control and Reclamation Act |
| SS | Suspended Solids |
| TRA | Tax Reform Act |
| TOX | Tijas Oil and Chemicals |
| TSCA | Toxic Substances Control Act |
| VAT | Value Added Tax |
| WDMRA | Waste Disposal and Materials Recovery Act |
| WMA | Waste Management Act |
| WAA | Wrecked Automobile Act |

# Deregulation and Environmental Quality

# 1
# Introduction

Environmental deterioration is not a recent phenomenon, even though the "environmental consciousness" of the past decade might lead the uninitiated to this conclusion. For example, Barry Commoner, who has been called the Paul Revere of ecology for his early efforts to alert people to the Environmental Crisis, has argued for years that most environmental problems first came to light, or became much worse, in the years following World War II.[1] In fact, the public's awareness of environmental problems did not begin until the late 1960s. This awareness probably crested shortly after Earth Week (April 1970), although it has remained a significant and enduring social concern.[2]

Many environmentalists have been concerned that "environmental consciousness" is on the wane because we have not even turned the corner in dealing with the problem of environmental deterioration. Only the naive, however, would have expected this "consciousness" to develop into an "environmental ethic" that would lead man to live in harmony with nature. An improvement in environmental quality in a mixed, capitalist economy can come only through public education and thoughtful government controls.

Public concern for the environment throughout the world probably peaked shortly after the end of the United Nations Conference on the Human Environment held in Stockholm, Sweden, in mid-June 1972. By the fall of 1973, worldwide public concern for the Environmental Crisis had been replaced by concern for the Energy Crisis. Even though temporary energy shortages had been experienced earlier, the Arab oil embargo between October 1973 and March 1974 hastened the inevitable. The public was forced to accept an end to the era of relatively inexpensive, abundant energy.[3]

The economies of the developed countries, which had been suffering from stagflation—stagnant economies with serious inflation—since the late 1960s, were temporarily paralyzed by the Arab oil embargo and by price-supply controls established by the Organization of Petroleum Exporting Countries (OPEC). Thus, the Energy Crisis triggered another crisis as all the developed countries fell into an economic slump. The worldwide recession was the primary focus of the public, even though most developed countries seemed to come out of that recession within two years. During the second half of the 1970s, most developed countries were experiencing some economic growth in spite of OPEC's continual threat to raise its prices.

At first, both stagflation and energy shortages were blamed on the missionary zeal with which legislative bodies established government programs for the regulation of water, air, and solid (and hazardous) waste pollution. The Arab oil embargo exonerated pollution control regulation as the primary cause of energy shortages. Moreover, most studies of the macroeconomic costs of pollution control, though suspect because of data limitations, have concluded that the cost of pollution control programs has not been unmanageable, nor will it be in the future.[4] Even though the costs of pollution control are material, the implications of these costs for economic growth, capital investment, unemployment, prices, and the relative supply of public and private goods have been, and will continue to be, limited.[5]

Inasmuch as neither economic problems, such as recession and inflation, nor the Energy Crisis can be blamed primarily on government regulatory programs, one would surmise that regulators, legislators, or the courts have made very few concessions in the implementation of environmental policies and standards. However, political pressure from industry, labor, and State (or provincial) and local government is mounting in an effort to delay the implementation of pollution standards or even to lower those standards because of their allegedly negative implications for economic growth, price stability, capital investment, unemployment, taxes (State or provincial and local), and the energy shortage.[6]

In theory, political economists have traditionally argued that public policymakers can control social behavior through three approaches: regulation, subsidization, and taxation (tax on pollution). None of the three is theoretically more efficient than the other two. However, political economists would classify most Federal legislation to control pollution as regulation, for example, the Federal Clean Air Act of 1963 and the Federal Water Pollution Control Act of 1956. Some Federal legislation to control pollution would be classified as subsidization, for example, Federal grants for the construction of municipal sewage treatment facilities, the rapid amortization of pollution control facilities, the investment tax credit, and industrial development bonds. Finally, there has been little, if any, reliance on taxation as a means of pollution control in the United States. The only examples are found at the State level, for example, beverage container taxes in Oregon, Vermont, Maine, and Michigan.[7]

Since regulation has been the primary method of implementing pollution control policy in the United States, one may ask how effective Federal pollution control legislation has been in improving environmental quality. In other words, what progress has been made in reducing the pollution of air, water, and land during the decade since the National Environmental Protection Act (NEPA) of 1969 created the Council on Environmental Quality (CEQ)? In December 1978, the ninth annual report of the CEQ suggested that, while significant improvements in air and water quality had occurred in several specific places and in respect to a few pollutants, there had been limited overall improvement in environmental quality, especially in view of the continued increase in the generation of solid, hazardous, and toxic wastes.[8]

As suggested above, political pressures from several interest groups— industrial, commercial, labor, and non-Federal government—support the antiregulation, anti-inflation movement with proposals to relax or delay pollution control standards. Even within the Federal government, the inflation- fighters seem to have singled out environmental regulation as a major cause of price instability inasmuch as the costs are alleged to be greater than the benefits.

The Federal government has decided to rely primarily on the regulatory process with limited subsidization as the means for controlling pollution. Nonetheless, some continue to propose the use of other means of social control, notably, subsidization and/or taxation, as complements or alterna- tives to the regulatory process.[9] The proponents of alternatives to regula- tion argue for taxation and subsidization because each method is alleged to be more efficient than regulation in practice. Based on economic theory, the superiority of these alternative methods of social control can be demon- strated only where suboptimal implementation (for example, uniform stan- dards that do not reflect each entity's marginal cost of pollution control are imposed) is assumed. Unfortunately, no empirical studies have been done demonstrating the superiority of any one of the three methods.

In the last few years, the public's dissatisfaction with Big Government and high taxes has precipitated a tax revolt in California (Proposition 13) and in other States and cities. Paradoxically, the public's antitax mood has not been reflected in its attitude toward the use of taxation to control pollution. In early 1978, a national survey of attitudes toward using taxes to encourage more "socially responsible" behavior by firms reported that 82 percent of those persons over eighteen who were surveyed favored a Federal tax system wherein companies that do not comply with some preconceived effluent or emission standards should pay higher taxes than those firms meeting standards for water and air quality. In summary, the public supports the use of tax policy to control pollution.[10]

## Objectives of the Study

Among political economists, the consensus is that taxation is the most cost-effective method for pollution control. Although the traditional objective of tax policy has been simply one of raising revenue to finance public goods without dissuading the private sector activity that produces tax revenue, the focus of tax policy today is on its potential for complementing all public policy objectives. Concern for neutrality has been replaced by a preoccupation with social and economic objectives. Thus, the tax system is often used to enhance regulation through tax policy provisions that lower tax liabilities to reward certain types of behavior. However, tax policy includes both taxation and indirect subsidies, such as exemptions, exclusions, accelerated deductions, and credits. Therefore, this study integrates tax expenditures or indirect subsidization and taxation of pollution, although it focuses on pollution taxation.

The objectives of this study are to (1) determine the state-of-art in using all three methods of social control to effect environmental quality in six major industralized countries in North America and Western Europe —the United States, Canada, France, the Federal Republic of Germany, Sweden, and the United Kingdom; (2) develop a comparative analysis of the use of tax policy (tax expenditures and taxation) as a complement to regulation and direct financial assistance in the control of pollution in the United States and the other five major industrialized nations; and (3) provide a foundation for further investigation of the political and economic problems inherent in operationalizing taxation as a complement to regulation and subsidization. These objectives are correlated with the following research questions: (1) To what extent has tax policy, especially pollution taxation, been implemented in any of the six countries being investigated? (2) Through comparative analysis, can any conclusions about the optimal mix of pollution control policy be drawn? (3) How can the United States benefit from the experiences of the other five countries included in this study?

## Tax Policy Defined

Before beginning the body of this study, the term "tax policy" should be defined so that there is no doubt about what the research objectives of this study are. Academics, professionals, politicians, bureaucrats, businessmen, investors, and a few politically sophisticated laymen continually use the term when they comment on public policy issues, especially economic policy issues such as inflation, recession, the Energy Crisis, the Environmental Crisis, and tax reform.

Explaining tax policy is not a simple task, for the term may have a different meaning for the particular professional using it. When used by an economist, political scientist, or sociologist, tax policy usually means changing the tax system to correct inequities or using the tax system as a vehicle for social control. When used by an attorney, businessman, investor, or accountant, the term usually means the effect the tax system has on the decisions of individuals and corporations. However, professionals from both camps are aware that tax policy has more than one meaning; one must, therefore, always be aware of the context in which it is being used.

Keeping in mind that there may be more than one meaning associated with any word or term, one can safely assume that the term "tax policy" is primarily thought of as an integral part of fiscal policy, which is an integral part of economic policy. Thus, an important element of tax policy is the non-taxability of selected transactions in an attempt to maintain the "neutrality" of the Federal income tax system, inasmuch as normal, non-income-producing commercial transactions are exempt.

The term is used to a lesser degree in a microeconomic context. For example, a corporation's tax policy may refer to the elections a corporation's management has made in complying with the Federal tax law. Alternatively, a corporation's tax policy may refer to the lobbying position management has taken on specific tax reform issues and proposals to change the tax law. Finally, tax policy may be synonymous with tax planning, that is, the minimization of an individual's, a firm's, or a fiduciary's taxes by carefully planning his or its economic affairs.

For purposes of this study, tax policy may be defined as

> the social and economic effects inherent both in financing the bulk of government activities through a system of taxes and in the continual amending of both the legal provisions and the administration of the tax system to complement the social and economic objectives of public policy, where "public policy" refers to the substantive programs of a national government, i.e., what is being done as opposed to how or why it is being done.[11]

In defining tax policy, both its static and dynamic dimensions are included. With respect to any tax—income, transaction, or wealth—presently being levied at any level of government, tax policy is implicit (a static dimension) in the legal provisions and administration of that tax. Moreover, when the legal provisions and/or administration of that tax are amended, the resulting change is an explicit expression (a dynamic dimension) of tax policy. Tax policy to control pollution includes both the imposition of taxes on the pollution of air, water, and land (solid waste) and the subsidy implicit in special provisions of the income, property, sales, and value added tax laws that lower the tax liability of those firms and consumers that reduce their pollution. Therefore, the concept overlaps the political economist's concepts of taxation and subsidization to the extent that the subsidization concept includes indirect subsidies or tax expenditures.[12]

## Organization of the Study

After this introductory chapter, a conceptual framework for analyzing environmental policy is developed in Chapter 2. The body of the investigation commences in Chapter 3 with a description of Canada's pollution problems, its pollution abatement programs for water, air, and solid waste, and its public assistance programs for polluters. Chapters 4 through 8 continue the same descriptive analysis for France, West Germany, Sweden, the United Kingdom, and the United States, respectively. Chapter 9 is a detailed analysis of capital cost recovery for pollution control investments in the United States. Chapter 10 is an explanation of the new U.S. environmental excise tax. Chapter 11 is a comparative analysis of the pollution control policy of the six countries as described in Chapters 3 through 8. Finally, Chapter 12 includes a summary of the results of the study along with conclusions as to its implications for the use of tax policy to control pollution in the United States.

## Study Methodology

Comparative studies are not commonly undertaken in the United States in academe, government, or industry. In recent years, however, comparative analysis has become a more popular research methodology, especially in the social sciences. For many years, the development of empirical research which utilized the comparative approach has been hindered by American feelings of superiority. Having experienced several public policy failures in recent years, notably, the War on Poverty, inflation, drug abuse, and the Vietnam War, American society seems to be willing to seriously consider what has worked in other developed countries. Awareness of what has worked should not impede the development of innovative public policy, and at the same time may reduce the probability of implementing public programs that have failed in a comparable environment.

In public affairs, the comparative approach usually deals with the identification, classification, measurement, and interpretation of similarities and differences between nation states. Comparative studies result from an awareness that many phenomena are both conceptually universal and operationally unique.[13] Therefore, in utilizing the comparative approach as a methodology, the researcher should differentiate between "the universal, the related, and the unique."[14] In other words, the researcher should distinguish between what is found everywhere, what falls into clusters of similarities, and what is the product of singular historical and environmental circumstances.[15]

In order to differentiate between the universal, the related, and the unique, one needs to know what to compare and how to compare. Boddewyn summarizes the problems inherent in differentiation as follows:

(1) We have some notion of what to compare . . . we need conceptual frameworks or constructs;
(2) We define these elements in a manner lending itself to empirical observation, measurement, and the subsequent ranking . . . we need operational variables; and
(3) We conceive of variations and combinations of the characteristics chosen to identify these systems, that is, we need classifications or typologies.[16]

Conceptual frameworks delimit what should be compared between systems, which enables the researcher to pursue an empirical inquiry comparing nations. For example, which attributes of each nation state—environmental, political, legal, social, economic, or fiscal—should be considered in a comparative analysis of the use of tax policy to control pollution? Note that a framework is not a theory because its function is not to explain but rather to specify what should be observed.[17]

Conceptual frameworks and theories are interrelated. This interrelationship has been summarized as follows:

Constructs and theories are thus in constant interchange, returning to one another for confirmation and revision. We ultimately need conceptual frameworks supplemented by hypotheses expressed in testable form and a series of verifying studies designed to test these hypotheses. Meanwhile, a construct is good to the extent that it catches all relevant systems in its net.[18]

Once one knows what to compare, one must deal with the problem of how to compare. It is hoped that empirical observations and measurement will enable the researcher to succeed in more than just determining and classifying similarities and differences. Ultimately, the researcher should be able to demonstrate the invariable agreement or disagreement between the presence, absence, or change of a phenomenon and the circumstances surrounding its appearance, disappearance, or change.[19] The comparative approach results in an explicit contrast of the similarities and differences of several comparable systems either within the body of the study or in the conclusions of the study. However, the contrasting should not be with the system of a particular nation state which is viewed as a norm or ideal reference point because such studies are more normative than comparative.[20]

## Study Limitations

The comparative approach is just one of a broad range of research methodologies available to the social scientist. This study goes beyond observation, description, and classification only to contrast the similarities and differences inherent in each nation's implementation or use of regulation, direct subsidization, indirect subsidization, and taxation to control pollution. No attempt is made to demonstrate that one nation is making better use of tax policy (indirect subsidization and taxation) than the others in solving pollution problems. Moreover, neither rigorous hypothesis testing nor complex model building is a possible outcome of this preliminary study.

As pointed out earlier, proponents of subsidies and/or taxation argue for these alternatives to regulation because of their alleged superior economic efficiency in practice. The superiority of one (or a specific combination) of these three methods of pollution control could be demonstrated empirically through the use of cost-benefit or cost-effectiveness analysis. Then the most efficient method or combination of methods would be determined through a comparative analysis of the measured economic efficiency of each nation's environmental policy. Such an empirical, macroeconomic study is a logical followup. However, the following study will only be a descriptive, comparative analysis of the use of regulation, direct subsidization, and tax policy to control pollution in the six countries.

Inasmuch as economic and environmental quality data for the past decade are either unavailable or incomplete, an empirical study demonstrating the superior economic efficiency of a specific method (or combination thereof) is not feasible. Ironically, by the time the data do become available, the results of any empirical studies will probably be moot. It is hoped that this preliminary study will lay the groundwork for future empirical studies of the pollution control policy of each industralized country included herein.

## Notes

1. Barry Commoner, *The Closing Circle* (New York: Alfred A. Knopf, 1971), pp. 140-77; and "The Paul Revere of Ecology," *Time*, February 2, 1970, p. 58.

2. Robert C. Mitchell, "The Public Speaks Again: A New Environmental Survey," *Resources*, No. 60 (September-November 1978), pp. 1-6.

3. U.S. Council on Environmental Quality (CEQ), *Environmental Quality: The Fifth Annual Report* (Washington, D.C.: U.S. Government Printing Office, 1974), pp. 94-98.

4. OECD Economic Policy Committee, *Economic Implication of Pollution Control* (Paris: Organization for Economic Co-operation and Development, 1974), pp. 9-11.

5. U.S. Congress, Joint Economic Committee, *Achieving Price Stability Through Economic Growth* (Washington, D.C.: U.S. Government Printing Office, 1974), pp. 36-42.

6. R. J. van Schaik, "The Impact of the Economic Situation on Environmental Policies," *OECD Observer*, No. 79 (January-February 1976), p. 25.

7. Frederick R. Anderson, et al., *Environmental Improvement Through Economic Incentives* (Baltimore: Johns Hopkins University for Resources for the Future, 1979), pp. 39-89.

8. CEQ, *Environmental Quality—the Ninth Annual Report* (Washington, D.C.: U.S. Government Printing Office, December 1978), pp. 1-3, 90-91, 159-60, 172, and 178-79.

9. Allen V. Kneese and Charles L. Schultze, *Pollution, Prices and Public Policy* (Washington, D.C.: Brookings Institute, 1975).

10. "Perspectives on Current Developments: Should Polluters Pay?" *Regulation*, Vol. 2, No. 4 (July-August 1978), p. 14.

11. Lewis A. Froman, Jr., "Public Policy," in *International Encyclopedia of Social Sciences* (New York: Crowell Collier & Macmillan, Inc., 1968), p. 14.

12. The concept of tax expenditures—the indirect subsidies implicit in the tax incentive provisions of a tax law—is explained in greater detail in Chapter 2.

13. S. F. Nadel, *The Foundations of Social Anthropology* (London: Cohen & West, Ltd., 1951), p. 229.

14. Clark Kerr, et al., *Industrialization and Industrial Man: The Problems of Labor and Management in Economic Growth* (New York: Oxford University Press, 1964), p. 10.

15. Jean Boddewyn, "The Comparative Approach to the Study of Business Administration," *Academy of Management Journal*, December 1965, pp. 263-66.

16. Jean Boddewyn, *Comparative Management and Marketing* (Glenview, Ill.: Scott, Foresman, & Co., 1969), p. 4.

17. F. J. Roethlisberger, "Contributions of Behavioral Sciences to a General Theory of Management," in Harold Koontz (ed.), *Toward a Unified Theory of Management* (New York: McGraw-Hill, Inc., 1964), p. 48.

18. Boddewyn, *Comparative Management and Marketing*, pp. 4-5.

19. Nadel, *The Foundations of Social Anthropology*, p. 323.

20. Boddewyn, "Comparative Approach to the Study of Business," pp. 265-66.

# 2

# Conceptual Framework for Analyzing Environmental Policy

As long as man could treat the earth as a purely open system, the "cowboy economy" flourished throughout the Western world. The term "cowboy" is symbolic of the reckless, exploitative, and romantic life of the independent frontiersman on the illimitable plains of the American West. In a cowboy economy, consumption and production are paramount; therefore, a successful cowboy economy maximizes the throughput of labor, capital, and natural resources. If there are infinite mines from which to extract raw materials and infinite "negative" mines into which to deposit pollutants, then throughput, as quantified in the gross national product, is an accurate measure of the success of an economy.

During the last decade, Western man has come to realize that we live on a spaceship, that is, in a closed system. Thus, our economic system is a "spaceman economy" without unlimited mines for either extraction or pollution. The environmental policy of Federal, State, and local governments is a manifestation of society's concern for the spaceship earth.[1]

Environmental problems have actually been with us for many years; however, because the air, water, and land can absorb large amounts of pollution for years, the environmental quality issue has, until the last decade, been viewed primarily as a local—intra-urban regional—problem. National and local parks, as well as the conservation of natural resources, were the environmental issues of the conservation movement at the turn of the century. During the past four decades, both population and economic growth have aggravated problems of environmental quality and raw materials supply and have thereby provided the impetus for the modern-day environmental movement.

Early legislation to control pollution was regulatory in nature. More

13

recent environmental legislation has generally expanded on this regulatory tradition. An understanding of pollution as both a technical and physical problem is a prerequisite to the economic analysis of pollution control policy. Moreover, an appreciation of political realities must precede an evaluation of the effectiveness of pollution control. Because of space limitations, the following explanation of the technical, physical, and political dimensions of environmental pollution will necessarily be cursory. The primary objective of this chapter is to develop an analytical framework based on macro- and microeconomic theory with which one can evaluate pollution control policy, especially the use of tax policy to control pollution. Finally, there is a commentary on the political economy of alternative strategies for implementing pollution control policy wherein the politics of environmental quality is implicitly analyzed.

## Environmental Pollution: Technical and Physical Concepts

The problem of environmental pollution has grown exponentially in the past few decades. What was once a local problem has now become a national and even a global problem. Technically and physically, pollution is perceived as an intra-urban regional, public problem. Therefore, it is probably most effectively solved by a regional governmental agency or authority with the guidance and financing of the national or Federal government.

Pollution (or environmental quality), like beauty, is "in the eyes of the beholder." The public defines pollution through the political process. Scientists cannot determine for the public the tolerable level of pollution or the proper use of the environment. Pollution is defined "by linking scientific knowledge with a concept of the public interest."[2] According to the Council on Environmental Quality (CEQ), "pollution occurs when materials occur where they are not wanted. Overburdened natural processes cannot quickly adjust to the heavy load of materials which man, or sometimes nature, adds to them. Pollution threatens natural systems, human health, and esthetic sensibilities. . . ."[3] In other words, pollution is a social cost of human activity that limits the usefulness of resources—air, water, and land—to man.

Each of the three major types of pollution—air, water, and solid waste—is explained briefly in this chapter. For a more detailed technical discussion of these as well as more esoteric forms of pollution—noise pollution, toxic and hazardous waste pollution, and visual pollution—one should consult the literature.[4]

Measuring Pollution

Some measurement of pollution is necessary for the development of effective pollution control policy. Such measurement should provide accurate and timely information on the status and trends in environmental quality. The two general problems faced by those attempting to measure pollution are "collecting accurate and representative data and presenting or analyzing the data so as to render it both comprehensible and meaningful."[5]

The most precise information on the environment would be raw data on particular environmental conditions. Even if raw data were available, they would have to be logically aggregated and summarized to make the information meaningful to technocrats, public policymakers, and the general public, all of whom are interested in both the status and trends in environmental quality.

One effective way of measuring environmental quality and the trends relating thereto is through the use of an index or indices. The CEQ describes an index as "a quantitative measure which aggregates and summarizes the available data on a particular problem."[6] An example of a simple environmental index is the ratio of the average ambient water pollution to some standard level of water pollution. A more complex index would involve the combination of a number of factors or ratios through the use of several different mathematical manipulations.

National economic policy is influenced and its effectiveness is evaluated by using economic indicators. If a national government can develop a series of environmental-quality indicators by aggregating and summarizing the available environmental data, then technocrats, public policymakers, and the general public could evaluate environmental quality trends and the success of national (or regional) pollution control policy. The national governments of most countries in North America and Western Europe are cooperating through their membership in the Organization of Economic Cooperation and Development (OECD) in the development of environmental-quality indices and standards for determining the pollution control costs of industry.

Air Pollution

Air pollution is "a phenomenon of urban living (concentrated populations, industrial growth, high motor vehicle usage) that occurs when the capacity of air to dilute pollutants is overburdened."[7] Fundamentally, atmospheric pollution results from the process of converting one form of energy into another. When the material by-products of energy conversion become airborne in sufficient concentrations to significantly affect the human use of the atmosphere, there is air pollution. Definition of an optimum level of air pollution "—the risks we are prepared to take for the benefits we enjoy—

is not a matter of scientific judgment, but rather one of moral and ethical judgment."[8]

Technically "pure," dry air consists of 78.09 percent nitrogen, by volume, 20.94 percent oxygen, and 0.97 percent of several other gases — carbon dioxide, helium, argon, krypton, and xenon. Moreover, water vapor in concentrations of 1 to 3 percent is normally present in air.[9]

In the urban areas of most developed countries, the five most common air pollutants by weight are carbon monoxide (CO), hydrocarbons (HC), sulfur dioxides ($SO_x$), particulates, and nitrogen dioxides ($NO_x$).[10] A ranking of air pollutants in the United States, shown in Table 2-1, by order of weight emitted probably does not parallel the ranking of air pollutants in order of damage or harmful effects. Moreover, the table does not include a measurement of photochemical oxidants (smog) effected by sunlight. Specific data on environmental phenomena in the United States are used throughout the following explanation of technical and physical concepts because they are so readily available.

**Table 2-1**
**Estimated Emissions of Air Pollutants**
**(In million tons per year)**
**by Weight Nationwide, 1971**

| Source | CO | PM[a] | $SO_2$ | HC | $NO_x$ |
|---|---|---|---|---|---|
| Transportation | 77.5 | 1.0 | 1.0 | 14.7 | 11.2 |
| Fuel combustion in stationary sources | 1.0 | 6.5 | 26.3 | 0.3 | 10.2 |
| Industrial processes | 11.4 | 13.6 | 5.1 | 5.6 | 0.2 |
| Solid waste disposal | 3.8 | 0.7 | 0.1 | 1.0 | 0.2 |
| Miscellaneous[b] | 6.5 | 5.2 | 0.1 | 5.0 | 0.2 |
| Total | 100.2 | 27.0 | 32.6 | 26.6 | 22.0 |

SOURCE: Council on Environmental Quality, *Environmental Quality: The Fourth Annual Report* (Washington, D.C.: U.S. Government Printing Office, 1973), p. 266.

Note: The table does not include data on photochemical oxidants because they are secondary pollutants formed by the action of sunlight on nitrogen oxides and hydrocarbons and thus are not emitted from sources on the ground.

[a]PM is an acronym for particulate matter.

[b]Miscellaneous includes forest fires, agricultural burning, and coal waste fires.

The actual air quality in a nation is also dependent upon "the geographic concentration of pollution sources and the dispersion of the pollutants"[11] by meteorological conditions once they leave the sources. Since the weight of air pollutants emitted nationally or regionally is only a crude measure of air pollution, the ambient air quality and the effect of air pollutants on human health, animals, vegetation, and property must also be considered in determining improvements in air quality. Thus, a regional air pollution index should adjust the amount of a pollutant emitted for the effect of meteorological conditions and damage caused by such a pollutant. In spite of these measurement problems, most developed countries have established emission standards, along with explicit or implied ambient air quality standards. Even so, we still have much to learn about both the long-term health effects of air pollutants and the actual damage to animals, vegetation, and property (in qualitative and economic terms).

The air environment, a dynamic system of nitrogen, oxygen, and water vapor, is constantly absorbing solids, liquids, and gases generated by both natural processes and human activity. Meteorological conditions, and to some extent geographical conditions, play a major role in the dispersion of emissions. Technically, the following four interrelated aspects of air pollution must be much better understood before air quality will be assured:

> The flow and dispersion of air contaminants and their degradation or conversion to other chemical and physical forms in the local, regional, and global atmospheres.
>
> The means of avoiding the generation of air pollutants or of abating pollution if it cannot be avoided.
>
> The effects of air pollutants on plant and animal life and on inanimate objects and materials.
>
> The means of detecting and measuring air pollutants and their effects.[12]

Air pollutants are eventually dispersed by wind movement—vertical and horizontal turbulence. The effect of such turbulence is somewhat dependent upon the origin—point-sources, line-sources, or area-sources—of the pollutant. The point-source is best illustrated by the smoke plume from the chimney of a factory or the tall stack of a utility. The freeway during rush hour is an example of a line-source of pollution. The most common characteristic of the origin of pollution is that it is generated by an area. An area-source can vary materially in size—an industrial park, a metropolitan area, or a megalopolis. Other meteorological conditions that may have an effect on atmospheric dispersion are the diurnal and seasonal cycles.[13]

The meteorological conditions that cause the dispersion of a point-source pollutant or a line-source pollutant differ significantly from those that diffuse emissions from area-sources. In the case of point-sources and line-sources, one would be concerned with either the behavior of wind with time at a single point or the increasing rate of wind dispersion with increasing scale. With area-sources or urban air pollution, there is concern for the replenishment rate of the air over an urban area since one must consider the total movement of a large volume of air as it ventilates the whole metropolitan area. Thus, any phenomenon that decreases the rate of ventilation in an urban area—whether it is a neighboring mountain range (geographical), a thermal inversion (meteorological), or low prevailing winds (meteorological)—intensifies air pollution.[14]

Air pollution affects human health, animals, vegetation, materials, climate, and visibility. Toxic levels of air pollution for humans and animals are relatively well understood, but the health and economic effects of low concentrations of heterogeneous mixtures of gaseous pollutants and particulates are less well understood.[15] We know that smog irritates the eyes and that at least four diseases—bronchitis, lung cancer, cardiovascular disease, and other respiratory disease—seem to be statistically related to higher levels of air pollution.[16]

Since plant damage caused by air pollution has been studied extensively, it is generally agreed that vegetation is more sensitive to atmospheric degradation than humans and animals are. Serious damage to plant tissue is the most common effect of air pollution on agricultural, forest, and ornamental vegetation. Very little information is available on the reduction in crop growth and productivity caused by air pollution, but studies show that material economic loss is correlated with air pollution.

The damage to materials (property) caused by air pollution is well known but not necessarily well understood. Air pollution cracks rubber, weakens synthetic fabrics, fades dyes, tarnishes silver, dirties laundry, corrodes metals, and damages exterior paint. The economic loss associated with the damage of air pollution to materials is difficult to determine.[17]

Air pollution affects the natural climate but not in a predictable manner. Visibility is also affected by air pollution. Smog conceals scenic views and hinders the safe operation of the air transportation system.[18]

There are five alternative technological strategies for controlling air pollution:

> Select process inputs, such as fuels, that do not contain the pollutant or its precursors.
>
> Remove the pollutant or its precursors from the process inputs.
>
> Operate the process so as to minimize generation of the pollutant.
>
> Remove the pollutant from the process effluent.
>
> Replace the process with one that does not generate the pollutant.[19]

Pollution abatement can, of course, be effectively implemented only by integrating one or more of the above alternatives with the pollution control policy for other forms of pollution so that a reduction in air pollution does not increase water or solid waste pollution.

Although the information is somewhat dated, Table 2-1 will help put the source of different types of air pollution in perspective. In North America and Western Europe, the major sources of air pollution are the by-products — carbon monoxide, hydrocarbons, and nitrogen oxides — of energy conversion from use of the internal combustion engine. This gasoline-powered engine is used in most motor vehicles which comprise a significant part of the transportation system in all developed countries. By any measure — weight emitted or damage of emissions — these by-products are the major source of air pollution.

Ambient air quality is a function of the weight of emissions from area-sources as well as meteorological and geographical conditions. Even though the air environment is not confined and randomly distributed like the water environment, meteorological and geographical factors create enough stability in the atmospheric conditions of a metropolitan area — that volume of air that covers an urban area — to create regional airsheds similar to the water basins that form the water environment.[20]

Awareness of the airshed phenomenon has resulted in the delineation of air quality control regions in North America and Western Europe. Ambient air quality standards for different pollutants can be set for and monitored by regions. Then emission standards for both mobile and stationary sources can be periodically adjusted to ensure that ambient air quality standards are met, even though the uncontrollable interaction of meteorological and geographical factors changes the replenishment rate of the atmosphere in the airshed.[21]

Water Pollution

Water pollution occurs when wastewater from urban and industrial sources, as well as agricultural run-off, flows into surface water systems in such quantities that the natural assimilative capacity of such systems is significantly decreased or destroyed until measures are taken to rejuvenate the water basin. Technically, water pollution occurs when the natural cleansing capacity of the oxygen in water is no longer capable of "breaking down" organic pollutants enough to maintain the requisite standard of water quality for generally accepted human use.[22]

Even though a substantial amount of data on water pollution has been collected, significant problems remain to be resolved in the development of an accurate measurement of water quality. Not only are there more water pollutants to measure, but also water can be put to more than one use — drinking, swimming, boating, fishing, agriculture, shipping, industry, and public wastewater disposal. Therefore, pollution levels cannot be compared to a single standard because the multiple uses of water necessitate that there be multiple standards for water quality.[23]

The main sources of water pollution in the order of their estimated significance are industrial wastewater, municipal wastewater, urban and agricultural run-off, soil erosion, and miscellaneous others—oil, watercraft wastes, thermal waste, and mine drainage.[24] The actual significance of these sources is not easily measured because of a lack of data on the amounts and composition of wastes generated periodically[25] and because of "the fact that contaminants often enter water in complex mixtures of many substances whose specific identities are largely unknown."[26] This practical problem is somewhat alleviated by describing the quality of a body of water in terms of its collective characteristics, for example, biological oxygen demand and suspended solids.[27]

The water environment, like the air environment, is a dynamic system that continuously absorbs a wide range of solids, liquids, and gases from both natural and man-made sources. Moreover, the ecosystem of natural waters is full of living organisms. All of these substances found in the water environment "flow, disperse, and interact chemically and physically before they reach a sink such as the ocean or a receptor such as a fish."[28] Technically, the following four interrelated dimensions of the water pollution phenomenon must be much better understood before an acceptable level of water quality can be sustained:

The flow and dispersion of water pollutants and their degradation or conversion to other chemical and physical forms.

The means of abating water pollution where generation of the pollutants cannot be avoided.

The effects of water pollutants on plant and animal life and on inanimate objects.

The means of detecting and measuring water pollutants and their effects.[29]

Water is a chemical compound of 100 percent $H_2O$; therefore, natural waters are never "pure" since they always contain living and nonliving substances, namely, decaying vegetation, animal wastes, soil erosion, and minerals. The bodies of water within a water basin have an amazing ability to purify themselves.[30] They utilize oxygen to degrade organic pollutants into relatively harmless and inoffensive forms.

The transport process through dispersion and the degradation process through oxidation affect water quality. Hence, they determine the amount of pollutants that the water environment can assimilate, while maintaining a predetermined water quality standard. In respect to water pollution that originates with a point-source—industrial wastewater discharge or municipal wastewater discharge that always enters the water environment at a specific location—information on the amount and composition of the point-source effluent when combined with knowledge of the effects of the transport and degradation processes should enable us to maintain the specific level of water quality related to the designated use of a body of water. Information on point-source effluent can be accurately measured by the polluter and the data reported to the government.

When water pollution originates from an area, it is said to be from an area source. Agricultural run-off, urban run-off, and soil erosion are a few examples of water pollution from an area source. Accurate information on the amount and composition of area-source effluent may never be available because of the inherent difficulty of collecting data.

Technically, the collective characteristics of a body of water can be used as a surrogate for the amount and composition of water pollutants. The two most common measurements of the collective characteristics are the biological oxygen demand (BOD) of a body of water and the suspended solids therein. BOD is a measurement of the weight of the oxygen dissolved in the water that is required both by microorganisms as they degrade and transform carbonaceous organic material and oxidizable nitrogen in organic material and by chemical reducing compounds—sulfide, sulfite, and ferrous iron. The chemical reactions effected by the microorganisms by utilizing dissolved oxygen occur over several days. This period of incubation differs somewhat for specific organic materials and mixtures thereof, but a five-day period of incubation is the accepted standard because the five-day BOD test measures the bulk of the oxygen demanded to stabilize most of the wastes discharged. In summary, the BOD test is a measurement of the gross amount of biodegradable waste being transported by a body of water as measured by its five-day oxygen demand.[31]

Suspended solids found in all waters—natural or polluted—vary considerably in their physical properties—size, shape, density, and concentration—and in their biological and chemical properties. Suspended solids or particles can be the pollutant, or they can effect the transport of other pollutants by absorbing, binding, and culturing pollutants. Even though centrifugal methods have been used both to separate suspended solids into their organic and inorganic components and to describe them in terms of their size, density, sedimentation rate, and contaminant content, there is no generally accepted surrogate—like BOD—which measures the effect of suspended solids on the collective or transport capability of a body of water.[32]

It should be emphasized that because water has multiple uses, the effects of water pollution depend upon how society decides to use its surface water—for drinking, recreation, marine life, wildlife, agriculture, industry, or wastewater disposal. In general, society is concerned with the effect of water pollution on human health, wildlife, marine life, recreational use, and aesthetic values. We may all agree that our goal should be zero discharge of pollutants into surface waters, so that all bodies of water are fit for human contact and minimal treatment is required for human consumption. But the cost to society for not utilizing the waste-assimilative capacity of selected bodies of water may be greater than the benefits derived from recreation and drinking. Therefore, zero discharge would be an inefficient use of a society's resources, and such inefficiency would be reflected in the higher cost of goods and services that could have utilized the waste-assimilative capacity of water.

In summary, there are three elements to water pollution control policy. First, the appropriate uses for each stretch of water in a country or region must be determined. Then the government must establish limits on the source and amounts of various pollutants allowed in the water environment. Finally, the government must establish a comprehensive plan to prevent or abate water pollution for each stretch of water in a country, including enhancement measures where necessary,[33] through the optimal mix of regulation, subsidization, and taxation.

### Solid Waste Pollution

Disposal of the solid wastes generated in the production and consumption of the "good life" requires more and more land. Such waste is the principal cause of "land" pollution. Litter—the illegal, careless disposal of waste by the general public—is also a cause of land pollution, as well as being aesthetically offensive. The main sources of solid waste are urban—residential, commercial, industrial, and institutional—refuse, including litter, open dumps, and sanitary landfills; junked motor vehicles; the solid waste materials of industry; and the solid wastes that are a by-product of the mining and processing of natural resources. Solid wastes are usually disposed of in sanitary landfills, in open dumps, by incineration, or by recycling.

The measurement of land pollution is much simpler than the measurement of air and water pollution. The amount of solid waste produced during a specific period of time as measured by volume and by weight is a useful, surrogate measure of land pollution. However, volume and weight are not precise measures because incineration and recycling reduce both. The amount of land (sanitary landfills and open dumps) being used to dispose of solid wastes is a more precise measure because, in the context of solid waste pollution, "pure" land is land that is not being used to dispose of urban refuse, junked motor vehicles, industrial waste materials, and mining or processing waste materials.[34]

The objective of any solid waste control program should be twofold: to reduce the volume and weight of waste materials and to convert the waste into a less offensive form. By reducing their volume and weight, solid wastes can be disposed of with the use of less land. The primary, practicable alternatives for reducing volume and weight are disposing of solid waste in sanitary landfills, which reduces volume by about one-third; incineration, which reduces volume by more than three-quarters; composting, which reduces organic matter by about 40 percent; and recycling or materials recovery.[35]

The urban refuse generated by the residential, commercial, and institutional sectors of a community is, by far, the most visible, tangible source of pollution to the layman. From a physical perspective, urban refuse is composed of cardboard, newspaper, miscellaneous other papers, plastic film, glass, textiles, ceramics, leather, molded plastics, rubber, wood, metallics, grass, dirt, and stones. In summary, urban refuse is a complex, heterogeneous mixture of solids. This heterogeneity has made it very difficult to develop measurements and standards for characterizing urban refuse.[36]

Many open dumps have been replaced by sanitary landfills, which are a short-term solution to the urban solid waste pollution problem as long as they are well planned and well operated. This solution depends upon the availability of suitable land—inexpensive and geologically sound—within economic range (minimal transportation costs) of an urban area. As the land that meets these requirements is exhausted, the use of alternative methods of solid waste abatement will grow.[37]

There are two somewhat overlapping alternatives to the sanitary landfill: incineration and resource recovery. Incineration has been of secondary importance for years in urban areas where land suitable for dumps and sanitary landfills has been scarce. This method can also alleviate the energy shortage at the cost of a potential increase in air pollution. The second alternative, resource recovery, includes in addition to energy recovery processes the following other major categories of resource recovery: *compost processes*, which produce a humus material useful as a soil conditioner from the organic matter in urban solid waste; *materials recovery processes*, which involve the separation and recovery of paper, glass, and metals from urban solid waste; *pyrolysis processes*, in which urban solid waste is thermally decomposed in controlled amounts of oxygen to produce gas, oil, tar, acetone, and char; and *chemical conversion processes*, in which urban solid waste is chemically converted into useful organic compounds, such as protein.[38]

The composting of urban solid waste to produce soil conditioner and low-grade fertilizer has been more successful in selected parts of Europe than in the United States. Inasmuch as compost is rich in organic matter, it can condition and enrich soil by improving the workability, structure, and resistance to compaction and erosion of that soil. However, farmers still find chemical fertilizers, which are more economical to buy and use, and animal wastes, which are more readily available, preferable to compost.[39]

Until the Energy Crisis of the winter of 1973-74, the most significant obstacle to the recycling of the paper, metal, and glass in urban refuse was not technology, but economics—the absence of markets for recycled materials. The huge rise in the price of energy has not only made incineration an attractive source of energy but has also increased the costs of production where virgin materials are used. Thus, post-Energy Crisis market forces have simultaneously reduced the problem of disposing of solid wastes and have provided needed resources in the form of energy as well as reusable raw materials. Moreover, the cost of the traditional methods—sanitary landfilling and incineration without energy recovery—of solid waste disposal has also been rising.[40]

The other two resource recovery processes—pyrolysis and chemical conversion—are still in the development stage. Refinements in the technology of both processes must occur before either will become an economic alternative. However, continued increases in environmental quality standards and in the price of energy, raw materials, and land should make both pyrolysis and chemical conversion viable, economic alternatives to landfilling, incineration, and other resource recovery processes in a few years.

Junked automobiles are another significant source of solid waste pollution. The junked automobile usually goes to the auto wrecker who strips it of those auto parts that can be sold for further use with or without rebuilding or that can be sold for their mineral (lead, copper, or iron) content. The auto wrecker then becomes a scrap processor, or he sells the discarded automobile to a scrap processor.[41]

Inasmuch as more cars are junked in North America and Western Europe annually than are processed as scrap, the stock of abandoned vehicles is increasing. In the United States, there is enough processing equipment to scrap all eight million automobiles junked annually as well as some of those twenty million abandoned in prior years. However, many junked automobiles (more than one million annually) are not inexpensively available to the scrap processor because of the prohibitive cost of transporting junked automobiles. The other problem facing the scrap processor results from the automakers' increasing the amount of nonsteel components in the automobile, including plastic, glass, rubber, aluminum, and zinc, in addition to copper.

Industry also generates large amounts of solid wastes in its production of goods and services. Industrial solid wastes are not to be confused with industrial scrap. Industrial scrap is that production material which has not been finally discarded by industry.[42]

The same environmental quality and economic trends that are affecting the disposal of urban solid waste are also having an impact on solid waste disposal by industry. As suitable land—inexpensive, unregulated, and geologically sound—within economic range of industrial sites is exhausted, on-site incineration and on-site landfilling, as well as materials recovery, will become more viable alternatives to the predominant disposal method—the sanitary landfill. Moreover, the high price of energy has significantly increased the cost of processing virgin materials and has stimulated the conversion of more industrial solid waste into industrial scrap.

The production of minerals and other virgin materials also generates annually over one billion tons of mine waste, mill tailings, washing plant rejects, processing plant wastes, and smelter slags. Over half of the mining and processing solid waste is recycled either for its mineral content or for use as building materials. Most of the remaining mining and processing solid wastes are not recyclable under present economic conditions. Consequently, these wastes are viewed as a public health and safety problem, as a poor use of the land where they are dumped, or as aesthetically offensive.[43]

## The Economy and Pollution: A Macroeconomic Perspective

Examination of environmental pollution as a materials balance problem for the entire economy is a logical extension of the above introductory analysis of environmental pollution from a technical and physical perspective. The materials balance approach is based on the notion that "at least one class of externalities—those associated with the disposal of residuals"[44] should be looked upon as a normal, inevitable part of production and consumption in a developed economy. As the economic development of a society proceeds, the economic significance of residuals disposal increases, that is, "the ability of the natural environment to receive and assimilate them is an important natural resource of rapidly increasing value."[45]

Traditionally, in economic theory the production and consumption processes have been conceptualized in a way that is inconsistent with the fundamental laws of nature in respect to the conservation of matter and energy. M. King Hubbert has succinctly explained the practical significance of these natural laws:

> The earth may be regarded as a material system whose gain or loss of matter over the period of our interest is negligible. Into and out of this system, however, there occurs a continuous flux of energy in consequence of which the material constituents . . . of the earth undergo continuous or intermittent circulation. The material constituents of the earth comprise the familiar chemical elements. These . . . may be regarded as being transmutable and constant in amount.[46]

In summary, nature does not allow the destruction of energy or matter. Thus, the residuals generated through economic activity must either be recycled or deposited in "negative" mines, that is, in an airshed, into a water basin, or in and on land.

Economists have traditionally classified air and water as free goods. Today, air and water are common property resources of significant and increasing value. Therefore, society must allocate such resources efficiently since voluntary exchange through private markets is impracticable. More-over, since abating—processing or purifying—one type of residual only alters its form, for example, from a liquid to a solid, residuals' problems should not be classified by type—air (gas), water (liquid), or solid waste pollution—and then be treated as separate, unrelated problems.

### The Materials Balance Approach

To speak of the consumption of goods is misleading since the material objects exchanged in markets are the vehicles through which services can be utilized both to satisfy consumer preferences and to add value in the production process. Not only does final consumption "never" occur, but also the residuals generated in the production and consumption processes can cause disutility to consumers and producers in the absence of voluntary exchanges in the market because the disposal options are pollution or recycling.

In any economic system, inputs of food, energy, and raw materials are converted into outputs of goods, services, and residuals. Goods and ser-vices are not "consumed" since consumption implies the destruction of matter and energy. Eventually, all goods also become residuals, even though they may be part of inventory and productive capacity for a short period of time. If we assume an economy is closed (no imports and exports) and there is no capital accumulation (that is, plant and equipment, inventory, buildings, and consumer durables), the amount of residuals generated in production and consumption should be approximately equal to the amount of food, energy, and raw materials as well as oxygen entering the system for a specified period of time.[47] In reality, a growing economy would be open—a net importer or net exporter on a national or regional basis—and capital accumulating or capital exhausting.

The overwhelming majority of the active inputs to a developed economy are eventually emitted to the atmosphere as $CO$, $CO_2$, and $H_2O$, as well as other gases, as a result of either the combustion of fossil fuels or animal respiration.[48] Of these gases, water vapor ($H_2O$) and carbon dioxide ($CO_2$), which is reabsorbed by vegetation and large bodies of water ("negative" mines), are harmless.[49]

The other gaseous residuals (emissions), like carbon monoxide ($CO$), sulfur dioxide ($SO_2$), and nitrogen oxides ($NO_x$), are potentially very harmful. For example, when sulfur dioxide (gas) and water (rain) are combined, the resulting chemical compound is sulfuric acid ($H_2SO_4$). Other residuals include liquid waste (effluents), like municipal and industrial sewage; urban and agricultural wastes suspended or dissolved in water; and solid waste such as urban refuse, discarded consumer and industrial durables, and industrial scrap. Ultimately, dry solid residuals are the irreducible, limiting form of waste. Technically, by utilizing the appropriate amount of energy as well as fixed capital, most harmful residuals can be transformed into harmless gases or dry solids, with the dry solids being disposed of as a solid waste or reused.

The materials balance approach underlines the importance of treating air, water, and land pollution as resulting from interdependent forms of residuals for purposes of planning and control policy. Figure 2-1 is a diagrammatic synopsis of the materials balance approach.

The environmental assimilation capacity (EAC) depends upon the waste-dispersal, waste-treatment, and waste-removal properties of the atmosphere (airsheds), bodies of water (water basins and the sea), and the land (including vegetation). From mankind's point of view, the utilization of EAC is an economical, nondisruptive method of transferring materials used by humans to other *loci* in the flow of materials between populations and environments in nature. In economic theory, EAC represents a link between the input-output flows of an economy and the input-output flows of natural ecosystems.[50]

Whenever the amount of residuals discharged overtaxes the natural waste-absorption capacity of the environment, the two choices available to society have always been either to collect, treat, and dispose of wastes or to recycle them for reuse in the production process. The materials balance approach facilitates comprehension of the notion that "the throughput of new materials necessary to maintain a given level of production and consumption decreases as the technical efficiency of energy conversion and materials utilization increases."[51] Moreover, *ipso facto*, increasing the longevity of producer (plant and equipment) and consumer (automobiles and appliances) durables decreases the throughput of new materials required to compensate for wear and tear as well as the obsolescence of such durables. However, the use of older durables tends to increase the discharge of other residuals because older plant, equipment, and motor vehicles are less efficient in converting energy. Completely efficient energy conversion would generate only $H_2O$, $CO_2$ and ash residuals; consequently, pollution would be limited to solid waste.

At the regional level for a given population, level of industrial production, transportation system, and standard of living, one can envision various combinations of socioeconomic policy, each of which would result in a different level of residuals discharge into the environment. In each region, the choice of a specific policy combination is a political decision and a value judgment by society wherein the public implicitly quantifies the social cost of the pollution they are willing to tolerate. The ideal choice would occur in the regions that "practiced high level recovery and recycle of waste materials and fostered low residual production processes to a far reaching extent in each ... economic sector"[52] to minimize the amount of residuals discharge into the air, water, and land. Moreover, the EAC of a region can be augmented through public investment (public goods) in facilities that can

Figure 2-1. The Flow of Materials in an Economy

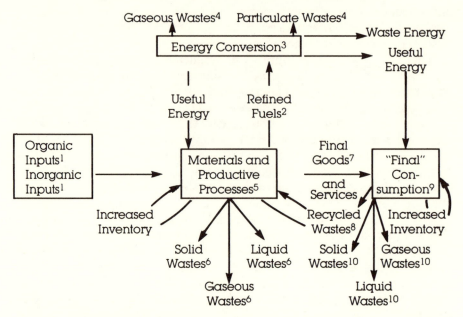

SOURCE: Adapted from Robert U. Ayres and Allen V. Kneese, "Production, Consumption and Externalities," *American Economic Review*, Vol. 59, No. 3 (June 1969), p. 285, and Allen V. Kneese, Robert U. Ayres, and Ralph C. D'Arge, *Economics and the Environment* (Baltimore: The Johns Hopkins University Press for Resources for the Future, Inc., 1970), pp. 18-68. Used with the permission of The Johns Hopkins University Press.

1. Inputs include products of photosynthesis (organic) such as fossil fuels (coal, petroleum, and natural gas) and agricultural products (food and forest products) and minerals (inorganic).
2. Refined fuels imply derivatives of crude oil (gasoline, kerosene, fuel oil, and other oils) as well as natural gas and coal.
3. Energy conversion refers primarily to power generation (electricity), transportation, industrial use, commercial use, institutional use, and household use.
4. Gaseous wastes include CO, $CO_2$, $H_2O$, $SO_2$, and $NO_x$ whereas the term "particulate wastes" refers to flyash and bottomash.
5. Materials and productive processes can be divided into four broad classes: direct and indirect products of photosynthesis; inorganic chemicals and products; primary metals and products; and structural materials.
6. Solid wastes include production-processing wastes, junked machinery and equipment, and mineral-processing wastes, whereas the liquid and gaseous wastes are primarily chemicals dissipated in processing such as solvents, neutralizers, and cleaners, but also include processing losses such as hydrocarbons and other organic wastes.
7. The primary, broad categories of final goods are food products, textiles, paper and wood products, petrochemicals (fibers, rubber and plastics), metal products, structural materials, and structures. Moreover, commercial and governmental entities produce "pure" services.
8. The recycling of wastes is actually a microcosm of the materials balance approach which should be categorized as residuals processing wherein organic and inorganic residuals

(inputs) are processed with useful energy to produce recycled inputs, final goods, or energy as well as residuals.

9. "Final" consumption can occur only in respect to "pure" services and the "services" (or utility) provided by final goods. Therefore, final goods become residuals after their "services" have been consumed.

10. Solid wastes include the urban refuse from households as well as discarded consumer durables, whereas the primary liquid wastes and gaseous wastes are sewage and the by-products of respiration ($CO_2$ and $H_2O$), respectively.

modify an environmental medium to maintain a floor on the level of assimilative capacity at all times or to increase the assimilative capacity during peak periods of residuals discharge. As an example, the construction of reservoir storage could be used to increase river flows, alleviating environmental problems that occur either because the flow periodically decreases while the discharge of effluent is constant, or because the discharge of effluent periodically increases while the river flow is constant.[53]

## The Materials Balance Approach and Macroeconomic Analysis

The conception of environmental pollution as a materials balance problem wherein external diseconomies (externalities) are inherent parts of the energy-conversion, production, and consumption processes requires relatively important changes in the formulation of the normal general-equilibrium mathematical model of an economy.[54] The main objective in such modifications would be to define a system of simultaneous equations in which flows of services ("pure" services and the services of goods) and materials are simultaneously accounted for and correlated with social welfare.

Several economists have analyzed environmental pollution in terms of a systems approach that integrates the externalities associated with residuals discharge with various simultaneous-equations models of the economy.[55] While these models are too complex to be explained in any great detail, the Kneese-Ayres modification of the Walras-Cassel model is explained below in general terms since it represents a rigorous consummation of their materials balance approach to environmental economics.

One can represent private exchange in a static competitive economy with a series of simultaneous equations by applying the concepts of Newtonian mechanics to the private market exchanges of an economy in which the following conditions and assumptions hold: (1) all markets are competitive, with no individual or producer affecting any market price; (2) all individuals and producers have perfect information about the quality and prices of commodities, goods, and services; (3) all participants in the economy are motivated by self-interest and economic gain in making economic decisions, that is, resource owners and producers maximize their returns and profits, respectively, while consumers allocate their income by acquiring that combination of goods and services which maximizes their well-being; (4) there is private, individual ownership of all resources and goods, and there are no common property resources; and (5) the quantity and quality of resources, the technology by which these are combined in production, the tastes and preferences of consumers, and the distribution of ownership of resources that determines the distribution of income and purchasing power among individuals are all fixed.

Since the assumptions provide that all utility functions of consumers and all production functions of producers are independent, the decisions of any consumer or producer will influence the decisions of any other consumers or producers only indirectly through the market mechanism, that is, through prices. Therefore, there is a single set of prices for resources, intermediate products, and final products/and a collateral distribution of goods and services which results in a "Pareto optimal" situation in which all possible trading gains have been exhausted. That is, any trading that would make one person better off would necessarily make another worse off. Thus, both individual welfares and social welfare are maximized.

For the purposes of this study, the most interesting assumption in models of this type is the limitation of all property rights to individuals. The problem of externalities has traditionally been perceived as a minor aberration of this assumption. Therefore, economists have treated externalities as a problem that could be satisfactorily dealt with in "partial" equilibrium analysis. If the spillover (or external) costs of production and consumption are inherent and pervasive because of the existence and scarcity of the common property resources into which residuals are being discharged, then social costs as well as prices affect the economic behavior of producers (production functions) and consumers (utility functions).

Therefore, the general equilibrium model must integrate the materials balance approach with the conventional economic models of production and consumption. That is, two sectors—an environmental sector and a "final consumption" sector—must be added to the conventional, two-sector Walras-Cassel model. The mathematical model must then be balanced in two dimensions since production (value) must equal consumption plus or minus changes in investment, and input (physical) from the environment plus recycled output must equal output from final consumption plus or minus changes in inventory. The model provides for the physical balance of materials, even though there is no "market" price for residuals. Thus, the Ayres-Kneese model implies central planning on a national or regional scale because of the necessity for centralized data collection and computation since there is no "market" for pollution or residuals discharge.[56]

## The Economy and Pollution:
## A Microeconomic Perspective

The general equilibrium, interdependency models of an economy have traditionally been formulated without allowing for the significance of residuals discharges for efficient allocation of resources. However, the external diseconomies associated with the production of firms and the consumption of households have been treated as material in respect to the functioning of the market under some circumstances, for example, by focusing on one industry or region at a time and then making aggregative extrapolations.

Two fundamental types of market failure are relevant to the problem of environmental pollution. The first is the lack of a well-defined and enforceable system of private property rights in respect to most environmental resources. Since no one owns environmental resources, there is no market price for them. Consequently, private economic decision makers do not receive complete information about the use of these resources; there are no proper economic incentives to place these common property resources in their highest valued use. The second market failure is related to the public goods nature of many common property resources. Private markets will always fail to allocate sufficient resources to the production of public goods. Hence, such goods require governmental intervention to ensure that they are produced in optimal quantities.

### Market Failure: The Problem of the Commons

For the market system to function properly, the ownership of every good or resource must be clearly defined and enforceable.[57] Ownership allows the owner to prevent others from using, benefiting from, or damaging a good or resource without making compensation. If such uncompensated benefits or damages occur, spillover effects or externalities result. The body of law in Western nations is based on the need to define the rights of ownership, protect them, and provide for their transfer. The control and allocation of resources, in any modern economic society, require a system of property rights.

Property rights may not be perfectly defined or fully enforceable. The person who owns personal property such as a sailboat can enforce his property rights in that sailboat. He can restrict its use to himself, exclude other people from receiving any benefits from the sailboat, and protect the sailboat from damage as he sees fit. However, with regard to common property resources, it is either difficult or impossible to assign enforceable property rights, as in the case of air; or enforceable property rights are not always assigned, as in the case of water. Similarly, the owner's use of his land may affect the welfare of others inasmuch as land use generally has effects that "spill over" to other people besides the landowner. Since these effects are external to the owner's economic decisions, he has no economic incentive to take them into account in his decision-making process.

For example,[58] Mr. Smith owns a house on land adjacent to a plot owned by Mr. Jones. Mr. Smith's satisfaction will be higher if Mr. Jones uses his land for a suburban residence rather than for a refuse dump. Mr. Smith's higher satisfaction has a monetary value—a willingness to pay. However, his neighbor's property rights exclude the power to require payment from Mr. Smith for the aesthetic value of the view which an attractive residence would create. Thus, Mr. Jones does not have an economic incentive to take Mr. Smith's wishes into account in deciding how to use his land. Alternatively, we could say that Mr. Smith's rights to his own property do not include the power to prevent Mr. Jones from imposing a loss of property value on him by creating a dump. No mechanism (market) exists through which Mr. Smith can make Mr. Jones compensate him for a loss of property value when Mr. Jones creates a dump.

For another example, assume there is a river from which a brewery takes water to make its beer and that a paper mill located upstream dumps its residuals into the river. The mill imposes damages or costs on the brewery, and these damages or costs are external. The mill has no incentive to economize on its use of the waste-assimilative capacity of the river.

Each producer feels that its location on the river gives it rights to use the water. However, these rights have not been clearly defined by the law. If the rights were originally owned by the brewery, then the mill would have to purchase the right to use the river water to dispose of its wastes in the river by offering to pay more than the value to the brewery of using the clean water for beer. The vesting in the brewery of the property right to the river, and the legal systems developed to protect property rights would effectively keep the mill from imposing external, nonmarket costs on the brewery without commensurate compensation.

In summary, the use of air, water, and, to some extent, land has in common the fact that one person's actions can affect a second person either favorably or unfavorably, even though there is no requirement that the compensation payments be made. Without these payments, or market prices, there are no incentives for the two groups to take the best action.

Where externalities (external diseconomies or economies) occur, there will be a divergence between social values or costs and private (market) values or costs. When a service or good is produced and sold in the market, its price can reflect only those costs that can be imposed on the producer or buyer when the good or service is produced or utilized. If third parties were harmed by the production of the good or are harmed by the particular use to which it is put, its social value—including the cost or value to third parties—is greater than its price or private value. Where such an external detriment is present, the producer and buyer who benefit do not have to pay for the benefit they receive in production or use. The result is a level of production of the good that is more than the optimum.

The concept of externalities, which arises whenever property rights are not clearly defined or enforceable, is almost universally applicable to environmental resources, for example, the waste-assimilative capacity of the environment. If the residuals discharge impairs other environmental services (such as recreation and EAC) or causes damages (specifically, health problems), there are third parties who are forced to bear the cost of residuals discharge. However, since their property rights in the environment are not enforceable, they will not be compensated. Since the price of the environmental services is zero to producers and users, they will expand the use of environmental resources to the level at which the marginal value of the service to them is zero. The marginal social cost, however, may greatly exceed zero. Consequently, their use of environmental resources for waste disposal will be more than optimal.

## Market Failure: Public Goods

Another source of market failure[59] results from the public-goods dimension of environmental resources. By definition, a public good is one that when supplied to one individual is available to all others. Because other users cannot be excluded for nonpayment of the price for the public good, the producers of a public good cannot collect revenues from these "free riders" or beneficiaries.

Most examples of public goods represent the extreme of a "purely" public good—the legal system, national defense, or an ocean lighthouse. In most other cases, the "publicness" of a public good is not an inherent characteristic, since it arises because it is costly or inefficient to exclude those who have not paid the price, for example, education, urban parks, and streets. Other goods possess an even lesser degree of publicness. There are public goods aspects to land use, inasmuch as Mr. Smith cannot exclude his neighbors (including Mr. Jones) from enjoying their view of his attractive suburban home. In general, whenever there is an element of publicness about the good, an insufficient amount of the good will be provided by the private sector.[60]

Public goods are characterized by their nonexhaustive consumption, whereas private goods are distinguishable therefrom because they can be exhaustively "consumed" by the final user thereof. As noted above, the "nonexhaustive-consumption" attribute of public goods allows individuals to consume such goods without revealing their true preferences therefor— the "free-rider" problem. Thus, if a public good is provided for one individual, it will be available to all whether or not the others have contributed to its production.[61] A public good is one that enters the utility of two or more individuals. Finally, some economists have suggested that all goods that are not "purely" private ought to be classified as public goods. An example would be Mr. Smith's attractive suburban home.[62]

If we utilize a line of reasoning similar to the one employed above in explaining the economic effect of a divergence between social and private values, public goods can be conceptualized as goods characterized by extreme external benefits. The social value of one unit of clean air is the sum of the private values for all individuals. If the owner of the exclusive property rights to the atmosphere could obtain payments from all users, his revenues would be equal to the social value of this public good. Therefore, he would have an incentive to produce clean air as long as the marginal social value equaled marginal social cost. But whenever "publicness" is a characteristic of a good or service, the producer cannot capture the full marginal value. Therefore, there would not be an incentive for him to maintain output at a socially optimal level.

All environmental resources possess the characteristics of a public good. The benefits of clean air represent an extreme case of "publicness." Similarly, the benefits of zoning are enjoyed by one's neighbors and other members of the community free of charge so that, from the public's point of view, very few landowners make the best (aesthetic) use of their land.

With regard to water, assume that a single firm that had been polluting a river was given exclusive property rights to the water basin. The firm might decide to reduce its waste discharges in order to sell "clean water." But if it reduced its residuals discharge for one customer, water pollution would be reduced for all, even if they had not paid. Since the firm would not be able to capture the full private value of the clean water, there would be no incentive for it to curtail pollution.[63]

## Economic Theory and Pollution Control Strategies

Based on the macro- and microeconomic analysis of environmental pollution described above, an economist would propose the following pollution control strategy: the government should create, through the use of property rights, a market for the residuals from production and consumption wherein materials balance is maintained through both recycling and disposal. In respect to both the EAC and the "public goods" dimensions of air, water, and land, the efficacy of giving selected natural and legal persons property rights in common property resources is suspect primarily because of the prohibitive transaction costs implicit in such a system of private markets.[64] However, if the government simulates the market through public ownership of common property resources, then a market solution is a valid strategy for pollution control.

### The Problem of Transaction Costs

Coase in his 1960 treatise on social costs[65] demonstrated the reciprocal nature of the pollution problem and effectively challenged the economists' conventional wisdom at that time through a comparative analysis of the treatment of externalities in the law and in economic theory. Contrary to the economists' notion that the fault must always lie with the polluter, he argued for the legal treatment of social costs wherein the issue is to avoid the more serious harm as between $A$ and $B$ when $A$ causes damage to $B$ and to allow (through public policy) $B$ to compel $A$ to cease production or pay damages also harms $A$. Coase also argued that the assignment of legal rights does not affect the economic efficiency of the resolution of the problem of social costs where transactions between polluters and "pollutees" are costless and where the persons involved can modify the initial legal determination of rights — either the polluter is liable for his damages or he is not liable — through market transactions.

Inasmuch as market transactions are not costless, however, economic efficiency through maximization of the value of production may not be identical regardless of the initial legal determination of rights. Therefore, society should prefer a legal system where the determination of rights reflects an underlying economic analysis that demonstrates the assignment of rights which maximizes the value of production. Moreover, to ignore the economic analysis of the "transaction costs" attributable to alternative assignments of rights would mean that "the costs of reaching the same result by altering and combining rights through the market may be so great that this optimal arrangement of rights, and the greater value of production which it would bring, may never be achieved."[66]

The Coase analysis raises two fundamental issues. First, can the market be modified to solve the problem of social costs? Second, how should society resolve the distributional problem implicit in its determination of who should pay for the external effects of production and consumption?[67] The first issue is implicitly discussed in the following text, and the second is deferred to the last section of this chapter.

### Pollution Control Strategies

For the government to simulate the market for pollution control, it must determine the cost of both pollution and pollution control. From a macro-economic point of view, the optimal level of pollution that society should allow firms, governmental units, and households to discharge is that amount of pollution wherein the sum of the social cost of pollution and the private and public cost of pollution control is at a minimum, as depicted next in Figure 2-2. With respect to the individual household, firm, or government entity, the socially optimal level of pollution exists when the marginal cost of an additional unit of discharge (pollution) equals the marginal cost of a reduction of one unit of discharge (pollution), as illustrated in Figures 2-2 and 2-3.

Figure 2-2. Optimal Level of Pollution:
Macroeconomic Perspective

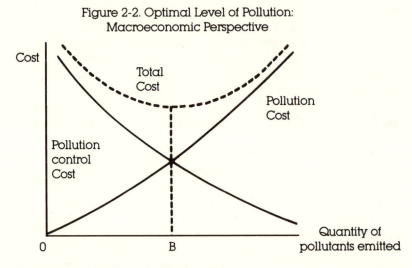

SOURCE: Reproduced from *Microeconomics: Theory and Applications*, Shorter Second Edition, by Edwin Mansfield, by permission of W. W. Norton & Company, Inc. Copyright © 1976, 1975, 1970 by W. W. Norton & Company, Inc.

The marginal cost of pollution, which is derived from a marginal social cost schedule or damage function, determines the price or pollution tax that is consistent with a socially desirable level of discharge by an individual household, firm, or governmental entity.[68] The marginal cost of pollution control is primarily the real resource cost of altering production and consumption to recycle and recover materials as well as collecting and treating wastes so as to reduce pollution to the socially desired level. In addition, there is an income cost because pollution control programs effect changes in relative factor prices as some resources become less valuable and their owners experience a decrease in income. An example would be the owners of high sulfur-content coal or crude oil after implementation of air quality standards which limit $SO_x$ emissions.

In analyzing Figure 2-3, note that an optimal level of pollution is generally not zero because of the EAC of common property resources. That is, the environment is a nondepletable or renewable natural resource. However, an individual household, firm, or governmental entity would not abate its pollution level to OB units because its marginal cost of pollution (zero) is less than its marginal cost of pollution control. The regional or national government, however, could cause an OB level of pollution by employing one or more of the following strategies in respect to each type of pollution:

Figure 2-3. Optimal Level of Pollution:
Microeconomic Perspective

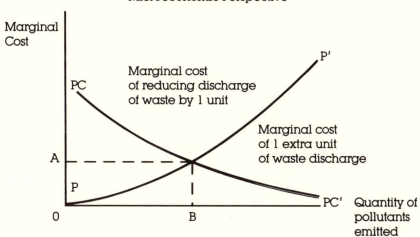

*SOURCE:* Reproduced from *Microeconomics: Theory and Applications,* Shorter Second Edition, by Edwin Mansfield, by permission of W. W. Norton & Company, Inc. Copyright © 1976, 1975, 1970 by W. W. Norton & Company, Inc.

(1) it could by law limit each household, firm, and governmental entity to OB units of polluting, thereby *regulating* the amount of pollution; (2) it could pay each household, firm, and governmental entity to limit its pollution to OB units, thereby *subsidizing* pollution; or (3) it could impose a price or user's charge of OA per unit on each household, firm, or governmental entity to limit its pollution to OB units, thereby *taxing* pollution.

## Taxation Versus Regulation

Legislators and the legal profession have traditionally preferred regulatory approaches to the solution of any public problem, whereas economists have generally argued for a user's charge (taxation), that is, a market solution through the simulation or reformation of a market mechanism. Subsidization in the form of direct grants and tax expenditures has traditionally been used to alleviate somewhat the inequities that are an inherent side-effect of the regulatory process and taxation.

In practice, economists prefer a user's charge or pollution for two reasons: the user's charge is often the least cost method for achieving a given amount of pollution abatement because regulation is usually implemented through imposition of uniform standards that do not reflect each firm's marginal costs; it is also the least cost method to administer, assuming voluntary compliance and minimal monitoring or auditing costs. As you may recall from Figure 2-3, a pollution tax is assumed to be uniform or is proportional; the benefit derived from reducing pollution by one unit is the same for all polluters; and a polluter reduces its discharge to the point where its cost of pollution control per unit is equal to the pollution tax rate. On the basis of this assumption, one would logically conclude that where all polluters experience identical marginal costs of pollution control for all levels of discharge, then a reduction to an OB level of pollution for each polluter (Figure 2-3) results in the same cost in real resources to society no matter which strategy the government employs to control pollution. However, it is unrealistic to assume that all polluters experience identical marginal costs of pollution control for all levels of discharge. One would therefore have to alter one's attitude of indifference toward the choice between regulation or taxation of pollution as the socially desirable (least cost) alternative for pollution control.

If one assumes that the marginal cost of any unit reduction of pollution differs for some polluters, one may conclude that the least cost method for reducing waste discharges is proportional taxation. By this method, polluters whose marginal cost of pollution is lowest are always able to reduce their waste discharge. At the same time, those with higher pollution control costs are allowed to increase their discharge so that total pollution is minimized at a minimum cost to society. Since uniform regulation would decree that all polluters generate the same amount of a specific pollutant, polluters would differ as to their marginal cost of pollution control at the regulated level of total pollution. Therefore, the desired level of pollution would be achieved at more than the optimal resource cost to society.[69] However, pollution control standards that reflect each firm's marginal cost of pollution control would be as efficient as a proportional tax.

As mentioned above, the second reason economists prefer user charges over direct regulation to control pollution is that the administrative costs for regulation are significantly greater than for the taxation approach. Since all pollution control strategies require some government action to ensure compliance—controlling discharges of pollutants at or below legal limits, installing pollution control equipment, or reporting the amount of discharge and paying a use tax thereon—each strategy or combination thereof can be evaluated in terms of the cost to the public of monitoring and enforcing compliance.

The government's monitoring and enforcement actions usually involve the comprehensive or selective investigation of compliance to detect violations so that penalties can be imposed administratively or through a judicial process. Even though an effective pollution control policy encompasses costly monitoring and enforcement activity, some policy strategies command fewer resources for such activity. The reason is that voluntary compliance is an integral part of the pollution control system, for example, a system of taxes on discharges with a level of voluntary compliance like that attributed to the national income tax systems of Anglo-American countries. Conversely, the level of policing activities needed is related to the tradeoff between the polluter's cost of compliance and the polluter's expected value (cost) for noncompliance—penalties for noncompliance combined with the probability of detecting a violation (noncompliance).

Assuming that environmental quality standards have been established, most economists support the market simulation or taxation strategy for controlling pollution on an *a priori* basis. Through this strategy, the cost of acquiring information about polluters, investigating them, and penalizing or prosecuting violators is minimized by virtue of the significantly greater voluntary compliance experienced by Anglo-American tax systems. In contrast, direct regulation encourages noncompliance because of the relatively greater role which monitoring and enforcement play in this approach. The regulatory process typically encourages noncompliance by providing for appeals of the decisions of both regulatory agencies and the courts.[70]

The cost of noncompliance—legal fees and lobbying expenses—with standards is often less than the cost of compliance—investment in pollution control equipment and technological innovations that minimize pollution. Moreover, there are over one hundred years of empirical evidence[71] to support the generalization that when regulatory agencies establish and enforce detailed standards to control individual and group behavior with the intent of correcting malfunctions in the socioeconomic system, the lobbying by those individuals and groups that will be affected most by the conscientious enforcement of rigorous standards usually results in a circumvention of the public interest.

If the regulators are not overwhelmed by the powerful special interest groups during the bargaining process, then the regulatees can move the conflict to the judicial system to reach a compromise solution. The malfunction (for example, pollution) usually continues during the delays that result from appeals within and without the regulatory agency. Even if a regulatory agency begins life with a missionary zeal, the overwhelming economic and political power of the regulatees eventually sabotages the regulatory process.[72]

Special interest groups would experience much greater difficulties "capturing" the government agencies that would administer a pollution tax system. In a pollution tax system, a government agency would levy taxes on each unit of discharge according to its nature and composition. Each polluter would have to monitor, measure, and report its discharges in order to determine its pollution tax liability for a specified period of time. The government agency's enforcement activities would include setting and periodic revision of pollution tax rates, when changes in pollution control costs or changes in pollution costs (damages) occur; promulgation of rules and standards of accuracy for measuring units of pollution; certification of monitoring devices; and periodic auditing of polluters to establish the accuracy (calibration) and continuous use of their monitoring devices.[73]

The pollution tax system outlined above would limit a government agency's administrative discretion. The extensive bargaining process which is an integral part of a regulatory agency's activities would be virtually nonexistent. In addition to policing compliance within this self-reporting tax system, the most visible, significant activity of a pollution tax agency would be the setting and periodic revision of tax rates—charges on pollution according to its nature and composition as well as its impact on overall environmental quality. There would be an objective criterion for judging the propriety of tax rates because if environmental quality standards are not met then tax rates need to be raised. Therefore, the tax system will remain effective as long as charges are revised to maintain environmental quality standards.[74]

### Taxation Versus Subsidization

Traditionally, the role of subsidization in public policy has been one of remedying inequities in regulatory programs and tax systems as well as compensating or rewarding individuals and institutions for their socially preferred behavior. Since we have assumed that the subsidy or benefit (see Figure 2-3) for reducing pollution by one unit is the same for all polluters, subsidization would be an incentive for a polluter to reduce his discharge to the point where the cost of pollution control per unit would be equal to the compensation for reducing pollution one unit. Thus, a subsidy of OA per unit of discharge should reduce the level of pollution to OB for each polluter, assuming all polluters experience identical marginal costs of pollution control for all levels of discharge (as in Figure 2-3).

The assumption of identical marginal costs of pollution control renders the choice of a pollution control strategy moot since regulation, taxation, and subsidization all result in the same cost in real resources to society. However, since such an assumption is unrealistic, the least real cost alternative for pollution control is the one that allows all polluters to reduce their discharges to a level where their marginal benefit (subsidy or tax) is equal to their marginal cost of pollution control. Thus, in the aggregate (see Figure 2-2), total pollution can be brought to an OB level if a total subsidy or tax of OA is realized (or incurred) proportionately by polluters. If the regulatory approach for control of pollution is adopted, then to effect an OB level of pollution through the use of discharge standards would only result in a least cost in real resources if all polluters experienced identical marginal costs of pollution control at all levels of discharge. In summary, uniform subsidization and proportional taxation are preferable to regulation through uniform pollution control standards.

In theory, one should be indifferent to the adoption of subsidization or taxation. However, detrimental equity implications are linked to the use of subsidies; and these implications are serious enough to render it a second best choice. The fundamental difference between a subsidy and a tax is that the subsidy is paid by taxpayers to polluters, whereas the tax is a cost of production and consumption. The pollution tax (penalty) ultimately affects the cost of production and consumption. In other words, pollution taxes can reduce the quantity of goods and services demanded and produced. Hence, charging individuals and institutions for their use of the environment can reduce monetary income for consumers and producers since environmental quality is paid for in higher prices and reduced output of goods and services. The environment is the common property of society.

In contrast, a subsidy presumes that individuals and institutions (and not society) own the common property resources and have the right to dispose of wastes therein. Consequently, society should pay consumers and producers not to discharge wastes. Thus, the subsidy is an opportunity cost because to forgo a reduction in discharges when the marginal cost of pollution control is less than the subsidy (marginal revenue) does not maximize profit. In other words, the polluter should conceptualize waste reduction as another marketable good or service; the polluter is selling society its right to use common property resources.

In summary, subsidies rely on normal profit-maximizing behavior to attain the optimal level of pollution consistent with environmental quality standards at a minimum real resource cost. Individuals and institutions will reduce waste discharges efficiently, until the marginal benefit of waste reduction equals the marginal cost of such behavior. However, what precludes firms and individuals from using production processes and goods that generate waste discharges just to receive subsidy payments? In fact, as long as the marginal revenue is greater than the marginal cost of that reduction, individuals and institutions will undertake pollution-generating activities.

A more fatal theoretical objection to a subsidy policy is related to the distributional differences between taxing pollution and subsidizing pollution reduction. Since a pollution tax is a charge for using EAC, the costs of using the waste-assimilative capacity of the environment are passed on to producers and consumers of goods and services. The incidence of the pollution or use tax is closely related to the amount of EAC utilized in the production and consumption of goods and services. With a subsidy, incidence of the costs of attaining an optimal level of pollution control— that is, a level consistent with environmental quality standards —is not directly related to the amount of EAC used to produce and consume goods and services. The public sector's purchase of reductions in waste discharge is paid for by increases in taxes and/or reductions in the production of other public goods and services. Thus, the incidence of pollution control costs will depend on the incidence of increased taxes and/or the incidence of reductions in the benefits derived from public sector activities resulting from a reallocation of public expenditures.[75]

Inasmuch as the subsidy approach to pollution control is a mirror image of a pollution tax system, the functions of a pollution subsidy agency would be similar to those of a pollution tax agency. The primary function of a pollution subsidy agency would be to make payments to polluters for each unit reduction in waste discharge. Each polluter would have to monitor, measure, and report its discharge reductions. The pollution subsidy agency's enforcement activities would include the setting and periodic revision of subsidy schedules, promulgation of rules and standards of accuracy for measuring reductions in waste discharge, certification of pollution monitoring devices, and periodic auditing of polluters to establish the accuracy (or calibration) of their monitoring devices. Thus, enforcement costs are coincident with those of a pollution tax system.

## The Problem of Imperfect Information

Throughout the above discussion, the problem of establishing the environmental quality standards necessary to develop detailed waste discharge standards, to set pollution tax rates, and to determine subsidy payments for reduction of waste discharge is seemingly trivial. The implicit assumption is no informational problems are associated with a technical solution to developing standards. In reality, the informational problems inherent in determining the marginal benefit curve for waste reduction, or conversely the marginal social cost curve for pollution (damage function), are significant and often prohibitive.

These prohibitive information costs are implicitly discussed in demonstrating the inferiority of the regulatory process in establishing, monitoring, enforcing, and revising standards (allowable waste discharges for individuals and institutions) consistent with the public interest (environmental quality). The superiority of taxation and subsidization is not derived from their having overcome the informational problems, but from the reliance in both approaches on a market mechanism wherein the tax rates or subsidies can be periodically revised in an effort to attain and maintain a specified level of environmental quality.

At present (and for the forseeable future), it is technically impossible to determine the social costs of most pollutants with any accuracy.[76] Without knowledge of the damages caused by different quantities and types of waste discharge, the marginal social cost curve for pollution (or the marginal benefit curve for waste reduction) necessary for technically establishing the optimal level of pollution control and the pollution tax rate or subsidy cannot be determined.

Throughout the world, regulatory agencies responsible for pollution control are establishing standards for the discharge of all residuals deemed to be pollutants and are prohibiting the discharge of certain extremely hazardous pollutants. In effect, standards for environmental quality not explicitly established by a national legislative body are implicitly determined by a regulatory agency because of the broad interpretive authority given to such agencies to determine discharge standards. Thus, discharge standards and environmental quality are the outcome of a political or bargaining process involving the technocrats of a regulatory agency, and the scientists and legal counsel of the polluters.

## The Political Economy of Environmental Policy

Pollution cannot be defined scientifically or physically because "pure" air, "pure" water, and "pure" land do not occur in nature. In fact, unpolluted air, water, and land are incompatible with the notion of an ecological system wherein the waste discharges of natural and technical processes are routinely assimilated by the environment. In economic theory, the optimal level of pollution or pollution control can be determined where it is technically feasible to measure the social cost of specific waste discharges, which implicitly reflects a prior social decision on how pure the environment should be. Thus, the determination of a tolerable level of pollution is essentially an issue of political economy.

Through a political-legislative process and a political-administrative process, business, labor, environmental, and consumer groups are forced to compromise their respective positions in agreeing on an equitable, economical level of pollution control. Inasmuch as very few members of any legislative body are knowledgeable enough to establish pollution (waste discharge) standards, pollution reduction (subsidy) payments, or pollution tax rates, pollution control statutes are written in broad, general language. Detailed interpretation of the statutes is entrusted the technocrats who are employed by the government agency charged with maintaining environmental quality. Thus, bureaucrats develop rules and regulations to specify discharge standards, subsidy payments, and tax rates for specific types of pollutants. In summary, the legislative body establishes a public policy of environmental quality, but the government agency that operationalizes that policy has been given such broad, interpretative authority that it in effect acts as the legislative body.

There is nothing inherently unjust in giving an essentially lawmaking authority to a government agency. Since scientists and economists are not equipped to define pollution, the definition depends on the two political processes—legislative and bureaucratic—explained above. This definition is based on a concept of the human use of the environment. Therefore, while the level of environmental quality necessary for a specific use as well as the most efficient (least economic cost) method for achieving that level can be determined scientifically, the definition of what constitutes pollution is dependent upon the public's decision as to what use it wants to make of the environment.[77] As long as the technocrats employed by the pollution control agency interpret the environmental quality statutes in a manner consistent with legislative intent, the public interest and political economy are served.

The role of economics in the legislative and bureaucratic processes is to design public programs that maximize decentralized public decision-making. Moreover, having exhausted the benefits inherent in decentralized public decision making, economic analysis can be used to improve the internal incentives of bureaucrats to implement public policy in a manner consistent with its enabling legislation, that is, the public interest. Since public policymaking is an adversary process, political economists can best participate by improving the quality of the debate.

Through reliance upon a common body of knowledge, political economists can rigorously analyze public policy issues. However, the political economist must translate the theoretical or conceptual aspects of this common body of knowledge for those involved in the public policy debate— legislators, lobbyists, bureaucrats, et al. In practice, transitional problems are critical to a public policymaker's decision-making process because the "losers" in respect to income distribution may be constituents or clients. Hence, the political economist should include equitable, efficient compensation plans in the design of public programs that optimize both decentralized public decision making and the internal incentives of bureaucrats to operationalize social control legislation in the public interest.[78]

### The Tax Expenditures Issue

The ultimate decentralization of public decision making is inherent in the concept of tax expenditures. As noted in Chapter 1, tax expenditures are the indirect subsidies implicit in the tax incentive provisions of a tax law. Therefore, such implicit subsidization is an element of tax policy.[79]

A tax system can be used to alter the allocation of resources for social purposes through provisions for tax incentives. Examples are a tax credit for investment in capital equipment, accelerated capital recovery through rapid depreciation or amortization, and low-interest loans for industrial development through exclusion of interest income from taxation. To the extent that individuals and private institutions avail themselves of tax incentives, there is a loss of tax revenue. The amount of the tax savings and tax deferral of individuals and private sector institutions is also the amount of the tax revenue loss of the government. Taxes deferred, of course, are not really lost in the long run, whereas taxes saved (tax credits) are permanent losses of revenue.

In summary, tax expenditures are an indirect government expenditure to the taxpayer who arranges its economic affairs to qualify for a tax incentive. Having developed a conceptual framework for analyzing pollution control policy, we can now turn to the six subjects of this study.

# Notes

1. Kenneth E. Boulding, "The Economics of the Coming Spaceship Earth," in Henry Jarrett (ed.), *Environmental Quality in a Growing Economy* (Baltimore: Johns Hopkins University for Resources for the Future, Inc., 1966), pp. 9-11.

2. J. C. Davies III, *The Politics of Pollution* (New York: Pegasus, 1970), p. 19. For an update of this treatise, see the 1975 revised edition and J. C. Davies III, "The Greening of American Politics," *The Wilson Quarterly*, Summer 1977, pp. 85-96.

3. U.S. Council on Environmental Quality (CEQ), *Environmental Quality: The First Annual Report* (Washington, D.C.: U.S. Government Printing Office, 1970), p. 8.

4. For example, John McHale, *The Ecological Context* (New York: George Braziller, Inc., 1970); *Cleaning Our Environment: The Chemical Basis for Action* (Washington, D.C.: American Chemical Society, 1969); Raymond F. Dasmann, *Environmental Conservation*, rev. ed. (New York: John Wiley & Sons, Inc., 1972); and CEQ, *Environmental Quality: Annual Reports* (Washington, D.C.: U.S. Government Printing Office, 1970-80).

5. CEQ, *Environmental Quality: The Third Annual Report* (Washington, D.C.: U.S. Government Printing Office, 1972), p. 4.

6. Ibid., p. 3.

7. CEQ, *Environmental Quality: The First Annual Report*, p. 62.

8. Air Conservation Commission, *Air Conservation*, AAAS Publication No. 80 (Washington, D.C.: American Association for the Advancement of Science, 1965), pp. 24-26.

9. Committee on Chemistry and Public Affairs, *Cleaning Our Environment: The Chemical Basis for Action* (Washington, D.C.: American Chemical Society, 1969), p. 24.

10. CEQ, *Environmental Quality: The Fourth Annual Report* (Washington, D.C.: U.S. Government Printing Office, 1973), p. 266.

11. CEQ, *Environmental Quality: The Third Annual Report*, p. 5.

12. Committee on Chemistry and Public Affairs, *Cleaning Our Environment*, p. 23.

13. Donald H. Pack, "Meteorology of Air Pollution," *Science*, November 27, 1964, pp. 1119-20.

14. Ibid., p. 1120.

15. Committee on Chemistry and Public Affairs, *Cleaning Our Environment*, p. 25.

16. Lester B. Lave and Eugene P. Seskin, "Air Pollution and Human Health," *Science*, August 21, 1970, p. 730.

17. Committee on Chemistry and Public Affairs, *Cleaning Our Environment*, pp. 78-80.

18. CEQ, *Environmental Quality: The First Annual Report*, pp. 70-71.

19. Committee on Chemistry and Public Affairs, *Cleaning Our Environment*, p. 46.

20. Philip A. Leighton, "Geographical Aspects of Air Pollution," *Geographical Review*, Vol. 56 (1966), pp. 151-64.

21. Ibid., pp. 170-74.

22. Daniel H. Henning, *Environmental Policy and Administration* (New York: American Elsevier Publishing Co., Inc., 1973), p. 117.

23. CEQ, *Environmental Quality: The Fourth Annual Report*, p. 279.

24. CEQ, *Environmental Quality: The First Annual Report*, pp. 30-39.

25. Committee on Chemistry and Public Affairs, *Cleaning Our Environment*, p. 96.

26. Ibid., p. 99.

27. Ibid., p. 96.

28. Ibid., p. 95.

29. Ibid., pp. 95-96.

30. CEQ, *Environmental Quality: The Fourth Annual Report*, pp. 29-30.

31. Committee on Chemistry and Public Affairs, *Cleaning Our Environment*, p. 101.

32. Ibid., pp. 102-3.

33. Henning, *Environmental Policy and Administration*, p. 120.

34. CEQ, *Environmental Quality: The First Annual Report*, pp. 106-10.

35. Committee on Chemistry and Public Affairs, *Cleaning Our Environment*, pp. 177-78.

36. E. R. Kaiser, "Refuse Reduction Processes," in *Proceedings: The Surgeon General's Conference on Solid Waste Management for Metropolitan Washington*, U.S. Public Health Service Publication No. 1729 (Washington, D.C.: U.S. Government Printing Office, 1967), p. 93.

37. Committee on Chemistry and Public Affairs, *Cleaning Our Environment*, pp. 168-70.

38. CEQ, *Environmental Quality: The Fourth Annual Report*, p. 97.

39. Wesley Marx, *Man and His Environment: Waste* (New York: Harper & Row, Publishers, Inc., 1971), pp. 128-31.

40. CEQ, *Environmental Quality: The Fifth Annual Report* (Washington, D.C.: U.S. Government Printing Office, December 1974), p. 131.

41. Committee on Chemistry and Public Affairs, *Cleaning Our Environment*, pp. 180-81.

42. Council on Economic Priorities, *Steel—The Recyclable Material* (New York: Council on Economic Priorities, 1973), p. 7.

43. Committee on Chemistry and Public Affairs, *Cleaning Our Environment*, p. 186.

44. Allen V. Kneese, Robert U. Ayres, and Ralph C. D'Arge, *Economics and the Environment* (Baltimore: Johns Hopkins University for Resources for the Future, 1970), p. 4.

45. Ibid., pp. 4-5.

46. M. King Hubbert, et al., *Energy Resources*, National Academy of Sciences (Washington, D.C.: National Research Council, 1962), Publication 1000D.

47. Kneese, Ayers, and D'Arge, *Economics and the Environment*, pp. 5-10.

48. Active implies chemically active inputs. Therefore, enormous amounts of inactive inputs—construction materials including stone, sand, gravel and other minerals used in construction—and inactive residuals—mine tailings and other solid wastes generated in the processing of minerals to develop raw materials for production—are excluded from the following description of materials flow in a developed economy because they undergo no chemical change, even though they are physically moved.

49. In the long term, $CO_2$ may be harmful if it accumulates in the atmosphere over time and thereby causes worldwide climatic changes.

50. Matthew Edel, *Economies and the Environment* (Englewood Cliffs, N.J.: Prentice-Hall, Inc., 1973), pp. 60-61.

51. R. U. Ayres and A. V. Kneese, "Production, Consumption, and Externalities," *American Economic Review*, Vol. 59, No. 3 (1969), p. 286.

52. Ibid., p. 287.

53. Kneese, Ayres, and D'Arge, *Economics and the Environment*, p. 13.

54. The analysis provided in this section is based on Kneese, Ayres, and D'Arge, *Economics and the Environment*, pp. 74-75; and A. Myrick Freeman III, Robert H. Haveman, and Allen V. Kneese, *The Economics of Environmental Policy* (New York: John Wiley & Sons, Inc., 1973), pp. 65-67.

55. For slightly different macroeconomic models, see Wassily Leontief, "Environmental Repercussions and the Economic Structure: An Input-Output Approach," in *The Review of Economics and Statistics*, Vol. 52, No. 3 (August 1970); Finn R. Forsund and Steinar O. Strom, "Outline of a Macro-Economic Analysis of Environmental Pollution: A Multisectoral Approach," in *Problems of Environmental Economics* (Paris: OECD, 1972); and William J. Baumol and Wallace E. Oates, "A Gallery of Externalities Models," in *The Theory of Environmental Policy* (Englewood Cliffs, N.J.: Prentice-Hall, Inc., 1975).

56. Ayres and Kneese, "Production, Consumption, and Externalities," *American Economic Review*, pp. 295-96.

57. For a more detailed analysis of the problem of "the Commons," see Garrett Hardin, "The Tragedy of the Commons," *Science*, December 13, 1968, pp. 1243-48.

58. The examples and explanation that follow are adapted from Freeman, Haveman, and Kneese, *The Economics of Environmental Policy*, pp. 72-77.

59. Other forms of market failure that are pertinent to the issue of pollution control include the problems of transaction costs and imperfect information, both of which are explained later in this chapter.

60. The above examples and explanation are adapted from Freeman, Haveman, and Kneese, *The Economics of Environmental Policy*, pp. 72-77.

61. Jared E. Hazleton, "Testimony of Economists Regarding the Proposed Rule-making," Report of Southwestern Econometrics, Inc., to U.S. Department of Labor (Austin, Tex.: Consultant's study, October 10, 1975), pp. 111-12.

62. See Paul A. Samuelson, "Pure Theory and Public Expenditure and Taxation," in J. Margolis and H. Guitton (eds.), *Public Economics* (New York: St. Martin's Press, 1969), p. 108.

63. Freeman, Haveman, and Kneese, *The Economics of Environmental Policy*, pp. 76-77.

64. Guido Calabresi, "Transaction Costs, Resource Allocation, and Liability Rules," *Journal of Law and Economics*, Vol. 10, No. 3 (April 1968), pp. 67-73.

65. Ronald H. Coase, "The Problem of Social Cost," *Journal of Law and Economics*, Vol. 3, No. 1 (October 1960), pp. 1-44.

66. Ibid., p. 16.

67. Jared E. Hazleton, "Public Policy for Controlling the Environment," *Journal of Urban Law*, Vol. 48, No. 3 (April 1971), p. 639.

68. The damage function, or marginal social cost schedule, is a term used to refer to increases in the cost of production and in the disutility to consumers generated by an externality or external diseconomy in a perfectly competitive economy. For a more detailed explanation of the price structure which can sustain a competitive equilibrium that is Pareto-optimal, see William J. Baumol and Wallace E. Oates, *The Theory of Environmental Policy* (Englewood Cliffs, N.J.: Prentice-Hall, Inc., 1975), pp. 14-55.

69. Edwin Mansfield, *Microeconomics: Theory and Applications*, shorter 2d ed. (New York: W. W. Norton & Co., Inc., 1976), pp. 367-68.

70. Freeman, Haveman, and Kneese, *The Economics of Environmental Policy*, pp. 102-4.

71. Robert Sherrill, "Nader's People Keep Shoveling It Out," *New York Times Book Review*, March 4, 1973, pp. 3-4. Also see Paul MacAvoy (ed.), *The Crisis of the Regulatory Commissions* (New York: Grossman Publishers, 1970); and David Zwick and Mary Benstock (eds.), *Water Wasteland* (New York: Grossman Publishers, 1971).

72. Joseph J. Seneca and Michael K. Taussig, *Environmental Economics* (Englewood Cliffs, N.J.: Prentice-Hall, Inc., 1974), pp. 206-7.

73. Donald N. Thompson, *The Economics of Environmental Protection* (Cambridge, Mass.: Winthrop Publishers, Inc., 1973), pp. 186-87.

74. Freeman, Haveman, and Kneese, *The Economics of Environmental Policy*, p. 106.

75. Seneca and Taussig, *Environmental Economics*, pp. 222-23.

76. K. William Kapp, "Environmental Disruption and Social Costs: A Challenge to Economics," *Kyklos*, Vol. 23, No. 4 (1970), pp. 833-48.

77. Davies, *The Politics of Pollution*, pp. 18-19.

78. Charles L. Schultze, "How Can Government Needs for Policy Research Be Better Met," 88th Annual Meeting of the American Economic Association, December 29, 1975.

79. Stanley S. Surrey, *Pathways to Tax Reform* (Cambridge, Mass.: Harvard University Press, 1973), pp. 1-7.

# 3

# Canada

Most of Canada's 23 million people live in urban centers located within a 100-to-200-mile-wide corridor that runs north of the Canadian-U.S. border.[1] Consequently, Canada's major pollution problems are very much like those experienced by the urban centers of the northern United States—air pollution from motor vehicles, utilities, and industry; water pollution from industry, agriculture, and municipalities; and solid and hazardous waste pollution from municipalities and industry. Other environmental concerns include the protection of the Atlantic and Pacific coastal zones and the fragile ecology of the Arctic areas of Canada, the restoration of the Great Lakes, "acid rain," and the control of pollution resulting from the extraction of natural resources. While the pulp and paper industry is generally considered to be the major polluter, several other major Canadian industries have significant pollution problems, namely, the chemical, petroleum, electric power, and food processing industries.[2]

## Pollution Abatement Programs

Until the early 1970s, legal and administrative responsibilities for pollution control in Canada rested almost entirely with provincial and municipal governments. In response to the concern for the environment, however, the Federal government has taken a much more active role in establishing national environmental policy and in coordinating provincial and municipal efforts to control pollution.[3] Since 1970, there have been numerous manifestations of the Federal government's more active role: the establishment of Environment Canada, a national ministry concerned solely with the environment, in June 1971; the passage of two significant pieces of

49

legislation— the Canada Water Act of 1970 and the Clean Air Act of 1971; and the amendments to the Fisheries Act of 1952.[4]

Under the Canadian system of government, the provinces have broad lawmaking authority in respect to civil and property rights,[5] including the authority to regulate the use of those natural resources lying within their jurisdiction. Therefore, the provinces have original jurisdiction over air and water quality as well as land use and solid and hazardous waste.[6]

The Federal government has the authority to regulate matters related to interprovincial trade; export trade; industrial development; navigation, shipping, and fishing on inland and coastal waters; and all powers that are not specifically granted to provincial governments, that is, residual powers. With regard to many public policy issues, the authority of the Federal government overlaps that of the provincial government. Consequently, many public programs are designed and operated cooperatively by both levels of government.[7]

Even though the provinces have original jurisdiction over property and civil rights, each has delegated much of the responsibility for local matters— such as water supply, solid waste collection and disposal, wastewater collection and treatment, land use planning in urban and rural areas, urban recreation, and public health—to municipal and/or regional governments. Each province has developed an extensive body of law in respect to water supply, public health and sanitation, power development, irrigation, land use and development, water reclamation, and recreation. In general, pollution problems are regulated under provincial legislation that deals with all types of environmental pollution, whether the degradation is of the water, air, or land, or by liquid, gaseous, or solid wastes. In carrying out their provincially specified responsibilities, municipal and regional governments are authorized to finance their activities by imposing property taxes, by charging fees for both services and licensing, and by borrowing, whereas the provinces rely on their share of Federal income tax collections and on both general and selective sales taxes.

All provinces maintain separate departments of public health and municipal affairs. The public health department protects human health by enforcing minimum standards of water quality and waste disposal. The municipal affairs department, relying on the province's power to approve and influence local government's activities, not only coordinates municipal and regional activity, but also subsidizes local revenue sources from its income, sales, and excise tax revenues. It also enforces minimum standards of quality and adequacy in respect to local government services, such as land use planning. All provinces assign electric power generation, distribution, and development to semi-autonomous, quasi-public corporations. Most provinces have separate ministries or departments with responsibility for tourism planning and development, and management of public lands and waters, and wildlife and fish. With regard to the management of other natural resources and the protection of the human environment, there are significant regional variations—the Maritimes, Quebec, Ontario, the Prairies, and British Columbia—in the structure of the provincial administrative machinery for environmental protection.[8]

The Federal government has *exclusive authority* to regulate only those aspects of pollution related to its jurisdiction over navigation, shipping, and fishing and over interprovincial and export trade. In practice, the Federal government's environmental policy is based on cooperation with provincial governments whenever practicable. Evidence of this strategy is provided below.

Under the Governmental Organization Act of 1970, the environmental protection activities of several Federal government agencies were reorganized in the creation of the Department of the Environment, which is popularly known as Environment Canada.[9] The Trudeau government's original Environment Minister, Jack Davis, delineated the general goals of Environment Canada:

(1) to maintain the capacity to meet historical and statutory responsibilities for research and management of air, water, fish, forest and wildlife resources; (2) to clean up and control pollution; (3) to assess and control the environmental impact of major new programs, projects, and developments; (4) to improve the understanding of long-term environmental phenomena; (5) to promote and support international initiatives; and (6) to create a better public awareness and understanding of environmental issues.[10]

Environment Minister Davis may have been too zealous in operationalizing the general goals he established. After Davis lost his House of Commons seat in the July 8, 1974, national election, Prime Minister Pierre Trudeau made no attempt to arrange for the continued eligibility of his defeated cabinet Minister. For most of his tenure as Trudeau's Environment Minister, Davis found himself faced with a classic dilemma, which he explained as follows: "You get it both ways. The businessmen think the government is doing too much. The environmentalists say you're not doing anything. The Environment Minister gets caught in the middle."[11]

In other words, the environmental groups naively expect the instantaneous clean-up of an environment damaged by one hundred or more years of industrial and urban development, whereas polluters realize that the costs of cleaning up will impede economic growth. During almost four years as Environment Minister, Davis effected the creation of a new Department of Environment, presided over the strengthening and enactment of environmental legislation, developed policy for requiring impact studies on major construction projects, and introduced legislation that would give Environment Canada the authority for screening new products and chemicals that might be harmful to human health and the environment.[12] Davis's proposed Environmental Contaminants Act was finally passed in 1975, and it became effective on January 1, 1976. This act "provides Canada with a new tool for protection of the environment, which is claimed to be among the most advanced of its type in the world."[13]

In executing its general responsibility for protecting Canada's environment and natural resources, the Department of the Environment has been organized into several subdivisions: the Environmental Protection Service (EPS); Lands, Forests and Wildlife Service; Fisheries Service; Atmospheric Environment Service; Policy Planning and Research Service; and Finance and Administrative Service. The EPS is probably the paramount subdivision because it not only regulates air and water pollution, solid waste management and resource recovery, toxic substances, environmental impact assessments and noise, but also promulgates and enforces environmental guidelines, regulations, codes, and protocols. In addition, EPS functions as the liaison between Environment Canada Department of Environment and provincial authorities, Federal agencies, industry, and the general public.[14]

Although neither the Federal government nor most of the provincial governments have enacted a comprehensive environmental impact assessment law (like the National Environmental Protectional Act enacted in the United States), both Federal and provincial environmental assessment programs or procedures have been established. During the first six years after the Trudeau government established an environment assessment program, more than 4,000 projects in which the Federal government had an interest were reviewed. Only about two dozen have required an environmental impact statement (EIS), including recent projects such as a proposed Alcan natural gas pipeline from Alaska through Canada to the United States, a proposal for oil and gas exploration in Canada's eastern Arctic waters, a nuclear power facility, a uranium refinery, an airport expansion, and a hydroelectric project.

In addition to the EIS requirement of the Federal government's Environmental Assessment and Review Process (EARP), there is a less stringent screening for projects referred to as an initial environmental evaluation (IEE). Thus, approximately 100 of the over 4,000 projects screened have undergone a required IEE. Finally, the remaining 3,900 projects were judged to have either no potential adverse environmental effects or minor effects that could be mitigated through environmental design changes. Therefore, over 90 percent of the projects reviewed were allowed to proceed without either an EIS or an IEE. Since 1977, any private company engaged in a project subject to screening for potential environmental effects has been incurring a major share of the required environmental study's cost, with the Federal agency involved therein incurring the balance.

While most of Canada's ten provinces have not enacted laws that explicitly require environmental impact assessment, some provincial laws provide their officials with the authority to require environmental impact statements or environmental studies. The scope of these provincial statutes varies, with some applying only to provincial government projects and others applying either to private projects or to cooperative provincial-private projects. In Alberta and Ontario, statutes provide specifically for required environmental impact assessments for both government and private projects. In general, such studies, whether required by law or by provincial officials, are paid for by the proponent of a project—either the private or public sponsor.[15]

## Water Pollution Control

Present legislation for controlling water pollution, provides for both direct regulation through the standard-setting authority of the 1970 amendments to the 1952 Fisheries Act and pollution taxation (or effluent charges) authorized by the Canada Water Act of 1970.[16] In addition, three pieces of legislation were enacted in 1970: the Arctic Waters Pollution Prevention Act, Oil and Gas Production and Conservation Act, and the Northern Inland Waters Act. These laws are designed to protect the water quality of Arctic waters north of the 60th parallel and the inland waters of the Yukon and the Northwest Territories. All three laws are being used to control northern development.[17]

The most unique aspect of Canada's water pollution control legislation is its agreement with the United States on the water quality of the Great Lakes. On April 15, 1972, Canadian Prime Minister Pierre Trudeau and U.S. President Richard M. Nixon signed the Great Lakes Water Quality Agreement which gives the International Joint Commission (established by a 1909 treaty that delineated boundary waters in the Great Lakes) much greater authority for studying and controlling pollution. The agreement is a landmark in international cooperation on the control of transnational pollution problems.[18]

As mentioned above, Canada's water pollution problems are similar to those experienced "south of the border." Canada's water problems include not only less-than-desirable water quality in the immediate vicinity of urban areas, but also flood plain management and localized water shortages. Therefore, conflicting demands for the urban, industrial, agricultural, and recreational use of water in the future will have to be resolved through regional water authorities that transcend municipal and provincial jurisdictions. Water pollution problems in Canada, as in the United States, are primarily attributable to concentrations of population and industrial activity. Consequently, the lower Great Lakes (Toronto, Hamilton, and Windsor), the Saint Lawrence River Valley (Montreal and Quebec City), and the lower Fraser River Valley (Vancouver) are the loci of the most severe pollution problems.[19]

The Fisheries Act, which was originally drafted at Canada's Confederation in 1867, has been amended several times recently both to strengthen and to expand its provisions. The law protects fish-inhabited waters from degradation by prohibiting the dumping therein of substances harmful to fish habitat, fish, or human use thereof. Thus, it is applicable to virtually all surface waters and, as such, functions as the principal water pollution control statute at the Federal level.[20] The most recent amendments to the Fisheries Act provide that anyone who owns waterfront property or proposes to build in an area that supports aquatic life must consult with Environment Canada's Fisheries Service before operating or constructing facilities that could disrupt or destroy fish habitat.[21]

Under the 1970 amendments to the Fisheries Act, regulations were promulgated in 1971 and 1972 to control the effluent both of the pulp and paper industry (Canada's largest industry) and of the chlor-alkali mercury industry. Pursuant to the Federal government's policy of working with provincial governments in resolving pollution problems, the aforementioned regulations were developed with the cooperation of provincial government representatives as well as representatives of both the pulp and paper and the chlor-alkali mercury industries.

The regulations to control water pollution in the pulp and paper industry were designed to substantially reduce discharges from old mills, whereas new mills have had to use the best practicable pollution control technology to minimize discharges. Compliance with these regulations is determined on a mill-by-mill basis, and the polluter and Environment Canada agree upon a time schedule for compliance. The regulations provide that only a specified amount of certain substances harmful to marine life can be discharged, namely, suspended solids, decomposable matter, and toxic wastes. For toxic wastes, a general test for the toxicity of effluent has been developed.

In addition to the regulations promulgated for the pulp and paper and chlor-alkali industries, Environment Canada either has developed or is developing discharge standards for municipal wastewater plants and several other industries, in cooperation with provincial governments and representatives of the industries affected. These industries include petroleum refining, mining, petrochemicals, steel, plastics, metal-plating, food processing, and animal feedlot operations.[22]

By the end of 1977, regulations and/or guidelines had been promulgated for the following industries: pulp and paper, chlor-alkali mercury, petroleum refining, food processing (including meat, poultry, fish, and potato), and the mining of uranium, iron ore, and other base metals. Moreover, regulations for the textile, metal finishing, and organic chemical industries were near completion. Regulations for all other major industries are being promulgated.[23]

The 1977 amendments to the Fisheries Act augment the Federal government's authority to regulate water pollution, especially for ongoing operations. The new amendments establish new procedures for reviewing any project that causes or may cause water pollution and new emergency clean-up authority wherein owners or carriers of a contaminant are required to clean up any spills. Now fishermen may take legal action to recover lost income attributable to water pollution from any source other than maritime vessels covered under the Maritime Pollution Claims Fund. Finally, fines have been substantially increased and penalties expanded, with the maximum fines for anyone discharging noxious substances now standing at $50,000 (Canadian) for the first offense and $100,000 (about US $80,000) for each subsequent offense. The actual amount of the fine remains at the discretion of the court,[24] but the increase in the maximum from $5,000 to $100,000 should materially augment the Federal government's enforcement power.

Under the 1970 amendments, the Federal government, with few exceptions, was only authorized to control the effluent from new, expanded, or altered industrial facilities. After several years of limited success in improving water quality through regulation of new operations, Federal environmental officials realized that effective water pollution control would be possible only when they had the authority to regulate the effluent of all operations—new and existing—and withdrawal of water from, discharge of effluent into, or any other use of water that would adversely affect fish-inhabited surface water. The 1977 amendments not only made effluent controls applicable to all operations, but also expanded the definitions of several key terms such as fish, deleterious substances, fish habitat, and fishing. Other revisions of definitions provide, in practice, for the control of indirect as well as direct sources of pollution. Another significant effect of the amendments will result from the authorization therein to convert existing guidelines (or standards) for the discharges of several industries to regulations. This conversion will occur over an extended period so that both provincial and industrial officials can be consulted.[25]

In summary, national effluent regulations and guidelines have been developed for most major Canadian industries under authority granted to the Federal government (in the Fisheries Act) to prohibit the discharge, without a permit, of deleterious substances into surface waters inhabited by fish. The regulations require a polluter to adopt the best practicable technology for pollution control and specify maximum discharges per unit of production, whereas guidelines establish interim discharge standards because pollution abatement technology is currently inadequate.[26] New facilities must be designed to comply with water pollution standards, and existing facilities must meet standards as soon as it is reasonably possible.[27]

There is still no explicit statement of ambient water quality objectives for the nation as a whole. Implicitly, one could assert that, based on the U.S.-Canada Agreement for Great Lakes Water Quality consummated in

1972 and renegotiated in 1978, the quality of Canada's surface water should be maintained at the level necessary to meet standards for all legitimate uses thereof, and there should be no degradation of unpolluted waters. Several provinces, including Ontario, have established ambient water quality guidelines. Eventually, there should be agreement on nationally applicable ambient water quality objectives, along with specific objectives to reflect the requirements of specific bodies of water or water basins.[28]

### Water Pollution Taxes

The mechanism for carrying out nationwide ambient water quality objectives became law on June 26, 1970, when Parliament enacted Bill C-144, popularly known as the Canada Water Act. The act provides for Federal water resource management in Canada with or without provincial initiative. Basically, the Water Act provides for Federal action in upgrading or maintaining water quality

1. Where there is a "national interest" in water quality or

2. Where, in the case of interjurisdictional waters, water quality is of "urgent national concern" and

3. Where a province refuses to upgrade water quality, the Environment Minister may establish either with the province or on his (or her) own a federal-provincial or federal agency to regulate the water quality within a specific basin or water quality management area.[29]

Moreover, with regard to strictly Federal waters, the Minister may establish water quality management agencies[30] and undertake agreements with provinces to set up intergovernmental committees to advise agencies on planning, management, and research.[31]

Water quality management authorities or agencies can delineate and enforce local water quality standards, design and operate facilities for wastewater treatment, and prosecute those who violate water quality standards. Upon summary conviction, violators would be liable for fines of up to $5,000 (U.S. $4,000) per day.[32]

The only regulations promulgated so far under authority provided in the Water Act were issued in 1970 to control the amount of phosphorus discharged into Canada's surface waters and thereby reduce the potential for eutrophication thereof. The Phosphorus Concentration Control Regulations set a limit on a detergent's phosphorus content at no more than 2.2 percent by weight, which is equivalent to 5 percent by weight as phosphorus pentaoxide.[33] In summary, the Water Act provides for "joint federal-provincial water basin planning, designation of water quality management areas, joint agencies for water quality management, and commissions to conduct water resource management programs."[34]

As an alternative to the regulatory approach—establishing standards, monitoring discharges, and prosecuting those violators which are detected with the intent of penalizing noncompliance, the Canada Water Act provides for the imposition of effluent charges or pollution taxes within a water quality management area. Section 8 of the act provides that "Except in quantities and under conditions prescribed . . . including the payment of any effluent discharge fee prescribed therefore, no person shall deposit . . . waste of any type in any waters comprising a water quality management area." In paragraph (c)(IV) of Subsection 13(1) of the act, a water quality management agency is authorized to recommend the imposition of effluent fees or pollution taxes to Environment Canada. Moreover, in paragraph (c) of Subsection 13(3), the agency is authorized to collect effluent charges. Paragraph (d) of Subsection 16(2) authorizes Environment Canada to prescribe effluent charges. However, the act does not provide a technical basis or guidelines for determining how effluent charges are to be established. Unfortunately, no water quality management authorities or agencies had been established as of June 30, 1982.[35]

Charges for the use of a public resource like water are not a recent development in Canada because in 1859 water fees were initially imposed in British Columbia. Inasmuch as local or regional government in eight out of ten Canadian provinces (all except New Brunswick and Prince Edward Island) and the Federal government in the Yukon and Northwest Territories assess fees for some (or all) uses of water, there exists a basis upon which to develop a comprehensive effluent charge or pollution tax system. However, only the charge systems in Saskatchewan and British Columbia were designed to reflect both the quality and quantity of water used. The other charge systems have been based on the quantity of water used.[36]

As noted above, all provinces have enacted omnibus environmental protection statutes that include provisions applicable to water quality protection. Generally, the provincial laws prohibit or regulate pollution. In addition, several provinces complement this regulatory approach by providing economic measures that discourage environmental degradation and by integrating the comprehensive management of natural resources, such as water, with their overall responsibility for the control of local and regional land use planning. Probably the most common form of water quality regulation is the requirement that all wastewater treatment systems be approved through issuance of a permit that specifies the appropriate design, construction, and operation thereof. Generally, there is a standardized set of technical requirements with which any domestic, municipal, or industrial wastewater treatment system must comply. If no standards have been established, the provincial authorities will advise the polluter as to the appropriate design for a permissible treatment system.

Another pervasive form of provincial regulation of water quality is the issuance of specific effluent regulations for those industries that are deemed to be major water polluters or the establishment of limits on the discharge of specific pollutants from any source. Other tactical approaches to the implementation of the regulatory strategy for controlling water pollution include the following: (1) under authority provided in provincial public health acts, prohibition by public health officials of activity, such as water supply sources and sewage treatment, which is or might be deleterious to human health; (2) under authority provided in provincial legislation regulating the extractive industries, imposition of standards for exploration, development, and production of natural resources; (3) under authority provided in fish and game legislation as well as pesticide legislation, control of activities harmful to marine and wildlife as well as public health; and (4) under provincial legislation establishing environmental impact assessment, control of the planning of major public and private projects.

In respect to the economic measures utilized in the provinces to minimize water pollution, the explanation of water use charges or pollution taxes that follows constitutes only one side of the "economic-controls coin." The other side—subsidization—will be covered in a later commentary on government assistance for pollution control. As discussed above, charges for the use of water were first imposed in British Columbia over a century ago. Currently, all municipalities collect funds from household, commercial, and industrial users of public sewage systems. The charges for water use and wastewater disposal are usually collected together, and the sewage charge component is based on the quantity of water use, on the value of the water user's property, or on some combination of water use and property value. Such sewage charge systems are not technically effluent charges or pollution taxes because the amount of a water user's charge is not systematically related to the quantity and quality of pollution he (she or it) discharges therein.

A few municipalities have instituted an effluent charge or pollution tax system regarding industrial discharges of wastewater therein. In at least ten municipalities encompassing about 20 percent of the Canadian population, there are effluent charge schemes wherein industrial wastewater discharges in excess of a minimum level trigger a service charge systematically related to the amount and type of waste therein.[37] Seven of the municipalities imposing effluent charges are large communities with populations of over 100,000—Toronto, Edmonton, Winnipeg, Calgary, London, Hamilton, and the Waterloo Region (Kitchener, Waterloo, et al.). Furthermore, ordinances providing for the imposition of effluent charges either have been enacted but not implemented, or are under consideration, in several other municipalities.

The charge schemes in all ten municipalities are similar in that industrial dischargers are regulated in terms of both the permissible content of their effluent and service charges imposed according to the quantity and quality of their effluent. In establishing charge levels, municipal officials consider neither the capital costs of the physical plant nor the excessive marginal costs attributable to the treatment of industrial effluent where household and commercial discharges are relatively constant in terms of their quantity and quality. Moreover, the charges are imposed only on the excessive quality of the industrial discharges over the normal quality of commercial-domestic sewage. Finally, municipal officials are generally lenient in establishing the level of normal quality for commercial-domestic sewage. In Toronto, for example, effluent charges are levied when an industrial discharger's effluent exceeds both 500 ppm (parts per million) of BOD and 600 ppm of suspended solids (SM), even though the average commercial-domestic load actually varies between 200 and 300 ppm for BOD and SM.

The objective of these effluent charge schemes is to compensate municipalities for their marginal cost of wastewater treatment for industrial discharges. Based on the above commentary, as well as on the fact that as recently as 1973 revenues from industrial effluent charges amounted to less than 1 percent of the public sewer and wastewater treatment costs, it could be inferred that the success of the municipal pollution charge system has been limited.[38] However, the following trends would seem to mitigate drawing such a hasty conclusion:

1.  A tendency to reduce the "normal strength" allowed without charge to a more realistic level, thus, narrowing or eliminating the free margin.

2.  Increases in charge rates (per unit of waste) at several locations. In part, this factor reflects the increasing tendency to recover the full amount of operating costs that can be allocated to industry (and, in several instances, to recover annualized capital costs as well).

3.  The tendency to include a wider range of wastes in the charge schedule, and to charge wastes separately instead of on the basis of the "highest concentration" of several wastes.

4.  Implementation over the near future of effluent charge by-laws as planned by other municipalities, such as Windsor, Ontario, and Oshawa, Ontario, and a number of other communities.

5.  A possibility of extending effluent charges to a "zero-basis," which would serve to convert effluent charges to a major means of municipal taxation.[39]

There is a wide variation in rates between municipalities. For example, the rates for BOD and SM vary between four hundredths of a Canadian cent per 1,000 gallons in Calgary to over 7¢ per 1,000 gallons in Winnipeg. In spite of this variation, many industrial concerns located within the ten municipalities have reportedly reduced their effluent discharges as a result of the effluent charge systems. The most damaging observation one can make on the efficacy of the continued proliferation of municipal effluent charge systems concerns the limits on their physical and legal jurisdiction where comparable limits do not exist for industrial polluters.[40]

Eventually, the Canada Water Act could lead to a new approach to the solution of public problems. In practice, water quality management studies—including water quality studies, water chemistry studies, hydraulic modeling, biological studies of the river basin, land use studies, and population and economic studies—must be made before comprehensive effluent charge systems can be established. Moreover, case studies of other effluent charge systems, such as the *Genossenschaften* in Germany and the French *Agence Financière de Bassin*, must be made and/or analyzed in order to develop standard-setting and enforcement processes and with a view towards minimizing institutional constraints to efficient, equitable water quality management. Since such a large investment of time and resources is required before a water quality management agency is operational, it will be some time before this novel taxation approach as authorized by the Canada Water Act can be assessed.[41]

## Water Pollution Control in the Great Lakes

In 1972, Canada and the United States agreed to a series of programs for ensuring the quality of the water in the Great Lakes: (1) Harmonization of water quality standards for phosphorus, coliform bacteria, dissolved solids, dissolved oxygen, pH, iron taste, and odor;[42] (2) completion or near completion by the end of 1975 of municipal wastewater treatment facilities, including phosphorus removal, in areas surrounding the Great Lakes; (3) reduction in the phosphorus discharged into Lakes Erie and Ontario in order to effect a natural restoration of their oxygen-deficient waters by 1975; (4) institution of wastewater treatment facilities by all industry around the lakes; (5) maintenance of a joint U.S.-Canadian contingency plan for oil spills, with an extension of regulations governing waste discharges from shipping and onshore facilities; (6) improved management of dredging operations on the lakes; and (7) two studies, one on pollution levels and remedial measures in Lakes Huron and Superior and the other on the reduction of water pollution from land drainage and agriculture, to be made by the International Joint Commission.[43]

The Great Lakes Water Quality Agreement is important because it is an example of how international cooperation in solving transnational pollution problems can work. The magnitude of the pollution problem is self-evident when one realizes that the Great Lakes are the largest single reservoir of fresh water in the world, constituting 20 percent of the world's fresh water; one out of every three Canadians (over seven million) and one out of every seven Americans (about thirty million) live around these lakes; and industrial activity around these lakes accounts for 50 percent of Canada's gross national product and about 20 percent of the U.S. gross national product. By the end of 1975, Canada had authorized the expenditure of most of the $250 million it had pledged to spend on municipal wastewater treatment facilities,[44] and over 95 percent of the population on the Ontario side of the basin were served by municipal sewage treatment facilities.[45] By mid-1978, 99 percent of Ontario's Great Lakes' water basin population were being served by such treatment facilities, whereas only 63 percent of the U.S. Great Lakes population were linked to municipal wastewater treatment facilities.[46]

After more than a year of negotiations, Canada and the United States, on November 22, 1978, consummated a new Great Lakes Water Quality Agreement[47] integrating many of the International Joint Commission's recommendations for improvement along with the revisions proposed by the respective governments of the two countries. The 1978 agreement, which took effect in July, covers all of the Great Lakes, including Lake Michigan even though it is not on the Canadian-U.S. border. In general, Canada and the United States have agreed to zero discharge of those toxic substances deemed persistent; more stringent and comprehensive pollution limits, especially for phosphorus, than those agreed on in 1972; a requirement that there be pre-treatment of industrial effluent to avoid overloading municipal wastewater treatment facilities; commitment to have all municipal wastewater treatment facilities operating by December 31, 1983; a requirement that there be an annual site-specific compliance report for each wastewater treatment plant; establishment of stringent radiological standards; tightening of controls on discharges from vessels, dredging operations, and transportation of hazardous materials; and a study of the effect of air pollution on the lakes.

Exceptions to the objectives of the agreement are permissible in "non-attainment" areas. These areas are "limited-use zones" wherein exemption from some standards is allowed, but there is no provision for automatic release from satisfying the objectives of the 1978 agreement. In addition, all decisions by domestic regulatory agencies to exempt "limited-use zones" must be reviewed by the International Joint Commission (IJC). The commission has the power to "go public" if it differs with an exemption decision. The two governments have also agreed to submit any information the IJC requests. The annual audits of the condition of the Great Lakes made by the IJC's Water Quality Board will now be made biannually. The governments must comment on each IJC audit and must review the agreement after three biennial reports have been made.

Even though the 1978 Great Lakes Water Quality Agreement is not a treaty between Canada and the United States and does not have the force of domestic laws, it will still have the effect of law when its provisions are implemented through the promulgation of regulations in each country. Finally, the countries are obligated to seek the necessary funding and State-provincial cooperation.[48]

Based on several measures made during the ten years since the 1972 Great Lakes Water Quality Agreement was consummated, there have been material improvements in water quality within the Great Lakes system. For example, phosphorus levels have decreased significantly in several areas along the Canadian side of Lake Ontario and Lake Erie. Algae growth on Lake Ontario has also decreased, resulting in clearer water and better swimming conditions. The concentrations of mercury in most species of fish in the western basin of Lake Erie have declined, with mercury levels in some species near or below the acceptable guideline for human consumption. The water quality of Lake Michigan, Lake St. Clair, and the Detroit River has also improved.[49]

Air Pollution Control

Environment Canada is responsible for administering the Clean Air Act of 1971. This act authorizes Environment Canada to

set national air quality objectives . . . ;

set national emission standards where health or international agreements are involved;

set national emission guidelines to assist provinces and local governments in developing uniform regulations;

set specific emission standards for all works or businesses under Federal legislative authority;

regulate composition of domestic and imported fuels; and

enter into agreements with individual provinces to combat air pollution within a province or in interprovincial areas.[50]

A court can assess penalties of up to a total of $200,000 (US $160,000) for violating national emission standards.[51] Penalties can also be assessed up to $5,000 per day both for importing or refining prohibited fuels and for violating other parts of the Clean Air Act.[52]

Environment Canada has established national air quality objectives for five major air pollutants: particulate matter, carbon monoxide (CO), photochemical oxidants (ozone), sulfur dioxides ($SO_2$), and nitrogen dioxide ($NO_2$).[53] The Clean Air Act is very similar to the U.S. Clean Air Act of 1963 in that both delineate different levels of air quality. The Canadian act designates three levels of air quality: desirable, acceptable, and tolerable.[54] The above national air quality objectives reflected only the "acceptable" and "desirable" levels of air quality until April 1978, at which time Environment Canada announced the maximum "tolerable" levels for the five contaminants.[55] These national objectives were developed as a cooperative effort of provincial and Federal authorities.[56]

About 90 percent of the air pollution in Canada can be attributed to the five pollutants for which maximum "tolerable" levels of air quality provide a frame of reference for measuring air quality throughout Canada. Therefore, the maximum "tolerable" level is used as a basis for determining priorities in solving air pollution problems, especially with respect to the amount of surveillance required and the enforcement programs of control agencies. The "desirable" and "acceptable" objectives serve as a basis for preserving cleaner air in those parts of the country that are still unpolluted. "Desirable" air quality standards are indicative of a concentration level wherein the environment is not affected by the contaminant. "Acceptable" is the level wherein the minimal effects of contaminants are considered within reason as far as public health is concerned. Finally, the maximum "tolerable" standard has been explained as when ambient air quality in a region deteriorates to a level where prompt abatement action will be necessary to protect the public's health. Since the air quality objectives for five contaminants were first established in 1974, Federal monitoring of ambient air quality has demonstrated that the maximum "tolerable" levels of air pollution seldom occur.[57]

This air quality control approach to air pollution control initially relies on standards for the quality of ambient air wherein numerical values for the concentration of specific pollutants are established for an airshed. Generally, air quality objectives are based on perceived or known effects of pollutants on public health. Such an approach presupposes a monitoring system to test compliance with the objectives. Eventually, to be effective the ambient air quality standards approach must be supported by some type of source emission standards (or emission taxes).[58]

Pursuant to the Federal government's authority to establish both national air quality objectives and national source emission standards where there are significant health hazards or transnational air pollution agreements, Environment Canada has been and will continue to draft both source emission regulations and source emission guidelines to aid provincial and local governments in the development of the uniform regulations necessary for attaining national air quality objectives. By June 1978, the Federal government had already published source emission guidelines for cement production, asphalt paving processing, coke production, and Arctic mining. In addition, the guidelines for incineration, petroleum refining, and natural gas processing were near completion. Guidelines for several other major industries—thermal power, pulp/paper, nonferrous metals, iron, and steel—were in the drafting stage. Some provincial governments have adopted these Federal source emission guidelines as their source emission standards.

The national source emission regulations in effect and in process have primarily been promulgated in response to public health risks associated with the use and production of certain substances. There are Federal regulations limiting emissions of lead in gasoline and from secondary lead smelters; of asbestos from the mining and milling thereof; and of mercury from chlor-alkali plants. Regulations for controlling vinyl chloride and arsenic emissions and for requiring information on fuel additives have also been developed.[59]

With regard to motor vehicles manufactured in Canada or imported therein since 1970, the Motor Vehicle Safety Act provides the authority for the Federal government's regulation of evaporative, crankcase, and exhaust emissions. In respect to gasoline-powered motor vehicles, the act stipulates that the gasoline engine be designed and maintained so that there are no crankcase emissions and so that exhaust emissions for hydrocarbons and CO not exceed specified limits for the type of vehicle. Finally, the exhaust emissions of diesel-powered motor vehicles are limited in terms of opacity during acceleration and during normal operation.[60]

The provinces still maintain the authority to administer and enforce the national air quality objectives and national emission standards. It was the intent of Parliament that Environment Canada would "provide leadership in the compilation of source emission data, prescription of national emission standards, strengthening of national air quality objectives, control of air pollution from all works under federal authority, and the control of the composition of fuels produced in Canada or imported."[61] Some provinces have already adopted the national air quality objectives as their own in the implementation of their provincial air quality control legislation. As mentioned earlier, the national source emission guidelines are also being adopted by some provincial governments.[62] For the most part, however, provincial air quality control resembles provincial water quality management in that a variety of laws and enforcement agencies are being used in the provinces to control the two types of pollution.

Solid Waste Pollution Control

Canada has no Federal legislation that specifically addresses the problem of solid and hazardous waste pollution. Generally, responsibility for control of this kind of pollution is vested in the provincial and municipal governments.[63] The existing set of municipal, provincial, and Federal laws has been effective in dealing with the solid and hazardous waste pollution problem by ensuring that waste is properly treated and disposed of without having a deleterious effect on public health or the environment. This essentially "defensive" approach is concerned primarily with preventing water pollution from waste disposal.

At the Federal level, numerous laws authorize to a limited extent the regulation of solid and hazardous waste, including the Fisheries Act and the 1970 and the 1977 amendments; Navigable Waters Protection Act of 1959; National Parks Garbage Regulations (1968); Canada Water Act of 1970; Arctic Waters Pollution Prevention Act of 1970; Northern Inland Waters Act of 1970; Territorial Lands Act of 1970; Animal Contagious Disease Act of 1970; National Parks Act of 1970; National Harbors Board Act of 1970; Oil and Gas Production and Conservation Act of 1970; Canadian-United States Great Lakes Agreement of 1972 (renewed with revisions in 1978); Ocean Dumping Control Act of 1975; Pest Control Products Act of 1970; and Environmental Contaminants Act of 1976. As noted above, most of the laws regulate solid and hazardous waste indirectly by ensuring that waste-disposal operations do not jeopardize water quality. The Pest Control Products Act and the Environmental Contaminants Act are concerned with toxic waste generated during the production or use of substances that are hazardous to human, plant, and animal life, even though present in the environment at very low levels of concentration.

In respect to packaging, the Federal government is authorized under the Consumer Packaging and Labeling Act of 1971 to regulate the size and shape of containers. This authority could be adapted through amendments thereto for solid waste management objectives. Canadian breweries have voluntarily agreed to a standard 12-ounce beer bottle in most parts of Canada to facilitate recycling. There is also a national organization, the Canadian Association of Recycling Industries, which acts as a trade association for approximately 250 firms in such industries as ferrous metals, nonferrous metals, paper, and textiles. The association has gained one concession important to the economic health of the recycling industry — preferential transportation rates for some recovered materials.

Although the regulation of solid and hazardous waste varies somewhat from one province to another, provincial regulation is usually authorized through both public health and water quality or omnibus environmental protection legislation. In addition, special legislation pertinent to sewage waste, to litter (beverage containers), to bulky wastes (such as tires, appliances, and motor vehicles) and to the wastes of the extractive industries (mining, oil, and gas) has been enacted in several provinces. Much of the responsibility for the collection and disposal of household and commercial wastes has been delegated to the municipalities. Novel solid waste management experiments at the municipal level include separate collection of glass and paper to aid recycling and incineration of wastes to generate heat and power.[64]

The traditional techniques for solid waste management, namely, sanitary landfills and incinerations, cannot adequately deal with the projected tripling in the amount of solid waste over the next thirty-five years. Hence, alternative methods of solid waste management must be developed with or without Federal government encouragement. One alternative would be to establish regional disposal operations. As a result of local government reorganization in Ontario, Quebec, and British Columbia, the regional approach to solid waste management is becoming more evident in those provinces. This regional approach should facilitate profitable resource recovery and recycling, as well as ease the financial burden on municipal refuse-collection operations.

In some provinces, resource recovery and recycling not only have been encouraged but also are mandated by legislation.[65] Several provinces, including Alberta, British Columbia, New Brunswick, Nova Scotia, Prince Edward Island, Sasketchewan, and Ontario, have legislation that provides for control of the disposal of litter, especially beverage containers. All require deposits on returnable beverage containers. Several prohibit non-returnables and/or tax beverage containers.

In Alberta, the Beverage Container Act of 1971, which was effective as of January 1, 1972, provides for the mandatory use of refillable containers for carbonated beverage containers except those containing beer and those of size 3 ounces or less and 60 ounces or more. The minimum deposits are 2¢ (Canadian) on soft drink containers and 5¢ on wine and liquor containers. In addition, the act provides for licensed recycling depots which are reimbursed 1½ cents plus the manufacturer's deposit for each wine or liquor container. Any violation of the act is punishable by a fine of up to $1,000 (US $800) upon conviction.[66] Both the Alberta Beverage Container Act[67] and the B. C. Litter Act explained below provide for deposits on those standardized refillable beverage containers for which throwaway containers are prohibited, licensed recycling deposits, and fines for violators.

In British Columbia, the Litter Act, which became effective on July 1, 1970, prohibits the use of throwaway bottles for beverages and establishes penalties for those convicted of littering on land or water. Initially, the act provided for a mandatory refund of 2¢ (Canadian) or more on soft drink or beer containers.[68] As of November 1, 1974, the mandatory refund was increased to a minimum of 5¢ per container for those 16 ounces or less in size, whereas for containers of less than 40 ounces but greater than 16 ounces, the refund must be at least 10¢ per container. The higher refunds do not apply to the metal containers with "flip top" or other mechanisms that are detached upon opening the can. In fact, on January 1, 1975, the sale of such "flip top" metal beverage containers was prohibited. For purposes of the B. C. Litter Act, "beverage" includes ale, beer, cider, and carbonated drinks.

These two amendments to the B. C. Litter Act were made in an effort both to ensure a more complete recycling of beverage containers and to reduce the amount of litter in British Columbia. During the first half of

1971, almost three million bottles were turned in at collection depots. During that same period, over eight million metal beverage containers were turned in for recycling, but that amounted to only about one-third of all such containers sold in British Columbia for the six-month period.[69]

In Ontario, regulations promulgated in April 1977 under the Ontario Environmental Protection Act require vendors of carbonated beverages (soft drinks) to package some of their products in returnable bottles, even though they would prefer to offer them only in nonreturnable bottles and cans.[70] The regulations were deemed necessary because voluntary industry cooperation with the provincial objective of significantly increasing the production of soft drinks in returnable containers during 1975 and 1976 proved fruitless. The breweries bottle beer in standardized containers.[71] In respect to soft drinks, there is a ban on all 10-ounce nonreturnable bottles, but other sizes—6½-, 26-, and 52-ounce bottles—are allowable. The Ontario Environment Ministry had originally planned to ban all nonreturnable bottles in anticipation of the Ontario Legislature's imposing a 5¢ tax on each can of carbonated beverage. However, a combination of both the inability of the ruling Progressive Conservative party to gain the support of the "liberal" legislators and lobbying by Ontario's soft drink industry, which had committed itself to package 75 percent of its beverages in returnable bottles by January 1, 1980, forced Environment Minister George McCague to cancel the planned banning of nonreturnable bottles.[72]

Hazardous wastes and toxic substances are controlled through provisions of the Environmental Contaminants Act (ECA), which was enacted in 1975 to take effect on April 1, 1976. In principle, the ECA provides for residual enforcement powers that are to be used only if other Federal or provincial environmental statutes and regulations are not adequate for dealing with a specific hazardous waste or toxic substances, such as polychlorinated biphenyls (PCBs), which were the first substances to be controlled under the act. The ECA is being used to restrict, but not to ban outright, hazardous wastes and toxic substances. Other substances that may be controlled include fluorcarbons, mercury, mirex, and polybrominated biphenyls (PBBs).[73] In June 1978, the Federal government published a list of substances delineated for priority investigation that "are either known to be harmful to human health or the environment or are suspected of being potentially harmful."[74]

## Government Assistance

As a member of the Organization for Economic Cooperation and Development (OECD), the Canadian government has officially adopted the "polluter pays principle" promulgated by the OECD Ministers in 1972. Therefore, the Federal government is precluded from providing financial assistance in the form of direct grants, tax expenditures, and preferential low-interest loans to producers and consumers that are required to implement pollution controls.[75] With regard to economic decisions made by producers and consumers either before pollution abatement legislation becomes effective or before more stringent pollution abatement standards take effect, there will always be some producers and consumers that suffer from very high short-term pollution control costs.[76] Temporary government assistance to such producers and consumers is not considered a violation of the "polluter pays principle."

Several provisions of the Canadian tax law offer tax incentives to industry to invest in pollution control facilities. Such tax provisions are supposed to make voluntary compliance with emission and effluent standards more attractive to the polluter than the delaying tactics that have proved so successful when a government employs direct regulation to alleviate public problems. In addition to using tax incentives, the Federal government subsidizes the construction of municipal wastewater treatment facilities.[77] Moreover, several provincial governments provide financial assistance for polluters. For example, Ontario, which is the locus of almost 40 percent of Canada's population and over 50 percent of its GNP, provides a loan program to assist small Canadian-owned companies in financing investments in pollution control equipment,[78] loans and grants to municipalities for sewage and waste-disposal investments, and rebates of the provincial sales tax on pollution control equipment.[79]

Paragraph (a) of Subsection 20 (1) of the Canadian Income Tax Act provides for the depreciation of the "capital cost" of assets used to earn income from business or property in accordance with regulations. Paragraph (t) of Subsection 1100 (1) of the Income Tax Regulations provides for the rapid depreciation of the "capital cost" of pollution control equipment and related structures at a rate of 50 percent per year. Pollution control equipment and structures are those assets acquired "primarily for the purpose of preventing, reducing, or eliminating pollution."[80]

Use of the word "primarily" means that in respect to new production facilities that incorporate the best practicable pollution control technology, taxpayers cannot depreciate pollution abatement equipment and related structures at the accelerated rate of 50 percent, and they must use the "capital cost allowance" rate provided for the "capital cost allowance" class into which their assets fall. If pollution abatement equipment and related structures effect recycling through materials recovery or by-products,[81] then a taxpayer can utilize the rapid depreciation rate only if such materials or by-products "were being discarded as waste by the taxpayer, or were commonly being discarded as waste by other taxpayers who carried on operations of a type similar to the operations carried on by the taxpayer."[82] Conversely, 50 percent accelerated depreciation is not available for investments in recycling equipment by waste or materials recovery firms.[83]

As regards water pollution control equipment and related structures, up to 50 percent rapid depreciation was authorized for assets acquired after April 26, 1965, and before January 1, 1974.[84] For air pollution control equipment and related structures, the rapid depreciation was available for assets acquired since March 13, 1970, and the incentive was to end on December 31, 1973. There may also be an additional allowance for property certified by the Minister of Supply and Services.[85] Both amortization provisions, which were originally to expire at the end of 1973, have been extended periodically in the Federal budget. Until the rapid amortization for pollution control investment was extended indefinitely for pre-1974 plants in a late 1978 Federal budget proposal,[86] there was a December 31, 1979, expiration date for the pollution control investment incentive in respect to industrial establishments existing or under construction as of December 31, 1973.[87]

The Income Tax Act provides for the current deduction of operating and capital expenditures—other than for land—for scientific research and development, which is defined as a systematic investigation or search carried out by means of experimentation or analysis in a field of science or technology to acquire new knowledge; to devise and develop new materials, products, or processes; and to apply newly acquired knowledge in making improvements to existing devices, products, or processes.[88] Moreover, expenditures to develop, test, and evaluate a prototype may be considered scientific research expenditures in certain cases.[89] Furthermore, the Department of Industry, Trade, and Commerce administers several direct grant programs for scientific research and development.[90] Both the current deduction for all expenditures on R and D and the subsidies for R and D are applicable to pollution control technology.

There is a 9 percent (which was 12 percent until November 15, 1978)[91] Federal sales tax on the manufacturer's selling price of Canadian manufactured or produced goods and on the duty-paid value of most imported goods.[92] Equipment utilized in reducing pollution, for sewage disposal, and in water purification systems is exempt from this tax.[93] This exemption, which is applicable to equipment that manufacturers or producers purchase to prevent, detect, reduce, or remove pollution, is an extension of the general exemption from the Federal sales tax provided for all production machinery and apparatus to pollution control equipment. In addition, Schedule III, Part XII (i) (e) of the Excise Sales Tax Act specifically exempts "goods for use as part of sewage and drainage systems."

With regard to reprocessed, used lubricating oil, there is an exemption from the Federal sales tax for the re-refined oil if the reprocessing refinery sells the oil back to the original user. All other sales of reprocessed waste oil are subject to the Federal sales tax.[94] There is also a special sales or excise tax on the acquisition of "gas-guzzling" automobiles manufactured in or imported to Canada. Although the objective of this special exaction is to discourage the purchase of heavy automobiles that burn more gasoline, there is also an incidental environmental benefit because fewer "gas-guzzling," heavy automobiles on Canadian highways should result in less air pollution. The "heavy-auto" excise tax is imposed on automobiles according to the following schedule:[95]

| Automobile Weight | Incremental Tax | Total Tax |
|---|---|---|
| Less than 4,425 pounds | $0 | $0 |
| From 4,425 to 4,525 pounds | 30 | 30 |
| Over 4,525 up to 4,625 pounds | 40 | 70 |
| Over 4,625 up to 4,725 pounds | 50 | 120 |
| Over 4,725 for each succeeding 100 pounds | 60 | 180+ |

In addition, since 1960 the Central Mortgage and Housing Corporation (CMHC), which was established by the National Housing Act, has been loaning municipalities at a preferential rate of interest approximately two-thirds of the capital cost of constructing wastewater treatment plants and the main sanitary sewers.[96] Originally, loans were not made to finance construction of storm sewer or water supply systems.[97] Amendments since 1960 have not only applied to storm sewers and water supplies, but they have also encouraged comprehensive land use and residential development in previously undeveloped areas.[98] If the approved project is completed within the agreed time period, the CMHC will forgive one-fourth of the loan[99] and approximately one-fourth of the interest accruing during the construction period.

The Federal government also subsidizes public water-supply and waste-water treatment systems through Department of Regional Economic Expansion (DREE) programs. These programs support infrastructure development in areas deemed growth centers and in agricultural service centers on the Prairies. DREE grants under the Regional Development Incentives Act serve as catalysts for employment in slow-growth regions. Grants are also available from DREE for pollution abatement investment by commercial and industrial enterprise.[100] Finally, all the provinces supplement Federal grants of low-interest loans.[101] The province of Ontario also subsidizes the construction of facilities for treating or recycling municipal solid waste.[102] To the extent that commercial and industrial enterprises use the public facilities subsidized by the Federal, provincial, and local governments, there is an indirect subsidy of the private sector's pollution abatement costs.

## Environmental Policy Developments, 1979-1981

With the exception of a nine-month period during 1979-80, the Liberal party of Prime Minister Pierre Trudeau has governed Canada since the late 1960s. Few developments in Federal environmental policy occurred during the 1979–81 period because of political turmoil over Federal-provincial relations, especially regarding national energy policy. Early 1979 was consumed by the politics of a national election that resulted in the formation of a minority government by the Progressive Conservative party of Joe Clark. Former Prime Minister Joe Clark was elected primarily because of his strength in resource-rich, sparsely populated western Canada. Consequently, he initiated reforms in the national energy policy by proposing the decontrol of the price of oil and gas, most of which is produced in the western provinces, and significant increases in excise taxes on motor vehicle fuels. Thus, the second half of 1979 was consumed by the formation of Clark's Progressive Conservative government and the debate over his controversial energy policy, which toppled the young government.[103]

The reelection of Trudeau's Liberal party in early 1980 did not end the Canadian preoccupation with energy policy. Although Clark's Progressive Conservative party was victorious again in western Canada, their losses in populous Ontario and in the Maritimes, along with Trudeau's Quebec support, gave the Liberals a majority in Parliament. Thus, the East-West dispute over energy policy continued to monopolize the efforts of the Federal government during 1980. When the Trudeau government proposed its national energy policy, including a continuation of price controls and "Canadianization" of the energy industry, in the fall of 1980, the western provinces screamed loudly about its constitutionality inasmuch as the Canadian British North America Act reserved the control of those national resources within the borders of each province to those provinces.[104]

The West's adamant opposition to the proposed national energy policy, along with the traditional secessionist attitudes of French Quebec, led Prime Minister Trudeau to propose a new Canadian Constitution that would replace the 1867 British North America Act. Thus, public attention in 1980 focused on the election and establishment of Trudeau's Liberal government, the development of a national energy policy, and early debate over a new Canadian Constitution, while 1981 was spent on negotiations with the energy-producing western provinces on the National Energy Program and on the development and enactment of a new Canadian Constitution. In summary, no significant changes or additions to Federal environmental law occurred during the 1979–81 period, although several developments occurred in the administration of environmental policy, especially at the provincial level. Since, an account of developments in environmental policy at the provincial level is beyond the scope of this text, references to such developments will occur only implicitly in the context of Federal environmental policy.

Environmental Contaminants Act

The Federal Environment Ministry (FEM), which is also known as Environment Canada, took several actions under the 1975 Environmental Contaminants Act (ECA) during the 1979-81 period. During 1979 and 1980, the FEM made several decisions to restrict the use of PCBs, which have been banned from use in the United States. The first challenge ever to the FEM's authority under the ECA occurred in 1979. However, in 1980 a Federal board of review supported the FEM's actions and recommended the eventual banning of *all* uses of PCBs.[105] Controversy also surrounded the proper method for disposing of PCBs that can no longer be used[106] and the clean-up of a massive underground spill of PCBs which occurred in 1976 and now threatens the water supply of Regina, Saskatchewan.[107] A partial ban on the use of chlorofluorcarbons (CFCs) as a propellant in aerosol products was also issued in 1980.[108] Finally, the December 1980 ban on the use of urea formaldehyde insulation was made permanent in 1981.[109]

Water Pollution Control

In late 1980, the FEM issued an interim report on water pollution abatement by the pulp and paper industry. The FEM report documents the marked decline in water pollution by that industry over the 1976-78 period as a result of 1971 regulations promulgated by the FEM under the authority of the Fisheries Act.[110] Similarly, two 1980 reports by the Ontario Environment Ministry indicate that the chemical contamination of fish in lakes and rivers has dropped significantly in recent years.[111] Conversely, the Fraser River in the Vancouver area of British Columbia continues to suffer degradation ostensibly because provincial environmental officials have relied too heavily on voluntary compliance by industrial and municipal polluters.[112] The emerging problem of "acid rain," as it affects the ecology of Canadian lakes, especially those in Ontario, is explained below, as are the progress and future agenda for controlling Great Lakes water pollution.

Environmental Protection Administration

Two changes in the organization of the FEM occurred in 1979. In March 1979, the status of the administrator of the Forestry Service was upgraded within the FEM. Later in April of that year, a new Department of Fisheries and Oceans was created by upgrading the staff of the old fisheries and oceans office within the FEM to a departmental level. Thus, the FEM is left with only water pollution control authority under the Fisheries Act, while the new department took the rest.[113]

## Future Environmental Policy

When the British Parliament approved the new Canadian Constitution in 1982, fundamental changes in the Federal-provincial governance of Canada could begin to occur. The resolution creating the new Constitution was introduced in Parliament in February 1981 and amended in April to satisfy provincial critics.[114] After the Canadian Supreme Court decided in September 1981 that the constitutional resolution being considered by Parliament would be legal if "provincial consent" were still necessary for any amendments thereof,[115] the Fourteenth Federal-Provincial Constitutional Conference was held in November 1981, where nine out of ten provinces (Quebec opposed) agreed to support the resolution if their proposed amendments were integrated therewith. These provincial amendments to the resolution were introduced in the House of Commons on the final day of the conference.[116]

When the House of Commons and Senate passed the constitutional resolution in December 1981, the final step in creating a new Constitution only needed the affirmative vote of the British Parliament in 1982 to end forever the right of that body to pass judgment on amendments to or creation of a Canadian Constitution. The new Constitution establishes rules that make the amendment thereof easier and requires two Federal-provincial conferences—one on the procedures for amending the new Constitution and the other on aboriginal (native indian) rights.[117] In addition, the new Constitution includes provisions dealing with civil rights, including human, mobility, legal, equality, language, and enforcement rights, some of which provincial law may overrule for five-year periods.[118] The Federal government's powers include the right to control trade and commerce, international affairs, banking, and criminal law, as well as the right to levy all types of taxes. A more nebulous, all-encompassing right to pass laws dealing with the "peace, order and good government of Canada" was also included. The provinces retain the power over highways, municipalities, property, and, for the most part, education which was usurped to the extent that linguistic freedom is guaranteed for minorities.[119]

The overriding issue for future Federal environmental policy is: how much more authority does the national government have now over natural resources inasmuch as the provinces retain their rights over property within their jurisdiction, while the Federal government acquired the right to pass laws dealing with peace, order, and good government of Canada? Resolution of this issue will occur over several years during which current Federal-provincial sharing in making environmental policy will remain essentially unchanged. More specific issues in the future development of environmental policy include solution of the transnational "acid rain" problem, development of comprehensive legislation on waste disposal (especially hazardous waste), further development of regulations on the production and use of toxic chemicals, the role of the Environment Ministry in the National Energy Program, and progress on the International Joint Commission's goals for improving the water quality of the Great Lakes.

## Acid Rain

The acid rain problem is transnational inasmuch as the Canadians claim that the burning of coal in the industrial Midwest (United States) emits sulfur dioxide ($SO_2$) and, to a lesser extent, nitrogen oxides ($NO_x$) which are carried north of the border and precipitated as "acid rain." This phenomenon has allegedly affected many lakes, especially in Ontario, threatening to destroy their ecology. Eventually, very little marine life will survive, and many lakes will be "dead." Acid rain may also cause damage to soil by leeching minerals and may accelerate the deteriorization of structures, especially those made of stone. The U.S. energy policy objective of switching from imported oil and valuable natural gas to coal has exacerbated the acid rain problem from the Canadian perspective.

Ironically, the two largest producers of $SO_2$ emissions in North America are in Ontario—the Inco Ltd. nickel smelting complex in Sudbury and Ontario Hydro's coal-fired, power-generating plants throughout the province. As explained above, the Federal government has established emission guidelines for specific industries, including ore-smelting and power generation, under the Clean Air Act. Based on these guidelines, provincial environmental agencies, such as the Ontario Environment Ministry (OEM), are supposed to enforce their air pollution control laws. OEM, in cooperation with FEM, has developed an extensive program for controlling the emissions of ore smelters and the publicly held Ontario Hydro power plants.[120] In fact, OEM has devoted considerable time and effort to controlling the emissions from the Inco smelting complex, which is the largest in the world, at the urging of the Federal government so that the United States cannot legitimately argue that Canada does not control emissions that create acid rain so why should it.[121]

Inco has also made considerable investment in pollution control equipment, and it has developed new pollution control technology that it hopes will ensure extension of an OEM pollution control order that expires in mid-1982. However, the OEM and FEM may insist upon even more stringent emission controls in an effort to impress the United States. If the control order is too restrictive, Inco has threatened to invest outside of Canada wherever pollution control standards are more compatible with its profitability insofar as it is not the only nickel producer in the world.[122] Only time will tell how seriously the Inco threat will be perceived by the OEM and FEM in their deliberations on a new pollution control order.

In early 1981, the OEM consummated an agreement with Ontario Hydro, the second largest producer of $SO_2$ and $NO_x$ emissions in the province, as part of the Canadian campaign to reduce acid rain sources and to set an example for the United States. In summary, the provincially owned power company agreed to the OEM's regulation to reduce its $SO_2$ and $NO_x$ emissions by 43 percent during the 1980s.[123]

The chronology of events leading up to the June 1981 negotiations on a Canada-U.S. treaty to control transnational air pollution is as follows: July 26, 1979, agreement to move from informal discussions on mutual problems to the more serious objective of a formal agreement; August 5, 1980, memorandum of intent to work toward a formal treaty; and June 1, 1981, negotiations on a formal treaty begin.[124] As noted above, the U.S. coal conversion policy, as well as the threatened attack on the U.S. Clean Air Act which must be reenacted in 1982, makes the Canadians very anxious about the chances of working out an effective solution to the acid rain problem through the treaty vehicle. Thus, the Federal government and the Ontario provincial government have funded a lobbying effort in Washington, D.C., to ensure that the U.S. Clean Air Act is not reenacted in a weakened form. As of mid-1982, it was not known when in 1982 the U.S. Clean Air Act would be passed or, therefore, when the treaty negotiations would be completed insofar as the details of an agreement are contingent upon reenactment of the Clean Air Act. Section 115 of that act authorizes the Environmental Protection Agency to control transnational air pollution where a reciprocal agreement can be consummated.[125]

There already is a precedent for a U.S.-Canada bilateral agreement on air pollution. Under the auspices of the International Joint Commission, a Memorandum of Understanding on transnational air pollution was consummated between Michigan and Ontario in 1974. The agreement probably contributed to a significant reduction in pollution from $SO_2$ and particulate matter between 1972 and 1975. Unfortunately, similar progress was not made between 1976 and 1978 when particulate matter remained generally at levels deemed "unsatisfactory," while sulfur dioxide emissions reached levels that were generally acceptable. Therefore, the goal of achieving acceptable levels of both types of pollutants by 1978 was not realized.[126]

### Great Lakes Pollution

Cooperation between the United States and Canada on reducing the pollution of the Great Lakes demonstrates that transnational pollution problems can be continually mitigated through bilateral agreements where both parties make "good faith" efforts to comply with the objectives of their agreement. As explained above, the IJC and the Water Quality Agreements of 1972 and 1978 have been the vehicles for improving the quality of water in Lake Erie and Lake Ontario. However, much more improvement is possible, and new pollution problems seem to emerge regularly.

In spite of provisions in the 1978 agreement to deal with chemical pollution of the Great Lakes, the problem of toxic pollution is far from being resolved. For example, the State of New York in the summer of 1980 approved a pipeline to discharge treated wastes through a pipeline into the Niagara River which runs into Lake Ontario, even though the 1978 Great Lakes Water Quality Agreement calls for zero discharge of toxic chemicals into the Great Lakes.[127] In early 1981, the Canadian government, relying on the U.S. Freedom of Information Act, discovered that during World War II the United States secretly dumped 37 million gallons of radioactive and toxic wastes from the Manhattan Project into the Niagara River. At about the same time, the IJC recommended that all discharges of chemical wastes into the Niagara River be stopped.[128] In late 1980, the United States and Canada agreed to conduct joint studies to determine the source of dioxin pollutants in the Great Lakes. The Canadian officials are also concerned that the temporary closing of two hazardous waste dumpsites in the Niagara Falls, New York, area in 1981 will result in the increased dumping of such wastes by Canadians in Ontario and that the wastes will eventually find their way into Lake Ontario.[129] Finally, one should not forget that Love Canal was formerly part of the Great Lakes water system.

In summary, progress on improving the water quality of the Great Lakes has been slow, in spite of the discharge limitations established for most pollutants from point-sources. Reductions in DDT, PCBs, and phosphorus have been dramatic in some areas of the lakes. Thus, some of the pollutants are entering the lakes from nonpoint-sources, such as urban and agricultural run-off. Moreover, many point-sources (especially sewage plants) are periodically or continually emitting contaminants at levels higher than allowed. Some pollutants are entering the lakes from precipitation because of air pollution. Finally, the Love Canal fiasco has exacerbated the toxic and organic chemical wastewater pollution problem because the shortage of legal waste dumpsites apparently has resulted in increased levels of those contaminants in the Great Lakes.[130]

## Waste Disposal

Public policymakers sometimes seem to forget that we live in a physical world where the reduction of air and water pollution does not eliminate the waste generated by industrial activity, but rather it changes the form of the waste. Consequently, every industrialized nation in the world seems to be facing the same problems related to waste disposal, especially hazardous waste disposal, because of successful efforts to reduce air and water pollution. Temporarily, industry can store waste at the production site. Eventually, however, so much waste accumulates that it must either be recycled or disposed of in an environmentally sound manner—bury, incinerate, or neutralize. Thus, Canada is facing a hazardous waste-disposal problem today because of its success in combating air and water pollution, and because such wastes can no longer be transported to the United States for disposal. Nor do communities in Canada seem to be willing to accept a proliferation of hazardous waste dumpsites when the Love Canal disaster is still on their minds.

Although the FEM has some authority to regulate hazardous substances, for example, the Environmental Contaminants Acts, the Transport of Dangerous Goods Act, the Fisheries Act, the Pest Control Products Act, and the Ocean Dumping Control Act, there is no Federal legislation that deals comprehensively with waste disposal inasmuch as each province retains the authority to regulate waste disposal. For example, Ontario relies primarily on its Environmental Protection Act and regulations applicable to waste management, and containers for carbonated beverages to control wastes. Other provinces have developed similar regulatory measures for controlling wastes. Because Ontario is the home of over one-third of the Canadian population and the locus of half the country's economic activity, its waste-disposal problems are most serious, especially with regard to hazardous wastes. In fact, about 200 abandoned dumpsites in Ontario are being investigated to determine their hazardous waste content. Those dumpsites nearest to or in urban areas are being checked first.[131]

Over the past few years, the FEM has worked with provincial environment ministries to develop comprehensive plans for the disposal of hazardous wastes. Most recently, the FEM and the western provincial environment ministries recommended that a single regional incineration plant in Alberta, along with physical-chemical treatment plants and a network of collection stations in all four western provinces (Alberta, British Columbia, Manitoba, and Saskatchewan), be established as the most cost-effective method for dealing with hazardous wastes in western Canada, including the Yukon and Northwest Territories.[132] Similar recommendations have been made for the Maritimes, Ontario, and Quebec.

### Environmental Contaminants Act

Although the FEM continues to implement the provisions of the ECA by reviewing old and new substances on its list of chemicals to be tested for safety in their production, importation, and use, the law does not include provisions that require cradle-to-grave accounting for those chemicals deemed carcinogenic or dangerous.[133] Canada has been very active in the OECD's efforts to harmonize the testing and regulation of toxic substances. Therefore, the Liberal government will propose amendments to the ECA to reflect the OECD's recommendations.[134] Eventually, cradle-to-grave accounting for hazardous substances could be integrated in the ECA. Consequently, the FEM would have the implicit authority to control the disposal of hazardous wastes as is the case in the United States under the Toxic Substances Control Act.

National Energy Program

In submitting the 1981 annual budget to Parliament in late October 1980, Trudeau's Liberal government included significant and controversial changes in energy policy, referred to as the National Energy Program (NEP). This program consists of three major components: subsidies to homeowners for conversion from oil to natural gas, electricity (nuclear power or coal), or wood for home heating except in Newfoundland and Prince Edward Island where subsidies are to be for conservation; "Canadianization" of the domestic oil industry, that is, 50 percent or more ownership of all companies that make up the domestic oil industry; and Federal tax incentives to stimulate frontier and offshore exploration for crude oil and natural gas. Thus, the NEP, which is legally known as the Oil and Conservation Act, encompasses the substitution of one nonrenewable resource (domestic and imported crude oil) for another (domestic natural gas), the Canadian control of the domestic oil industry, and the production of more nonrenewable energy supplies.[135] Price controls were continued with some revisions that hold natural gas prices below crude oil prices.[136] In summary, the Canadians are buying time and are allowing energy demand to grow by stimulating domestic energy production while controlling prices.

Environmentalists are concerned about the NEP's emphasis on energy production for several reasons. The growth of nuclear and coal-burning power generating capacity in Ontario increases nuclear waste-disposal problems and air pollution, especially emissions of $SO_2$ and $NO_x$. The development of synthetic crude oil and exploration for oil and gas in territories controlled by the Federal government, that is, offshore territories and the Yukon and Northwest Territories, may involve significant environmental costs. Because production of synthetic crude oil from tar sands and heavy oil in Alberta is assured of world market prices, its development is limited primarily by energy economics, especially after the Federal-Alberta agreement in the summer of 1981 on modifications in the NEP. If Federal and Alberta pollution control regulations are not strictly enforced, the synthetic crude projects may generate significant emissions and pollute scarce surface and ground water. Similarly, nonenforcement of Federal environmental regulations applicable to oil exploration and production on Federal lands and offshore, especially in the Arctic region, could result in significant damage to marine life and the fragile Arctic ecology.[137]

Although the Federal government's agreements with Alberta, Saskatchewan, and British Columbia have reduced resistance in the western provinces, the environmental provisions of the bill have been controversial because the FEM is not granted authority to regulate the exploration and development of oil on Federal lands and offshore in respect to environmental considerations. The Federal ministries with responsibility for administering all aspects of oil exploration and development are the Ministry of Energy Mines and Resources in respect to the offshore zone south of the 60th parallel, and the Ministry of Indian Affairs and Northern Development in respect to all other offshore territory and Federal lands. Not only does the current legislative proposal preclude the role of the FEM, but it also gives considerable authority to the two ministries in environmental considerations. For example, each minister may order a public inquiry if an oil spill occurs, but she/he is not required to do so. Moreover, "authorized" oil spills are defined by the minister, and the Federal government has no liability for "authorized" spills. Needless to say, amendments to the NEP enabling legislation that would give the FEM the authority under the Oil and Conservation Act for environmental matters were proposed. However, it is unlikely that the Liberal majority in the House of Commons would allow the amendments to pass.[138]

Another environmental problem that may result from the NEP results from the price and export controls on natural gas that are continued under the proposed Oil and Conservation Act. The petrochemical industry plans to more than triple its capacity and double its share of the world market within ten years because of the current difference between natural gas prices in Canada and the United States. If the United States accelerates the decontrol of natural gas prices, the differential will increase and even more expansion of the Canadian petrochemical industry may occur. Unfortunately, the air and water pollution problems associated with the petrochemical industry are myriad. Thus, the provincial environment ministries in Alberta, British Columbia, and Saskatchewan will have to be vigilant in the enforcement of their air and water pollution control laws and monitor closely the disposal of hazardous wastes from the petrochemical plants.[139]

### Water Resources

In spite of an apparent abundance of water, Canada may be facing serious water shortage problems within two decades. Thus, the Canadians have the opportunity to manage their water resources to mitigate this impending water scarcity. In addition to the acid rain problem, pollution of surface water by agricultural run-off, industry, municipalities, and urban run-off will contribute to the water shortage. Currently, there is no general shortage of water, but in some areas of the country either water is scarce for periods of time, as during drought periods in the agricultural Prairie provinces of the West and southern Ontario, or there is not as much water as is wanted with the right quality, for example, recreational-use water in eastern Canada near urban areas.

While acid rain destroys lake ecology, thereby killing fish, reduces forestry growth, damages crops, and contaminates ground water by leeching heavy metals out of the ground as it percolates to the water table, other types of water pollution increase the cost of water treatment and reduce the availability of water for human consumption and recreational use as well as for agricultural and industrial use. Major diversions of water to alleviate periodic drought conditions in western Canada and southern Ontario will eventually affect climate and the availability of water in the areas from which the surplus water is diverted. In addition, low water levels accelerate the deterioration of water quality by increasing the water treatment costs of industry and municipalities, as well as adversely affecting the tourist and recreation industry.[140]

Although Canadian environmental and public health officials have been preoccupied with contamination of ground water in urban areas by toxic chemicals from leaking hazardous waste dumpsites, agricultural pollution has emerged as a potential threat to ground water in rural farming areas. For example, poorly stored animal wastes and excessive use of fertilizers with high nitrogen content have contaminated ground water in the Kingston agricultural region of Ontario.[141]

Over the next two decades, the urbanization of Canada should accelerate so that 90 percent of the population will inhabit urban areas near surface water. Thus, the current 50-50 split between urban and rural population will become a 90-10 relationship. Although overall population will also increase, the use of water by urban areas and for agriculture will continue to increase more rapidly than population growth. In summary, not only must current water pollution control regulations be enforced, but also more stringent measures to control water pollution must be enacted.[142]

In recognition of these long-term problems of water resource management, the FEM is conducting several research projects aimed at determining the causes and effects of water pollution (especially acid rain), establishing water quality objectives and monitoring procedures, and delineating wastewater treatment and disposal practices, especially innovative methods for controlling municipal and industrial pollution. Currently, over $50 million is spent annually on water resources research.[143] Because of the severity of the acid rain pollution threat, the Federal research budget to study that problem is to be more than doubled (from $4 million to about $10 million annually) for a four-year period (1980–81 through 1983–84), with the research being conducted by three ministries—Environment, Health and Welfare, and Fisheries and Oceans). The acid rain research will focus on scientific, engineering, social, and economic problems, including the effects on human health and on fish, wildlife, forests, and crops; predicting the effects of acid rain; and protecting endangered areas from acid rain.[144] Finally, Canadian and U.S. environmental officials are working on a joint research project to develop a mutually acceptable method of measuring acid rain.[145]

## 1982 Addendum

The two most significant developments during the first half of 1982 only implicitly affect environmental policy. On April 17, 1982, Queen Elizabeth II completed the patriation of Canada by proclaiming the new Canadian Constitution, a document to which she had given Royal Assent on March 29, 1982. As explained above, the ultimate impact on the Federal-provincial sharing of authority in environmental matters will not be known for years. Section 31 provides that the Constitution does not extend "the legislative powers of any body or authority," which means that neither the Federal government nor the provincial government gained power when the British North America Act was superseded.[146] While the Constitution is explicit in providing for retention of provincial power over highways, municipalities, and property, including natural resources, the Federal government was extended the right to enact laws dealing with peace, order and good government.[147]

The second development relates to a severe economic recession that threatened to continue throughout 1982. Canadianization of domestic industry and the National Energy Program for energy independence have discouraged foreign investment, delayed or terminated "mega" investments in energy development (including tar sands in Alberta), and triggered capital flight. In addition, high interest rates and economic recession in the United States have further depressed the Canadian economy.[148] Ironically, the severe recession in both the U.S. Midwest and eastern Canada have probably reduced the $SO_2$, $NO_x$, and particulate matter emissions that induce acid rain.

More permanent reductions in acid rain are, however, still being negotiated. At the third negotiating session between Canada and the United States in late February 1982, Environment Minister John Roberts submitted a draft agreement providing for a 50 percent reduction in $SO_2$ emissions for both countries by 1990. The 30-page draft agreement, which is modeled after the U.S.-Canadian Great Lakes Water Agreement, was not received enthusiastically by the U.S. delegation, but the U.S. State Department officials were pleased to have a comprehensive draft with which to work.[149] At the New York State Acid Rain Conference in early March 1982, New York Governor Hugh L. Carey proposed a joint U.S.-Canadian program to control acid rain. Carey told the conference, which was attended by State and provincial officials, that Congress should amend the U.S. Clean Air Act to provide for greater control of $SO_2$ emissions.[150] At the tenth annual conference of New England governors and eastern Canada premiers in June 1982, the governors and premiers attending unanimously adopted a resolution advocating the implementation of a regional program to reduce $SO_2$ emissions in eastern North America. However, Canadian frustrations with alleged "foot dragging" by the United States on consummating an agreement to establish a joint program on acid rain prevention has resulted in a threat by Environment Minister Roberts to withdraw from negotiations with the United States.[151]

In a related matter, environmentalists and Parliament members are pressuring the Federal government to veto a National Energy Board decision to allow the export of electric power to the United States because coal-fired plants to be built by Ontario Hydro would significantly increase $SO_2$ emissions. During the NEB export hearing, Environment Canada intervened to argue that Ontario Hydro should only be allowed to export power if it is willing to improve pollution control. To allow Ontario Hydro to increase $SO_2$ emissions in producing electric power for export would be inconsistent with Canada's willingness to reduce $SO_2$ emissions by 50 percent. The Ontario government, which controls the provincial crown corporation Ontario Hydro, has already relaxed $NO_x$ emission standards for the utility.[152]

With regard to the Great Lakes transboundary pollution problem, Canadian officials in a March 1982 diplomatic note to the United States formally expressed concern that the Reagan administration's budget cuts will make it impossible for the United States to meet the guidelines for reducing industrial effluents by December 31, 1983 as provided in the 1978 Great Lakes Water Quality Agreement. As mentioned above, the other major concern that has emerged relates to toxic wastes pollution of the Great Lakes. To the extent that Reagan budget cuts hamper research on the type, quantity, and source of toxic chemicals, the International Joint Commission will have difficulty in developing a program for controlling such pollution.[153]

With respect to a recent U.S. Environmental Protection Agency report on cleaning up hazardous waste dumps in the Niagara Falls border area (including Love Canal), Environment Minister Roberts expressed general agreement with the report's priorities. However, he reserved final judgment until detailed plans for action can be reviewed. Consultation with U.S. officials on the impact of Reagan budget cuts is expected later in 1982.[154]

## Notes

1. Touche Ross & Co., *Business Study—Canada* (New York: Touche Ross International, 1975), p. 8.

2. U.S. Department of Commerce, *The Effects of Pollution Abatement on International Trade* (Washington, D.C.: U.S. Government Printing Office, 1973), p. A-8.

3. Ibid., p. A-9.

4. Canadian Department of Industry, Trade, and Commerce, "Canadian Cures and Preventions: Measures to Clear Pollution Haze," *Canadian Courier*, Vol. 2, No. 8, (1972), p. 1.

5. Canadian Department of Secretary of State, *Canadian System of Government* (Ottawa: Queen's Printer for Canada, 1970), p. 30.

6. Canadian Department of Industry, Trade, and Commerce, "Canadian Cures and Preventions," p. 1.

7. Canadian Department of Secretary of State, *Canadian System of Government*, pp. 22-23.

8. OECD Environment Directorate, *Economic and Policy Instruments for Water Management in Canada* (Paris: Organization for Economic Cooperation and Development, 1976), pp. 7-10.

9. Canadian Department of Industry, Trade, and Commerce, "Canadian Cures and Preventions," p. 1.

10. Jack Davis, "What Environment Minister Plans," *Financial Post*, October 16, 1971, p. A-9.

11. Stephen Duncan, "What Went Sour for Jack Davis?" *Financial Post*, August 24, 1974, p. A-11.

12. Ibid.

13. "Tough Laws to Clean Up Environment," *Canada Weekly*, January 21, 1976, p. 5.

14. *International Environment Reporter* (Washington, D.C.: Bureau of National Affairs, Inc., 1978), p. 51:0101.

15. *International Environment Reporter—Current Report*, April 10, 1978, pp. 112-15.

16. U.S. Department of Commerce, *Effects of Pollution Abatement on Trade*, pp. A-14 & A-15.

17. Canadian Department of Industry, Trade, and Commerce, "Canadian Cures and Preventions," pp. 1-2.

18. Francis T. Mayo, "An Important Step Forward in International Pollution Control," *Environment Midwest*, October-November 1975, p. 2.

19. OECD Environment Directorate, *Water Management in Canada*, pp. 4-5.

20. *International Environment Reporter*, p. 51:0101.

21. "New Fish Habitat Laws," *Canada Weekly*, July 12, 1978, p. 6.

22. Canadian Department of Industry, Trade, and Commerce, "Canadian Cures and Preventions," p. 1.

23. *International Environment Reporter*, p. 51:0102.

24. Ibid., p. 51:0101.

25. *International Environment Reporter—Current Report*, February 10, 1978, pp. 40, 45.

26. OECD Environment Directorate, *Water Management in Canada*, pp. 32-34.

27. *International Environment Reporter*, p. 51:0102.

28. OECD Environment Directorate, *Water Management in Canada*, pp. 32-34.

29. Leonard Waverman, "Fiscal Instruments and Pollution: An Evaluation of Canadian Legislation," *Canadian Tax Journal*, Vol. 18, No. 6 (1970), pp. 508-9.

30. Ibid., p. 509.

31. U.S. Department of Commerce, *Effects of Pollution Abatement on Trade*, p. A-14.

32. Waverman, "Fiscal Instruments and Pollution," pp. 508-9.

33. *International Environment Reporter*, p. 51:0101.

34. *International Environmental Guide—1975* (Washington, D.C.: Bureau of National Affairs, Inc., 1975), p. 51:0101.

35. *International Environment Reporter*, p. 51:0101; and *International Environment Reporter—Current Report*, July 10, 1978.

36. Waverman, "Fiscal Instruments and Pollution," p. 510. For a detailed analysis of the British Columbian experience with water use charges, see Anthony H.J. Dorcey (ed.), *The Uncertain Future of the Lower Fraser* (Vancouver: University of British Columbia Press, 1976); and James B. Stephenson (ed.), *The Practical Application of Economic Incentives of the Control of Pollution: The Case of British Columbia* (Vancouver: University of British Comumbia Press, 1977).

37. OECD Environment Directorate, *Water Management in Canada*, pp. 19-23.

38. Ibid., pp. 52-61.

39. Ibid., p. 55.

40. OECD Environment Directorate, *Pollution Charges: An Assessment* (Paris: Organization for Economic Cooperation and Development, 1976), pp. 14, 19, 21, 26, and 30.

41. Environment Canada, *Inland Waters Directorate* (Ottawa: Queen's Printer for Canada, 1973), p. 36.

42. U.S. Department of Commerce, *Effects of Pollution Abatement on Trade*, p. A-10.

43. Charles Davis, "Detail Work in Pollution Control Just Beginning," *Financial Post*, October 28, 1972, p. P-1.

44. Leonard Zehr, "U.S. Falls Behind in Doing Its Share to Carry Out Agreement With Canada to Clean Up Great Lakes," *Wall Street Journal*, January 16, 1974, p. 24.

45. William Omohundro, "Great Lakes Water Quality: 1974 Annual Report to IJC," *Environment Midwest*, October-November 1975, p. 21.

46. "Good News About Great Lakes," *Canada Weekly*, September 20, 1978, p. 3.

47. "News Pact to Clean Up Great Lakes Highlights Visit of U.S. State Secretary," *Canada Weekly*, December 6, 1978, pp. 1-3.

48. *International Environment Reporter—Current Report*, June 10, 1978, pp. 168-69.

49. "Good News About Great Lakes," p. 3.

50. U.S. Department of Commerce, *Effects of Pollution Abatement on Trade*, p. A-13.

51. Canadian Department of Industry, Trade, and Commerce, "Canadian Cures and Preventions," p. 2.

52. U.S. Department of Commerce, *Effects of Pollution Abatement on Trade*, p. A-13.

53. *International Environment Reporter*, pp. 52:1941 and 1961.

54. U.S. Department of Commerce, *Effects of Pollution Abatement on Trade*, p. A-10.

55. *International Environment Reporter—Current Report*, May 10, 1978, p. 126.

56. U.S. Department of Commerce, *Effects of Pollution Abatement on Trade*, p. A-10.

57. *International Environment Reporter—Current Report*, p. 126.

58. Swedish Delegation to the OECD Environment Committee, *Environment Policy in Sweden* (Paris: Organization for Economic Cooperation and Development, 1977), p. 137.

59. *International Environment Reporter*, pp. 51:0101 and 51:0102.

60. *Ibid.*, p. 51:2301.

61. Canadian Department of Industry, Trade, and Commerce, "Canadian Cures and Preventions," p. 2.

62. *International Environment Reporter*, p. 51:0102.

63. U.S. Department of Commerce, *Effects of Pollution Abatement on Trade*, p. A-10.

64. OECD Environment Directorate, *Waste Management in OECD Member Countries* (Paris: Organization for Economic Cooperation and Development, 1976), pp. 5, 16, 17, 22, 23, 37, 44, 45, 55, 56, 64, and 69.

65. Davis, "Detail Work in Pollution Control Just Beginning," p. 1.

66. *International Environment Reporter*, pp. 51:5121-5127.

67. Richard Grace and Jonathan Fisher, *Beverage Containers: Re-Use or Recycling* (Paris: Organization for Economic Cooperation and Development, 1978), p. 106.

68. Associated Press-Vancouver, "Recycling Plan Aims at Bottles," *Vancouver Sun*, August 11, 1971.

69. British Columbian Department of Lands, Forests, and Water Resources, "Pop Container Refunds Are Officially Increased," *Vancouver Sun*, p. 1.

70. *International Environment Reporter—Current Report*, March 10, 1978, p. 84.

71. Grace and Fisher, *Beverage Containers*, pp. 146-47.

72. *International Environment Reporter—Current Report*, April 10, 1978, p. 115.

73. Ibid., February 10, 1978, p. 34.

74. Ibid., July 10, 1978, p. 205.

75. U.S. Department of Commerce, *Effects of Pollution Abatement on Trade*, p. A-11.

76. Canadian Department of Industry, Trade, and Commerce, "Canadian Cures and Preventions," p. 2.

77. Waverman, "Fiscal Instruments and Pollution," p. 508.

78. U.S. Department of Commerce, *Effects of Pollution Abatement on Trade*, p. A-11.

79. OECD Environment Directorate, *Waste Management in OECD Countries*, p. 24.

80. Canadian Income Tax Regulations, Schedule B, Class 24.

81. Waverman, "Fiscal Instruments and Pollution," p. 507.

82. Canadian Income Tax Regulations, Schedule B, Class 24.

83. OECD Environment Directorate, *Waste Management in OECD Countries*, p. 10.

84. Canadian Income Tax Regulations, Schedule B, Class 24.

85. Ibid., Class 27.

86. "Budget Measures Aim at Holding onto Gains and Stimulating Growth," *Canada Weekly*, November 29, 1978, pp. 1 and 2.

87. "Tax Incentives for Pollution Control," *Canada Weekly*, December 7, 1977, p. 7.

88. Canadian Income Tax Regulation 2900.

89. Canadian Department of Industry, Trade, and Commerce, *Doing Business in Canada: Taxation—Income, Business, Property* (Ottawa: Queen's Printer for Canada, 1972), p. 7.

90. Canadian Department of Industry, Trade, and Commerce, *Doing Business in Canada: Federal Incentives to Industry* (Ottawa: Queen's Printer for Canada, 1972), pp. 6-9.

91. Arthur Andersen & Co., *Tax and Trade Guide—Canada* (Chicago: Arthur Andersen & Co., 1973), p. 72.

92. "Budget Measures," *Canada Weekly*, pp. 1 and 2.

93. U.S. Department of Commerce, *Effects of Pollution Abatement on Trade*, p. A-11.

94. OECD Environment Directorate, *Waste Management in OECD Countries*, p. 56.

95. Martin J. Shannon, "Business Bulletin: Canadians Tax Auto Air Conditioners and Heavy Cars in Bid to Save Fuel," *The Wall Street Journal*, April 14, 1977, p. 1.

96. Waverman, "Fiscal Instruments and Pollution," *Canadian Tax Journal*, p. 508.

97. David Carlson, *Revitalizing North American Neighborhoods: Comparison of Canadian and U.S. Programs* (Washington, D.C.: U.S. Government Printing Office, 1978), p. 3.

98. "CMHC Makes Generous Loan," *Canada Weekly*, November 30, 1977, p. 8.

99. Waverman, "Fiscal Instruments and Pollution," p. 508.

100. OECD Environment Directorate, *Water Management in Canada*, p. 38.

101. Ibid., p. 23.

102. OECD Environment Directorate, *Waste Management in OECD Countries*, p. 23.

103. John Urquhart, "Liberals' Victory in Canada Could Mean More Activity by Government in Economy," *The Wall Street Journal* (Southwest Edition), February 20, 1980, p. 16.

104. Robert Paehlke, "Canada—The National Energy Program," *Environment*, June 1981, pp. 4-5.

105. *International Environment Reporter—Current Report*, February 10, 1979, p. 516; June 13, 1979, pp. 723-33; and May 14, 1980, pp. 177-78.

106. Ibid., February 10, 1979, p. 516; March 10, 1979, p. 584; and January 9, 1980, p. 26.

107. Ibid., March 12, 1980, p. 93.

108. Ibid., May 9, 1979, p. 666.

109. Ibid., January 14, 1981, p. 604, and May 13, 1981, p. 845.

110. Ibid., December 10, 1980, p. 551.

111. Ibid., April 9, 1980, p. 141, and June 11, 1980, p. 264.

112. Ibid., February 13, 1980, p. 56, and March 12, 1980, pp. 99-100.

113. Ibid., April 11, 1979, p. 622, and May 9, 1979, p. 671.

114. "Federal Government and Provinces Come to Consensus on Constitution," *Canada Weekly*, November 25, 1981, p. 1.

115. "Supreme Court Brings Down Divided Decision on Canadian Constitution," *Canada Weekly*, October 14, 1981, pp. 1-2.

116. "Government and Provinces Come to Consensus," p. 1.

117. "Canadian Constitution Awaits Passage by British Parliament," *Canada Weekly*, December 23 and 30, 1981, p. 3.

118. "Government and Provinces Come to Consensus," p. 1.

119. United Press International, Ottawa, September 9, 1980.

120. "How Many More Lakes Have to Die?" *Canada Today*, February 1981.

121. *International Environment Reporter—Current Report*, June 11, 1980, pp. 232-34; July 9, 1980, p. 279; August 13, 1980, pp. 346-47; October 8, 1980, pp. 474-75; April 8, 1981, pp. 792-93; and June 10, 1981, pp. 871-72.

122. Ibid., March 12, 1980, p. 93, and June 11, 1980, p. 234.

123. Ibid., February 11, 1981, pp. 620-21.

124. Ibid., August 8, 1979, p. 798; August 13, 1980, pp. 320-21; and July 8, 1981, pp. 920-21.

125. Ibid., March 11, 1981, p. 684; April 8, 1981, p. 802; September 9, 1981, pp. 1008-9; October 14, 1981, pp. 1039-41; and November 11, 1981, p. 1075. (In December 1980, the Parliament amended the Canadian Clean Air Act to authorize Federal controls for transnational emissions.

126. Ibid., July 9, 1980, p. 287.

127. Ibid., September 10, 1980, p. 408.

128. Ibid., March 11, 1981, pp. 691-92.

129. Ibid., January 14, 1981, pp. 602-3.

130. Ibid., December 10, 1980, pp. 543-44, and May 14, 1980, pp. 183-84.

131. Ibid., October 8, 1980, pp. 484-85.

132. Ibid., May 13, 1981, p. 832.

133. Ibid., February 10, 1982, p. 77, and June 10, 1981, pp. 876-77.

134. Ibid., April 8, 1981, p. 807.

135. Paehlke, "The National Energy Program," pp. 4-5.

136. "Canada: A Petrochemical Boom That Threatens the U.S.," *Business Week*, February 8, 1982, p. 49.

137. Paehlke, "National Energy Program," p. 5.

138. *International Environment Reporter—Current Report*, November 11, 1981, pp. 1094-95.

139. "Petrochemical Boom Threatens U.S.," *Business Week*, p. 49.

140. *International Environment Reporter—Current Report*, October 14, 1981, p. 1055.

141. Ibid., May 14, 1980, p. 184, and April 8, 1981, pp. 802-3.

142. Ibid., October 14, 1981, p. 1055.

143. "Study Looks at Impact of Environment on Water," *Canada Weekly*, October 14, 1981, p. 3.

144. *International Environment Reporter—Current Report*, November 11, 1980, p. 513.

145. "Joint Acid Rain Research," *Canada Weekly*, October 21, 1981, p. 5.

146. Tom Kelly, "The Constitution Comes Home," *Canada Today*, April 1982, pp. 3-11.

147. United Press International, Ottawa, September 9, 1980, and "Canadian Constitution Comes Home," *Canada Weekly*, April 28, 1982, p. 6.

148. Frank J. Comes and Thane Peterson, "The Canadian Economy is in Crisis," *Business Week*, June 28, 1982, pp. 80-83.

149. *International Environment Reporter—Current Report*, March 10, 1982, pp. 122-23.

150. *Ibid.*, April 14, 1982, pp. 146-47.

151. Ibid., July 14, 1982, pp. 280, 281 and 288. See also Lee Walczak, "Washington Outlook: The Environment," *Business Week*, December 20, 1982, p. 91, and Henry G. DeYoung, "Acid Rain Regulators Shift into Low Gear," *High Technology*, September-October, 1982, pp. 82-86.

152. Ibid., June 9, 1982, pp. 219-20, and May 12, 1982, pp. 189-90.

153. Ibid., April 14, 1982, pp. 135-36.

154. Ibid., May 12, 1982, pp. 192.

# 4

## France

As in all other developed countries since World War II, rapid industrialization and urbanization have been the major causes of France's pollution problems. This situation may be further aggravated in the future because France is less urbanized[1] and industrialized[2] than the United States, Canada, West Germany, Sweden, and the United Kingdom, and is striving to achieve an industrial and urban growth at a rate greater than that projected for those five countries.

Through the impetus of a unique form of national economic planning in which industrial and urban growth has been encouraged in urbanized regions other than metropolitan Paris, the French have made substantial progress in modernizing their industrial base and in urbanizing the provincial cities. Nevertheless, metropolitan Paris, which has almost 20 percent of the country's over 50 million people, remains France's political, industrial, social, and cultural center.[3] Therefore, it is also the center of some of the country's most serious pollution problems.[4]

Even though France has experienced an unprecedented industrial and urban growth in the post-World War II era, it is still the most agricultural country in the nine-member-nation European Economic Community (EEC) or Common Market, as well as the least densely populated country therein. However, France's population density is much greater and its agricultural land per capita is much less than that of the United States and Canada.[5]

During the quarter century following World War II, an increasing proportion of France's water resources has become heavily polluted. Those rivers flowing through industrialized areas—the lower Seine (Paris and Rouen), the Moselle-Rhine (Metz and Strassbourg), the Meuse (northeastern France), the Rhone (Lyons and Grenoble), and the Loire (Orleans, Tours, and Saint

Nazaire)—are especially polluted.[6] In 1971, the French Economic and Social Council estimated that about 1,600 miles of French rivers were polluted, but only 350 miles were continuously polluted, with the remaining 1,250 miles being polluted only during low water flows.[7] Serious water pollution problems are also found along the Mediterranean coast, especially between Nice and Marseille. Both the Rhine River and Mediterranean Sea pollution problems are transnational and, therefore, insoluable without cooperation from neighboring members of the EEC and neighbors in Europe, Africa, and the Middle East.

Air pollution is a serious problem in those regions characterized by heavy industrialization, urbanization, and unfavorable geographical and climatic conditions. The cities of Paris, Lyon, and Marseille, as well as the industrial regions north of Paris and in the northeast (Lorraine), suffer from acute air pollution most of the year.

The French public's awareness of the severity of environmental degradation has been a significant social and political force since the late 1960s as a result of the publicity given to the *Torrey Canyon* oil spill in the English Channel, the acute pollution of the Rhine River, industrial pollution incidents on several French rivers, and the wreck (March 1978) of the *Amoco Cadiz* supertanker on the Brittany coast which resulted in the second worst oil spill in history. However, even though the French national government has had the legislative authority to control both air and water pollution since the early 1960s,[8] the preoccupation of the prior two Gaullist governments of Charles de Gaulle and Georges Pompidou with the continued modernization of the French industrial state necessitated a cautious approach to the implementation of environmental policy.

The personal commitment of the French president to the support of any particular facet of public policy, such as environmental policy, has been essential to its successful implementation since the French abandoned their traditional form of parliamentary government and established a strong executive branch in 1958.[9] Subsequent to the untimely death of President Georges Pompidou on April 2, 1974, the government of President Valery Giscard d'Estaing, who was elected in May 1974, has established a national commitment to environmental quality on a par with the nation's thirty-year commitment to the creation of a modern industrial state.[10]

Substantive manifestations of President Giscard's commitment occurred early in his administration when he vetoed the super highway planned for the Left Bank, the international trade center to be constructed on the site of the *Les Halles* vegetable market, and the sale of the *Cite Fleurie*— a private garden area in Montparnasse—to property speculators.[11] All three projects would have contributed to the further deterioration of central Paris. This deterioration began under Pompidou with the building of several skyscrapers in desecration of one of the world's finest skylines which begins at the Louvre and stretches from there through the

*Tuileries* gardens and the *Champs Elysées* to the *Arc de Triomphe*.

Confident of public support, Giscard has continued to support environmental quality in spite of the Energy Crisis and subsequent recession and stagflation.[12] In fact, under the leadership of Giscard, the French have entered a new phase in their post-World War II development wherein there often seems to be more emphasis on urban amenities (infrastructure or public services) and the environment than on either industrialization or the French heritage. Thus, Giscard's new outlook seems to have coincided with a shift in public feeling, which supports environmental quality at the expense of economic growth.[13]

## Pollution Abatement Programs

The French national government has had the legal authority to control air pollution since 1961 and to control water pollution since 1964.[14] Periodically, these pollution control laws have been strengthened by the addition of amendments and decrees (regulations) thereto and by the enactment of new laws, as follows:

(1)  Law on Waste Disposal and Recovery of Materials in July 1975.

(2)  Environmental Impact Review Law in July 1976.

(3)  Law on Installations Classified for Purposes of Environmental Protection in July 1976.

(4)  Chemicals Controls Law in July 1977.[15]

(5)  Decrees implementing both the Laws on Installations Classified for Purposes of Environmental Protection in September 1977[16] and the Environmental Impact Review Law in October 1977.[17]

In addition, the Law on Energy includes provisions that are pertinent to the regulation of air pollution.[18]

Thus, the French have developed a body of environmental law which gives the central government extensive authority to deal with the most important pollution problems both nationally and regionally. However, the more recent laws, amendments to existing laws and decrees have not been fully implemented. Over the next few years the pollution control directives of the European Economic Community will also have to be implemented.

As a result of the authority given to him in the Decrees of February 2, 1971, and April 2, 1971, President Georges Pompidou created the *Ministère Chargé de la Protection de la Nature et de l'Environnement* (Ministry of the Environment). In February 1974, the Ministry of the Environment became a "Secretariat of State," but, its organization and functions were not affected. As provided in the Decree of February 2, 1971, the Minister of the Environment is generally responsible for coordinating interministerial matters related to the protection of nature and the environment, in addition to having powers of his or her own regarding (1) establishments engaging in dangerous, noxious, and unsanitary trades; (2) national and regional parks; (3) monuments and scenic areas; (4) hunting and fishing; and (5) interministerial coordination in water management, which includes the formulation and execution of policy in water resource management, making recommendations to the National Water Board, supervising and making recommendations to the river or water basin agencies, and Chairmanship of the Interministerial Commission (or Committee) on Water Problems and supervision of the Permanent Secretariat for the Study of Water Problems.[19]

The Environment Minister is something more than a cabinet minister in that he or she also exercises broader powers like a Prime Minister through his or her responsibility for coordinating, controlling, and promoting the environmental policy and actions of other ministries. To effect such high-level supervision and coordination, an Interministerial Action Committee for Nature and the Environment (CIANE) was established with the Prime Minister as its Chairperson. This committee of fourteen (or more) principal ministers or secretaries is authorized to allocate monies from a fund,[20] the Fund for Intervention and Action for Nature and the Environment (FIANE), for the purpose of assisting in the financing of operations deemed necessary to improve environmental quality,[21] including environmental projects involving actions by several ministries. Inasmuch as the CIANE cannot meet often, the Environment Minister is authorized to monitor the accomplishments of his ministerial colleagues, and he may urge them to greater efforts by subsidizing their activities. The following scenario illustrates how the Environment Minister can use FIANE as a "carrot": "For example, if the Minister of Transport has planned a new freeway, the Minister of the Environment can allocate some of his own money to see that flowers line its borders."[22]

In May 1974, additional responsibilities for cultural matters were transferred to the Ministry when it was renamed the Ministry for the Quality of Life (later redesignated Culture and Environment) by President Giscard.[23] The most recent change, in April 1978, effected a diminution of the Ministry for the Environment by spinning off responsibility for cultural matters so that the Environment Minister can concentrate on protecting and enhancing the environment. As a result, the former Ministry of Culture and Environment was split into a new Ministry of Culture and Communication and a new Ministry of the Environment and the Lifestyle (*Ministère de la Environnement et Cadre de Vie*).[24]

As the most powerful member of the French government and the chief executive of the central government, the President presides over the Council of Ministers, which is responsible for determining and implementing national policy through submission of legislation to Parliament, by promulgating regulations or decree-laws where the Parliament—National Assembly and Senate—has authorized the interpretation of law by administrative agencies, and through the initiation of a vote of confidence in the Assembly. The President appoints, but cannot dismiss, the Prime Minister. The Prime Minister, along with the President and the cabinet ministers, who are also appointed by the President, make up the Council of Ministers. The President, who is elected directly by the French voting populace for seven-year terms, has been given the following additional powers under the Constitution: (1) He may dissolve the Assembly but only in specific situations; (2) he can escape the Assembly's legislative authority by submitting specified types of significant legislation to a referendum; (3) he can, when emergency conditions exist, assume plenary powers, thereby circumventing the operation of a constitutional government; and (4) he appoints the Prefects (Administrators) of the ninety-five Departments into which provincial France is subdivided for the implementation of central government policy and programs.[25]

### Licensing of Installations for Environmental Protection

Since 1917, the central government has had the authority to classify industrial, governmental, and commercial installations as dangerous, unsanitary, or a nuisance and to regulate those installations so classified for the protection of the health, safety, and well-being of the public. Amendments and decrees (regulations) have been promulgated over the years to strengthen the 1917 law.[26] However, because the 1917 law and subsequent amendments and decrees thereto were not effective in dealing with the environmental degradation inherent in the location of industry, commerce, and government during the late 1960s and the early 1970s, the French National Assembly and Senate enacted a Law on Installations Classified for Purposes of Environmental Protection (Law No. 76-663) in July 1976 to supersede the Classified Establishments Act of December 1917 (as amended), effective January 1, 1977.[27] In addition to strengthening the 1917 law for registering "classified installations," the National Assembly enacted an Environmental Impact Review Law in July 1976 to take effect on January 1, 1978, in an effort to provide for the assessment of potential environmental impact during the planning stages of selected public and private projects.[28]

What follows is a synopsis of provisions of the Law on Installations Classified for Environmental Protection. The general provisions of Title I provide for the act's applicability to "classified installations" as defined in a subsequent decree (regulation) and for the integration of registration—authorization or declaration—with the building permit application process administered by the Prefect (Administration) in each Department (Region). Title II is a general explanation of the procedures for registration of those installations subject to authorization wherein the locating and operating conditions (best technology economically achievable) for each "classified installation" are specified by a Prefectoral Order subject to the review and consultation of the public (through a public inquiry), Municipal Councils within the Department, the Health Council of the Department, and central government ministers responsible for classified installations, industry, environmental protection, health, and agriculture (especially in respect to wine-producing areas). Title III provides for the regulation of those installations which, while not likely to threaten any danger or nuisance to neighborhood amenities, public health, or safety, agriculture, the environment, or conservation of sites and monuments, must still comply with the regulations promulgated by the Prefect for protecting neighborhoods, public health and safety, and so on. Titles IV and VII deal with procedural matters related to the registration process. Finally, Titles V, VI, and VII include financial provisions, administrative penalties, and criminal penalties, respectively.

The financial provisions in Title V delineate two types of exactions. One is a nonrecurring charge or tax payable upon authorization or declaration, and the other is a recurring charge payable by those installations that require detailed regular inspection because the nature or volume of their activities implies unique risks for the environment. The one-time charge for an authorization is 3,000 francs (approximately $450). For a declaration, the charge or tax is 1,000 francs ($150). Much lower charges (750 and 250 francs) are applicable to "small business." The charge is doubled when the information requested is either not given or given incorrectly. The penalty for late payment is 10 percent.

In addition to the nonrecurring charges, there may be an annual charge or tax for those installations deemed to be engaged in high-risk activities. The basic charge or tax rate is 500 francs ($75) per activity. In respect to each activity, there may be a multiplier of from 1 to 6; therefore, the actual charge will vary between 500 and 3,000 francs per year per activity. The penalties for incorrect or omitted information and late payment are the same as for nonrecurring charges. Finally, an extensive system of criminal penalties is provided for in Title VI.

Independent of any criminal action, Title VII provides that whenever an inspector or appraiser determines that the operating conditions delineated in an authorization are not being complied with, the Prefect shall notify the violator that compliance must occur within a specific period. If compliance is not achieved within that specified period, the Prefect may either force compliance at the operator's expense or suspend operations. Finally, where a classified installation is being operated without having been declared or authorized, the Prefect shall notify the operator to declare or apply for authorization within a specified period. During the interim, the Prefect may suspend the installation's operations. Moreover, if compliance is not forthcoming within the specified period, the Prefect may either force compliance at the operator's expense or suspend operations.[29]

Since the 1917 law for registering classified installations was repealed by the 1976 law explained above, there was a need for transitional provisions regarding those natural and legal persons in compliance with the old law. As provided in Decree No. 77-1133, which was promulgated in September 1977, those installations registered under the old law are in compliance with the new law. However, Section 4 (Title I) of the 1976 law stipulates that a person must renew his or her authorization or declaration both when there is a transfer, extension, or transformátion of an installation and where there is any change in a manufacturing process which gives rise to a danger or nuisance as delineated in Section 1 (Title I).[30] As decreed in 1964 and 1965, the Classified Establishments Act of 1917 delineates 420 types of industrial and commercial activity divided into three categories according to the potential of each for causing serious air, water, and noise pollution in adjoining communities.[31]

In addition to the registration of classified installations for environmental protection, Law No. 76-629 of July 1976, along with Decree No. 77-1141 of October 1977, requires an environmental impact statement for all private and public projects that may have a detrimental impact on the harmonious balance of the inhabitants of urban or rural areas and on the natural surroundings or resources. The law not only establishes basic principles for environmental protection, but also specifies the procedure whereby all interested groups and persons may evaluate and object to projects. Consequently, the long-term environmental impact of major public and private projects may be mitigated during the planning thereof.[32]

Inasmuch as the 1977-81 Environmental Program of the EEC provides for the introduction of procedures for assessing the environmental impact of projects, the French environmental impact assessment procedures may have to be modified somewhat to conform to any EEC directive on environmental impact surveys. The environmental impact evaluation procedures of West Germany and the United Kingdom may also be affected by any EEC directive thereon.[33] The EEC's Environment and Consumer Protection Service has been responsible for drafting a directive on environmental impact assessment which will probably resemble the U.S. environmental impact statement in content and procedure at the national level.[34]

## Water Pollution Control

France's water pollution control legislation is neither as old nor as well-developed as the general pollution control legislation. However, the Water Management Act of 1964, which is designed to protect and enhance the country's water resources, gives the central government very broad authority to manage surface and subsurface water, including the authority to establish water basin agencies or authorities throughout the country.

Therefore, the development of water pollution control policy has not been piecemeal. The water basin authorities, officially known as Water Basin Finance Agencies, are public administrative and financial institutions that govern water management, the distribution of water resources, and water pollution control.

Six water basin authorities have been set up based on the major French hydrographic or river basin boundaries: (1) Artois-Picardy Agency in northern France with headquarters in Donai; (2) Rhine (Moselle)-Meuse Agency in northwestern France with headquarters in Metz; (3) Seine-Normandy Agency in north central France with headquarters in Paris; (4) Loire-Brittany Agency in north central France south of Paris with headquarters in Orleans; (5) Adour-Garonne Agency in southeastern France with headquarters in Toulouse; and (6) Rhone-Mediterranean-Corsica Agency in southwestern France with headquarters in Lyons. The antipollution activities of these agencies, which did not actually begin their operations until January 1, 1969, include charging for the use and pollution of water, the financing of pollution control facilities, and the monitoring of public and private pollution control facilities within their jurisdiction.[35]

In general, water is not privately owned under French law, even though there are a few restricted exceptions to the State ownership of water resources. Individuals and institutions can claim only the right to use State-owned waters, and that right to use always reverts to the State.[36] Thus, the French have established a very progressive system for controlling water pollution in which "Every major French user of water must pay for authority to pump water out of wells or rivers. Towns and cities pay, as well as industry, and they must all pay again if, after usage, they have contaminated the water."[37]

In France, each case of water pollution is dealt with separately. In response to a polluter's application for licenses to abstract and discharge water, the Prefecture—the departmental administrator for the Environment Ministry—will normally issue a water use license and a license stipulating conditions as to the quality and quantity of waste which the polluter may discharge after a public inquiry and consultation with competent authority within and without the Environment Ministry. In addition, industrial establishments classified as Category 1 or 2, as is provided in the 1976 Law on Installations Classified for Environmental Protection, must apply for a second license as part of the registration process explained above.[38]

Thus, direct control or regulation is the basic technique for controlling effluent discharges. Consequently, the effluent tax system implemented through the Water Basin Finance Agencies has been designed to facilitate the enforcement of effluent standards inasmuch as the taxes function only as a method for reallocating the cost of financing public and private pollution abatement investment, and not as an incentive for a polluter to discharge less effluent than that prescribed in its license.[39] With the possible exception of the effluent standards which are contracted by the central government and selected industries, there are still no explicit effluent standards at the national level,[40] even though the water pollution inventory mandated by the Water Management Act of 1964 has been carried out twice, in 1971 and 1976.[41]

Implicit in the national economic planning process explained below are water quality standards that have been established for bodies of water within the twenty-two planning regions and that are consistent with the short-term objectives of each Basin Finance Agency and its Committee. The information and data acquired during the survey of water pollution every five years are to be used not only to monitor progress in the planned effort to improve water quality, but also to develop national effluent standards for industry and the public. Standards will be based on the level of discharge necessary to produce water quality consistent with each body of water's specified use. There already are recommended discharge guidelines for both "registered installations" and publicly owned wastewater treatment facilities, but the long-term preferred use for each significant body of water in France has, as yet, to be determined. Therefore, more stringent national effluent standards will apparently not be delineated for several years.[42]

Thus, French water quality is maintained and improved through the use of effluent charges or taxes to reinforce existing effluent standards. Each Water Basin Finance Agency prepares and subsidizes from effluent tax revenue an "action" program covering five years to be coincident with national economic planning periods. The purpose of the program is to develop water resources and control pollution within its jurisdiction. The main effluents subject to taxation are oxidizable substances, suspended solids, and toxic pollutants. A more detailed explanation of the pollution tax base and the determination of the pollution tax rate is provided next.

Water Pollution Taxes

The six Water Basin Finance Agencies exact effluent charges or water pollution taxes on a regional basis. The Water Development Act of December 16, 1964, was passed "to supplement the specific controls on each individual consumer of water resources by adding an organization adapted to the natural environment as a whole and designed for economic planning."[43] Each water basin covers a separate hydrographic area, which is independent of the other basins. Hence, each basin can effect a homogeneous form of water resource management suited to its particular characteristics and to the area's economic demands.

Each basin has a Basin Committee which serves as an advisory body. The water users and governmental authorities are given equal representation on the Basin Committee. The most important function of the Basin Committee is the advice it gives the Basin Finance Agency on the Agency's (1) proposed program of action for water resource development and water pollution control; (2) proposed basis for calculating effluent charges; and (3) proposed level of such charges. In other words, program objectives and effluent charges are somewhat negotiable.

The Basin Finance Agency acts as the executive authority for water management in the water basin region. Among the Agency's technical functions are (1) preparation of a pluri-annual action program for developing water resources and controlling pollution within the hydrographic area; (2) conducting engineering studies and research (for example, inventorying and forecasting utilization of water resources and preparing pollution control programs) necessary for preparation and implementation of its action program; and (3) consulting in the public interest on matters related to water pollution, especially consultation on the design, construction, and operation of community or private projects which help improve water resources. The Agency's economic role is holistic because it levies effluent taxes and then disburses the tax revenues either to finance community projects or to make loans and grant subsidies.

In summary, a Water Basin Finance Agency's functions are limited to those technical and economic roles explained above. Moreover, an Agency's "technical and economic role" is an indirect one in that it does not commission or construct pollution control facilities, nor does an Agency have any regulatory powers to determine or enforce pollution standards.[44]

Inasmuch as many different polluting substances are discharged into the aquatic environment and a separate charge for each pollutant would be too complicated and costly, two fundamental problems arise in delineating a tax base: (1) Which polluting substances are most typical of overall pollution; and (2) which polluting substances can most easily be measured.[45] Since the effluent tax base is the weight of the pollution discharged, the French charge incorporates the following parameters: (1) The weight of suspended materials denoted $SM$; (2) the weight of oxygen needed for oxidizable materials denoted $OM$; and (3) the weight of toxic substances denoted $TS$. The pollution weight and tax base, which is denoted $P$, is determined by the following formula. Because the oxidizable materials ($OM$) are decomposed both through biological oxygen demand or bacterial action (denoted $BOD$) and chemical oxygen demand (denoted $COD$), they are assigned weighting coefficients of $2/3$ and $1/3$, respectively. The weight of toxic substances is adjusted by a toxicity factor (denoted $r$).[46]

$$P = \frac{COD + 2\,BOD}{3} \times OM + SM + rTS$$

In the French charge system, the common pollution measure used is known as a "per capita population equivalent," and it represents the highest average daily discharge by weight of pollutants for each inhabitant of a hydrographic basin. The rate is actually derived from projections of pollution and of revenue required to finance water quality management within a water basin. Since an estimate of the highest average daily discharge of pollution within a water basin can be made, a rate $(R)$ of taxation can be calculated by dividing total basin pollution $(P)$ by revenue required $(T)$, as follows: $R = P \div T$. Then the rate and pollution are divided by the basin's population. In 1974, the annual effluent tax rate ranged between 3.60 and 5 francs (90¢ to $1.25) per capita population equivalent of discharge, depending upon the water basin region in which the pollution occurs.[47] By 1976, the tax rate was to have increased to more than 10 francs ($2.50) per capita population equivalent of discharge.[48]

## National Economic Planning and Water Pollution Control

Before they can establish effluent tax rates, the Basin Committee and the Basin Finance Agency must determine regional objectives for water resource development and water pollution control.[49] As a part of the French process of national economic planning wherein the National Planning Commission "attempts to combine the dynamic forces of a market system with explicit consideration of the ways in which markets can be used to serve collectively determined social goals,"[50] the Basin Finance Agency prepares a pluri-annual program of action outlining the steps that must be carried out to effect water resource development and water pollution control for both short-term (five years) and long-term (twenty years) planning periods.[51] The Seventh State Plan, which covers the period 1976–80, established the following objectives for French society through the end of 1980: (1) Achievement of full employment and creation of 1.1 million jobs; (2) maintenance of a reasonably fast annual rate of economic growth of 5.5 to 6 percent; (3) achievement of a rate of inflation (6 percent) one-third lower than the 1975–76 rate of 9 percent; (4) reduction of inequalities in income earned by those gainfully employed; (5) improvement of the quality of life through policies that promote family-owned housing, check the demographic deterioration of rural areas, and better control the rapid urbanization of the past thirty years; and (6) delineation of twenty-five separate high-priority programs, including water resource development, and estimates of the financial resources required to operationalize each program, with a guarantee that each program represent essential fields of action that should be protected from phenomena like the Energy Crisis, stagflation, and recession.[52]

A desirable target has been set for water resource development and water pollution control under the Seventh State Plan, and the financial resources required to attain that target during the five-year period have been calculated. These specific objectives were determined through the consultation of the Basin Committees, Basin Finance Agencies, and Delegations for Basin Affairs with planning authorities (the General Commissariat and planning commissions for environment, public health, industry, interior, and so on) and governmental authorities.[53] In addition to the six water basins, there are twenty-two Program Regions for economic planning and over ninety Departments. In other words, there are three overlapping levels of regional government at which water management policy must be reconciled.

To effect regional cooperation, twenty-two committees on water technology have been established in the Program Regions.[54] Moreover, each water or river basin has a *Mission Délégués de Bassin* (Delegation for Basin Affairs) made up of representatives from those offices within each Department concerned with water management. The *Secretariat Pour l'Eau* (Secretariat for Water), which was originally known as the Permanent Secretariat for the Study of Water Problems, is the direct link between the Interministerial Committee or Commission on Water Problems, six Basin Finance Agencies, and twenty-two technical committees in the planning regions. The Secretariat also acts as the basic technical and research agency for the central government, including the Interministerial Water Commission and the National Water Board.[55]

The short-term water pollution control and water resource development objectives are actually based on pollution trends and needed action over the long-term (twenty years), as follows:

> goals are determined in terms of a *minimum objective*, which is to maintain pollution at its present level, i.e., to eliminate any fresh pollution; and of a *desirable objective*, which is to treat 80% of all effluent so that 80% of all gross pollution can be wiped out by 1985 or 1990.[56]

Having developed a multiyear (pluri-annual) action program that must be financed primarily by effluent tax revenues, the Water Basin Finance Agencies must determine the appropriate effluent tax rate structure. That is, it must decide how much polluters should contribute to the financing of a Water Basin Agency's program of action. The desirable objective or "maximum contributive capacity" is based on the most effective effluent treatment technology currently available. It is the maximum amount of a Water Basin Agency's expenditures if the Agency finances the cost of implementing the most advanced technology throughout the Basin Finance Agency's jurisdiction during the pluri-annual period of the action program. The expenditures of those users of water resources who have developed the most advanced effluent treatment technology are assumed to be maximal. Therefore, the highest effluent tax rate in terms of francs per kilogram of pollution is that rate which will generate sufficient effluent tax revenue to finance that ideal program of action drawn up by an Agency. The objective of such a program is to continuously reduce water pollution to the lowest level technologically possible.

The "minimum objective" is the minimal constraint on the effluent tax rate structure wherein not only the relatively less-polluting flows of newly created pollution but also effluents from earlier sources are taken into account. Once the five-year investment expenditures for the water basin program of action have been determined, an effluent tax rate must be established such that the water resource user is indifferent to the alternatives of paying the tax or reducing pollution.

As noted above, the Basin Committee approves effluent tax rates and action program objectives. Therefore, the rates and objectives proposed by the Basin Finance Agency are approved after negotiation has occurred within the Basin Committee between governmental authorities and the representatives of water users. Once the Basin Committee has agreed upon the pollution tax base and effluent tax rate, pollution discharges by each industrial firm are measured by determining the daily average weight of pollution for the month in which the largest amount of pollution occurs, as follows:

$$Daily\ average\ pollution = \frac{Weight\ of\ discharge\ during\ month\ of\ greatest\ pollution}{Number\ of\ days\ in\ month}$$

Then an annual tax rate is applied to the average daily pollution to compute the annual effluent tax liability for a polluter in an industry.

In practice, flat rate schedules of discharges covering pollution by type of industry have been developed in cooperation with water resource users so that all discharges by industrial firms will not have to be measured. However, if an industrial firm feels that such a flat rate is discriminatory, it has the right to request that the actual amount of pollution discharged be measured. If the measured daily average pollution is less than the flat rate, the Basin Finance Agency pays the cost of measuring the discharge, whereas if the flat rate is less than the daily average, the industrial firm pays measuring costs.

Within municipalities, a flat per capita rate that varies only somewhat with community size is applied to a community's population, and the appropriate flat rate is applied to industrial pollution discharged into municipal wastewater treatment facilities. The municipality pays the full amount of effluent tax to the Basin Finance Agency, and the industries reimburse the municipality for their share of the tax burden.

Because the amount of pollution does vary significantly with the size and location of the community, "community coefficients" are applied to adjust the flat per capita rate levied per unit of population.[57] The following is a summary of other characteristics of the French effluent tax system: (1) Charges are collected annually; (2) a straight-line rate is charged, meaning that the scale does not vary in terms of the weight of pollution discharged; (3) rates are applicable to the net amount of pollution discharged = gross pollution × treatment − premium coefficient; (4) the charges are calculated without reference to standards of discharge or of quality (as by payment of the charge so long as a standard fails to be reached); (5) the threshold for the collection of charges is, however, set at 30 kilograms per day per polluter; and (6) the charges are limited to 2.5 percent of the value added of any industrial firm.[58]

In general, the permanent Secretariat for the Study of Water Problems, under the Ministry for the Quality of Life (Environment), is responsible for the coordination of water resource management and water pollution control policy, especially supervision of the Water Basin Agencies. The Secretariat is also responsible for (1) the nationwide monitoring of water pollution for the over 1,000 areas that have already been identified and analyzed or studied; (2) the preparation and monitoring of the execution of each five-year State Plan; (3) the development of both a network for measuring pollution (through installation and operation of experimental stations) and a national water index; and (4) the promotion of joint research on water resources management and water pollution control.[59]

As part of the French government's 1978 environmental policy, CIANE announced several measures for implementing major planned projects, including a long-term clean water program to be carried out over a fifteen-year period. Based in part upon recommendations from the Environment Ministry's Secretariat for Study of Water Problems, the clean water program would include a series of studies and specific antipollution programs for specific bodies of water, such as rivers, lakes, harbors, and bays. Such an extensive program will, of course, have to be approved by Basin Finance Agencies, Basin Committees, Regional Planning Councils, and the National Water Committee.[60]

## Transnational Water Pollution Control

The French government did not carry out the 1976 EEC directive on the purity of drinking water immediately because of a reluctance to enforce lower standards than already exist domestically. Similarly, the government postponed implementation of the EEC Council's 1973 water quality objectives on detergents and on testing biodegradability because the country already had more stringent standards. All three objectives have now been implemented however.[61]

The French have two major water pollution problems, the Rhine River and the Mediterranean Sea. They cannot resolve these problems on their own because both involve transfrontier pollution. In the case of the Rhine, the four riparian countries France, West Germany, The Netherlands, and Switzerland agreed in September 1972 to adopt common water pollution standards. However, responsibility for applying the common standards remains with the authorities of each country. The four countries also agreed to establish a program of protection for the Rhine: (1) To implement the French system for water resource use and pollution, that is, you pay for the water you use and also for the water you pollute; and (2) to clean up the French (Alsatian) potash plants that have been notorious polluters of the Rhine for years.[62]

In October 1975, the Council of Ministers of the EEC adopted two proposals relating to transfrontier pollution: (1) the EEC became a party to a convention limiting the chemical pollution of the Rhine, and (2) the EEC became a party to a framework convention related to pollution of the Mediterranean Sea.[63] In February 1976, France, Italy, and fourteen other countries bordering the Mediterranean Sea negotiated a convention for controlling pollution of the sea from land-based sources, hydrocarbons (oil from tankers), and the dumping of the waste of aircraft and ships. Because the EEC has sole authority for some facets of its members' environmental policy, it signed the convention along with France and Italy.[64] However, as of mid-1978 the French National Assembly had not ratified the convention signed over two years earlier by the government.[65] An update on the status of ratification appears at the end of this chapter.

## Air Pollution Control

Law 61-842, passed on August 2, 1961, provides for government authority to control air pollution from home heating, industry, and motor vehicles through regulation of emissions, construction of new industrial facilities, use of fuels, and the designation of "special protection zones." In addition, the law authorizes local officials to exercise emergency powers to control air pollution where public health is endangered. The "special protection zone" designation is to be used to establish more stringent air pollution control regulations for an urban area suffering from seriously deteriorated atmospheric conditions.[66]

Heavily polluting productive facilities such as electric power, steel, pulp and paper, cement, sugar, distilling, petroleum refining, and petrochemical plants are required to operate air pollution monitoring equipment. Moreover, the emissions of such heavily-polluting plants are checked by Departmental (regional) government inspectors. The central government, local governments, and quasi-governmental and private institutions also operate an extensive system of air pollution monitoring stations.[67] Most of the main industrial and urban centers are now equipped with air pollution measuring and warning networks financed and maintained by the central government.[68]

Based on the central government's 1973 environmental program, a system of monitoring within all industrial zones and all urban areas with more than 100,000 inhabitants should have been operative by 1978. By 1975, there were 800 monitoring systems in eighty places of varying importance, with the most notable being located in thirty-three urban areas and at other principal sources of pollution, including electric utilities, refineries, cement factories, petrochemical plants, pulp and paper mills, and steel plants. The monitoring systems are continually being modernized and reorganized in the larger urban and industrial areas to effect an optimal arrangement for measuring emissions. Data are collected at the Departmental (regional) level to be used both to determine compliance by registered installations and to monitor ambient conditions in anticipation of an air pollution alert. In addition, Departmental authorities transmit interpreted and sample data on average emissions to the Environment Ministry for use in investigating the problem and effects of air pollution.[69]

The air pollution control legislation has been implemented in a piecemeal fashion in France beginning with the 1917 Classified Establishments Act, followed by the 1961 Air Pollution Control Act with implementing decrees. For example, there are still no comprehensive regulations establishing national emission standards such as those promulgated in Canada and the United States. Heating plants and steam generators are regulated according to decrees promulgated in 1964 and 1967, respectively. Under a second 1964 decree, two categories (out of the three delineated) of industrial development are prohibited in certain developed or developing urban areas. Department and local authorities are required to regulate the construction of productive facilities and to establish the maximum emissions therefrom[70] through administration of the 1976 Law on Installations Registered for Purposes of Environmental Protection and Law No. 61-842 Air Pollution Control Act enacted in August 1961.[71]

Emission standards have been established for certain metropolitan regions. Metropolitan Paris, established as a specially protected zone by a decree promulgated in 1963, has been successful in controlling air pollution. Over a recent six-year period,[72] $SO_2$ emissions and CO emissions were reduced 40 percent and 30 percent, respectively, while energy consumption increased significantly. This success has prompted the central government to extend the status of special protection zone to Lyon and the Lille region. In addition, there are three "alert zones" where ambient air quality can deteriorate to dangerous levels during periods of peak activity and/or when climatic conditions are optimal. During a period of very poor air quality, emissions by industries in the "alert zone" must be curtailed. Alert zones are located at Rouen, at Havre, and in the new port and industrial center west of Marseille on the mouth of the Rhone—the Fos-Berre region.

The major sources of air pollution can be summarized as follows: (1) home heating systems which produce $SO_2$, $NO_x$ and particulate matter (smoke); (2) motor vehicles which emit CO, hydrocarbons, particulate matter (smoke), and $NO_x$; and (3) industrial facilities that generate several types of emissions, including $SO_2$ and particulate matter. The control of air pollution from home heating systems has resulted both from improved efficiency in combustion producing more $CO_2$ and water vapor, along with less smoke, and from regulation of the sulfur content of fuels used for domestic heating.[73] The French standard for the sulfur content of domestic fuels is consistent with the EEC directive thereon. Thus, there is presently an 0.5 percent limit on the sulfur content in gas oil and home heating oil. In 1980, the limit was to be lowered to three-tenths of a percent.[74]

There are also regulations for new and used transportation equipment. The central government has established emission standards[75] for toxic gas (hydrocarbons), particulate matter (smoke), and CO pursuant to regulations established by the Commission of the EEC in 1970 and under the authority provided in the 1961 Air Pollution Control Act. As a result of these emission standards, automobiles in 1972 discharged 50 percent less pollution in their exhaust than 1962 models did. By 1974, the discharge rates were to be reduced by an additional 20 percent.[76] During the first two years (1970-72) of EEC auto emission standards, the reduction of CO and hydrocarbon emissions from new European cars just about offset the pollution increase caused by the growth of auto use. Before an overall reduction in automobile exhaust emissions could occur in France (or the other EEC nations), the EEC would have to establish more stringent standards.[77]

In May of both 1974 and 1975, the EEC adopted more stringent auto emission standards, but the 1975 standards were not implemented by the French until October 1977, which is more than the two-year grace period usually allowed for implementing an EEC policy.[78] In view of the positive effect on air quality caused by the Energy Crisis of 1973-74 and the subsequent recession of 1974-76, the delay probably had little effect on air quality.

The permissible lead content of auto fuels used within the EEC has not, as yet, been reduced to that allowable in the United States because the Commission's environmental scientists suspect that significant reductions in the lead content of gasoline would cause a commensurate increase in emissions of noxious gases, especially nitrogen oxides, which might be just as detrimental to human health. With regard to establishing more stringent standards for auto exhaust emissions, an Organization for Economic Cooperation and Development report on automobile pollution completed in 1972 supported the EEC's contention that not enough was known about the role of automobile emissions in air pollution for a common policy to be formulated.[79] As of January 1, 1976, the maximum permissible lead content of gasoline must be 1.81 grams per gallon in France as compared to a maximum of 1.7 grams per gallon (with 0.05 grams per gallon being available) permissible in the United States at that time.[80] In May 1978, the EEC adopted a directive on the lead content of gasoline which provides for a limit between 1.81 and 0.67 grams per gallon.[81] France now is complying with this EEC directive. For a more detailed analysis of the lead in gas issue, see the update in Chapter 7.

## Solid Waste Pollution Control

In July 1975, the French National Assembly enacted omnibus legislation to deal with the problem of waste disposal and materials recovery—Law No. 75-633. The general provisions in Title I broadly define the terms "waste" and "waste disposal" and explicitly delineate the law's applicability to any person producing or disposing wastes harmful to air, water, land, flora, fauna, and humans. Title III further stipulates that the Law on Installations Registered for Purposes of Environmental Protection applies to all waste-disposal installations, both private and public. Even though the 1975 law established a National Agency for Recovery and Disposal of Wastes, Title IV allows local government to retain its traditional responsibility for municipal waste collection and disposal.

The provisions of Title II, which stipulates that both producers and importers must be able to prove that their products generate only disposable wastes,[82] have been designed to facilitate the regulation or prohibition of products that generate cumbersome or hazardous wastes in production or consumption. In addition, there are provisions requiring that waste disposal be organized so as to facilitate either materials recovery or energy conservation. In anticipation of the comprehensive control of hazardous waste, the 1975 Waste Disposal and Materials Recovery Act (WDMRA) stipulates that entities producing waste must provide environmental authorities with complete information at any time requested and that all waste-treatment facilities must be licensed as provided in the Law on Installations Registered for Environmental Protection.

In principle, the WDMRA authorizes the central government to use its powers to ensure that recycled materials are an economic alternative to new materials in the production of goods and services. For example, the government can require that minimum amounts of recycled materials be included in the production of certain goods and services. Generally, the law provides for the elimination of any fiscal or juridical policy that discriminates against recycled materials except where there is technical or economic justification therefor. Moreover, the central government can impose taxes to encourage recycling (or materials recovery) and waste reduction. Pollution taxes which have been considered include those on (1) production waste to be disposed of by industry; (2) products, such as tires and motor vehicles, the disposal of which is difficult; and (3) packaging, the disposal of which is also difficult. Finally, in carrying out its urban waste-disposal responsibilities, municipalities can rely on pollution taxes, service charges, or general tax revenue to finance such public services.[83]

As a complement to the 1975 Waste Disposal and Materials Recovery Act, the National Assembly enacted legislation to control toxic chemicals in July 1977 (Law No. 77-771). Implementing regulations (decrees) were promulgated in 1978, and they became effective in early 1979. In addition, there will eventually be an EEC directive on toxic chemicals with which France and other EEC nations will have to comply.[84] For an update on EEC regulation of toxic substances, see the discussion thereof at the end of the next chapter. In addition, Law No. 76-1106, enacted in December 1976, protects employees from toxic substances in the workplace. The implementing regulations (decrees) for Law No. 76-1106, which were promulgated in mid-1978, went into effect in 1979.[85]

Although the French drafted an implementing decree for the 1975 Directive on Waste Oil some time ago, the motor vehicle repair and maintenance industry has lobbied against the draft decree because it would require "uneconomical" recycling of oil similar to the German system explained below. The 1975 Directive on Wastes, which was modeled to a large extent on the French law, will be implemented as soon as the EEC Commission makes detailed recommendations on its implementation.[86]

### Government Assistance

The French government, as well as the governments of the other eight nations of the EEC, have adopted the OECD's "polluter pays principle." Therefore, there is very little government assistance in the form of direct grants or subsidies, tax expenditures or incentives, and loans to producers and consumers who are forced to invest in pollution abatement. With respect to investments in effluent treatment facilities, industrial firms are eligible for subsidies (direct grants from Water Basin Finance Agencies and indirect grants through use of subsidized municipal effluent treatment facilities) and low-interest loans from both Water Basin Finance Agencies and a central government lending institution. Investments in emission control facilities are normally eligible only for the low-interest loans of a central government lending institution. Waste-disposal facilities may benefit from both loans and grants. In addition, pollution abatement investments are eligible for accelerated depreciation. Subsidies are available for the development of new products and processes that result in pollution abatement.[87]

Section 14 of the Water Management Act of December 16, 1964, provides for the following financial role of the Water Basin Finance Agencies:

> The *Agence* shall contribute, by means of funds jointly provided by the *Agence* and the State budget, to studies, research and installations in the common interest of the basins and towards its operating costs. The *Agence* shall allocate grants and loans to corporate bodies

and private persons for carrying out any work in the common interest of the basin or group of basins which they directly perform whenever such work is of a kind which will reduce the financial costs of the *Agence*.[88]

The amount of direct grants or subsidies to industrial firms varies between 30 and 50 percent of water pollution abatement investment, depending on the specific basin and area within the basin. Direct grants may be converted into loans on a 1.4 to 1.5:1 basis or into advances on a 1.2 to 1.4:1 basis. In addition, an industrial firm may be given loans or advances in addition to grants, with the amount varying as between basins and areas within a basin.

In order to encourage the efficient use of municipal and industrial water pollution abatement facilities, certain direct grants of limited duration are also awarded: (1) Up to 50 percent of the expenses of technical assistance services, training operating staff, and testing the efficiency of operation; and (2) reimbursement for a portion of operating expenses. In general, the more efficient the operations of the water treatment facilities, the greater the percentage amount of the "efficiency" grant.

Financial assistance from Basin Finance Agencies is awarded on the basis of an industrial firm's detailed application, along with a survey by the Agency to determine both the usefulness of the proposed pollution abatement investment and the technical conditions required for obtaining such assistance. Generally, assistance is awarded for construction of water pollution abatement facilities, but certain Basin Finance Agencies also provide assistance for altering the production process to reduce pollution. The upper limit on financial assistance to a firm making technological improvements in its production processes is the cost of the "public" pollution abatement facility that would have been necessary had the pollution-abating change not been made.[89]

The Water Basin Finance Agencies collected almost 1 billion francs (about 150 million) in 1977 when the tax rate was 10 francs ($1.50) per capita population equivalent. About half of the revenue came from the taxes paid by industry. Thus, financial assistance from the Basin Finance Agencies in the form of grants, loans, and advances to *existing* industrial establishments and municipalities should be about 1 billion francs annually.[90]

In addition to the aids granted by Basin Finance Agencies, supplemental grants and advances are available from the central government. These supplemental aids are limited to 40 percent of a Basin Finance Agency's investment, and they cannot be used to increase the total financial assistance package to more than 80 percent of an industrial firm's or industry's pollution abatement investment. The principal source of the aid from the central government is the "ecretement" or "tax ceiling" scheme wherein the State subsidizes the Agency's investment expenditures to the extent that the effluent tax liability of an industrial firm(s) exceeds 2½ percent of the value added by the industrial firm(s) concerned.[91]

The above financial assistance is available for existing industrial establishments that are investing in water pollution control. For new establishments, an operating permit will be issued only after the environmental impact survey, a technical and economic review, and a public inquiry have been carried out. Thus, the "best technology economically achievable" is incorporated in the design of the planned installation. The central government provides the primary source of financial assistance for new investments through its regional economic development program. These grants from the *Delegation à l'Amenagement du Territoire et à l'Action Regionale* (DATAR) are made to firms that establish operations in those regions delineated in the national economic plan as meriting economic support to encourage growth therein. Moreover, these subsidies may be increased when the amount of a firm's investment in pollution abatement is greater than 5 percent of the total investment in its industrial facility.[92]

Another source of central government aid is the exceptional transitional assistance provided for several heavily polluting, industrial sectors in the form of *contrats de branches* (individual industry contracts). In July 1972, the first contract was consummated with the French pulp and paper industry, which accounted for 20 percent of all water pollution and is the largest pulp and paper industry in the EEC. As a result of the first contract, the industry agreed to reduce pollution discharged by about 80 percent in return for subsidies amounting to 80 percent of the industry's projected investment in pollution abatement over a five-year period. Similar contracts have been concluded with other major industries, including beet sugar refineries, distilleries, breweries, and the producers of starch, yeast, and woolen clothing. Finally, agreements have been negotiated with several other polluting industries,[93] including cement manufacturers, metal finishers, gypsum quarries, tar-coating works, and chlorine-electrolysis works.[94]

In addition to direct forms of financial assistance to industry, the Basin Finance Agencies make indirect grants to industrial firms by financing municipal treatment plants and intermunicipal sewage collection systems. Capital grants to municipalities are awarded on the basis of a detailed application made to the Agency and following the allocation of central government grants prepared in cooperation with the Prefect (or regional administration) for the Ministry for the Environment. The rate of assistance may vary between 15 and 40 percent, with the average being 25 percent. Moreover, municipalities are eligible for loans and advances from the Basin Finance Agencies or central government.[95]

In early 1978, the central government proposed legislation that would provide for air pollution/emission taxation through creation of an "Air Agency" system similar to the existing water pollution effluent tax system operated through the Water Basin Finance Agencies. Initially, the Air

Agency (*L'Agence de l'Air*) would conduct an intensive investigation of air pollution abatement problems. Having developed an adequate understanding of the abatement problems, the Air Agency would provide financial and technical assistance for air pollution control programs by industries. The revenue from taxes imposed on polluters based on the amount and content of their emissions would provide much of the funding for the Air Agency's financial assistance program.

Probably the most serious flaw in the logic of those asserting that the success of the Water Basin Finance Agencies in alleviating water pollution problems is easily transferable to the Air Agency relates to the different perceptions that water polluters and air polluters have of their common interests. The effectiveness of the water pollution control system through effluent taxation and Basin Finance Agencies has been aided by a commonality of interest of all those using or living in a water basin area. Inasmuch as air pollution is more diffuse and its impact may vary widely in relation to atmospheric and topographic conditions, there is much less likelihood that all those living in an airshed will perceive a common interest.[96]

CIANE's 1978 environmental program included creation of an Air Agency to be established in eastern France at Metz, with an initial budget of 5 million francs (about $1 million). This Air Agency's three major short-term responsibilities are (1) measuring and abating air pollution within its jurisdiction; (2) aiding industry's air pollution abatement research and development programs; and (3) providing the public with air pollution information. In the long run, Air Agencies will be imposing emission taxes on polluters and financing abatement investments, as well as research and development.[97]

Several government assistance programs are applicable to the disposal and recycling of wastes. As noted above, municipal governments can finance their urban waste-disposal services through taxation, service charges, or general revenue. In addition, there is a central government grant and loan program for municipal investments in household waste-treatment facilities, with grants ranging from 10 to 40 percent of cost. Central government assistance, necessary for the timely establishment of collective treatment centers, includes grants to installations to provide collective treatment services for industrial wastes and reimbursement for part of the cost of treating industrial waste when subsidized treatment centers are inaccessible or inappropriate. Other government assistance includes the exemption from taxation of recycled oil and central government grants to Departmental officials for one-half the cost of initial clean-up of abandoned automobiles.[98]

The Ministry for Industrial and Scientific Development provides grants of up to 50 percent of research and development expenditures for new products and processes utilized for pollution abatement. However, if the new product or process becomes commercially viable, the grant must be repaid. In addition, many central government assistance programs that encourage technological development are applicable to research and development expenditures on pollution abatement.

French tax law provides tax incentives to industry for pollution abatement investments,[99] including accelerated depreciation (exceptional amortization) of investments in pollution abatement facilities made before January 1, 1982, for industrial installations existing on December 31, 1976, at a rate of 50 percent during the first year.The remainder of the investment is recoverable over the balance of its useful life at an amortization rate that is applicable under the general provisions for depreciation of tangible personal property used in a trade or business. The same exceptional depreciation is applicable to investments in scientific research and development. Under normal circumstances, such investments in pollution control and research would be subject to either straight-line or declining-balance amortization, with rates for the latter beginning at 50 percent for investments with a three-year useful life. The declining-balance accelerated depreciation method is not allowed for most industrial and commercial buildings, except for hotels and light buildings with a useful life of fifteen years or less.[100]

The provisions for rapid amortization of air and water pollution abatement investments were enacted as temporary measures in the mid-1960s to offset a portion of the cost of compliance for industrial installations in existence when more stringent effluent and emission standards take effect. The provision applicable to water pollution control facilities was enacted in 1965, whereas rapid amortization for air pollution control equipment was not authorized until 1967. In both cases, the original expiration date has been extended and may be extended again.[101]

In respect to the 17.6 percent value added tax levied on all industrial and commercial transactions, selected noncommercial transactions, and the provision of services at the taxpayer's option, there is no special treatment for purchases of pollution control facilities or for investments in scientific research. If the tax is not recoverable through the credit mechanism, then it can be capitalized and amortized over the useful life of the investment as explained above. The value added tax is imposed at a reduced rate of 7.0 percent on the provision of water supply and sewage services; the sale of new cars for private use is subject to a punitive tax rate of 33⅓ percent with no recoverable credit. Finally, no recoverable credit is applicable to the sale of petroleum products.[102]

As a result of government assistance programs, an industrial firm could secure government financing of 90 percent of its investment in pollution abatement. For example, the central government's first *contrat de branche* with the pulp and paper industry, which was negotiated during 1972, provides for an 80 percent subsidy of their pollution abatement costs as follows: 50 percent financed from the effluent tax revenues of Basin Finance

Agencies; 20 percent financed from tariffs on paper imports; and 10 percent financed from direct government grants. The remaining 20 percent of the cost of pollution abatement is supposedly the financial responsibility of the pulp and paper industry. However, favorable income tax treatment reduces the actual financing of the industry to less than 10 percent of the total pollution abatement capital expenditure. Subsequently, similar agreements were signed with other heavy-polluting industries, such as sugar refining and distilling.[103] As an alternative strategy for expediting industrial clean-up, the government has also negotiated antipollution contracts directly with private companies.

In August 1975 after more than a year of negotiation, the Minister for the Quality of Life (Minister for the Environment) signed the first antipollution contract with a private company, Pechiney-Ugine-Kuhlmann (PUK). This company is a major French conglomerate with sales of almost $5 billion in steel, chemicals, aluminum, mining, and other metal alloys. Not only does the contract specify the company's level of pollution abatement expenditure (initially $45 million) for a seven-year period, but it also stipulates in some detail the steps PUK must take now to clean up its operations. Moreover, the company and the Ministry will agree upon more money for pollution abatement at a later date.[104] In September 1977, the Environment Ministry consummated a similar agreement with Societé Crusot-Loire (SCL).[105]

The most interesting facet of the government's agreements with PUK and SCL is that all financing will be borne by a private company. In contrast, the *contrats de branches*, signed with entire industrial sectors, provide for substantial governmental assistance in the form of direct grants, low-cost government loans, and favorable income tax treatment for pollution abatement equipment. The government has assumed a more aggressive posture in negotiating pollution control by industry for those installations that existed when more stringent pollution control standards were implemented because of the substantial popular support for tough environmental policy and a quality environment, even at the expense of higher prices for the goods and services of heavily polluting industries.[106] As discussed above, any new facilities or expansion of existing ones must be licensed and, therefore, must comply with the pollution control standards that were in force at the time of registration, that is, the "best technology economically achievable."

## Environmental Policy Developments, 1979–1981

During 1979, the French government completed much of the implementation of the Chemical Controls Law of 1977 (Law No. 77-771). In January 1979, a decree was issued to establish premarket notification for the production and distribution of toxic substances.[107] Later in April, the Environment Ministry promulgated a chemical substances file,[108] but another two years would pass before a file was published listing the effects of chemical substances on man and the environment. The April 1981 document included 270 entries, with additional substances to be added on an annual basis.[109]

In the summer of 1979, a special tax on waste-lubricating oils was promulgated.[110] An order patterned after a similar law enacted in Germany several years ago was issued in November 1979 to require compulsory recycling of used motor oil, as well as the imposition of a temporary tax on new motor oil.[111] The tax and regulatory provisions of this order are explained in greater detail below.

Several other changes or additions to French environmental law occurred in 1979. For example, public notification rules were established in February for authorizations of classified plants under the 1976 Law on Installations Classified for Purposes of Environmental Protection (Law No. 76-663). Similarly, a new decree establishing restrictions on newly constructed facilities near shore areas was promulgated in August 1979.[112] The National Assembly also enacted a new law in April 1979 to tighten government controls on the use of fertilizers.[113] Finally, December 1979 saw the issuance of an order specifying the conditions that must be met by those discharging polluting substances into surface waters and ground waters.[114] Details of this order are provided below.

The year 1980 witnessed few new developments in environmental law. By far the most important development of the year occurred in February when the Environment Ministry issued a "technical instruction" on dumping procedures for nontoxic and toxic industrial wastes. The objective was to specify situations in which expensive waste treatment could be avoided through controlled dumping.[115]

Inasmuch as the first few months of 1981 were consumed by the national campaign for the Presidency and the National Assembly, very few developments in environmental law occurred. In fact, at a press conference Giscard's Environment Minister Michel d'Ornano announced a pause in the adoption of new environmental law and the resolve to improve the enforcement of existing environmental law. He also summarized the accomplishments of the Giscard government in the field of environmental law.[116]

Environmental Protection Administration

From a policy point of view, the most important development probably occurred in July 1980 when Robert Toulemon, Chairperson of the Environment Intergroup (a group of experts on regional planning, the environment, industry, energy, housing, and agriculture) announced the environmental guidelines for the Eighth Five-Year Plan. This plan both establishes general goals of social and economic development and describes the probable consequences of the plan's stated objectives as guidance for public policymakers. Although the guidelines are not binding, Toulemon's stature as a top civil servant and the past history of the acceptance of such guidelines by central government planners and administrators mean that the Toulemon report will most likely establish the framework for environmental policy for the 1981-85 period. Details of these guidelines are explained later in the analysis of future environmental policy.[117]

Another indicator of trends in environmental protection administration is the growth of the budget for the Environment Ministry. The Ministry's budget for 1979 was 30 percent larger than that for 1978.[118] However, the increase in the 1980 budget was slightly less than the increase necessary to keep up with inflation.[119] As might be expected, the Ministry's 1981 budget as proposed by the Giscard government reflected a significant increase (18 percent) over the 1980 budget for environmental protection. As is explained below, the Mitterrand government increased the 1981 Ministry budget twice more. The most significant aspect of the 1981 budget was the proposal to increase appropriations for promotion of clean technologies by almost 200 percent over the amount budgeted for 1980, which was the first time any such appropriations had been made. The emphasis on promotion of clean technologies was consistent with the Giscard government's new emphasis on pollution prevention as the preferred strategy for environmental protection.

The 1981 budget also included significant increases for subsidies to the private sector (23 percent), prevention of pollution and nuisances or enforcement (26 percent), and research (27 percent), while appropriations for operating expenses (8 percent) and the protection of nature (4 percent) were less than the inflation rate (13 percent). By far the largest portion of the budget (71.5 percent) is appropriated for the antipollution programs of the Water Basin Agencies.[120]

The Giscard government's increased emphasis on environmental research was further strengthened in late April 1981—a few days before the presidential elections—when Environment Minister d'Ornano announced the creation of three centers for such research. The Parisian center is to be devoted to the study of nature conservation and urban environmental problems, such as air pollution and noise. Another center, which will be located in Montpellier, Aix-en-Provence, and Marseilles, is to study the problems unique to Mediterranean areas, namely, water management and urban or industrial development in coastal lagoons. The third center, which will be based in the Breton cities of Rennes, Nantes, and Brest, is to specialize in the study of seashore protection, especially marine pollution. In reality, these three new institutions will be established through reorganization of existing research facilities. Moreover, there will be greater coordination between academic, government, and private sector research teams, insofar as the research effort will be operationalized in 1982 jointly by the Environment Ministry, the Ministry of Universities, and the Research Secretariat. The "hidden agenda" lies in France's intention to promote its environmental expertise in foreign trade.[121]

The Giscard government also consummated a sectoral program with the asbestos concrete industry in late 1980. The industry agreed to reduce by at least 30 percent, or completely wherever possible, water pollution from its plants by 1984. This contract is the most recent of the sectoral programs (*programmes de branche*) and sectoral contracts (*contrats de branches*) through which industries agree to control the pollution of their existing facilities by a specified amount within a specified period of time. Under such agreements, the government promises not to impose more stringent pollution control standards on the industry in return for industry's pledge to abate emissions and/or effluent. However, sectoral contracts differ from sectoral programs insofar as the contracts provide financial assistance from the government, for example, a reduced effluent tax on discharges into water.

This sectoral program applies only to water pollution inasmuch as the asbestos concrete industry has already reduced its air pollution significantly since 1977. Other measures have been agreed upon. Asbestos dust and asbestos concrete waste are to be recycled where possible; waste from the fabrication of finished products is to be taken to controlled dumps and buried; and sludge from effluents is to be either recycled or condensed to a concentration of greater than 30 percent for transport to controlled dumps. Compliance at each asbestos concrete plant is to be controlled through the Prefectoral (Departmental) Order required of every classified installation as provided in Law No. 76-663 and its 1977 implementation decree.[122]

Although the asbestos concrete industry accounts for about three-fourths of asbestos consumption in France, another 5 percent is attributable to brake lining producers. Consequently, new plants manufacturing brake linings or pads were included on the supplementary list of classified installations requiring operating permits in a June 9, 1980, decree. Later, in April 1981, a "technical instruction" applicable to such new plants was issued by the Environment Ministry to the Prefects to guide them in their efforts to reduce air, water, and solid waste pollution therefrom. A similar "technical instruction" for new asbestos concrete facilities was issued in February 1981. It should be noted that French and European environmentalists are adamantly opposed to sectoral agreements insofar as they often violate the "polluter pays principle."[123]

Waste Disposal and Recycling

As noted above, in February 1980, the Environment Ministry issued a "technical instruction" on dumping procedures for toxic and nontoxic industrial wastes. The text of the instruction was drafted as a complement to the March 9, 1973, circular on urban (household) waste elimination and dumping.

This "technical instruction" delineates two types of industrial wastes depending on their toxicity. It also deals with dumping site selection, criteria for acceptance therein, the operating regulations for dumpsites, classification of dumpsites, materials that may not be dumped, and dumpsite management. The instruction allows dumping of nontoxic industrial waste with urban (household) waste. However, to avoid a proliferation of sites, such industrial waste must be dumped at sites with a capacity of at least 30,000 tons.

"Special" industrial wastes, which are defined as wastes likely to have a specific environmental effect, may only be dumped under limited conditions, including obtaining an operating permit and the provision of long-term controls. At least one dumping site for special industrial wastes is supposed to be opened in each of the twenty-two economic regions in France. Most toxic substances may not be dumped, with the exception of some mineral muds.

The operating rules and controls for dumpsite management are the responsibility of the Departmental-level inspection services for classified installations, along with the cooperation of the water management service. The same control procedures that apply to city waste dumps are applicable to industrial dumpsites.[124]

In March 1981, the National Agency for the Recycling and Elimination of Waste (NAREW) announced that it would begin the job of cleaning up hazardous industrial waste dumpsites *à la* Love Canal. The dumps to be cleaned up will be identified by regional authorities.

NAREW also decided to allocate funds for the development of urban (household) waste sorting and recycling plants. The Agency has already subsidized the construction of two urban waste-recycling installations. Similarly, it approved funding for recycling nonferrous metals, paper, wood pellets, and a variety of organic materials, including slaughterhouse blood and sludge from urban and industrial sources.[125]

### Tax on Recycled Waste Oil

As already mentioned, compulsory recycling of used motor oil began on November 23, 1980. Essentially anyone who possesses used motor oil must store the substance until it can be delivered to a registered collector. The collector must pick up all amounts over 200 liters (about 53 gallons). It must also advertise its business as a registered collector. For amounts of used oil between 1,000 (about 265) and 5,000 liters (about 1,323 gallons), the collector must pay 3 centimes (about ½ cent) per liter. For amounts over 5,000 liters, the charge is 5 centimes (almost 1 cent) per liter, while there is no required payment for amounts below 1,000 liters.

A registered collector is licensed for a three-year period. Each Prefecture (or Department) compiles a list of registered collectors based on the list thereof maintained by NAREW.

Waste oils are to be delivered by registered collectors to registered eliminators who are to recycle the oil whenever possible. Alternatively, burning should only be authorized in unusual circumstances, and such burning must be carried out in certified facilities that are equipped with both pollution control and heat recovery equipment. Insofar as burning waste oil in facilities that are not certified or the dumping thereof is strictly forbidden, there are criminal penalties; two months to two years in prison and fines of 2,000 (about $300) to 100,000 francs (about $1,500) are associated with violations of the waste oil recycling provisions.

To aid oil waste holders, collectors and eliminators in financing the equipment for recycling of oil, a special tax on new motor oil was created. Both compulsory recycling and the tax on motor oil are authorized by the 1975 Waste Disposal and Materials Recovery Act (Law No. 75-633). The tax is to be phased out progressively in three years. The revenues, which should amount to 35 million francs (about $5 million) annually, are to be used to finance additional storage facilities for holders and collectors, improved efficiency of recycling installations, and improvements in anti-pollution equipment. In summary, compulsory recycling should serve the dual purpose of protecting the environment from a polluting substance and of recovering a valuable energy resource.[126]

## Air Pollution Control

As explained above, the central government has had the legal authority to control air pollution since 1967, and there have been emission standards for the Paris urban area since 1963. Thus, air quality in metropolitan Paris and most other major urban areas has been improving.[127] However, air pollution in Paris is still a serious problem, especially during periods in the winter when atmospheric conditions reduce the natural capacity of the airshed to assimilate emissions. Consequently, the network of automated stations that monitor air pollution, including sulfur dioxide, nitrogen oxide, ammonia, hydrocarbons, and particulates, in the metropolitan Paris area was significantly extended in 1980. If climatic conditions precipitate serious air pollution in the Paris metro area as measured by the expanded network of automated air pollution monitoring stations, industrial fuel users can be ordered to close down their operations. Because sulfur dioxide is the most serious air pollution problem in the Paris urban area, the sulfur content of fuel oil used therein is limited to 1 percent within the city and to 2 percent in the "special zones of protection" in suburbs surrounding the city.[128]

In one of the last acts of the Giscard government, Prime Minister Raymond Barre signed the May 13, 1981, decree (Decree 81-593) that created an Air Agency which is under the authority of the Environment Minister. As provided in Article 2 of the decree, the new Agency has been given responsibility to "initiate, stimulate, coordinate, facilitate and possibly implement activities aiming at development and demonstration of technology for the prevention of air pollution, reinforcement of air quality monitoring, and information . . . for the prevention of air pollution."[129] The Agency's board will have twenty-one directors with eight representatives from the central government, eight environmental experts or environmental associations' representatives, and five from local government. Finally, the agency's Chairperson is to be appointed by the Environment Minister for a three-year term.

The Agency will function as a clearinghouse and coordination center for the hundreds of stations monitoring air quality in the most sensitive air pollution areas in the country. It will also accumulate data on air quality control obtained from other government agencies as well as from local governments, public establishments and the private sector. In summary, the Agency's information services are intended for the use of the public, business, and local governments.[130]

A bill enacted in the summer of 1980 authorized the creation of the Air Agency by adding a new Article 9 to Law 61-842, that is, the Air Pollution Control Act. Article 9 provides in part as follows:

> An air quality agency, a government public establishment having an industrial and commercial character, is created for the purpose of contributing to and conducting monitoring and prevention of, and informing on atmospheric pollution matters.
>
> The Agency is empowered to conduct any research, study and work related to this purpose, or to contribute to this purpose.
>
> The Agency's Board of Directors is composed one third of State representatives, one third of local government representatives, and one third of qualified people and representatives of associations or interested groups.
>
> An information report presented (by the Agency) as an annex to the budget of the Environment (Ministry) will enable Parliament members to follow in detail the (Agency's) orientations and activities.
>
> In the course of its activities the Agency is authorized to grant subsidies and loans.
>
> The Agency, among other income, may receive royalties for inventions and processes to which it has contributed, fees for services rendered, and the proceeds of special taxes.[131]

## Water Pollution Control

Rather than national effluent standards, France has recommended discharge guidelines for both "registered installations" and publicly owned wastewater treatment facilities. The central government also has established industry-specific effluent standards in its sectoral programs and sectoral contracts with a number of industries. However, a system of national effluent standards is being developed as a result of a December 1979 order describing specific conditions that must be met for effluent and waste discharges that are likely to influence surface and ground water quality as well as the French territorial sea water. The order, which serves to implement Decree 73-218 (February 23, 1973), was agreed to by the Ministries of Environment, Interior, Health, Agriculture, Industry, Transport, and Budget. It requires that all discharges affecting water quality must be approved by the government based on the criteria below.

The permit or authorization to discharge effluent must quantify the maximum discharge of pollutants for any two-hour and twenty-four-hour period as well as the minimum quality of the effluent discharged. Of course, discharge quantities will be allowed to vary in concert with seasonal conditions that affect the assimilative capacity of water. Moreover, the use to be made of the water into which the discharge occurs will affect the permit conditions. Similarly, the existing amount of pollution in the water source, its natural assimilative capacity, and the need or desire to protect environmental balance affect permit conditions. Determination of the effluent's minimum quality is to be based on both the characteristics of the water source into which the effluent is discharged and the opportunity for treatment thereof. The permit is also to specify the effluent's maximum temperature, as well as its maximum and minimum acidity so that the temperature is compatible with both the aquatic environment and the EEC's Directive on Fishing Water Quality.

With regard to liquid discharges on land, the permit must specify conditions to be met to avoid prolonged stagnation of the discharge, overflow from the area designated for the discharge, and contamination of ground or surface water. In addition, the permit may specify the minimum quality of the liquid discharge, the maximum area of the land upon which the discharge is allowed, the maximum level of pollutants to be discharged, the irrigation conditions where the discharge contains fertilizing substances, and the discharge process to be used. Finally, the permit must require the monitoring of underground water quality.

Similarly, approval is required for underground discharges of effluent. The permit must specify maximum quantities that may be injected both at any one time and for any twenty-four-hour period, as well as the maximum average flow of pollutants for any twenty-four-hour period.

A permit is also required for the deposit of solid wastes either on land or underground. Such a permit must specify the maximum area in which waste dumping may occur while taking into consideration the nature of the ground, its elevation, the vulnerability of underground water to contamination, and the proximity of any surface water or sea that may be contaminated. In addition, the permit must specify the maximum capacity of the dumping area. Finally, the permit must delineate the wastes that may and may not be dumped, the type of containers (if any) to be used, and the type of treatment (if any) to be applied to the wastes before dumping.[132]

Near the end of 1980, President Giscard d'Estaing raised an issue that is not unique to the French experience with publicly supported wastewater treatment facilities when he questioned the cost-effectiveness of allocating billions of public monies to the construction of these facilities in times of restrictive budgets and conservative fiscal policy. In order to make up for past underinvestment, a new wastewater treatment facility is operationalized in France at a rate of one per day. However, in late 1980 existing facilities only had a capacity to handle the wastewater from the equivalent population of 30 million out of a total of 67 million. Insofar as the French population is about 53 million, the wastewater of industry is equivalent to a population of 14 million additional inhabitants.

At present, the French spend over a billion dollars on the construction of wastewater treatment facilities annually. About 8 percent is funded by the central government, and another 12 percent by water basin finance agencies. Thus, the balance, or about 80 percent, is financed by business and industry, as well as individuals through local government. Unfortunately, the large government subsidies may often act as an incentive to local governments to construct plants that they are not able to operate effectively because either sufficient operating expenses cannot be budgeted or sufficient trained maintenance personnel are not available.

Estimates indicate that many wastewater treatment facilities may be operating at only 40 to 50 percent of their capacity. To increase the cost-effectiveness of government subsidies, current policy is being reappraised, with other treatment techniques being given careful attention. Suggestions for improving cost-effectiveness include increased reliance on the "lagooning" process for resorts and other communities, with significant seasonal variations in population and increased use of septic tank equipment in areas with low population density. The lagooning process takes advantage of bacterial purification in natural basins during periods of peak wastewater discharge. Therefore, a wastewater treatment facility's capacity and the commensurate investment therein can be reduced to that necessary for more normal wastewater discharge.[133]

Transnational Pollution Control

Although the pollution of the Rhine River and the Mediterranean Sea have been major transnational environmental problems for a number of years, the most significant development in transnational pollution control during the last two and a half years of the Giscard Presidency occurred in the summer of 1980 when the Prefect of Lorraine agreed to apply German air pollution control standards to a battery plant being constructed in France on the border with the German State of Saar. More specifically, the Lorraine Prefect agreed in negotiations with the economics and environment ministers of the Saar to include existing German standards of air quality control technology and ambient air quality for lead emissions in the permit which it authorized for the battery plant. Even though such negotiations are not formally permissible in any bilateral or multilateral agreements, the Lorraine-Saar cooperation was made possible by a November 1979 agreement to establish a permanent Governmental Commission of Germany, France, and Luxembourg. The commission meets as a regional consultation body on matters of siting installations on their borders that have potential environmental consequences for their neighbors, but are not otherwise covered by multilateral or international agreements.[134]

The problems of the multination pollution of the Rhine River have been somewhat of an embarrassment for the French inasmuch as they are signatories to a five-nation agreement thereon, the Convention for the Protection of Rhine Against Chemical Pollution. They did not ratify the agreement, however, because of political pressure from Alsatian public officials and the potash industry in Alsace. The Mitterrand government, as is explained below, is renegotiating the previous agreement so that it will be able to convince the National Assembly and Senate to ratify such a multilateral agreement.

On the issue of Mediterranean Sea pollution, the French position has been much more positive. Beginning with the first United Nations-sponsored meeting in 1975 and in the subsequent ten meetings on Mediterranean pollution, the French have been adamant in their support for abating pollution of the sea. They signed the Mediterranean Action Plan (MAP), also known as the Barcelona Convention on Mediterranean Sea Pollution, in 1976. Seventeen of the eighteen nations (Albania excepted) on the sea agreed to jointly monitor pollution and to take actions to abate their pollution, beginning with a halt to all dumping by ships and aircraft and the regulation of oil pollution.

In May 1980, they signed a third agreement (the Athens Protocol) on cleaning up land-based pollution. This agreement is essentially a commitment to abate pollution from rivers that flow into the sea, including the French Rhone River. The French claim to have reduced by 90 percent pollution from the highly industrialized Gulf of Fos area west of Marseille on the Rhone River since 1975.[135] In addition, the French government has invested heavily ($1 million in 1980) in the start-up costs for major wastewater treatment facilities in Marseille, Toulon, and Nice, protection of lagoons along the southwestern seashore near Spain, and, of course, pollution abatement of the Rhone River-Gulf of Fos industrial region. Although the United Nations financed all of the MAP budget during its first three years, France financed almost 50 percent during 1979 and 1980. The French agreed to a similar level of funding for 1981 and 1982.[136]

### EEC Environmental Policy

Under the Gaullist party governments of Georges Pompidou and Valery Giscard d'Estaing, the French government has been intransigent on several proposed EEC directives on environmental policy because of their preference for bilateral arrangements for settling disputes between neighboring nations. However, French behavior on the salt pollution of the Rhine caused by Alsatian potash mines indicates that they have lacked the resolve to solve disputes on even a bilateral basis. In recent years, the French have opposed directives on environmental impact assessments, toxic chemical production, industrial accidents, and the siting of nuclear power plants.[137] As is explained below, the Mitterrand government intends to be more cooperative on working out an acceptable proposal for such directives rather than taking unyielding positions as was common during the previous two administrations.

### Environmental Protection Accomplishments

In announcing a change in emphasis in central government policy from legislative action to the enforcement of environmental protection legislation in January 1981, Giscard's Environment Minister Michel d'Ornano cited the progress made during the past few years of legislative action and enforcement. The following is a summary of the accomplishments cited by D'Ornano:

> Water pollution has stopped increasing since 1976 and is now decreasing by 5 percent per year on average as a result of: 1) Growing pollution control investments by local governments now reaching an average of 3.6 billion francs (approximately $540 million), 20 percent to 50 percent of which is financed by the Central Government; 2) The collective financing system of the six water basin agencies, which received 1.5 billion francs ($225 million) in subsidies in 1980 and 6 billion francs (over $1 billion) since their creation; and 3) Controls by classified installations.
>
> The sectoral agreements concluded since 1971 with 17 industries have been implemented, sometimes with spectacular results, in particular in the field of cement, oil refineries, and yeast and sugar.
>
> With a few exceptions—including the city of Marseilles—air quality

has improved in the last seven years in all major cities, in particular in Paris, which has 20 percent less pollution, Lyons, Lille, and Rouen, where peak pollution has been cut by 60 percent. Similarly the situation is under control in such large industrial areas as Fos, the Seine estuary, and the Lyons area.

Domestic waste is collected in 90 percent of communities, and treatment conditions are improving in 70 percent of them. Selective recovery of domestic waste is in place in 10,000 communities out of a total of 30,000, concerning 15 million people or nearly 30 percent of the French population. Industrial recycling of recovered material is starting.

Elimination of toxic industrial waste is now under control, and the elimination of some 30 dangerous material dumps is underway. Financing is provided by individuals when identified, and if not by public funds.

Sewerage systems are being constructed in major cities on the Mediterranean shore in compliance with the Athens Protocol to the Barcelona Convention.[138]

## Eighth Five-Year Plan

In July 1980, the Environmental Intergroup announced the environmental policy guidelines for the Eighth Five-Year Plan, which will be applicable to the 1981-85 period. Speaking for the Intergroup, Robert Toulemon pointed out that rapidly increasing energy prices and stagnant economic growth should not erode concern for environmental protection because it is best served by waste reduction, recycling, and energy conservation. The plan describes the probable consequences of the specified objectives as guidance for government policymakers responsible for various facets of social and economic development.

On the matter of air quality, the Intergroup projected that emissions of sulfur dioxide should be reduced from 3.4 million to 2.4 million tons insofar as the sulfur content of home heating oil and gas oil should be lower and the national electric company *Électricité de France* should reduce consumption of heavy oil. It also projected a comparable 25 percent reduction in emissions of particulate matter inasmuch as coal will be used primarily in power plants and other large facilities that should be equipped with efficient emission collectors. As a consequence of these reductions in emissions, ambient acid and particulate content in urban areas should be reduced by 10 to 20 percent over the 1981-85 period. Additional improvements in urban area air quality should result from the creation of more "special protection areas" and "alert areas."

The Intergroup recommended that additional research be conducted on developing substitutes for lead additives in gasoline. It recommended that research be conducted by the EEC so that a more stringent directive on lead additives can be developed.

Although the Intergroup stresses the development of new industrial technology which is less polluting, it recognizes that special care must be given to preventing accidental pollution and, in particular, inadvertent emission of toxic, metallic, nitrogen, and phosphorus pollutants. Thus, special attention must be given to the elimination and treatment of toxic and hazardous waste. Considerable progress was made during the Seventh Five-Year Plan in developing fifteen waste-treatment centers with incinerators, chemical treatment facilities, and oil emulsion treatment facilities. Furthermore, seven controlled dumpsites equipped with disposal facilities for special wastes were developed between 1976 and 1980. Finally, many industrial establishments are equipped with treatment facilities for toxic and hazardous waste.

The Eighth Five-Year Plan should provide for a two-pronged attack on toxic and hazardous waste. First, Love Canal-type hazardous waste dumps should be detected and cleaned up. Second, industrial and hazardous waste-treatment and disposal facilities should be expanded. More specifically, the Intergroup recommends the creation of new centers for gathering and sorting industrial wastes, including dumpsites for special wastes and the development of new treatment equipment, including new detoxification systems. In addition, incineration capacity, physiochemical treatment, and soluble oil treatment should be increased 20 percent, 100 percent, and 300 percent, respectively.[139] As mentioned above, NAREW began the removal of hazardous dumpsites during the summer of 1981.[140]

The Intergroup emphasized the need to reduce noise pollution from automobiles, trucks, motorcycles, and aircraft. Furthermore, they recommended more stringent carbon dioxide and monoxide control requirements, along with more efficient antipollution equipment for motor vehicles.

On the subject of the prevention of catastrophes and accidental pollution, the Intergroup report included suggestions applicable to the shipping and road transport of hazardous and toxic substances, as well as to accidents in industrial plants. There should be plans for accidents involving toxic substances, and these plans should be well known and understood by central and local government officials who would play a role therein. In addition, downstream alarm systems, like those already in place along the Rhine, the Moselle, and the Saar, should be installed along other rivers downstream from industrial development areas. Similarly, the Intergroup gave special consideration to nuclear safety in its recommendations.

The Intergroup also proposed new indemnification procedures for accidental pollution incidents. An indemnity fund for compensating people and property harmed by the most dangerous pollution accidents might be established. Improved insurance coverage and financial guarantees for accidental pollution incidents should also be discussed.[141]

In an apparent eleventh-hour effort to drum up support for the reelection of Giscard D'Estaing as President, his cabinet approved waste-recycling objectives for the Eighth Five-Year Plan in late April 1981. Although the subsequent election of François Mitterrand may affect the details of these recycling objectives as well as the environmental policy guidelines explained above, there would seem to be a national consensus on the need for a recycling program designed both to protect the environment and to decrease dependence on imported basic materials. The following is a summary of the waste-recycling objectives approved for the 1981-85 planning period:

Paper recycling in 1986 should account for 42 percent of total consumption, up from 35 percent presently, resulting in an additional 300,000 tons of recycled paper in that year.

The number of reusable bottles will double, through an addition of 100 million units per year.

Production of compost, using household refuse as a basic material, will grow from the present 400,000 tons to 700,000 tons.

Recycled plastic will triple in the next five years, through a yearly increase of 100,000 tons.

Processing of slaughterhouse blood for the manufacture of proteins will grow from 40 percent of available supplies to 70 percent.

Recycled solvent production will increase from 200,000 tons (25 percent of current consumption) to 400,000 tons.

Motor oil recycling plants will be used at full capacity, namely 200,000 tons, while present operations total some 130,000 tons.

An additional one million tons of recycled rubber will be used for tire recapping.

It is noteworthy that NAREW had already announced funding in March 1981 for: development of garbage sorting and recycling plants; construction of two garbage recycling facilities; and programs for the recuperation of nonferrous metals, recycling paper and wood pellets, and recycling of various organic wastes.[142]

As noted above, the Intergroup recommended the development of less polluting technology by industry. In fact, it drew a parallel between good management policy and pollution control. Therefore, the objectives of the Eighth Five-Year Plan vis-à-vis industry are environmental protection and profitability. For new production facilities, the operator should be responsible for measuring discharges at regular intervals. The Intergroup explained its support for this proposal:

This policy is based on improvement of the public decisionmaking process, to avoid unjustified constraints on industry, simplification and decentralization of decisionmaking, formulation of quantified objectives by industry, definition of quality objectives, international comparison of technical and economic data, and adoption of Community (EEC) directives.[143]

With respect to the development of less polluting industrial production, the Environment Ministry has been conducting periodic surveys of industry to identify modes of clean technology. A total of 150 modes had been discovered by the summer of 1980, while only about twenty were identified three or four years earlier. In addition, the 1980 survey discovered extensive recycling, especially in the reuse of cooling water.[144]

After a delay of many months caused by publishing problems, the results of the Environment Ministry's survey of clean industrial technology were published in December 1981 in a book entitled *Les Techniques Propres Dans L'Industrie Francaise*. The work lists seventy-three processes or types of technology in use by 121 industrial companies as well as twenty-four processes or types of technology currently developed for use by industry to control pollution. Although the list is not final or complete, it is representative of the industrial sector's diversity in respect to types of pollution problems and solutions. The book is actually a series of files classified by type of industry, with each file containing a concise description of type of clean technology and a comparison in detailed illustrative form of the old and new processes. In addition, comparative data on emissions, energy consumption, and selected costs, that is, limited to capital investment and direct operating expenses, are provided for the old and new processes. Finally, the book emphasizes the following caveat:

> The development of a clean technology does not guarantee its success. Once the decision is reached to modify the manufacturing process everyone at the plant must be associated with its implementation. This is because [the new process] will result in a modification of production mechanisms and, as a consequence, the habits of workers whose conscious and active participation is obviously required.[145]

### Future Environmental Policy

The election of Socialist François Mitterrand to a seven-year term as President of the French Republic in May 1981 marked the beginning of many policy changes, including environmental policy changes. Of course, his nationalization of many major industries and the banking system, as well as higher taxes for those with large incomes and property, has received the most media attention, but significant changes in environmental policy are also anticipated. Nationalization of major industries implicitly results in a change of environmental policy inasmuch as government assistance for investments in pollution control will change in form, if not in substance.

The environmental policies of the new government will focus on decentralization of decision making and on transnational decision making within the EEC. The new government also promised to deliver on its campaign pledge of a moratorium on construction of nuclear power plants until a national referendum can be held in about a year after debate in the National Assembly and public debate has occurred. The cabinet headed by Pierre Mauroy, Mitterrand's successor as head of the Socialist party following the election, intends to conduct international environmental policies in unison with its European partners to a greater extent than under former President Valéry Giscard d'Estaing. Mauroy also plans to deal with transnational problems in a European framework.

Succeeding Michel d'Ornano as Minister of Environment is Michel Crepeau, leader of the *Mouvement des Radicaux de Gauche*, a minority party that contributed to Mitterrand's victory in the May presidential election. The fifty-year-old Crepeau was a candidate in the first round of the election in which he received about 2 percent of the vote. Crepeau's nomination was seen as a reward for his personal achievements as the Mayor of La Rochelle, a medium-sized Atlantic seaport, where he was active in promoting pedestrian streets, a pool of bicycles freely available to the public, household waste recycling, public transportation, solar heating for residential construction, and atmospheric and marine pollution control. A lawyer, Crepeau allegedly prefers common sense to theory, as well as the company of ordinary citizens to that of technocrats.

Dr. Alain Bombard replaces François Delmas as Environmental Undersecretary. Bombard is remembered for a 1952 survival experiment in crossing the Atlantic. He demonstrated that a man aboard an eighteen-foot rubber raft can survive on rainwater and plankton alone. Since then, he has been engaged in other research, including studies on radioactive materials, marketing rubber boats, and political activities for the Socialist party.

The cabinet includes for the first time a Ministry of the Sea. Maritime affairs have been the responsibility of the Ministry of Transport's Merchant Marine Division. The new Ministry head is Louis Le Pensec, a forty-seven-year-old Breton militant. Educated as a teacher, Le Pensec is equally committed to his province's culture, as he speaks Breton and French fluently, and to the preservation of the marine environment. It is not clear, however, whether the new Ministry will have broader environmental responsibilities than the old Merchant Marine Division.

As was true during the second half of the Giscard Presidency, no major environmental legislation is anticipated immediately by the Mitterrand government. Thus, environmental policy will emphasize specific regional (Departmental) and local enforcement of existing environmental law.

The new government will use a "European" emphasis for international issues broader in scope than European, including seeking increased compensation funds for oil spill damages and control of substandard ships in

European harbors. For European matters specifically, the government will continue its regional approach within the EEC context but will put new emphasis on a "Europe of regions" concept. For example, Mitterrand will probably solicit co-financing of the cost of the Mediterranean effluents control. Therefore, the French will also be willing to contribute to expenses incurred by one of their EEC partners in programs involving France.

On Rhine pollution, the local opposition in Alsace to the salt injection technique called for by the Bonn Convention will require a renegotiation of the treaty. Shortly before the election, President Giscard d'Estaing announced that the government would study the marketing of the salt presently dumped in the river. Although the new government will seek a solution retaining the co-financing feature agreed to by the convention's five signatories (France, West Germany, The Netherlands, Luxembourg, and Switzerland), it will work for an elimination process other than injection, for example, evacuation by barge or pipeline.

In another concession to environmentalists as well as farmers, the Mitterrand government announced on May 29, 1981, that a decision would be forthcoming on the controversial Lazac army base, which was to be expanded from 7,500 to 40,000 acres as a result of a December 26, 1972, Executive Order. The enlargement of the base, which is located in the central province of Aveyron, has been fought by environmentalists and local farmers since the order was issued. Mitterrand left little doubt during the presidential campaign that he would eventually rescind the order that occurred during the late Georges Pompidou's Presidency.[146]

In a July 1, 1981, interview, Environment Minister Crepeau listed the following issues among his highest priorities:

> (v)igorous action to prevent marine disasters such as the *Amoco Cadiz* and breakthroughs in issues deadlocked at the European Community, such as the proposed "Seveso" directive on industrial accidents. . . .
>
> As part of a "good neighbor" policy with respect to the environment, Crepeau said France would seek further EEC action on noise and waste, stricter enforcement of existing EEC legislation, further harmonization of legislation on toxic substances, progress on dealing with transfrontier air pollution, and increased attention to changes in climate.
>
> "International relations in the first place must be relations of good neighbors," Crepeau said. "Commissions covering practically all our border areas have responsibilities in the field of environment. I will try to facilitate their action, among other things through a better exchange of information with our partners and helping find a solution to difficult problems still unresolved, in particular the Rhine pollution problem."[147]

Because many French pollution problems are particularly appropriate for EEC action, Crepeau is very supportive of any opportunity for EEC resolution of transnational and interregional pollution problems. However, his support is contingent upon EEC actions being compatible with the Mitterrand government's intention to implement the decentralization of national environmental policy.[148]

Proof of Crepeau's resolve on European pollution problems occurred on November 17, 1981, at a Paris press conference attended by his colleagues from the Bonn Convention signatories. He announced that a French-Dutch compromise solution had been reached on the problem of salt pollution by potash mining on the Rhine River in Alsace. Although France, West Germany, The Netherlands, Luxembourg, and Switzerland signed the Bonn Convention on December 3, 1976, France failed to ratify the agreement because of fierce opposition from the Alsace populace and its elected officials.[149] A few days later, the European (EEC) Parliament adopted a resolution calling for the rewriting of the agreement to regulate pollution of the Rhine River and chastising France for not taking any steps to ratify the 1976 Bonn Convention. In addition to the five original signatories to the Rhine River pollution treaty, the Parliament proposed that the EEC be a signatory and that it, therefore, be prepared to take part in the solution of the Rhine salt problem.[150]

The Mitterrand government's resolve on limiting the future development of nuclear power slipped somewhat during the latter half of 1981. Although the new cabinet decided on July 30 to freeze construction on eighteen nuclear reactors planned for five sites, it only decided to scrap the Brittany site and five of the fourteen units at the other five sites.[151] Moreover, the Mitterrand government decided not to hold a national referendum on the nuclear energy program. Unfortunately, the future energy production lost by scrapping must be made up by further development of lignite at the probable cost of increased sulfur emissions.[152] When the Socialist members of the National Assembly overwhelmingly approved the cabinet's energy program on October 6, 1981, the government was ensured approval of the program, which includes the construction of six new nuclear power plants to begin in 1982 and 1983.[153]

In August 1981, Environment Minister Crepeau announced publication of a working paper proposing both "framework" laws on air, water, noise, and hunting and fishing and the reorganization of his Ministry. Before any further action is taken on these proposals, the working paper is to be discussed at regional meetings attended by officials of the central and local governments as well as civic group representatives.

Crepeau's proposed reorganization of the Environment Ministry would add four new divisions (water, air, wildlife, and education), and information and innovation to the three current divisions (for the protection of nature, pollution prevention, and quality of life). Earlier in the year, the Mitterrand government implicitly reorganized the Environment Ministry through budget revisions that removed monies for housing and construction from the Ministry's budget.

The Air Division would be given authority over the Air Agency, which was created at the end of Giscard's tenure as President. The Air Agency, which is headquartered in Nancy, would require an increase in financial resources. In conjunction with the Urban Planning and Housing Ministry, the Air Division would pursue a vigorous anti-noise policy, especially regarding existing homes and motorcycles.

Similarly, the Water Division would be given the authority over the six water basin finance agencies, and the responsibilities of the agencies would be expanded so that they would be comparable to the "basin authorities" of the United Kingdom. Thus, the agencies would be responsible for water resource management and distribution, in addition to their current responsibility for water pollution control activities. In addition, water distribution companies could be nationalized, but the working paper does not propose such a government takeover. Finally, Water Division staff members would provide technical assistance to the expanded basin agencies.

National parks and reserves, regional parks, and hunting activities would be managed by the Natural Areas and Fauna Division. A portion of the budget for this division would be raised through higher taxes on hunting society shares, some of the proceeds of which would be used on increased research in game behavior.

In addition, the National Forest Office, as well as selected units of the Agriculture and Interior ministries, would be transferred to the Environment Ministry. Conversely, the establishment of the Ministry of the Sea should result in a transfer of the Environment Ministry's research institute on oil spill control to the new Sea Ministry. Crepeau would have to ask for a substantial increase in his Ministry's budget, which was set at about 370 million francs (about $55 million) or less than one percent of the central government's budget for 1981, to effectively carry out the expanded activities and responsibilities of the Environment Ministry after reorganization.[154]

On November 10, 1981, the National Assembly approved the Environment Ministry's 1982 budget of 504 million francs (about $75 million). The budget, which includes current expenses and investment funds, represents a 30 percent increase over the original 1981 budget of 370 million francs. However, the final 1981 expenditures should amount to 551 million francs (about $83 million) inasmuch as 181 million francs were added for economic stimulation purposes. Therefore, the 1982 Environment Ministry budget, which amounts to less than 1 percent of the total budget, will actually amount to an 8.5 percent reduction from 1981 spending.

The total resources spent by the central government for the environment should include revenues from special taxes earmarked for such purposes. When the water basin agencies' revenues from effluent taxes and fishing and hunting permit fees are included, central government expenditures for the environment increase to 2.476 billion francs (about $375 million). Furthermore, when local government spending for the environment is included, total spending increases to 7.7 billion francs (about $1 billion).

More than half of the Ministry's budget (51 percent for current expenses and almost 54 percent for investment funds) will be allocated to the Pollution Prevention Division. More specifically, the division's budget will be allocated as follows: 46.3 percent for water control programs, with 30.6 percent for large dam projects; 21.6 percent for safety; 18.8 percent for waste control and clean technologies; 9.7 percent for air quality control; 1.9 for noise control; and 1.6 percent for economic and statistical research.

About two weeks after his Ministry's budget was approved, Crepeau presented his 1982 program to the cabinet. He emphasized the adoption of a more systematic monitoring procedure for inspecting oil and chemical facilities, new measures for the protection of air and water quality, and new initiatives to mitigate noise pollution. Moreover, enforcement of existing regulations would be strengthened by the addition of eighty-seven inspectors of classified installations, thereby increasing the total number to 487. This increase is deemed necessary because the number of inspectors has not kept pace with industrial growth.

Measures for the improvement of air quality will focus on the creation of an air agency in the Paris area. Actions for the protection of water quality will encompass financial assistance by water basin agencies to local governments so that the local governments can develop pollution-free water resources, improved enforcement of regulations designed to prevent the contamination of ground water, and measures to restrict the development of new gravel pits.

With regard to noise pollution, a noise specialist will be appointed for each of the ninety-four Departments (*Prefectures*). The Ministry intends to sign voluntary agreements for establishment of noise control programs with twenty local governments during 1982 and 1983. In addition, the nuisance tax on airline companies will be modified so that it is based on landing and takeoff noise rather than passenger traffic. The tax proceeds will be used to purchase 10,000 homes near the Charles de Gaulle and Orly airports. Finally, new motorcycles and mopeds will be required to have sealed mufflers, and 1,500 of 300,000 homes especially vulnerable to vehicular traffic noise will be soundproofed.[155] In summary, the Mitterrand government's Environment Ministry under the leadership of Michel Crepeau plans to be much more aggressive than it was during Giscard's Presidency.

## 1982 Addendum

Although there were some minor developments in environmental policy during the first half of 1982, the Mitterrand government was concerned primarily with national economic and administrative policy. The National Assembly finally enacted the Socialist-Communist coalition's nationalization program in early February 1982. The nationalization plan, which will cost the government an estimated $7 billion (about 50 billion francs), includes the takeover of France's military industry, five industrial conglomerates, and 39 domestic banks. The government also proposes to nationalize at least three major foreign companies by the end of 1982.[156]

Mitterrand plans to use the nationalized companies to reduce unemployment, while stimulating increased economic growth, technological development, and capital formation. To do so, Mitterrand revised the Eighth Five-year Plan developed during the last year of Giscard d'Estaing's presidency. However, the environmental goals in the 1981-85 plan were not affected because the stopgap 1982-83 plan was primarily concerned with economic growth and investment.

To reemphasize the national planning system that the Gaullists had ignored for a decade in deference to a freer market, Mitterrand proposed a new law in May 1982 to provide participation by consumers, lobbyists (including environmentalists), and politicians in addition to bureaucrats, industrialists, and trade union leaders. The new law will both allow local governments to make their own plans and authorize planners to make contracts with public and private companies. When the law is enacted, planning for the Ninth Five-year Plan, which is to cover the period 1984-88, will begin in earnest with government subsidies offered to public and private companies in return for commitments to promote exports, increase employment and production, invest at specific locations and in the amount specified, conduct research, and introduce new technology with the overall objective of recapturing domestic markets in critical industries, for example, electronics, information processing, energy, and transportation.[157]

In its first year, the Mitterrand government's stimulation of demand through increased transfer payments reduced unemployment somewhat, but it also aggravated the inflation problem and precipitated two currency devaluations that reduced the value of the franc by almost 20 percent in eight months. As a result, Mitterrand instituted wage and price controls in mid-June 1982, only thirteen months after his election.[158] Later in the same month, Mitterrand was forced to replace several cabinet members who resigned to protest wage-price controls that were initially to be imposed through October 1982. The 1983 budget, however, will be austere.[159]

With regard to his campaign promise to decentralize political power, Mitterrand has initiated measures to simplify and ease the bureaucratic controls for which the national government has been famous (or infamous) since Napoleon's reign in the late eighteenth century. The Parliament enacted Mitterrand's public administration reform package transferring economic power from the central government to departmental and regional assemblies, which will now be elected. Thus, the 95 *Prefects* (except Paris) have lost some of their power. The *Prefects* retain police and political powers, but they no longer have the authority to make key economic decisions in their departments or in the 22 administrative regions. Thus, the *Prefects* should retain in the short-term much of their responsibility for licensing and regulating installations that are classified as dangerous, unsanitary, or a nuisance, that is, *Prefect* environmental bureaucrats are still responsible for the protection of the health, safety, and well-being of the public.[160]

One can certainly envision situations in which the Mitterrand government's economic policy and administrative reforms would be in conflict with controlling pollution. For example, unemployment may result from increased costs incurred to control pollution because the resulting inflation would reduce the demand for and ultimately the supply of goods and services. Thus, the Mitterrand government might decide to relax or defer any pollution abatement guidelines that could precipitate increased unemployment. Alternatively, if local and regional assemblies, which have traditionally been concerned with economic development,[161] are given greater autonomy in environmental protection matters, then they may occasionally decide that the benefits of economic growth are greater than the costs of more environmental pollution.

Such scenarios are, however, unlikely because of the continuing efforts of Environment Minister Michel Crepeau to implement Mitterrand's campaign promise to consult with the people most interested in the environment to determine policy thereon, that is, to decentralize public policymaking. In May 1982, Crepeau forwarded a compilation of the White Papers from the regions to the Parliament. These regional reports represented the position and opinions of about 10,000 individuals representing themselves and/or over 4,000 groups of environmentalists, and hunting and fishing societies. The reports were used to draft a national White Paper on the Environment to be used as a working document for environmental policymakers.

The regional White Papers were generated from testimony by individuals and groups at meetings held in January and March of 1982. This consultation process is designed to develop popular support for Mitterrand's "chart of the environment" which was announced during his campaign and adopted by the Cabinet in late November 1981. As noted above, such a consultation process to enlist broad-based public support is also envisioned for the national planning process, that is, the Ninth Five-year Plan.

Although Crepeau has successfully carried out the Socialists' policy of effecting greater public involvement in the public policymaking process, other ministeries which are members of the Quality of Life Interministeral Committee (including industry, agriculture, health, interior, and marine affairs) have been guilty of "foot-dragging" in the development of Socialist government environmental policy. The necessary cooperation by the aforementioned ministeries is inevitable insofar as President Mitterrand is very enamored with his campaign promise to develop a "chart of the environment,"[162] which became officially known as the "Estates General of the Environment" when confirmed by the Cabinet at its November 25, 1981, meeting.

In early June 1982, the Ministry of the Environment officially published its 800-page White Paper on environmental policy, that is, a compilation of the regional White Papers, as the first stage of the Mitterrand government's new environmental policy. The next step will be consultation with elected officials at the departmental and regional levels of government.[163] However, this second step will be delayed since Parliament has only recently enacted the legislation to reform the electoral system so that regional assemblies can be elected. Traditionally, department assemblies have been elected directly, but such elections at the regional level will not be held until late 1983 or early 1984.[164]

As one might expect, the unique problems of each region are stressed in the Environment Ministries White Paper. In general, the policy positions in the White Paper are consistent with those of environmentalists, including the substitution of energy conservation for increased production through nuclear energy development, opposition to programs for systematic economic growth, general opposition to the concentration of power with the central government (especially with regard to energy policy), and an emphasis on prevention over cure as the objective of pollution control policy. Any proposed reforms pertinent to water pollution, forestry, wildlife and wilderness areas, agriculture and rural areas, flora and fauna preservation, tourism, urban living, human health, and energy conservation will, of course, have to be enacted by Parliament, but that would seem to be a foregone conclusion given the Socialist majority in the National Assembly and the limited power of the Senate to impose its will on the lower house.[165]

Earlier in 1982, the Quality of Life Interministerial Committee, which is now chaired by Prime Minister Pierre Mauroy, met for the first time since the Socialists took power. In addition to affirming the policy of "redistributing power" in environmental matters, it adopted a major program for the control of noise, including the creation of both a national noise council and a commission on airport noise, the promulgation of regulations to effect control of the noise emissions from two-wheeled motor vehicles, the establishment of a noise complaint bureau in each of the *Prefects*, and a procedure for contracting with larger municipalities to subsidize local noise control programs.[166] Some aspects of the new noise control program were initiated by the Environment Ministry in early June, including appointment of the National Council on Noise members.

In respect to rural environmental protection, the Committee adopted a program that includes creation of cattle manure depositories, limitations on the construction of small hydroelectric power installations, and agreements to be arranged between the central government and the nationalized utilities (electricity and telephone) to better integrate transmission facilities with the natural landscape. Moreover, the Committee decided to fund the development of "ecological balance sheets" by nationalized and private business to provide information to the public on progress in environmental protection.[167]

For industries experiencing the most serious odor problems, the Committee adopted a policy of financial aid. Thus, the chemical, petrochemical, and agribusiness industries will receive subsidies, as well as technical assistance, for control of noxious odors. The Committee also plans to publicize those preventive and curative techniques used in France and foreign countries for the control of industrial odors.[168]

Prospects for solving the Mediterranean Sea pollution problem were enhanced when representatives of 18 nations on the sea (all except Albania which has a very small coastline) signed a United Nations-sponsored treaty on specially protected areas.[169] The treaty, which is part of the UN Environment Program's action plan for the sea, commits the 18 nations to take "all appropriate measures with a view to protecting those marine areas which are important for the safeguard of the natural resources and sites of the Mediterranean Sea and of their cultural heritage." The treaty provides for the preservation of "sites of biological and ecological value, the genetic diversity of species, satisfactory population levels, their breeding grounds and habitats, representative samples of eco-systems ... sites of particular importance because of their scientific, aesthetic, historical, archeological, cultural, or educational interest."[170]

One setback in environmental policy became apparent in early 1982 when a trade group for the lubricating oil industry (*the Centre National des Lubrifiants*) reported that recycling of such oil decreased for the second year in a row. The amount recycled in 1981 was 15 percent less than that recycled in 1980. The decrease in recycling between 1979 and 1980 had only been about 13 percent. The reductions occurred in spite of the excise tax on new motor oil that took effect in November 1980. As explained above, the revenues from the tax are being used to subsidize the collectors of used oil in an effort to stimulate recycling.

Business representatives argue that the tax and subsidy system is failing because the setting of a maximum purchase price has restricted profits for oil producers, collectors and processors. The regulated price along with the exclusion of producers delivering less than 1,000 liters from the subsidy program has encouraged small businesses (especially garages and service stations) to burn the oil themselves in spite of the legal requirement that burning only occur in registered installations. In addition, collectors are increasing their exports of used oil to West Germany and Belgium. Finally, many collectors and processors were denied licenses and thus were forced out of business.[171]

## Notes

1. OECD Environment Committee, "Urban Environmental Indicators," *The OECD Observer*, No. 78 (November-December 1975), p. 26.

2. John Sheahan, "Planning in France," *Challenge*, March-April 1975, pp. 15-17.

3. G. K. Dykes and E. Tomsett, *France: Business Study* (New York: Touche Ross International, 1979), pp. 1-5.

4. U.S. Department of Commerce, *The Effects of Pollution Abatement on International Trade* (Washington, D.C.: U.S. Government Printing Office, 1973), p. A-23.

5. Philip W. Whitcomb, "France Explores Ways to Grow Soybeans," *The Christian Science Monitor*, August 9, 1973, p. 9.

6. U.S. Department of Commerce, *Effects of Pollution Abatement on Trade*, p. A-23.

7. Counseil Economique et Social, "Les Problemes de l'Eau en France," *Problemes Economiques*, May 6, 1971, pp. 7-10.

8. U.S. Department of Commerce, *Effects of Pollution Abatement on Trade*, pp. A23-A25.

9. *France: A National Profile* (New York: Ernst & Ernst, 1975), pp. 11-12.

10. U.S. Department of Commerce, *Effects of Pollution Abatement on Trade*, pp. A-23 and A-24.

11. Margot Lyon, "Crusader for Better Quality of Life," *The London Times*, A Special Report on France, November 1974.

12. "French Antipollution Contracts to Proliferate in the Future," *Business Europe*, August 15, 1975, pp. 258-59.

13. John Ardagh, "Coastlines and City Centres in Danger," *The London Times*, A Special Report on France, November 25, 1975.

14. U.S. Department of Commerce, *Effects of Pollution Abatement on Trade*, pp. A-29 and A-31.

15. *International Environment Reporter*, pp. 231:2001, 5001, and 7001; and *International Environment Reporter—Current Report*, January 10, 1978, pp. 4-5.

16. Service de l'Environnement Industriel, *Installations Registered for Purposes of Environmental Protection* (Paris: Ministère de l'Environnement et du Cadre de Vie, 1978), p. 15.

17. *International Environment Reporter*, p. 231:2201.

18. Service d'Information et de Diffusion, *Des Actions Pour la Qualité de la Vie* (Paris: Premier Ministre, 1976), p. 7.

19. Ambassade de France, *Protection of the Environment of France* (New York: Service de Presse et d'Information, 1974), p. 4.

20. Philip W. Quigg, "How France Attacks Its Environmental Problems," *World Environment Newsletter*, February 13, 1973, p. 3.

21. Service de l'Information, *Des Actions Pour la Qualite de la Vie*, p. 9.

22. Quigg, "How France Attacks Its Environmental Problems," p. 3.

23. Ambassade de France, *Protection of the Environment of France*, p. 5.

24. *International Environment Reporter—Current Report*, April 10, 1978, p. 97.

25. Bureau d'Etudes Fiscales et Juridiques Francis Lefebrre, *Business Operations in France*, Tax Management Portfolio No. 39-4th (Washington, D.C.: Bureau of National Affairs, Inc., 1972), pp. A-1 and A-2.

26. U.S. Department of Commerce, *Effects of Pollution Abatement on Trade*, p. A-29.

27. Service de l'Environnement Industriel, *Installations Registered for Environmental Protection*, p. 14.

28. *International Environment Reporter—Current Report*, January 10, 1978, pp. 4-5.

29. Service de l'Environnement Industriel, *Installations Registered for Environmental Protection*, pp. 1-14.

30. Ibid., pp. 1, 3, and 37.

31. OECD Environment Directorate, *Study on Economic and Policy Instruments for Water*

*Management in France* (Paris: Organization for Economic Cooperation and Development, 1976), p. 3.

32. *International Environment Reporter—Current Report*, January 10, 1978, pp. 4 and 5.

33. "File: EEC," *World Business Weekly*, January 15, 1979, p. 60.

34. *International Environment Reporter—Current Report*, April 10, 1978, pp. 109-110.

35. Ambassade de France, *Protection of the Environment in France*, p. 15.

36. OECD Environment Directorate, *Water Management in France*, p. 3.

37. Quigg, "How France Attacks Its Environmental Problems," p. 3.

38. OECD Environment Directorate, *Water Management in France*, pp. 7-8.

39. OECD Environment Committee, *The Polluter Pays Principle* (Paris: Organization for Economic Cooperation and Development, 1975), p. 97.

40. OECD Environment Directorate, *Water Management Policies and Instruments* (Paris: Organization for Economic Cooperation and Development, 1977), p. 81.

41. Secretariat General du Hait Comite de l'Environnement, *L'Etat de l'Environnement: Rapport Annual 1976-77*, Tome 1 (Paris: La Documentation Française, 1978), p. 10.

42. OECD Environment Directorate, *Water Management in France*, pp. 5-8.

43. Ibid., p. 99.

44. Ibid., pp. 99, 100, and 110.

45. OECD Environment Directorate, *Water Management Policies and Instruments*, p. 66.

46. Ibid., pp. 101 and 102.

47. OECD Environment Directorate, "Pollution Charges: An Initial Assessment," *The OECD Observer*, No. 78 (November-December 1975), pp. 23-24.

48. OECD Environment Directorate, *Water Management Policies and Instruments*, p. 66.

49. Ibid., pp. 101 and 102.

50. Sheahan, "Planning in France," *Challenge*, p. 15.

51. Environment Committee, *The Polluter Pays Principle*, p. 102.

52. Charles Hargrove, "France Sets the Priorities for a New Economic Plan," *The London Times*, May 12, 1976.

53. Environment Committee, *The Polluter Pays Principle*, p. 102.

54. OECD Environment Directorate, *Water Management Policies and Instruments*, pp. 51-52.

55. OECD Environment Directorate, *Water Management in France*, pp. 14-18.

56. Environment Committee, *The Polluter Pays Principle*, p. 102.

57. Ibid., pp. 102-6.

58. Ibid., pp. 106 and 112.

59. Ambassade de France, *Protection of the Environment in France*, p. 16.

60. *International Environment Reporter—Current Report*, March 10, 1978, p. 63.

61. Ibid., January 10, 1978, p. 7.

62. Quigg, "How France Attacks Its Environmental Problems," p. 3.

63. Stanley Johnson, "EC Policy Reports: Environment," *European Community*, March 1976, p. 45.

64. Community News Editor, "Environment: Mediterranean Pollution," *European Community*, April-May 1976, p. 47.

65. *International Environment Reporter—Current Report*, March 10, 1978, p. 60.

66. U.S. Department of Commerce, *Effects of Pollution Abatement on Trade*, pp. A-29 and A-30.

67. Ibid., p. A-25.

68. Ambassade de France, *Protection of the Environment of France*, p. 17.

69. Service d'Information et de Diffusion, *Des Actions Pour la Qualite de la Vie*, pp. 5 and 6.

70. U.S. Department of Commerce, *Effects of Pollution Abatement on Trade*, pp. 29 and 30.

71. Service de l'Environnement Industriel, *Installations Registered for Environmental Protection*, pp. 1-14.

72. Ambassade de France, *Protection of the Environment of France*, p. 19.

73. Service d'Information et de Diffusion, *Des Actions Pour la Qualite de la Vie*, pp. 5-6.

74. *International Environment Reporter—Current Report*, January 10, 1978, pp. 5-7.

75. U.S. Department of Commerce, *Effects of Pollution Abatement on Trade*, p. 30.

76. Ambassade de France, *Protection of the Environment of France*, p. 20.

77. Philip W. Quigg, "Auto Standards Emit Confusion," *World Environment Newsletter*, January 30, 1973, p. 2.

78. *International Environment Reporter—Current Report*, January 10, 1978, pp. 5-7.

79. Quigg, "Auto Standards Emit Confusion," p. 2.

80. Ambassade de France, *Protection of the Environment of France*, p. 21.

81. *International Environment Reporter—Current Report*, June 10, 1978, p. 162.

82. *International Environment Reporter*, pp. 231:7001-7002.

83. OECD Environment Directorate, *Waste Management in OECD Countries*, pp. 7, 8, 10-12, and 15.

84. *International Environment Reporter—Current Report*, June 10, 1978, p. 174.

85. Ibid., pp. 204-5.

86. *International Environment Reporter—Current Report*, January 10, 1978, p. 7.

87. U.S. Department of Commerce, *Effects of Pollution Abatement on Trade*, p. A-26.

88. Environment Committee, *The Polluter Pays Principle*, p. 110.

89. Ibid., pp. 111-13.

90. Service de l'Environnement Industriel, *Industrialization and Environmental Protection in France* (Paris: Ministère de l'Environnement et du Cadre de Vie, 1979), pp. 4 and 6.

91. Environment Committee, *The Polluter Pays Principle*, p. 110.

92. Service de l'Environnement Industriel, *Industrialization and Environmental Protection in France*, pp. 4 and 6.

93. Service d'Information et de Diffusion, *Des Actions Pour la Qualite de la Vie*, pp. 19 and 20.

94. Service de l'Environnement Industriel, *Industrialization and Environmental Protection in France*, p. 7.

95. Environment Committee, *The Polluter Pays Principle*, pp. 110 and 111.

96. *International Environment Reporter—Current Report*, January 10, 1978, p. 9.

97. Ibid., March 10, 1978, p. 63.

98. OECD Environment Directorate, *Waste Management in OECD Countries*, pp. 26, 47, and 58.

99. U.S. Department of Commerce, *Effects of Pollution Abatement on Trade*, p. A-11.

100. Jean-Claude Goldsmith, *Business Operations in France*, Tax Management Portfolio No. 39-6th (Washington, D.C.: Bureau of National Affairs, Inc., 1981), pp. A-71 and A-72.

101. Jean Lamarque, "Aides Publiques et Environnement: La Situation en France," *Environmental Policy and Law*, Vol. 1, No. 1 (1975), pp. 24-26.

102. Goldsmith, *Business Operations in France*, pp. A-56, A-57, and A-58.

103. Lamarque, "Aides Publiques et Environnement," p. 26.

104. "French Antipollution Contracts," *Business Europe*, pp. 258-59.

105. Haut Comité de l'Environnement, *L'Etat de l'Environnement: 1967-1977*, Tome 1, p. 12.

106. "French Antipollution Contracts," pp. 258-59.

107. *International Environment Reporter—Current Report*, February 10, 1979, pp. 503-4.

108. Ibid., June 13, 1979, pp. 711-12.

109. Ibid., August 8, 1979, p. 846.

110. Ibid., May 13, 1981, pp. 829-30.

111. Ibid., July 9, 1980, pp. 298-99.

112. Ibid., March 10, 1979, pp. 562-63, and September 12, 1979, p. 846.
113. Ibid., May 9, 1979, p. 670.
114. Ibid., January 9, 1980, pp. 7-8.
115. Ibid., March 12, 1980, pp. 97-98.
116. Ibid., February 11, 1981, p. 643.
117. Ibid., August 13, 1980, pp. 318-19.
118. Ibid., April 11, 1979, p. 606.
119. Ibid., October 10, 1979, p. 888.
120. Ibid., November 12, 1980, pp. 521-22.
121. Ibid., May 13, 1981, p. 830.
122. Ibid., November 12, 1980, pp. 504-5.
123. Ibid., May 13, 1981, pp. 830 and 840.
124. Ibid., March 12, 1980, p. 97.
125. Ibid., May 13, 1981, p. 831.
126. Ibid., July 9, 1980, pp. 298-99.
127. Ibid., February 11, 1981, p. 643.
128. Ibid., February 13, 1980, p. 52.
129. Ibid., June 10, 1981, p. 877.
130. Ibid.
131. Ibid., August 13, 1980, p. 332.
132. Ibid., January 9, 1980, pp. 7-8.
133. Ibid., December 10, 1980, p. 538.
134. Ibid., September 10, 1980, pp. 403-4.
135. Mark J. Kurlansky, "Saving the Mediterranean," *Environment*, June 1981, pp. 2-4.
136. *International Environment Reporter—Current Report*, March 11, 1981, p. 697.
137. Ibid., August 13, 1980, pp. 317.
138. Ibid., February 11, 1981, p. 643.
139. Ibid., August 13, 1980, pp. 318-19.
140. Ibid., May 13, 1981, p. 831.
141. Ibid., August 13, 1980, pp. 318-19.
142. Ibid., May 13, 1981, pp. 830-31.
143. Ibid., August 13, 1980, p. 319.
144. Ibid.
145. Ibid., January 13, 1982, pp. 16-17.
146. Ibid., June 10, 1981, pp. 868-69.
147. Ibid., July 8, 1981, p. 924.
148. Ibid.
149. Ibid., December 9, 1981, pp. 1136-37.
150. Ibid., pp. 1116-17, and Mark J. Kurlansky, "International: Who Is Killing the Rhine?," *Environment*, September 1982, pp. 41-42.
151. Ibid., August 12, 1981, p. 965.
152. Ibid., September 9, 1981, pp. 1011-12.
153. Ibid., October 14, 1981, p. 1057.
154. Ibid., September 9, 1981, p. 1011.
155. Ibid., December 9, 1981, p. 1137.
156. Felix Kessler, "France Enacts $7 Billion Nationalization of Five Industrial Groups and 39 Banks," *Wall Street Journal*, February 12, 1982, p. 26.
157. "France's Planners Spell Socialism With a Big S," *The Economist*, April 10, 1982, p. 72.
158. "The First Year of Mitterrand," *The Economist*, May 8, 1982, pp. 62-63; and "Je Reviens," *The Economist*, June 19, 1982, pp. 14-15.
159. "French Cabinet Members Resign," United Press International, Washington, D.C., June 29, 1982, and "France Concedes Failure, *New York Times*, September 2, 1982.

160. "The First Year of Mitterrand," *The Economist*, May 8, 1982, pp. 62-63. See also "Judgment on Paris," *The Economist*, July 10, 1982, pp. 17-18.

161. "A Strange Sense of Proportion," *The Economist*, June 26, 1982, p. 51.

162. *International Environment Reporter-Current Report*, May 12, 1982, p. 195.

163. Ibid., July 14, 1982, p. 284.

164. "The First Year of Mitterrand," *The Economist*, May 8, 1982, pp. 62-63.

165. *International Environment Reporter—Current Report*, July 14, 1982, p. 284.

166. Ibid., March 10, 1982, p. 104.

167. Ibid., July 14, 1982, p. 291.

168. Ibid., March 10, 1982, p. 104.

169. Ibid., April 14, 1982, p. 141; and "Good News for Midwife Toads," *The Economist*, April 10, 1982, p. 57.

170. *International Environment Reporter—Current Report*, April 14, 1982, p. 141.

171. Ibid., July 14, 1982, p. 301.

# 5

# West Germany

The rapid urbanization and industrialization that has characterized the post-World War II economic development of North America and Western Europe has not bypassed the Federal Republic of Germany. In fact, West Germany's redevelopment, which is known as *Wirtschaftswunder* or the "economic miracle," has not only jettisoned the Federal Republic into the dominant economic position in the EEC and Western Europe, but it has also significantly aggravated traditional pollution problems in this country of over 60 million. West Germany is twice the size of New York State and has twice its population.[1]

Air and water pollution are especially serious in the large industrial centers located in the Ruhr and in the major cities such as Berlin, Hamburg, Munich, Cologne, and Frankfurt. The major urban areas suffer from air pollution caused by industrial plants, electric power plants, home heating systems, motor vehicles, and the public transportation system.[2] Air pollution is worst when adverse weather conditions impede the normal air exchange in the atmosphere.

The largest sources of water pollution are the wastewater of Germany's large chemical industry, the wastewater of the other industrial users, such as the iron and steel industry, domestic sewage, chemicals in detergents and fertilizers, and solid wastes dumped into surface waters. By the early 1970s, the Rhine was more than twenty times as contaminated as it was in 1949 and several of its tributaries were biologically dead. Lake Constance, which is the country's only "Great Lake," had probably lost its capacity for self-purification. Both the North Sea and the Baltic Sea have been heavily polluted by Germany and the other countries bordering the seas.[3]

The level of solid waste pollution has been steadily rising because increases in population and the standard of living have been accompanied by both more packaging for each product and shorter-lived products.[4] Moreover, since solid waste pollution in North America measured on a per capita basis is at least twice that of West Germany and the other countries of Western Europe, the continued adoption of North American packaging methods should cause solid waste pollution levels to rise much faster in Germany and the rest of Europe.[5] The solid waste pollution problem is further aggravated by an antiquated solid waste-disposal system:

> For something like one quarter of the population, a plant for the systematic treatment and disposal of refuse is lacking. Almost half of the inhabitants in West Germany have to put up with the impossibility of their refuse being carted away regularly. Many waste materials are still being dumped as they were in the last century, contrary to sanitation, anywhere in the open and occasionally in the middle of densely populated areas.[6]

Domestic refuse—18 million tons in 1974—is about equal in importance to industrial solid waste—20 million tons in 1973. Other significant waste items include animal wastes and soil erosion from agriculture, sewage sludge, commercial refuse, wrecked and abandoned motor vehicles, debris and refuse from construction, and chemical wastes.[7]

### Pollution Abatement Programs

Since 1969, the coalition governments of Chancellors Willy Brandt and Helmut Schmidt have made a concerted effort to modernize West Germany's legislation, regulation, and technology in the field of pollution control. Traditionally, most Germans have considered themselves friends of nature. The effect of this *Naturfreunde* attitude is reflected in Germany's well-cared-for countryside which may top the list of the tidy, pleasant landscapes of Europe. Forestry management has been practiced since the eighteenth century. Water and air pollution have been curbed by legislation since the nineteenth century. In Germany, urban planning and land use control have never suffered from the *laissez-faire* diseconomies of North America or France. In summary, the German preoccupation with order and cleanliness has been instrumental in preserving a relatively attractive and apparently healthy environment.

During the summer of 1969, the West Germans were rudely awakened to the Environmental Crisis in their own country when the dumping of toxic chemicals into the Rhine caused the death of millions of fish. In July 1970, Chancellor Willy Brandt created a Cabinet Committee for the Environment with the Minister of the Interior—Hans Dietrich Genscher—as Chairperson. Then emergency legislation was proposed, and a long-term legislative program, the Federal Government's Program for the Protection of the Human Environment, was developed jointly with State government authorities and groups of scientific experts from industry and academe. The German State authorities were included in environmental policy determination because traditionally they have had much greater authority for environmental protection than their U.S. counterparts.

Since 1970, the Social Democrat-Free Democrat coalition under the leadership of Chancellors Brandt and Schmidt has been implementing its legislative program for protection of the environment. The Brandt government experienced many legislative accomplishments during the 1970-73 period: (1) A 1972 constitutional amendment transferring concurrent authority to the Federal government in matters of air, noise, and solid waste pollution; (2) the "polluter pays principle" implicit in a number of new laws; (3) a system for determining the environmental impact of all government decisions; (4) a comprehensive law dealing with waste disposal; (5) stringent new limits on the lead content of gasoline, even stricter than those recommended by the EEC; (6) prohibition of certain household detergents; and (7) a law providing for the reduction of noise from aircraft.[8]

Because the Environmental Crisis was replaced by an Energy Crisis and worldwide stagflation during the winter of 1973-74, Chancellor Helmut Schmidt, Willy Brandt's successor, has not been as successful as his predecessor in effecting the environmental legislation necessary to complete the Social Democrat-Free Democrat coalition's program for environmental protection and improvement. It is noteworthy, however, that the environmental protection policy established through the impetus of the Brandt government has been and will continue to be implemented without delay, in spite of the industrialized world's current focus on the Energy Crisis and related economic difficulties.

Even in late 1975, the worldwide recession was uppermost in the minds of those involved in the German economic policy debate at the last "Concerted Action" conference convened by the Federal government that year. The participants in that conference agreed not to delay the implementation of environmental policy, even though it might impede economic recovery. Therefore, after extensive review by Chancellor Schmidt, his Interior Minister (along with several other ministers), industry and trade union representatives, environmental specialists from national political parties, State government officials, and the scientific community, the Federal government officially declared "that, even in the present circumstances, Germany will continue to pursue its environmental goals."[9] As further evidence of the German government's concern for environmental quality, it was reported in early 1977 that about 25 billion marks (about $10 billion) of fixed investment that was already financed and set to go was being delayed because of government restrictions. Consequently, the standards-permit-enforcement pollution control strategy relied on by the Federal government was delaying construction of productive facilities at a time when real economic growth in West Germany had dropped to about zero.[10]

In general, prohibition and regulation, the social control measures that are most effective in initially reducing pollution, are constitutionally within the jurisdiction of the eleven German States (*Länder*) rather than the Federal government. Therefore, the environmental program of the Federal government has institutionalized four basic principles to be followed in reducing pollution, although its present power is still somewhat limited. (1) Programs must be holistic, encompassing land, air, water, noise, and animal and plant life; (2) precautionary action is better than corrective action, and standards should be applied to dissuade all from harmful activities; (3) social and economic forces should be relied upon to discourage polluters, and governmental action should be reserved for use only if socioeconomic strategies fail; and (4) costs of pollution abatement should be borne by the polluter, and the government should complement environmental policy with tax, budgetary, and infrastructure policies. These four principles have received widespread publicity in the schools and industry, and through the mass media. However, these principles may not be fully operationalized because of the relative weakness of the Federal government in environmental matters, especially in the area of water pollution control.[11]

### Federal-State Sharing
### of Pollution Abatement Responsibility

Germany is a Federal state wherein the eleven German States (*Länder*) are constitutionally equal with the Federal government (*Bund*). Legal authority is shared by the Federal and State governments as follows: (1) the Federal government has exclusive jurisdiction for selected responsibilities such as national security, customs duties, and foreign relations; (2) the Federal and State governments have concurrent jurisdiction for many responsibilities, such as air and noise pollution control, solid and hazardous waste control, control of drugs and poisonous substances, consumer protection, and trade regulation, but the States may only enact legislation where the Federal government has not passed laws; and (3) the Federal government has only broad "framework" jurisdiction with respect to several traditionally State responsibilities such as water use, land use, and nature and landscape protection, but the State must implement the Federal "framework" legislation by enacting detailed laws with those adaptations appropriate for each State's unique conditions.[12] There are three city-states (Berlin, Bremen, and Hamburg) among the eleven German *Länder*.

Local government includes counties and municipalities. There are two types of municipalities—towns and villages, and "free" municipalities. Towns and villages are actually administered by counties, whereas "free" municipalities administer their own affairs. In addition, there is a district or

regional level of State government that supervises the local governments—
counties and "free" municipalities.[13]

The Federal government's legislative branch consists of a popularly
elected *Bundestag* (lower house of Parliament) and a *Bundesrat* consisting
of appointed representatives of the *Länder*. The *Bundestag* enacts Federal
legislation within the aforementioned jurisdictional constraints on a simple-
majority-rules basis, except in the case of constitutional amendments which
require a two-thirds majority vote. The *Bundesrat*, which represents the
interests of the States, serves primarily in a advisory role. However, inasmuch as it must approve specified types of Federal legislation, regulations,
and standards (including those affecting the environment), the *Bundesrat*
has veto power over many legislative and executive decisions.[14]

Although the Federal government has concurrent authority to set standards for air, noise, and solid and hazardous waste pollution, enforcement
of any Federal regulations is generally the responsibility of the States. Moreover, each German State has its own idea on the importance of pollution
control, and often these ideas differ both from each other and from the
Federal government's policy objectives.[15] The responsibility for enforcing
pollution control laws generally falls on the shoulders of the environmental
section of each State's public prosecutor's office. The public prosecutor
must also rely on information provided by police and factory inspectors.[16]

The Ministry of the Interior has been given the primary responsibility
for carrying out the environmental protection program of the Federal
government. At least ten other ministries continue to have limited, and
often concurrent, responsibility for specified aspects of environmental
protection. Before the Federal government was reorganized to exert more
independent, centralized control over its environmental protection program,
there were over fifty Federal offices, institutes, establishments, and societies authorized to deal with various facets of the environmental quality
issue. Moreover, there were (and still are) institutions in each State, including several centralized State ministries for environmental protection, with
responsibilities for maintaining environmental quality.[17]

Reorganization of nineteen Federal ministries to bring most of the environmental protection functions under an office of the environment within
one Ministry required an amendment to the organization act governing the
proceedings of the Federal government.[18] The Federal Environmental
Office (BUA) was modeled on the U.S. Environmental Protection Agency.
BUA, however, was designed to be subordinate to the Ministry of the
Interior because the German Federal system does not permit independent
agencies or commissions. The BUA was designed to

> have a great deal of practical autonomy and authority. However, a
> wide range of environmental powers and programs will remain spread
> among ten other federal ministries. (It)... will be mainly a policy,

planning and coordinating authority that will have direct responsibility for a few key areas, such as pollution control.[19]

Thus, in July 1974 when the Federal legislature (*Bundestag*) amended the Federal organizational law establishing a Federal Environmental Office in the Ministry of the Interior, the new office's responsibilities included providing technical support in the form of drafting regulations and guidelines on ambient air quality and solid and hazardous waste management, as well as effecting maximum cooperation among those Federal agencies with an impact on the environment policy and coordinating research on the environment. In addition, there is an Environmental Department in the Interior Ministry. This department, which is also involved in drafting regulations and legislation, is divided into three units: Subunit I, which has primary responsibility for general environmental planning, economics, and enforcement policy; Subunit II, which is responsible for water resources and supplies, waste removal, and transnational water law; and Subunit III, which is responsible for air quality and noise control.

The Interior Minister can also call upon the expertise of the Council of Experts for Environmental Questions (CEEQ) and selected nongovernmental groups. The twelve-member German equivalent of the U.S. Council on Environmental Quality consists of specialists from various disciplines related to the environmental sciences field who advise the Interior Minister. There are also several professional groups which advise the Ministry on standards and guidelines for pollution control. For example, the Association of German Engineers, through its Air Quality Maintenance Commission, advises the Interior Ministry during the process of establishing legally binding standards and guidelines for ambient air concentrations of various gases and particulate matter. Three professional groups—the German Association of Gas and Water Management Experts, the Sewage Technology Union, and the Water Division of the German Standards Committee—consult with the Ministry during the process of establishing nonbinding standards and guidelines for water quality.

Several other Federal ministries have maintained specified responsibilities in the area of environmental protection. Thus, the Transport Ministry maintains the primary authority to control the air and noise pollution caused by motor vehicles. The Ministry for Food, Agriculture, and Forestry has been given the authority to control pesticide use,[20] as well as responsibility for landscape management, nature conservancy, and the pollution of air, water, and land by agriculture and forestry. The workplace environment is the responsibility of the Ministry of Labor and Social Affairs.[21] Protection from radiation is primarily the responsibility of the Ministry for Education and Science, with support from the Ministry of Youth, Family, and Health. The Ministry of Youth also has been given some responsibility for protection from pesticides and for waste disposal,[22] including toxic substances, inasmuch as it is concerned with general health care. Responsibility for Federal policy in land use and regional planning, urban renewal and the development of human settlements, and modernization and construction of housing has been given to the Ministry for Regional Planning, Building, and Urban Development. There are other ministries with more limited responsibilities for environmental protection. Finally, the Interior Ministry, Ministry for Research and Technology, and other ministries with environmental protection responsibilities provide financial support for and/or conduct research on the problems and issues related to environmental quality.

Federal environmental protection programs are coordinated through a cabinet-level Committee for Environmental Questions, which is chaired by the Prime Minister and which consists of the Interior Minister, the Prime Minister, and representatives from the eleven other ministries with environmental protection responsibilities. In addition, a Permanent Board of the Heads of Divisions for Environmental Questions, which is chaired by the Interior Minister and which consists of the senior officials for environmental questions from twenty-one Federal agencies, is responsible for coordinating environmental policy implementation both in the Federal government and between the States and the Federal government. Finally, there is a Conference of State Ministers for the Environment which provides a major forum wherein the Federal Interior Minister and his State counterparts can coordinate Federal and State environmental policy. In conjunction with the Conference, there are Federal-State working committees for coordinating most facets of environmental protection, including air quality and emission control, water and waste management, nature conservancy and landscape management, chemical control, and atomic energy.[23]

In conclusion, a 1972 amendment to Article 74 of the West German Constitution gave the Federal government concurrent authority with the States for control of air pollution, waste disposal, and noise control.[24] Therefore, responsibility for the control of air pollution, solid and hazardous waste, and noise pollution are shared by the Federal and State governments with the Federal laws, regulations, and guidelines taking precedence, while the State governments along with the local governments are responsible for most of the administration and enforcement.

Environmental Planning

Although Germany has no U.S.-style environmental impact assessment law, it has several Federal laws pertinent to land use, government planning, and pollution control. The Germans, like their European neighbors, have elected to integrate environmental planning with those traditional planning and licensing procedures employed for the review and control of development programs or construction projects. Because of constitutional limitations, the Federal Regional Planning Act of 1965 provides only a framework for comprehensive development planning. There is a conference of Federal and State officials that coordinates national and State growth and investment plans.

Each State is in the process of preparing and implementing a State development program that encompasses regional and local development as well as zoning plans. A comprehensive evaluation of the social, economic, and environmental impact of major public and/or private development projects is usually an integral part of the regional planning procedures of State land use planning statutes. All government agencies and interested groups may participate during public hearings and through review of pertinent documents.

The major Federal legislation providing for environmental impact assessment of major public and private facilities is the Emission Control Law which requires each new or expanded installation to apply for a site permit. The Federal Emission Control Act, in concert with the Federal Waste Disposal Act, mandates that permits may be issued only if environmental damage can be avoided or held to a minimum in the construction and operation of industrial plants, highways, railways, and other polluting establishments. The environmental impact assessment occurs only after a site decision has been made and is applicable only to the specified site. Hence, the permit-issuing agency, which is usually a local pollution control authority, has no perspective or power from which it can review the broader issues of more preferable sites or whether a proposed project should even be constructed. Similarly, the Federal Atomic Law provides for permit issuance and environmental impact assessment in respect to nuclear power plant, waste-processing, and waste-disposal sites. In most cases, there will be public participation through a public hearing. Finally, the Federal Highway Act requires planning authorities to integrate pollution control deliberations with the other more traditional planning and cost considerations pertinent to the decision to construct new highways.

A U.S.-style environmental impact assessment model law has been proposed by the German Conference of Environment Ministers. Constitutional limitations prevented mandatory adoption of the model law in all States, although three *Länder* have voluntarily enacted environmental impact assessment laws. In 1974, the Federal cabinet in its Resolution on Adopting Environmental Assessment Principles established in principle a National Environmental Protection Act-type policy for all Federal planning and projects. In practice, the German bureaucracies, which have traditionally been highly independent, hierarchical, secretive, and proprietary, have experienced difficulties in implementing the Federal cabinet order, that is the order's alleged "loopholes" have enabled Federal agencies to avoid more restrictions on the environmental impact of their plans and projects.[25] Inevitably, the proposal of the Commission of the EEC for a directive on environmental impact assessment procedures patterned after the U.S. statement will result in the introduction of EEC measures for assessing environmental impact.

## Water Pollution Control

The Federal government has much less power to control water pollution because the constitutional amendments that gave it concurrent authority with the States to control air, noise, and solid waste pollution have not been complemented by an amendment providing for Federal-State sharing of the power to control water pollution. For many years, such a constitutional amendment has been pending, but through their legislative power in the *Bundesrat* the State governments have resisted any further erosion of their authority to determine the quality of the environment within their jurisdictions. The proposed constitutional amendment and its related Federal enabling laws would require all water users, public (towns and cities) and private (industry), to restore water to the degree of purity in which they originally found it or pay a tax or effluent charge to a State.

Until the German Constitution is amended, the Federal government's authority to control water pollution generally is limited to "framework laws" establishing binding principles to serve as the "framework" for State legislation. Of course, the Federal government's superior fiscal resources from income, value added, and net worth taxation can usually be used to encourage State legislative action where substantial intergovernmental transfers from Bonn are contingent upon enactment of the appropriate State law.[26]

Inasmuch as the Constitution provides for concurrent authority on matters of taxation, the Federal government has exercised its broad taxing powers so that all major sources of tax revenue—income, value added, net worth, and trade—are regulated by Federal laws, while the States have retained the responsibility for administering most taxes, except income tax on nonresidents, customs duties, and the value added tax. The States have also retained some legislative authority over minor taxes, such as the real estate transfer tax. The fiscal authority of municipal (local) government is limited primarily to the determination of the tax rate for their major source of revenue—the municipal trade tax. Although there are separate legislative and administrative responsibilities, revenue which a State collects from the income tax on individuals and corporations is shared equally with the Federal government. There is also a revenue-sharing arrangement between the municipalities and the States.[27]

Because the German Constitution limits the Federal government to framework competence in matters of water quality, the primary authority for water pollution control remains with the sovereign States. Thus, there are various State laws governing water quality. Although the State laws conform to the Federal model, each also reflects to some extent a State's unique conditions and objectives. Since the Federal Water Resources Management Act (WRMA) was enacted in 1957, Federal law has provided general guidelines for the abstraction of surface, ground, and coastal waters, both by prescribing conditions for the use of water resources and by establishing permit requirements. In addition, the Federal government has the authority to specify the technical requirements for all water pipelines, even though the States retain the authority for regulating the construction and operation of such pipelines.

Under a 1961 Federal statute[28] amended in 1975, the Bonn government has established minimum standards for biodegradability in washing and cleaning agents. The law not only requires detergent producers to inform the Federal government of their product's composition, but also authorizes the Federal government to prohibit the use of such cleaning agents if they are a threat to water quality. Detergent producers must also inform consumers as to how to use their products with a minimal effect on water quality.[29] The 1975 amendment brought the German law into compliance with the 1973 EEC directive on detergents.[30]

In early 1976, the WRMA was amended and strengthened to provide for national minimum requirements as to the granting of licenses for effluent discharges into public waters. As a result of this amendment, subsequent licenses or permits to discharge effluents are granted only if the discharger can comply with the aforementioned minimum effluent standards established by the Federal government.[31] Thus, the Federal government's licensing or permit authority under the Emission Control, Waste Disposal, and Water Resource Management Acts enables it to control the environmental impact of any new, or substantially altered, industrial or public facility. In addition, the 1965 Federal Regional Planning Law as explained above provides for comprehensive land use planning at the State level.[32]

The WRMA not only prescribes minimum requirements for treatment of wastewater from industrial and public facilities, but also mandates preparation of Water Use Plans by the States to ensure the supply of water. Private and public institutions that store, load, and transport materials that could endanger the water supply are required to use the best practicable technology. The permit or licensing procedure, which is based on the Federal Emission Control Act, provides for the appointment of "water protection officers" in industrial and commercial installations making use of the water supply. State and local governments have the authority (within their jurisdiction) to prohibit construction, or later the land use of areas, where such actions are deemed necessary to protect the water supply. Finally, the penalty for violating the WRMA is a fine of up to DM 100,000 (about $40,000).[33] Although only one German State (Bavaria) had established a schedule of appropriate fines for violators of the WRMA who pollute rivers and lakes within the State's jurisdiction by 1978,[34] most states now have developed fine schedules like Bavaria. In summary, the day-to-day responsibility for water supply and water quality management continues to be a State activity, although the Federal government's superior technical and fiscal resources, along with its framework competence, will enable it to continually usurp more responsibility.[35]

The *Länder* exert their primary responsibility for water quality through regulation of the use of surface and subsurface water supplies. A State can carry out its responsibility in several ways, including municipal and county agencies and water authorities. These authorities may prescribe effluent standards that are more stringent than the Federal minimum, grant permits, and enforce Federal and State water laws. The water authorities regulate the taking, diverting, and dredging of, as well as the discharging of wastes into, surface and subsurface water resources within their jurisdiction.[36]

Water authorities, popularly known as water associations, are created by a legislative act wherein a State transfers the responsibility and management of water quality to an association. The primary function of such associations is to execute and finance water management policy. Most of the several thousand water associations are small, rural institutions involved in the supply and treatment of domestic water supply, drainage, irrigation, and the maintenance and improvement of waterways. Some are also involved with regional flood protection and dike maintenance, as well as wastewater treatment. Finally, there are a few large water associations (called *Genossenschaften*) which are authorized through special legislation to manage water quality for its members throughout a water basin. All associations receive financial aid for construction costs from the States according to their size and responsibilities.[37]

The water pollution control/water quality policy in effect in the Federal Republic since 1956 can be characterized best as a standards-permit-enforcement system with standards prescribed and permits issued at both the Federal and State levels, but enforcement occurring only at the State and local levels.[38] This system has not halted the decline of water quality that has occurred in all of West Germany except the Ruhr Valley, which is the locus of several large, independent *Genossenschaften*[39] and over two-thirds of the country's iron and steel production.[40]

As a consequence of the success of the *Genossenschaften* in mitigating somewhat the continual decline in water quality, the Federal Interior Ministry has been proposing since 1973 an innovative statute that would establish an effluent tax or charge system much like that developed by the *Genossenschaften* in the Ruhr. The proposed constitutional amendment giving the Federal government concurrent authority with the States for water quality has not been enacted so that the above enabling law can take effect.[41] Even so, in 1976 the *Bundestag* enacted a Federal Waste Water Charges Act that will require wastewater dischargers to pay an effluent tax or fee per unit of pollutant introduced into public waters after 1980. The base for determining the effluent charge will be equal to the unpurified waste of one person, and the rate will begin at DM 12 (about $5) per unit (base) in 1981 and then increase periodically to DM 40 (about $16) per unit in 1986.[42]

## Water Pollution Control in the Ruhr Basin

The *Genossenschaften*, which are still the best known effluent tax systems in the world, were initiated in the Ruhr Valley seventy years ago as river-basin water quality management associations. They are quasi-governmental agencies that have been granted broad authority. The principles of water quality delineated by the States in their basic charter encompass a variety of technological options which they systematically integrate according to economic criteria.

The water quality management program developed by these agencies combines objective policies and physical facilities with economic incentives to effectively and efficiently maintain water quality within a river-basin region. Thus, a *Genossenschaft* resembles a hypothetical, basin-wide firm charged with the responsibility for maximizing the benefit-cost ratio of the services it provides for society within that basin.[43]

Membership in a *Genossenschaft* is mandatory for any public or private entity with "material" discharges into the river basin for which the association has been given responsibility for water quality. Each *Genossenschaft* not only constructs and operates water treatment facilities, but also engages in any other collective control measures on a regional basis that aid in cost-effective water quality within the river basin. Effluent "taxes" or charges are levied on member-users based on their proportionate waste load. The revenue from these charges covers the expenses of operating the *Genossenschaft*'s facilities, including the cost of long-term financing for construction of treatment facilities. However, member-users do not pay for all the costs of constructing treatment facilities because State governments have been granting subsidies to the water management association for such construction.[44]

In addition to their regional treatment facilities, *Genossenschaften* construct and operate systems that improve river flow and installations which augment the dissolved oxygen in the river. Some rivers actually function as open sewers carrying wastewater to central treatment facilities. The water quality in other rivers is maintained in accordance with specifications in regional water use plans, so that the water can be utilized for both recreation and drinking (after treatment).[45]

A discharger's effluent "tax" or fee is not systematically related to the private treatment costs it would incur at a particular location. Charges actually amount to the discharger's pro rata share of the operating costs for the whole river-basin's water quality management system. Economic efficiency is served by this management system, both through integrated planning for and operation of water treatment facilities and through an effluent tax system that internalizes much of the water disposal costs of industrial dischargers.[46]

Inasmuch as the effluent tax is a flat-rate charge related to an approximation of the damaging effect of the discharger's waste water, it violates the principle of marginal-cost pricing. It does so by ignoring differences in damage-avoidance costs and marginal damages associated with variations in factors such as the locus of the waste source, river flow, ambient waste concentrations in the river, and water temperature. However, as Kneese and Bower pointed out in their definitive study of the *Genossenschaften*:

> Methods that are less than theoretically ideal may be optimal in practice, since an important element in determining the best method for actual use is the cost of making marginal refinements. A comparatively crude method that is generally correct in principle will often realize the major share of the gain that could be achieved by more complex and conceptually more satisfying techniques.[47]

The *Genossenschaften* minimize their monitoring or auditing costs by relying on crude measurements that are easily administered. Levels of discharge are developed by relying on the production figures of established production functions. In the case of a brewery, there might be a determination that the brewing process being used would generate five population equivalents of wastes per barrel of beer. If the discharger's production process changes or improvements in wastewater treatment reduce effluent, the polluter has the burden of proof with regard to its lower discharges. The staff of a *Genossenschaft* occasionally does sampling to evaluate the accuracy of the system. The appeal process that exists for resolving disputes over discharge figures seems to work smoothly inasmuch as the agency levying the charges is not an adversary government institution.[48]

The water management associations in the Ruhr do not have the regulatory or enforcement powers that municipalities, counties, and water authorities do. Moreover, they have no monopoly over discharges into the river basin that defines their jurisdictional authority. Consequently, they cannot prevent a municipal, county, or State agency from issuing a discharge permit for a new industrial facility, even though its discharges will adversely affect water quality in the river basin.[49] The result is increased pollution control costs and/or effluent taxes for all other polluters where the ambient water quality can only be maintained by more treatment and/or less pollution by each facility.

The membership of a *Genossenschaft* meets at least annually. At the annual meeting, the members elect directors and decide upon an annual budget. Voting power in an association is directly related to a member's wastewater load and represents an extension of the stockholder concept to the public domain. This voting arrangement has allegedly guarded against the incurring of unnecessary expenditures for wastewater treatment.[50]

### Water Pollution Taxes

The limited success of the *Genossenschaften* has served as a model both for the French water or river-basin authorities and for the 1976 revision of the German water quality management program embodied in the Waste Water Charges Act (WWCA). These associations have been somewhat more effective and efficient by relying on an effluent tax system—although the tax rates have been too low—rather than on the standards-permit-enforcement regulatory system like the Federal government has been utilizing since 1957. However, as long as the effluent taxes of the *Genossenschaften* and similarly organized water management associations are insufficiently large to effectively dissuade wastewater dischargers, ambient water quality throughout Germany will not improve enough to attain the Federal Water Program's goal of 90 percent of the desired water quality levels.

Although approximately one-half of industrial wastewater is discharged into the public sewage system where it receives more treatment, the balance of industrial wastewater is discharged directly into natural waters with little or no treatment. Of course, some water management associations treat the water in rivers and lakes after its members have discharged therein. The construction of wastewater treatment facilities by public sewage systems often has been deterred inasmuch as the Germans decided in 1945 to reconstruct their cities and industries with their capital. Therefore, the Environmental Crisis precipitated the recognition that public facilities were insufficient to properly handle the domestic, commercial, and industrial wastewater flowing therein.[51]

The 1976 WWCA mandates the imposition of charges or taxes on the discharge of wastewater or effluent into waters, which are defined in the 1957 WRMA as amended. The charge or tax has been levied by the States beginning on January 1, 1981, although the law took effect on January 1, 1978.[52] The revenue from the effluent taxes or wastewater charges is used both to administer the tax or charge system and to finance the construction, maintenance, operation, and technical development of facilities and other measures for enhancing water quality.[53]

Article 2 of the WWCA defines wastewater discharging and the wastewater treatment plant. The wastewater itself is "water changed in its properties by domestic, commercial, agricultural or other use and the water running off in conjunction therewith in dry weather (polluted water), as well as water running off from builtup or paved surfaces following precipitation (hereinafter referred to as rainwater)."[54] Discharging is defined to encompass both the direct conveyance of wastewater into bodies of water and the indirect conveyance into water bodies through initial disposal of wastewater in subsoil except such disposal as part of agricultural soil treatment. Finally, the wastewater treatment plant includes both the facilities used to reduce or eliminate the pollutants in wastewater and the facilities that mitigate in full or in part the generation of wastewater.[55]

In general, the wastewater charge or effluent tax is assessed on the noxiousness of wastewater or effluent. Noxiousness or damage is determined on the basis of the wastewater volume, the suspended solids and oxidizable substances therein, and the toxicity of the wastewater.[56]

The major exceptions to the general system for assessing wastewater charges relate to water management associations, settling ponds connected to wastewater treatment plants, persons generating wastewater with large amounts of settleable solids, urban run-off from rainwater, and the wastewater from households and other small dischargers. For example, water management associations such as the *Genossenschaften* which operate river sewage plants to purify bodies of water may determine their assessment by measuring the number of units of noxiousness or damage in the water bodies downstream from their river sewage plants. The *Länder* may exempt from the wastewater charge assessment the noxiousness or damage of discharges that is dissipated through use of secondary settling ponds connected to a wastewater treatment plant. The *Länder* may also allow the noxiousness of suspended matter and suspended solids to be determined by the weight of such solids upon the request of a discharger or liable person when the number of cubic meters of solids is greater than five times the number of tons of dry substance generated annually.[57]

The exceptions related to urban run-off and domestic sewage are unique because they provide that a community's polluted rainwater and its households generate units of noxiousness in relation to the number of inhabitants connected or not connected to the public sewage system. More specifically, Article 7 deems that the units of noxiousness resulting from urban run-off shall be 12 percent of the population connected to the public sewage system. As provided in Article 8 and unless otherwise specified by a *Länd*, the units of noxiousness of household and commercial sewage for which a public corporation is liable shall be one-half of the population not connected to the sewage system. In both cases, the population may be estimated where it is difficult to determine or where the determination would be unreasonably expensive. In respect to polluted rainwater that is either retained or treated in a sewage treatment plant, each *Länd* determines how much of an adjustment should be made to allow for the resulting reduction in the noxiousness of urban run-off. Thus, a *Länd* may decide to exempt a community from the wastewater charges related to urban run-off because of either retention or treatment of polluted rainwater.[58]

Since December 31, 1980, wastewater dischargers have been liable for wastewater charges or effluent taxes, with the following rates per unit of noxiousness applicable on a calendar-year basis:[59]

| | |
|---|---|
| DM 12 as of 1/1/81 | DM 30 as of 1/1/84 |
| DM 18 as of 1/1/82 | DM 36 as of 1/1/85 |
| DM 24 as of 1/1/83 | DM 40 as of 1/1/86.[60] |

Inasmuch as these rates were established in 1976, they may have to be periodically increased or indexed to reflect the effect of inflation. The rates were originally set at a level so that wastewater dischargers would prefer the cost of operating a wastewater treatment plant with an efficiency of 90 percent to paying the effluent tax or wastewater charge on their discharges.[61] Based on the German experience with the rate-setting practices of water management associations, the producers of wastewater have no serious interest in reducing the noxiousness of wastewater discharges as long as the wastewater charges or effluent taxes are less than their cost of water pollution. In principle, the 1986 rate of DM 40 per unit of damage or noxiousness has been set to approximate the average purification cost of an average-size wastewater treatment plant designed for full biological purification. Such a plant has at least 90 percent efficiency, and the average cost includes both operating costs and amortized capital costs.[62]

As noted above, several types of wastewater dischargers are not liable for regular wastewater charges, even though they do pay for their water pollution. With regard to the river sewage plants of water management associations, a *Länd* may allow a water management association to assume the wastewater charge liability of dischargers in its catchment area, which is to be delineated in advance. The *Länder* are to determine the proper method which water management associations must use to exact wastewater charges from the dischargers in its catchment area. As explained earlier, these associations may determine their assessment for wastewater charges by measuring the noxiousness in a water body downstream from their river sewage plants.

In respect to public corporations that provide wastewater treatment services, the effluent tax or wastewater charges for households and other small dischargers not connected to the public sewage system are transferred to the local authorities as explained above. In addition, a *Länd* may allow domestic households or other polluters who discharge on an annual average less than 8 cubic meters (283 cubic feet) of wastewater per day into the sewage system to transfer their liability to such public entities providing abatement services. The *Länder* are also responsible for determining the method to be used in exacting wastewater charges and/or other financial support from households for the wastewater charges paid on their behalf by public corporations providing wastewater treatment services.[63]

Article 10 of the Waste Water Charges Act provides for six additional exemptions from liability to pay wastewater charges. Such exemptions are granted for the discharge of (1) water that is already polluted when abstracted for use and does not exhibit any additional noxiousness upon its return to the body of water from which it was extracted; (2) polluted water into surface water that is used in natural resource extraction, but only if such water is used for washing the minerals extracted, does not contain any pollutants other than those extracted, and is always isolated from other bodies of water; (3) sewage from watercraft at the locus of generation; and (4) rainwater if a public sewage system is not used for collection and treatment thereof, inasmuch as there is a special provision in Article 7 for levying a wastewater charge on polluted rainwater or urban run-off. Exemption from liability is also allowed when the *Länder* determine that wastewater discharges into the underground will not be harmful because the ground water's natural properties make it unsuitable for use as drinking water even after conventional treatment.

Exemption from liability for an antecedent period of three years are granted to any wastewater treatment plant under construction, provided the anticipated decrease in units of noxiousness discharged into the body of water concerned is at least 20 percent. The number of units of noxiousness is reduced in an amount corresponding to the anticipated decrease in discharges. If the plant does not become operative, retroactive liability for the exempted wastewater charges is imposed. Moreover, if the actual degree of purification effected by the new treatment plant falls short of its projected efficiency, a retroactive liability for wastewater charges is imposed to the extent of the shortfall.[64]

The Waste Water Charges Act should cause an estimated increase in costs and prices of only 0.5 percent as measured by the wholesale price index. In those few sectors of the economy where the increases in production costs are fully passed on to consumers, such costs could rise 10 percent or more. Even in extreme cases of passing on increased production cost, however, there should not be more than a 1 percent increase in overall prices.[65] To avoid any significant economic disruptions, the Federal government is authorized to exempt certain parties through statutory ordinances approved by the *Bundesrat*. Both persons liable to pay wastewater charges and regional or sectoral groups of such parties that operate systems for reducing noxiousness of wastewater from the liability for wastewater charge may be exempted. No exemption from liability may remain in force, in whole or in part, beyond December 31, 1989.[66]

With certain exceptions, the value to be used in determining the number of units of noxiousness is obtained from the official permit licensing wastewater discharge. These exceptions are polluted rainwater (Article 7), small wastewater discharges by households and others (Article 8), members of those water management associations operating river sewage plants, and domestic and commercial discharges into public sewage systems. Each permit provides data on the (1) maximum amount of annual wastewater discharge, (2) suspended solids, (3) oxidizable substances, and (4) the degree of toxicity, according to the mean values to be maintained (*standard values*) and the values that may not be exceeded under any circumstances (*maximum values*). Either the standard values or at least 50 percent of the maximum values are used as a basis for determining the number of units of noxiousness (*reference values*).[67]

Except in the case of rainwater-induced urban run-off (Article 7) and small discharges by households and others (Article 8), these wastewater charge rates are reduced 50 percent for those units of noxiousness that cannot be avoided while complying with the minimum requirements for the quantity and noxiousness of discharge as established in Article 7a of the Water Act. Article 7a provides that a wastewater discharger must comply with the minimum requirements of those general administrative regulations applicable to the process(es) generating wastewater. These regulations are based on the commonly accepted rules of engineering applicable to each case. If a permit specifies more stringent requirements than are provided by Federal regulations in respect to the standard and maximum values, the 50 percent rate adjustment is available only where the more stringent requirements are observed.[68]

There are two types of penalties for noncompliance with the WWCA provisions. One penalty results from a breach of regulations, and the other relates to underpayment of the wastewater charge or effluent tax. A person who conducts himself or itself in the following manner, either intentionally or negligently, commits a breach of regulations: (1) submits measuring values that do not conform to the measuring program delineated in Article 5; (2) fails to submit calculations or documents or does not submit calculations or documents in a complete or accurate form; or (3) fails to provide the liable person with the necessary data or documents in an accurate or complete form. Any breach of regulations may result in a penalty not exceeding DM 5,000 (about $2,000).[69] In addition, the penal clauses of Article 370 (Paragraphs 1, 2, and 4) and Article 371 of the Federal Fiscal Code are applicable to any act constituting evasion of the wastewater charge. The penalty provisions of Article 378 of the same Fiscal Code are also applicable to any unlawful reduction of the liability for wastewater charges or effluent taxes.[70]

### Transnational Water Pollution Control

As explained in Chapter 4, the Rhine River's water quality problem is transnational, and accordingly, can be solved only by multinational cooperation through the Rhine Commission and the European Economic Commission. As a member of both commissions, the Federal government is legally committed to ameliorating the pollution in the Rhine

River. Thus, the government's program for future investments includes a financial assistance program for the construction of public and private wastewater treatment facilities in areas along the Rhine and near Lake Constance.[71]

## Air Pollution Control

Prior to the recent constitutional amendment, the Federal government's authority to control emissions was limited to its constitutional authority to regulate interstate transportation, which was prescribed in the Federal Road Transport Licensing Ordinance. Moreover, there were, and are, State laws for control of air pollution. Since these State laws vary considerably, they have never resulted in national ambient air quality standards.[72] The only significant air pollution control legislation enacted under the Federal government's authority to regulate interstate transportation is the 1971 Federal Petrol Lead Law which limits metal additives in motor vehicle fuel. The law limits the lead in gasoline to 0.15 grams per liter as of January 1, 1976, with penalties of up to DM 50,000 (about $20,000) for violators.[73] As discussed in Chapter 4, the EEC has promulgated a very liberal directive on the lead content of motor vehicle fuels. Inasmuch as the directive thereon, reducing the content from 1.81 to 0.67 grams per gallon, is less stringent than that provided in the Federal Petrol Lead Law, Germany is already in compliance therewith.[74]

The Federal Emission Control Act (ECA) of March 1974 is the primary statutory authority for the Interior Ministry's air pollution control program. This law authorizes the Interior Ministry to promulgate regulations to minimize or eliminate noxious environmental influences on human, animal, or plant life. The law is applicable to the construction and operation of industrial and commercial facilities, and to the production, sale, and importation of transportation equipment, as well as fuels, propellants, and related products made for vehicles.[75]

Inasmuch as the ECA is generally applicable to private and public facilities that are likely to cause material levels of air and noise pollution, the construction of public roads and railways for intra- and intercity transportation is also subject to Federal-State regulation.[76] Thus, the air pollution from most stationary and mobile sources can be controlled through enforcement of the ECA, wherein willful violations are crimes punishable by up to ten years in prison and/or a fine of up to DM 100,000 (approximately $40,000).

A permit program has been initiated to aid in controlling emissions from those stationary sources that are likely to cause material levels of air pollution.[77] Although there are now provisions for open hearings with citizens and third parties,[78] the permit must still be obtained from the State in which the facility will be constructed. Both the State and Federal governments may impose emission limits and technical requirements on the planned facility upon issuing the permit. Technical requirements may include standards for plant equipment and installation of scrubbers. The permit may also mandate regular monitoring of emissions by the operator. Each State must monitor the ambient air quality within its jurisdiction and, based on its periodic measurements, must develop a Clean Air Plan to improve air quality within its borders. While some facilities are exempted from the permit process because they are unlikely to cause significant levels of air pollution, they are still legally required to utilize the best technology available for controlling emission of air pollutants.[79]

The ECA requires those facilities that operate under permit to employ a sufficient number of pollution control experts or "emission control officers" to ensure that the facility is in compliance with Federal emission regulations and orders. In addition, the "emission control officer" is responsible for developing and/or introducing equipment, processes, and products to decrease pollution. Finally, an institution must not only appoint those who qualify as pollution control experts because they satisfy guidelines established by the Interior Ministry, but it must also provide enough equipment and personnel so that the emission control officer can fulfill his or her responsibility.[80] This officer must report annually to a facility's operator(s) and must supply the State and Federal governments with information about a facility's pollution control efforts and emissions.

The *TA Luft*, which are the General Administrative Regulations interpreting the ECA, were promulgated in August 1974.[81] These regulations have provided more specific air quality standards and emission guidelines for issuing permits to both new stationary sources of air pollution and old stationary sources materially altered in size, location, or process. Emissions of pollutants, including noxious gases, dark smoke, and particulate matter, are regulated by specifying the maximum allowable concentrations and the prescribed ambient air quality levels for each noxious substance, as well as the preferred techniques for monitoring and sampling air pollution.[82] The *TA Luft* mandates that the air pollution control system of every new or expanded facility conform to the most advanced technology available at the time of construction. Therefore, the General Administrative Regulations include special air pollution control requirements for more than forty types of facilities. Finally, the *TA Luft* establishes the maximum daily level of concentration for the ten most significant sources of air pollution in Germany.[83]

More specific regulations on the design, installation, and operation of specified furnaces utilizing liquid and solid fuels were also promulgated in August 1974. Furnace operators must install equipment for monitoring emissions, and they must allow government pollution inspectors to sample emissions therefrom. In addition, there are regulations that establish standards for determining the minimum height of chimneys for furnaces and other equipment that produces emissions from combustion.[84]

A second set of regulations promulgated in August 1974 limits the emission of chemicals by dry-cleaning establishments. In January 1975, another set of regulations were issued to set the maximum sulfur content for imported and domestic light fuel oil and diesel oil. More recently, regulations have been issued to control air pollution from wood dust and to limit selected emissions in those geographic areas suffering from unusual air pollution problems.[85]

As stated above, the ECA authorizes the Interior Ministry to test and regulate not only the production, sale, and importation of vehicular equipment that generates noxious emissions, such as air and noise pollutants, but also the fuels used in such transportation equipment. Exhaust emission standards for vehicular equipment powered by the internal combustion engine required the German auto industry to produce an essentially pollution-free motor vehicle by 1980. More specifically, the noxious emissions of motor vehicles manufactured for 1980 must be less than 10 percent of the 1969 level for such pollutants, that is, a 90 percent reduction in 1969 emission levels by 1980.[86] An update on government-auto industry agreements appears at the end of this chapter.

In May 1978, a landmark decision by the Federal Administrative Court reversed the decision of two State courts and reaffirmed the use of administrative regulations instead of laws to establish environmental quality standards. Specifically, a legal challenge to the *TA Luft* was precipitated by the issuance in 1974 of a permit for the significant enlargement of a coal-fueled power plant in Voerde, North Rhine-Westphalia. The State issued a partial construction permit to prepare the site for building two 700-megawatt plants as provided in the Federal Emission Control Law. In addition, the State relied on the *TA Luft* to specify various air pollution control requirements that would have to be met before either construction or operating permits could be issued. A suit filed by a person who lived within 4 kilometers of the site challenged the validity of the *TA Luft* air quality standards and emissions on the basis that they were not stringent enough to protect the plaintiff's interests. The North Rhine-Westphalia administrative court and its court of appeals agreed with the plaintiff that the *TA Luft* standards and guidelines should be used only as minimum criteria for State officials issuing permits consistent with their obligation to protect the health and safety of persons living near any installation with air pollution potential. The plaintiff is appealing his case to the Federal Constitutional Court.

Concern for the outcome of the Federal Constitutional Court's decision spurred the Federal cabinet to order the Interior Ministry to review and revise the Emission Control Act and *TA Luft*. The Interior Ministry anticipates submitting proposals to the cabinet in the near future to revise both the law and its regulations, so that more stringent standards and guidelines can be imposed where essential to protect the health and safety of persons living in areas adjoining industrial, commercial, and public installations.[87] In addition, the Interior Ministry's proposed amendments to the August 1974 regulations that control emissions from furnaces or heating plants were not approved by the *Bundesrat* in June 1978. However, the Interior Ministry was instructed to revise the amendments in anticipation of *Bundesrat* approval by the end of that summer. The proposed amendments would not only strengthen standards for emissions of air pollutants from furnaces fueled by oil and gas, but would also increase the efficiency of oil-and-gas-fired heating plants consistent with the objectives of the Federal Energy Conservation Law.[88] An update on the *TA Luft* controversy appears at the end of this chapter.

Ironically, significant improvements in ambient air quality and emissions of air pollutants are being reported at a time when air pollution control standards and guidelines are being tightened under authority provided by the Federal Emissions Control Act and the Federal Petrol Lead Law. Particulate matter, $SO_2$ and lead emissions, and the ambient air quality levels related thereto all dropped dramatically during the periods 1965 to 1970 and 1970 to 1975.[89]

### Solid Waste Pollution Control

The constitutional amendment that gave the Federal government concurrent authority with the States for control of air and noise pollution created a similar sharing arrangement for control of solid and hazardous waste pollution. As in the case of air pollution abatement, the Federal government's previously limited authority to control solid and hazardous waste pollution meant that State and local governments generally monopolized public policy on waste collection and waste disposal prior to April 1972. However, several special Federal laws had an impact on the collection and disposal of a variety of waste materials. The scope of these laws can be summarized as follows: (1) meat and other animal parts (Animal Carcass Disposal Act); (2) plants and vegetables which must be disposed of because of their danger to human health; (3) radioactive waste materials; (4) waste materials produced by activities designated by State law as mining; (5) any liquid waste materials dumped into surface waters or channeled into sewage plants; (6) waste oil and oil residues; and finally, (7) gaseous wastes that are regulated partly by Federal law (Federal Emissions Control Act) and partly by State law as explained above. While the Federal Waste Disposal Act of June 1972 did not consolidate all Federal legislation regulating waste collection and waste processing in West Germany,[90] the law did substantially fill the interstices that had existed. For example, there are now Federal regulations under the Waste Act specifically applicable to Mining Wastes.[91]

The Federal Waste Oil Act was enacted in December 1968 and took effect on January 1, 1969. This law provides for a levy or "tax" of DM 7.50 (about $3) per 100 kilograms of lubricant, which is payable by the manufacturer or importer and is integrated with the Federal excise tax on mineral oil. The revenues from the "waste oil tax" finance the collection and disposal or recycling of used oil with impurities of 10 percent or less free of charge. Those firms licensed to dispose of or recycle waste oil must charge for collecting used oil with impurities of more than 10 percent because they are not subsidized.[92] Thus, "taxes" or fees are levied on new and used oil, respectively, to pay for the collection and disposal (or recycling) of waste oil. This Federal law requires any natural or legal person that accumulates more than 500 kilograms of waste oil annually to maintain accurate records on both the accumulation and disposal thereof.[93]

The "waste oil tax" is collected by the Federal government and is redistributed at the regional, or district, level as a subsidy to private firms that collect and recycle used oil. The subsidies are as follows: (1) DM 12 (about $5) per 100 kilograms of oil reprocessed into lubricating oil; (2) DM 10.20 ($4) per 100 kilograms of oil reprocessed into heating oil; and (3) DM 10 per 100 kilograms of oil burned coupled with a DM 2.60 ($1) additional subsidy for fuel gas purification. As is explained below, German States are divided into districts or intra-State regions in which there are one or more private firms licensed to collect and reprocess waste oil.

This German pollution, or waste oil, tax system seems to work well. The efficiency of the system is allegedly attributable to both a regulatory system and the organization and development of networks for the collection and reprocessing of waste oils, rather than to the system of pollution charges or taxes that partially finances the networks. An EEC directive authorizes the implementation of such a "waste oil tax" system in other Common Market countries. However, one can argue that such an excise tax levied on lubricating oils is a beneficiary or service charge and not a "pollution tax" because, while the revenue is redistributed for control of waste oil pollution, the levy is based on neither pollution nor damage.[94]

The Federal Waste Disposal Act (WDA) of 1972 prohibits the disposal of any wastes in a manner that could endanger human health or damage the environment. Such wastes can be disposed of only in licensed private facilities or in public facilities, both of which enjoy virtual monopolies in respect to the residential and/or commercial waste processed in their disposal plants. All commercial enterprises must provide local (city or county) authorities with data on both the amount and the properties of specified types of wastes they generate.[95] Through the WDA, the Federal government has exercised its power to regulate solid and hazardous waste pollution. Nonetheless, the *Länder* have been left with considerable discretion to enforce this Federal law and its related regulations. Therefore, the WDA's effectiveness is dependent upon how zealously each State carries out its policing responsibilities.[96] In summary, the Waste Disposal Act delineates how and by whom wastes are to be disposed of, prescribes what authority supervises waste-processing or -disposal facilities, and provides generally for the manner of disposal.

The WDA requires the *Länder* to draw up mandatory waste-disposal plans for each region within a State. These plans specify not only the location of waste-disposal facilities, but also the appropriate design for the waste-processing plant. The WDA does not specify how the responsibility for waste disposal is to be allocated by regions and within regions. Bavaria was the first State to remodel its system for waste disposal in response to the general provisions of the Federal legislation. Except for some variations as to the State governmental subdivision with responsibility for the waste-collection and-disposal system, the other German States have been remodeling their waste-disposal systems by adopting an allocation system identical to the Bavarian model. Licensing for, and the subsequent policing of, the collection, transporting, storing, and/or processing of refuse are the most important continuing responsibilities of the designated State or local government authorities.

The responsibility for solid waste collection and disposal traditionally has been deferred to the local authorities—counties and municipalities. The Bavarian State "model" law stipulates that counties and "free" municipalities have the responsibility for waste collection and disposal. The waste disposal, and coincidently most other administrative, responsibilities of the cities and counties are supervised by the district government—an intra-State, regional level of public administration. The Bavarian law for implementing the WDA also provides for delegation of waste-collection and -disposal tasks by the counties to the villages and towns they administer.

The WDA provides for the use of private contractors where a local authority decides and the district administration approves. However, the county or city that decides to delegate all or some of its waste-collection and -disposal responsibility to a private contractor must maintain sufficient ownership interest therein so that the local authorities will always be able to vote on the governing bodies of the contractor's company. Moreover, such public involvement in the private contractor's management may result in State subsidies where the State legislation provides for financial aid to local authorities responsible for waste collection and disposal.

The Bavarian "model" law legitimizes various degrees of cooperation between municipalities and private contractors as to both the legal form of municipal participation and the scope of the private contractor's waste-collection and -disposal operations. The local authority's responsibility encompasses all types of refuse, no matter what their origin.[97]

With the approval of the district administration, local authorities may delegate the collection and disposal of "such refuse as, because of its nature or quantity, may not be disposed of together with residential refuse."[98] Thus, the municipalities can continue to collect and dispose of residential solid waste, while the commercial and industrial wastes are disposed of either by the enterprises generating such waste or by utilizing the services of enterprises specializing in such waste collection and disposal. In both cases, the State must license the private firms at the district level, and the firms must maintain accurate records for local authorities on the nature and kind of refuse they generate and/or dispose.[99]

The charges for local waste collection and treatment vary according to the amount or weight of waste generated.[100] Since rates for waste collection are set by elected city or county councils, private contractors to whom responsibility for waste collection and disposal has been delegated depend on fixed-fee arrangements as provided in their contracts with such local authorities. These fees are usually collected by the local authorities for transfer to the private contractors. Where the wastes of certain industrial and commercial enterprises are excluded by statute from the Federal-State waste-disposal system, for example, the recycling of waste paper or scrap metal, there are no legislatively determined, contractually fixed rates. Therefore, such waste-collection and -disposal services by a private contractor are limited only by market conditions. However, if the private contractor develops a monopoly position, its pricing may also be subject to the antitrust laws.[101]

The Federal law specifies that the penalties for violations of the WDA include both prison sentences (up to five years) and fines (up to DM 100,000)[102] where life or health is endangered, but only fines for other violations. With regard to the importation of waste materials, the WDA provides for regional approval thereof upon determination that such waste can be disposed of harmlessly.[103]

The WDA also authorizes the prohibition or regulation of the sale of throwaway packaging or containers for which harmless disposal is either technically difficult or extremely expensive. This provision (Section 14) of the WDA was aimed at those packaging materials made of plastics, especially polyvinylchloride (PVC) plastics, and was inspired by the success of the Waste Oil Act in that it authorizes penalty levies or "pollution taxes" on manufacturers of consumer goods for which high disposal costs are later incurred. However, implementation of an excise, or "pollution," tax on such solid waste must by law await the development of economically feasible technology for recycling.[104]

In West Germany, the glass beverage container is a serious solid waste problem because about 15 percent (by weight) of household and commercial waste is composed of glass. Metal and plastic account for 4 percent and 2 to 3 percent, respectively, of solid waste. The Federal government's Waste Management Program of October 1975 provides for regulation and/or taxation of glass containers if conversion to lightweight bottles, recycling, and other measures do not lead to a decrease in the quantity of glass in solid waste. The threat of direct government action has resulted in the rapid expansion of a voluntary recycling scheme by industry, with the cooperation of local governments responsible for household and commercial waste collection and disposal. By 1976, 16 percent of the glass containers, amounting to 260,000 tons, were being recycled. The recycling program's rapid growth has been attributed to both an effective advertising campaign and a system of waste-disposal service charges wherein most Germans pay extra for generating more than a minimal amount of household refuse. To avoid extra charges, Germans take their no-deposit, nonreturnable bottles to recycling centers.

In an effort to improve the returnable bottle scheme, the German brewers have also agreed that each beer producer will accept returnable beer bottles from retailers, with redemption of the deposit thereon, even though the bottle may not have originated with the brewer. The adoption of the standard "Euro bottle" by German brewers has facilitated the reuse of returnable bottles. The program has been much more successful in States, such as Westphalia, where there are a small number of breweries, and relatively unsuccessful in States, such as Bavaria, where voluntary cooperation between a large number of highly competitive brewers has failed to produce more reuse of returnable bottles.[105] Although the Federal government has the authority under the WDA to prohibit and/or tax the use of glass bottles if the beverage industry does not continue to reduce the amount of its containers which end up in solid waste, there are no immediate plans to ban nonreturnable bottles or impose a mandatory deposit on all glass beverage containers.[106]

The WDA is applicable to the storage and disposal of discarded tires and junked motor vehicles in that it provides for the licensing of facilities that store and dispose of tires and vehicles. Shredding plants for junked vehicles are more closely scrutinized under special licensing procedures. The private organizations that are licensed to handle used tires recover or export about 40 percent of the tires discarded annually, while the private associations that handle abandoned and wrecked automobiles scrap almost all (97 percent) such vehicles in eighteen shredding plants.

In addition to glass, tire, and motor vehicle recycling, there is a well-developed system of wastepaper recycling organized by a national trade association, which provides collection, transportation, and pre-treatment services for its members. About one-third of all wastepaper is reused as a result of this well-organized, voluntary system of recycling. Preferential freight rates for wastepaper have also been established.

Even though State and local governments seem to have practically all of the authority to control solid and hazardous waste collection and disposal under the Waste Disposal Act, the Federal government does rely on its authority under the Federal Emissions Control Act to implicitly control the solid and hazardous waste pollution of new industrial facilities. Thus, the Federal government will not grant a construction permit to an industrial firm until the firm can assure the government that proper incineration, appropriate composting, or other harmless disposal of its wastes will be technically possible at its new or altered production facilities.[107] Under authority provided by the WDA, three Federal regulations have been promulgated to establish a permit system for the transportation and disposal of solid and hazardous wastes, to require recordkeeping by waste-disposal enterprises, and to regulate the importing of wastes.[108] The August 1974 Law of Environmental Statistics provides the Federal government with additional authority to collect data about waste quantities and qualities, including that in wastewater and emissions.[109]

Although the public control of such wastes is generally very stringent, with several States planning and/or maintaining disposal facilities, there have been practical problems related to the precision with which hazardous wastes are classified. Thus, the States have enlisted the support of the Federal government in the development of legislation that delineates a more detailed classification system for hazardous wastes so that Federal-State regulation and State institutions for control of such wastes could be more effectively developed and maintained.[110] There is a central salt mine in which highly toxic industrial wastes can be disposed of safely underground—40,000 tons in 1974. However, the charges imposed for disposal of hazardous wastes in the mine are very high to encourage recycling.[111] In addition, there is the related problem of regulating the use of toxic substances from which many hazardous industrial wastes originate.

In early 1978, the Interior Ministry submitted a proposal for toxic substances control legislation to the Federal cabinet in anticipation of an advisory review by the *Bundesrat* before submission to the *Bundestag* by the end of 1978. The proposed legislation would conform with the proposed EEC directive on toxic substances.[112] The German law would require much less testing, however, than the American EPA has proposed in its implementation of the U.S. Toxic Substances Control Act.[113] Inasmuch as the EEC directives on motor vehicle emissions and wastes have been based on the German law or German initiatives and in view of the German interest in developing a toxic substances law appropriate for both the international and domestic regulation thereof, it is highly probable that the German legislation to regulate toxic substances will be a model for any future EEC directive thereon.[114] An explanation of the German toxic chemicals law and the related EEC directive appears at the end of this chapter.

## Government Assistance

As a member of the EEC and the OECD, the Federal Republic of Germany has adopted the "polluter pays principle,"[115] but the Federal government recognizes that there should be a sharing of the costs of pollution abatement among the producer, the consumer, and the public, especially in view of the increasingly stringent standards being imposed on industry. The Federal government and the States share the cost of pollution with the producer and consumer by providing tax incentives, direct grants, loans and loan guarantees to effect improvements in pollution control.[116] The public sector also provides information and research services.

Under Federal tax law, tangible personal property is usually eligible for both straight-line and accelerated depreciation. Depreciable real property is generally eligible for only straight-line depreciation, although buildings and building components may on occasion be eligible for accelerated depreciation. With regard to acquisitions before January 1, 1975, an installation for air or water purification or for the prevention of air or water pollution was eligible for additional accelerated depreciation of 50 percent during the first five years of its useful life. To qualify for the accelerated depreciation of 50 percent in addition to regular depreciation, the owner of a pollution control installation had to elect to compute regular depreciation under the straight-line method.[117] The undepreciated cost of the asset that remained after the initial five-year write-off period was depreciated on a straight-line basis over its remaining useful life. Thus, the original tax incentive for investing in air and water pollution control facilities provided for heavy depreciation deductions during an initial five-year period without permitting total depreciation — accelerated and straight-line — to exceed the basis (usually cost) of the fixed asset.[118]

An initial depreciation allowance of 60 percent and a subsequent annual allowance of 10 percent were available for tangible personal property utilized to protect the environment from air, water, waste, and noise pollution and acquired or manufactured after December 31, 1974, and before January 1, 1981.[119] The maximum depreciation allowance available for tangible personal property acquired or manufactured after August 31, 1977, is 3 times the declining balance (straight-line rate) at an annual rate not to exceed 30 percent.[120]

In addition, the Federal Ministry of Finance recommends the following rates to local finance offices, which can deviate in individual cases: 10 to 12 percent for machinery; 20 to 25 percent, automobiles and trucks; and 2 to 4 percent, industrial buildings, such as factories and warehouses. With the exception of the provision for rapid amortization of pollution control investments, the maximum depreciation allowance available for tangible personal property used to control pollution is 25 percent.[121] Traditionally, the Federal government's fiscal policy has included periodic suspensions of accelerated depreciation and the investment credit in an effort to reduce aggregate demand by discouraging capital investment. However, such periodic suspensions have not been applicable to the rapid amortization provisions for pollution control.[122] The special provision could only be temporary since a rapid amortization incentive for pollution control investments is technically a violation of Article 92 of the Rome Treaty of the European Economic Community which prohibits competition-distorting tax incentives in member-countries.[123]

Although the *Kraftfahrzeugsteuer* (automobile tax) is a minor tax in terms of its revenue, it can be a material expenditure for a firm in which transportation services are either a major or an integral part of its economic activity. This State tax,[124] which is based either on the size of a motor vehicle's engine or on the total weight of a motor vehicle,[125] will not be imposed on the sale of a motor vehicle if it is going to be used in the collection and transportation of waste by private firms licensed to provide such services.[126]

In addition to tax relief, the Federal government guarantees loans for water, air, and noise pollution control installations. Direct grants are made to higher education and research institutions for the investigation of pollution problems. At the Federal level, however, there are no direct grant programs for investments in pollution control facilities.[127] The possible exception is the use of revenue from the waste oil (excise) tax to finance the recycling of waste oil with impurities of less than 10 percent.[128]

At the State level, there are various direct-grant type subsidies for capital investment in pollution control facilities, as well as loan guarantee and loan financing programs. Inasmuch as State government assistance varies from one *Land* to another, its impact is difficult to evaluate. Examples of such State financial aid include (1) direct grants and preferential loans to water management associations[129] and local government[130] to aid in financing wastewater treatment facilities; (2) lending programs to aid in financing air quality maintenance facilities;[131] and (3) direct grants[132] and preferential loan programs[133] to aid in financing waste-disposal and -recycling facilities. In addition, State financial aid has been provided for energy recovery in the disposal of wastes.[134]

Federal and State governments cooperate closely in providing financial assistance for the construction of wastewater treatment facilities operated by local government and water management associations. Because of its superior fiscal resources, the Federal government has always been able to make the larger contribution, but State and local governments are typically required to make significant public investments and loans on a matching-fund basis. The Federal government has made loans at preferential interest rates through the European Recovery Program (ERP) Special Fund since 1950. In 1965, the Federal government began to supplement the ERP loans with long-term credit guarantees for financially weaker, medium-sized private enterprises. In addition, the Federal government has provided funds for groups, localities, regions, and industries with special economic or water pollution problems.

ERP loans are made to municipalities, rural communities, water management associations, and industrial firms at the lower-than-prime-rate interest rate—6½ percent in the mid-1970s. Since 1959, such preferential loans have been available only for investments related to the sewer treatment of wastewater, although funds were originally available for water supply systems. The ERP loans are, in principle, only to be used for financing the "net remaining balances" of funds required for a project. Consequently, ERP funds seldom provide more than one-third of a project's capital investment. An ERP loan is repaid over a period of eighteen years or less, with the first payments to begin within two to four years after consummation. The amount of ERP loans granted increased significantly in 1970 in concert with the Federal Government's Program for Environmental Protection. Moreover, the annual amount of loans has continued to increase modestly each year since 1970. Finally, since 1965 the Federal government has supplemented ERP loans with loan guarantees for long-term financing provided by commercial banks for construction of the wastewater treatment facilities of medium-sized, financially strapped firms.

In addition to these loan and loan guarantee programs, the Federal government has established special programs of financial assistance as part of its environmental protection and economic growth-and-stability responsibilities. An EEC or OECD member-country does not violate the "polluter pays principle" by including grants and loans for public sewage systems as part of its fiscal policy for the country and/or depressed regions and sectors of the economy. Inasmuch as private firms only benefit to the extent that they utilize public sewage systems for treating their wastewater, industry usually pays for its water pollution abatement.

One special funding program is administered by the Federal Ministry of Agriculture and Forestry. Both loans and grants are provided to country districts wherein nonmunicipal communities are not large enough to support their own sewage system and wastewater treatment facilities, but several communities could economically share a wastewater treatment plant with financial assistance from the Federal government. Limited amounts of Federal funds are also available for water pollution control projects from the rationalization program in agriculture wherein small farms are amalgamated to ensure proper drainage facilities; from the regional development program which is concentrated along the border with Eastern Europe; and from the *Arbeitsverwaltung* or employment fund program. In all of these Federal programs, there are always State, and often local, matching funds.

The *Bodensee* Project was initiated as part of the Federal government's detailed program for environmental protection in 1971 to clean up both the Rhine River and Lake Constance. It has been the largest single funding program for water pollution control in Germany. As noted above, Germany has also agreed to control its pollution of the Rhine River in cooperation with the other signatories to the conventions limiting the pollution of the Rhine. The Federal government agreed in the *Bodensee* Project to fund over a five-year period (1972-76) the public wastewater treatment investments along the Rhine and on Lake Constance that benefit more than the localities in which they are located. The Federal government's commitment of DM 150 million (over $60 million) was supplemented by State and local funding for wastewater treatment facilities as formalized in a 1971 agreement between the riparian States and the Bonn government. The Federal-State agreement stipulated that (1) wastewater treatment facilities should utilize the most advanced technology available; (2) such facilities should be operated by well-trained technicians; (3) each facility should treat an annual wasteload equivalent to at least DM 35 per inhabitant within the facility's service area; and (4) the owner—the local government or water management association—of each facility should provide a reasonable level of capital investment for its facility.[135]

In anticipation of the 1981 imposition of wastewater charges and subject to the vagaries of the Federal government's fiscal policy, a National Program for Future Investments was initiated by the Federal government with the cooperation and financial support of the *Länder* and municipalities. The total funding for construction of wastewater treatment facilities was shared as follows: DM 1.7 billion (about $700 million) in Federal funds; and DM 1.3 billion (over $500 million) in State and municipal funds. In addition, the *Bodensee* Project was extended and expanded as part of the federally sponsored National Program of Future Investments with the following arrangement for sharing the costs of wastewater treatment facilities along the Rhine and on Lake Constance: DM 800 million (over $300 million) in Federal funds; DM 800 million in State funds; and DM 400 million (over $150 million) in municipal funds. During 1977, the Federal government awarded DM 450 million (about $180 million) to local government, water management associations, and private firms for construction and planning of wastewater treatment facilities.[136]

As part of its fiscal policy program, the Federal government periodically relies on deficit spending through tax reductions and temporary expenditure increases to spur economic growth and full employment. Following the recession of late 1973 through 1975 and the stagnant period of early 1977 to mid-1978, the Federal government's fiscal stimulus included, in addition to personal income tax cuts, the following financial assistance which to some extent finances wastewater treatment facilities: (1) During 1974-75, DM 1.8 billion (over $700 million) to alleviate the unemployment situation through the funding of approximately DM 1.2 billion (about $500 million) of building projects in depressed regions, as well as a temporary reinstatement of the 7½ percent investment credit and DM 1.13 billion (about $450 million) to encourage investments, especially in energy projects;[137] and (2) for 1977-81, a DM 8.7 billion (about $3.5 billion) medium-term investment program[138] which encompassed the aforementioned programs for the Rhine and Lake Constance and wastewater treatment facilities in anticipation of the January 1, 1981, effective date of the wastewater or effluent tax system.

After 1980, State financial aid should have been provided with revenues from collection of wastewater charges or effluent taxes imposed on persons discharging noxious wastewater. Revenue from wastewater charges was to be used for maintenance or improvement of water quality. The *Länder* were also authorized to use the revenue to finance the administrative expenditures attributable to the enforcement of both the Waste Water Charges Act and the States' own supplementary regulations related thereto. More specifically, the revenues should be used (1) to construct wastewater treatment plants; (2) to construct rain retention basins and facilities for purification of rainwater; (3) to construct ring-shaped, holding canals at and along lake shores, sea shores, and dams; (4) to construct main connecting sewers so that jointly operated sewage treatment facilities can be utilized; (5) to construct facilities for disposal of sewage sludge; (6) to finance measures in and at bodies of water to improve their water quality — for example, augmenting the low-water flow of a river and providing oxygen enrichment for water bodies — and to maintain the water quality of such bodies; (7) to finance research on and development of technology for water quality improvement and wastewater treatment; and (8) to finance fundamental training and continuing education for the operating staff of water treatment facilities and other installations designed to improve and maintain water quality.[139]

When the wastewater charge or effluent tax system is operational, an indirect subsidy will be in effect until 1990. Firms or industries will be allowed to reduce their charge or tax payments because to pay their actual liabilities would be an economic hardship to them, to consumers in a price increase of more than 1 percent, or to workers in higher unemployment.[140] Another even more indirect subsidy will result from the three-year antecedent period of reduced wastewater charges during construction of wastewater treatment plants that will reduce the noxiousness of wastewater at least 20 percent.[141]

One might conclude that inasmuch as relatively less government financial assistance has been available to industry in West Germany than in the other countries of Western Europe and North America, German producers have been operating at a comparative disadvantage. Such a conclusion would not be true if producers in both Germany and the other countries were being required to internalize the same pollution costs because standards and pollution control costs are the same in all countries.[142] Nevertheless, the Federal government is contemplating the liberalization of its tax concessions for future investments, in spite of EEC Treaty Article 92[143] and OECD policy which mandate the "polluter pays principle." It is also remotely possible that the Federal government will subsidize investments wholly designed to protect the environment for those industries that cannot afford to internalize in the short run all their pollution control costs without disrupting economic stability, employment, and growth.

## Environmental Policy Developments, 1979–1981

The most important recent development in German environmental policy occurred in the summer of 1980 when the Federal Chemicals Law (FCL), also known as the Act on Protection Against Dangerous Substances, was enacted to establish controls on the production, use, and importation of new toxic chemicals.[144] Developments applicable to air, water, noise, and waste pollution were of an administrative nature, rather than statutory, and minor in scope, although several legislative proposals were introduced during the period.

### Federal Chemicals Law

The FCL, which became effective on January 1, 1982, was enacted to be consistent with the EEC's "Sixth Amendment" to its 1967 Directive on the Classification, Packaging, and Labeling of Dangerous Substances. As already noted, the 1972 Waste Disposal Act along with several other environmental laws, authorizes the Federal government, in cooperation with the *Länder*, to control the disposal and emission of hazardous wastes. The FCL provides the Federal government the authority to control the production and importation of toxic substances, especially toxic substances that are "new" on the market.

Although the FCL was first proposed in 1978, its controversial nature and a delay in approval of the EEC's Sixth Amendment resulted in a lengthy debate on its provisions. The law was opposed by the powerful German chemical industry (Europe's largest, which includes three of the world's five largest chemical producers).[145] Support came from both the environmentalists and the powerful German labor unions. Thus, politics again makes for "strange bedfellows" inasmuch as environmentalists and the unions have fought bitterly over the regulation of nuclear power and the expansion of industry in congested areas. The unions were not willing to allow economic development to take a "back seat" to worker safety in exposure to toxic substances, whereas on the nuclear power issue, construction jobs and economic growth have been deemed more important than absolute environmental protection.[146]

In addition, the public has been concerned about the regulation of toxic chemicals because of recent incidents, notably:

> The slow poisoning of the surroundings of a cement factory in North Rhine-Westphalia by the metal thallium;
> A noxious odor in parts of Bavaria, identified by the Bavarian government as harmless only after an anonymous phone call identified the source of the smell;
> Discovery of long-stored World War II poison gases in Hamburg,

which caused two deaths and evacuation of the surrounding area;

Thousands of dead fish in the Main River caused by liquid waste emissions from the Hoechst Chemical Company; and

Discovery of high levels of cadmium in solid wastes around the city of Munich.[147]

Alarming levels of cadmium and lead from the emissions of lead factories have also been discovered in the blood of children and adults in an agricultural area of Lower Saxony inhabited by 27,000 people and producing fruit, vegetables, grain, and foodstuffs. At the time of the incident, there were no national standards for the emission of lead and cadmium. The State of Lower Saxony has provided financial assistance for the relocation of families with children.[148] Later, medical researchers discovered kidney disorders in those living closest to the lead factories.[149] Finally, in at least two of the incidents, in Hamburg and in Hesse (location of the Hoechst plant), it was discovered that local government officials had long-standing knowledge of the potential danger of an accident before it occurred.[150] In reality, the FCL was enacted in the legislative rush to complete parliamentary business before the recess for national elections called for October 10, 1980.

The major provisions of the FCL dealing with notification of new and imported chemicals are congruent with the provisions of the EEC's Sixth Amendment, which does not specify mandatory methods for controlling the production, importing, use, or disposal of toxic chemicals. The FCL does, however, go beyond the notification and labeling requirements of the EEC directive by subjecting existing chemicals to notification and testing procedures, which are to be promulgated for both new and old chemicals through future regulations. Thus, an inventory of the 45,000 chemicals currently being marketed in Germany will have to be developed by the Interior Ministry, and extensive regulations will have to be written.

Other pertinent provisions include an exemption from the notification requirements for production of less than one ton of a particular chemical substance and a forty-five-day period between notification and the initial marketing of the chemical. Moreover, the General Accounting Office (GAO) was assigned the task of recommending the appropriate Federal agency for handling notification responsibilities because legislators could not agree on to whom they should give the authority.[151]

Following the GAO's recommendation (after much debate thereof), the Federal cabinet named the Federal Ministry of Labor's Office for Occupational and Safety Policy (BAU) as the notification authority, that is, the agency to which notifications of new or imported chemicals are to be sent. However, the BAU will review the notifications to ascertain which government scientific experts can best conduct an evaluation of the notification and the chemical substance related thereto. If the BAU does not have the appropriate expertise on its staff, then it must transfer the evaluation responsibility to one of two other Federal offices with scientific expertise — either the Interior Ministry's Federal Environment Office (BUA) or the Youth, Family and Health Ministry's Federal Health Office (BGA).

As mentioned earlier, no single Federal ministry has been assigned exclusive responsibility for environmental protection. Thus, the BAU (Ministry of Labor) must share FCL notification responsibilities not only with the BUA (Ministry of Interior) and BGA (Ministry of Youth, Family, and Health) but also with the Office of Substances Testing (Ministry of Economics), which is concerned with water quality programs and the transportation of hazardous substances, and the Office of Biological Research (Ministry of Agriculture), which is responsible for selected aspects of nature protection.

Since the Ministry of Youth, Family, and Health has been given formal authority for implementing the FCL, it is the only Ministry with responsibility for coordinating the law's implementation. Of course, environmentalists argue that the powerful chemical industry created the five-agency conundrum to make control of the production of toxic substances unworkable, especially in view of the forty-five-day period for evaluation.[152] Another complicating factor is the fact that all four offices may make proposals to the BAU for controlling a new chemical.

In the case of disagreement among the five offices on a proposal to control the production or importation of a chemical, the BAU is authorized to mediate, but it can only reject a proposed action if it can demonstrate that the action is "disproportionate" to the identified hazard or risk. Even when there is agreement on a proposal to control a chemical, the proposed action must be submitted to the *Bundesrat* because such regulatory actions are the joint authority of the States and the Federal government. However, the Federal cabinet makes the decision to submit the proposed control action to the *Bundesrat*.

The BAU has exclusive authority over at least two aspects of the control of chemicals. First, it can require a firm to correct a deficiency in its notification of a new chemical when the BAU deems that notification to be inadequate or incomplete. In addition, the BAU can regulate the use of selected chemicals in the workplace environment as authorized by the 1971 Worker Protection Regulation incorporated in the FCL when it was enacted in 1980. The regulation allows the BAU to impose requirements that are necessary to protect workers, including labeling, the handling of hazardous chemicals, and production procedures.[153]

The general form and content of the notification on a new chemical which a manufacturer or an importer submits to the government are specified in the FCL regulations. The regulations also define the tests and requirements of notification. The chemical manufacturer or importer must prepare a summary of the tests and results so that a quick review of the notification documentation is possible, thereby facilitating the review process. The notification information must be filed in quadruplicate to further expedite the review process.

The BAU is also the liaison office for relations with the EEC in respect to the Sixth Amendment on notifications for new chemicals. This role is especially important because the EEC seems to be more serious about monitoring compliance among its ten members on the Sixth Amendment than it has been in the case of other environmental directives. Furthermore, the EEC is developing an inventory of chemical substances marketed commercially as of September 18, 1981. Those substances not on the list will have to comply with the notification requirements of each member's law implementing the Sixth Amendment of the EEC Directive on Classification, Packaging, and Labeling Dangerous Substances. Although the EEC will not be developing testing procedures for chemicals in the near future, the OECD of which Germany and the other EEC-member countries, non-EEC European countries, Canada and the United States are members, has been working on such procedures as well as guidelines for divulging information on the importing and exporting of chemicals.[154]

### Other Developments During 1979–1981

Although legislative changes were proposed in 1979 for revision of the *TA Luft* regulations under the Federal Emissions Control Act (ECA)[155] and in 1980 for the deferral of the January 1, 1981, effective date of the wastewater charge system under the Federal Waste Water Charges Act (WWCA), neither legislative proposal was enacted during the 1979-81 period. With regard to implementation of the wastewater charge system at the *Länder* level, only the heavily industrialized State of North Rhine-Westphalia located in central Germany had enacted detailed enabling legislation for the WWCA by March 1980. As explained above, the *Länder* must enact such legislation because the Federal Constitution only permits the Federal government to enact general, "framework" laws in respect to water quality and resource management. Insofar as the *Länder* were unsuccessful in delaying the effective date of the WWCA to January 1, 1983, the rest of the States should have enacted enabling legislation either in 1980 or early 1981.[156]

In early 1979, the Federal government established limits on municipal wastewater discharges from sewage treatment facilities under the WWCA.[157] Similarly, in the summer of 1980 the Federal government finally promulgated regulations on the permissible levels of phosphate in soaps and detergents as authorized by the 1975 Detergents Act (Act Concerning the Environmental Compatibility of Washing and Cleansing Agents). The regulation requires that phosphate levels be reduced in two stages with 25 percent reductions by October 1, 1981, and by January 1, 1984.[158] Finally, there was an historic agreement between the State of the Saar and the French Prefect of Lorraine wherein the French agreed to apply German air quality standards in permitting the construction of a battery factory 3 kilometers from the German border.[159] The details of this agreement are provided in Chapter 4.

## Future Environmental Policy

The reelection of Helmut Schmidt as Chancellor in the October 5, 1980, national elections generally reaffirmed the environmental policy of the past decade.[160] Even though several other issues, including world peace, the economy, energy, and nuclear power, were more important, environmental protection was still a major campaign issue because of the emergence of an environmental political movement. It was feared that the new movement, known throughout Europe as Green Parties or the Greens, would erode support for the Free Democrats enough to preclude the majority coalition of Social and Free Democrats.[161]

### Environmental Policy Agenda

Less than two months after the election, Chancellor Schmidt and those government ministers with environmental protection responsibilities were expounding upon the Federal government's environmental policy objectives for the early 1980s. The highest priority is to be given to the reduction of both emissions and noise from motor vehicles. Consequently, the proposal to control traffic noise, which failed to pass in the *Bundestag* in 1980, was to be reintroduced in 1981. Previous policy on toxic chemicals, solid wastes, water pollution, and emissions will generally be unchanged. However, amendment of the nature protection law has been proposed to allow a limited number of environmental groups to sue the government and/or industry to ensure implementation and enforcement of environmental laws. Although conversion from oil and gas to coal as the source of energy for generating electric power will continue, the solution to nuclear waste-disposal problems will also be emphasized. In this way, further development of the nuclear power industry will be possible. In the construction of highways, more funds are to be spent on the control of traffic noise and environmental protection.

Inasmuch as controlling motor vehicle pollution and energy consumption is a controversial issue in the Social Democratic party, the Schmidt government did not anticipate any proposed legislation on the issue in 1981. However, a proposal to abolish the motor vehicle tax and to replace it with a significant increase in the Federal gasoline tax could improve air quality, while reducing energy consumption and the Federal budget deficit. Unfortunately, the *Länder*, as well as local governments and many members of the *Bundestag*, have been opposed to this proposal.

Other environmental policy objectives outlined by Gunter Hartkopf, State Secretary and Minister of the Interior, provide more insight as to the specific proposals anticipated during the next few years. In addition to supporting the reintroduction of legislation which will allow certain environmental organizations to sue the government and industries to ensure proper enforcement of environmental protection laws, Hartkopf noted that all national political parties support the introduction of an "environmental protection clause" into the German Constitution (Basic Law).[162] Bipartisan support for such a constitutional amendment was confirmed in early 1981 when all major German political parties approved in principle the proposed language of such an amendment to Article 20.

The proposed amendment, which would be a human right, would be worded as follows: The protection of the natural foundations of life is a responsibility of the government. Such an amendment would not, however, create any new procedural or substantive rights for either groups or individuals, but it would mandate that environmental protection policy be on a par with any other governmental policy responsibilities specified in the Constitution, for example, protection of the family and health. In fact, such an amendment would aid the Federal government in its efforts to balance environmental protection issues with related issues such as economic development, energy supply and conservation, and government austerity.[163] Of course, an environmental group's right-to-sue law would still be necessary.

Another legislative goal of the highest priority will be the establishment of air quality standards for industrial plants.[164] Unfortunately, this legislative issue has been lingering since late 1978.[165] Such amendments to the Federal Emission Control Act (ECA) were proposed as far back as August 1979.[166]

Water pollution control efforts will continue as scheduled insofar as *Länder* efforts in the *Bundesrat* to delay the January 1, 1981, effective date of the WWCA were unsuccessful. Hartkopf predicted that approximately thirty regulations establishing wastewater discharge limits for a variety of industries under the WWCA would be developed in 1981.[167] Wastewater discharge limits for municipalities under the WWCA were adopted in early 1979.[168]

Amendments to the Waste Management Act (WMA) are to be proposed to encourage more recycling of urban and industrial wastes (especially packaging), as well as waste dumping on the high seas. When the Federal government's proposed amendments of the WMA are effected, it will present the *Bundestag* and *Bundesrat* with a regulation concerning the disposal of sewage sludge.[169] The proposed amendments would

> Empower the Minister of Interior, who is primarily responsible for environmental protection, to adopt regulations limiting the presence

of toxic chemicals in sewage sludge used for fertilizer;

Exclude household and non-toxic industrial garbage, excavations, and construction rubbish from the waste disposal permit requirements insofar as it is not contaminated by other substances; and,

Delegate the authority to grant permits for the disposal of these wastes, in particular animal fecal matter, to the state governments [and empower the states] . . . to restrict or ban disposal of these wastes on agricultural lands "if the concentration of harmful substances in the soil could lead to injury of the public welfare."[170]

The *Bundesrat* modified the Federal government's proposals somewhat with proposals to exempt both mining wastes, where such wastes are already regulated under mining laws, and substances used in military weapons. These modifications were included when the amendments were sent back to the government in May 1981.

The Federal government's proposals were devised to deal with two particular problems that have recently concerned the public: insufficient legal authority to control the use of sewage sludge that contains harmful levels of organic chemicals and heavy metals; and excessive permit requirements on nonhazardous wastes when authorities do not have sufficient resources to properly regulate toxic wastes. Heretofore approximately 60 percent of the waste-disposal permits have been issued for the disposal and/or transport of nonhazardous urban (household and construction) waste. Logically, exempting such waste from the permit process should allow authorities to allocate more resources to the regulation of hazardous waste disposal. Unfortunately, sewage sludge sold by the larger, heavily industrialized cities, such as Hamburg and Munich, to nearby farmers has contained harmful levels of the metal cadmium. In fact, some acreage has had to be removed from agricultural production.[171]

With regard to both improvements in emissions and noise from motor vehicles and the control of toxic chemicals, the Federal government intends to cooperate with automotive and chemical industries and other private institutions to develop voluntary compliance. If such efforts are not fruitful, then regulations will have to be adopted.

Some regulations are necessary to implement the FCL. Thus, the Federal government either has adopted or is developing regulations under the FCL to provide an inventory of chemicals marketed commercially on September 18, 1981, testing procedures for chemicals, and standards for evaluation of the risk to humans and the environment from the development and use of new chemicals. Similarly, implementing regulations must be devised for compliance with the EEC's "Seveso Regulation," the directive that requires all EEC members to establish emergency procedures in the case of chemical plant accidents,[172] especially explosions like the one that occurred in Seveso, Italy, a few years ago.

In an effort to encourage voluntary compliance by the chemicals industry, the Interior Ministry has already endorsed testing procedures being developed by the OECD and the EEC. Specifically, a set of tests proposed by the OECD for initial assessment of the impact of chemicals on human health and the environment was deemed proper as a result of research sponsored by the Interior Ministry. Likewise, the binding "basic dossier" of testing procedures required by the EEC's Sixth Amendment was investigated by Federal government scientists who confirmed that the procedures were sufficiently accurate to identify potentially dangerous chemicals before their marketing is permitted.[173] Such endorsements are consistent with both Germany's responsibilities as a member of the EEC and the Federal government's commitment to the development of testing procedures in the context of the toxic chemicals control program of the OECD.[174]

Air Pollution Control

In its 1979 annual report, released in August 1980, the BUA concluded that overall air quality had not improved in spite of successful programs for regulating industrial emissions because air pollution from the growth in motor vehicle ownership had increased.[175] Government studies indicate that motor vehicles produce one-third of all emissions, including carbon monoxide, nitrogen oxides, and hydrocarbons. Unfortunately, there is considerable political resistance to any further tightening of motor vehicle emission standards through legislative action. Therefore, the Federal government with *Länder* support has temporarily adopted a strategy of using "moral suasion" to enlist voluntary reductions in emissions for motor vehicles produced by German industry. In addition, the continuing conversion from imported oil and gas to domestic coal as a source of energy, along with the limited growth of the nuclear power industry, aggravates air quality because burning more coal increases emissions of sulfur dioxide and soot (particulate matter) as well as cadmium.[176] Thus, the Schmidt government has been attempting to impose more stringent emission control standards on industrial polluters through amendments to the ECA since 1979.

The most recent proposals (September 1981) to revise the 1974 *TA Luft*, the General Administrative Regulations interpreting the ECA, seem to be as controversial as the 1979 proposals were. Just as before, the environmentalists argue that the revisions do not go far enough to protect either plant and animal life or physical structures, while industry claims they are too stringent and they would discourage new investments, increase unemployment, and hamper economic growth. In addition to reducing many existing ambient air quality standards, the revisions include new standards for measuring sulfur dioxide emissions by reducing the area for which it is measured, as well as new ambient air quality standards for hazardous or carcinogenic substances, including lead, cadmium, and asbestos. The existing ambient standard for sulfur dioxide remains unchanged, although many German *Länder* are growingly concerned about the detrimental effect of acid rain on their soil and forests, which cover 29 percent of the country.[177]

The proposed revisions would allow special emission controls to be established in areas that are either particularly clean or polluted to prevent increases in air pollution. Furthermore, the siting of new industries in heavily polluted areas would be permissible only if total emissions of existing industrial facilities located therein were reduced and overall air quality improved.

In the summer of 1981, the Federal government made some progress in obtaining the voluntary cooperation of the domestic automobile industry to reduce motor vehicle emissions, noise pollution, and energy consumption. The automobile industry agreed to a five-point program: (1) short-term reduction in motor vehicle emissions of CO, nitrogen oxides, and hydrocarbons by 20 percent; (2) longer term reduction in emissions of 50 percent by 1985-86, while at the same time conserving fuel; (3) fuel consumption savings of at least 12 percent by 1985 as compared with 1978 consumption levels, with further reductions of up to 15 percent; (4) reduction of noise pollution to levels established by the Federal government in its earlier recommendations to the EEC; and (5) a promise to support the Federal government's dissemination of information on the noise, fuel consumption, and emission levels of various makes and models of motor vehicles marketed in Germany.[178] Earlier in the summer, the Federal government published a list of automobiles that satisfied its specified noise standards.[179]

The 20 percent reduction of motor vehicle emissions is consistent with the recommendations of the U.N. Economic Commission of Europe which includes EEC member-countries. However, the Federal government's longer term goal of 50 percent reduction is a uniquely German position. Moreover, if the EEC fails to adopt the 50 percent reduction policy in amending its directive on motor vehicle emissions, the German government has threatened to unilaterally adopt such a standard. The adoption of such a stringent standard is allowable under Article 36 of the Treaty of Rome which established the EEC; in the treaty, barriers to trade are permissible if deemed necessary to protect human health and life.[180]

## Water Quality and Water Resource Management

Even though the Federal government has made considerable progress in controlling the pollution of surface water, two new water pollution problems have emerged—deterioration of ground water supplies and rapid degradation of the North Sea bordering the German coast. Several public and private institutions have expressed concern about not only the pollution but also the depletion of ground water. Agricultural, urban, and industrial pollution and use may be endangering the quality and supply of ground water for human consumption.

The two major causes of the ground water pollution and supply problem are intensive agriculture that relies on fertilizers, pesticides, and herbicides and the use of large amounts of ground water. The chemical-intensive agriculture is contaminating ground water because residue from fertilizers, pesticides, and herbicides seeps down to the water table. The use of sewage sludge which was contaminated with heavy metals, such as cadmium, may also result in the pollution of ground water. Similarly, industrial emissions, effluent, and solid waste find their way into ground water. In addition, profligate use of ground water by industry rapidly depletes its supply, allowing unpurified, surface water to be absorbed by aquifers.

The Federal Constitution provides that a landowner's property rights may be limited only by the public interest which may not always include the natural environment. Consequently, changes in Federal laws will probably be required to ensure that the protection and preservation of ground water is in the public interest.[181]

In response to the problem of ground water and North Sea pollution, the Social Democrats' Environment Commission has proposed fundamental changes in water resource management programs which will be considered at the Social Democrats congress in 1982. The Environment Commission's recommendations include

> Limiting industrial use of groundwater for non-essential purposes, such as cooling and industrial purposes;
>
> Improving and expanding biological waste treatment facilities;
>
> Increasing protection of watershed areas by providing long-term protection through regional plans and bans on construction; and
>
> Banning new construction in already burdened coastal areas and near the Dutch Wattensee.[182]

Although pollution of the Rhine River has been alleviated somewhat through the efforts of Germany and the other four signatories to the 1973 Bonn Convention, pollution of the North Sea has worsened and there is currently no multilateral agreement devised to mitigate its pollution of. In September 1981, all four major German political parties agreed in a *Bundestag* debate not only on the problems of North Sea pollution, but also on the priorities for mitigating those problems. The debate was precipitated by the issuance during the summer of 1980 of a report to the Interior Ministry by the Council of Experts for Environmental Questions on the environmental problems of the North Sea.

The actual sources of North Sea pollution cannot be precisely delineated because much of the waste is carried into that body of water by rivers that run through several countries. Nonetheless, the specific types of pollution being dumped therein are well understood. In addition to industrial and urban wastes emitted into rivers flowing into the North Sea, they include the dumping of chemical wastes by industries located in North Sea coastal countries, the dumping of sewage sludge, poorly treated municipal wastes by cities on the North Sea coast, and oil and chemical spills by ships passing through the North Sea. The most serious pollution threat is in the low-tide area of the North Sea; almost two-thirds of that area is within the jurisdiction of Germany.[183]

Although the low-tide area near Germany's coast is most threatened by waste dumping in the North Sea, most of the dumping areas are near the coast of Great Britain (United Kingdom). The Council's report indicated that at least 88 million metric tons of waste were approved by seven Western European nations for dumping in eighty-six areas of the North Sea during 1978. About 70 million tons of the waste was excavated material. About 8 million tons each of industrial waste and sewage sludge were also dumped. Finally, 67,000 tons of waste were incinerated at sea. Almost 2.5 million tons of the more harmful industrial wastes were from Great Britain; 1.5 million tons from The Netherlands; almost 1.4 million tons from France; and almost 800,000 tons from West Germany. However some of the industrial wastes passing through Dutch ports for dumping originated in Germany. It is also noteworthy that European industries have been flocking to the North Sea coasts where fewer effluent discharge controls are imposed to build plants rather than comply with increasingly restrictive water quality regulations applicable to waste disposal on inland waters.[184]

In summary, the *Bundestag* agreed that a seven-nation multilateral agreement on protecting the North Sea should be the vehicle for controlling pollution from and by the North Sea coastal nations. West Germany has also been negotiating with East Germany and Czechoslovakia to control the pollution from rivers running through those countries into the North Sea. Unfortunately, the other six North Sea coastal nations seem unwilling to discuss a common action program for protection of the North Sea similar to the 1980 Baltic Protection Convention. Of course, Rhine River pollution control as agreed to by Germany, France, The Netherlands, Luxembourg, and Switzerland should reduce North Sea pollution.

West Germany has taken unilateral actions to reduce North Sea pollution. Sewage sludge dumping in the North Sea will be terminated within a few years because the sludge has been found to contain heavy metals such as cadmium. Likewise, the controversial dumping of titanium oxide wastes should be terminated in 1982 or 1983, inasmuch as the two major dumpers, Bayer Chemical Company and American-owned (National Lead) Kronos Titan, have been pressured by environmentalists and the Federal government to eliminate their offshore dumping.[185]

Without a North Sea Protection Convention signed by West Germany as well as Belgium, Denmark, France, The Netherlands, Norway, and the United Kingdom, the prospects for reducing North Sea pollution are still good because of EEC directives on the discharge of hazardous substances into water (including one applicable to titanium oxide) and several international or European agreements. Among these agreements are the 1974 Paris Convention for the Prevention of Marine Pollution from Land-Based-Sources, which was not ratified by Germany until the summer of 1981, the 1977 Oslo Convention for the Prevention of Marine Pollution by Dumping from Ships and Aircraft, the 1977 London Convention on Civil Liability for Oil Pollution from Exploration, and the 1977 Bonn Agreement on Cooperation in Dealing with Pollution of the North Sea by Oil. If the EEC directives are implemented and the Paris, Oslo, and London conventions are ratified and enforced, there is still hope for the North Sea.[186]

### Environmental Protection Administration

Although West Germany's "economic miracle" had evolved into an "economic mess" by 1979 primarily because of the high cost of oil imports, spending abroad,[187] and antinuclear power sentiments, popular support for pollution control measures has enabled the Schmidt government to increase expenditures for environmental protection in the face of austerity Federal budgets to hold down deficits. However, the 1981 budget for the BUA was held at about 1980 levels, with "substantial cuts" in its research budget and the budget for nuclear reactor safety and radiation protection being used to fund the activities necessary to implement the Federal Chemicals Law which took effect on January 1, 1982.[188]

Funding for environmental research has been a "political football" for several years.[189] The BUA's research capacity and laboratories have always been limited. Therefore, much environmental research has been conducted under contract by nongovernmental public and private organizations with BUA supervision. Research related directly to regulatory initiatives was the only aspect of BUA research activity that was spared in the budget cuts, while the following types of research suffered cutbacks: biological and medical effects of air pollution, implementation of aircraft noise protection measures, and water quality and waste disposal. Generally, any basic research not directly related to regulatory initiatives was affected by the austerity budget.

BUA officials have expressed concern that the cutbacks would hamper future environmental policy initiatives. Historically, the BUA research has been devised to develop alternatives to industrial technology that pollutes and to evaluate the implementation of environmental programs. Based on this research, it has accumulated the expertise necessary for achieving voluntary compliance by industries with environmental goals.

The BUA's success with this strategy is reflected in a reduction in the use of chlorofluorcarbons, the recycling of waste oil and hazardous industrial wastes, the use of returnable beverage containers, and the effort to control traffic noise and motor vehicle emissions.[190] Although environmental research may continue to suffer from budget cuts, thereby hampering the BUA's development of long-term environmental protection objectives, the Schmidt government anticipates no drastic Reagan-type cuts in the overall BUA budget.[191]

### 1982 Addendum

In spite of a two-year economic recession, the Social Democrat-Free Democrat coalition through which Chancellor Helmut Schmidt governs survived a late June battle over a proposed budget for 1983, thereby increasing the chances for survival of the Schmidt government until the 1984 national election.[192] However, the Social Democrats have been faring poorly in recent state and local elections, while the conservative Christian Democrats and the Green or Alternative party has been gaining strength in such elections. In fact, the environmentalist party (often referred to as "the Greens") has elected representatives in five states (*Länder*), and it has supplanted the Free Democrats, traditional coalition-makers or breakers, as the third largest party in three states. Public opinion polls indicate that the Greens could win over 10 percent of the vote in the state election in Hesse in September; and, consequently, it would be the third largest party in that state. A late spring poll indicated that the Greens would win as much as 7 percent of the votes in a national election, giving them an influential role in a new *Bundestag*.[193]

The recent growth of the Green or Alternative Party is primarily attributable to the growing anti-nuclear, peace movement in West Germany, but the party also appeals to the young and liberal because of its "alternative" position on the environment, health, energy, defense, and the Third World. Unfortunately, the Greens often act their age; that is, the immaturity of the Green party has often led its elected state representatives to react negatively to any legislation which is not consistent with the party's alternative ideology. In fact, the Greens are not a homogeneous party at the national level, and, consequently, most Social Democratic party leaders do not consider the Greens a reliable coalition partner to help a government after the 1984 national election. Based on a poll conducted during 1982, the Christian Democratic party, which has been out of power for 16 years, would win a majority in any national election; moreover, the centerist Free Democratic party leadership seems to be itching to join the Christian Democrats in a new coalition government if the conservatives do not win a clear majority in the 1984 election.

In spite of the Greens' apparent inclination to play the role of obstructionists in influencing state and national politics, the prospect of a more conservative government after the 1984 elections may accelerate their maturity as a political force. In other words, if the Green party's success in the 1984 election were to enable it to extend the life of a Social Democratic government more favorably disposed to the Green's ideology, then a coalition with the liberal Social Democrats would certainly be better for the ecology-peace movement than several years of a Christian Democrat-Free Democrat coalition, especially if the conservatives also end up controlling the states and the *Bundesrat*.[194]

Progress on Emissions Control

Although the emerging political power of the Green party in a coalition government with the Social Democrats could bring about more stringent environmental policy (especially with regard to nuclear power, toxic wastes, and air pollution), the first half of 1982 saw few developments related to the environment. The best news for environmentalists occurred in May when the Federal Environmental Office (BUA) issued its second report under the Emissions Control Act (ECA). The report, which covered the period since the first report issued in 1978, disclosed that levels of $SO_2$ and hydrocarbons have remained constant in spite of constant economic growth and a continuous increase in energy consumption. The report also indicated that noise pollution has remained unchanged since 1978 with the exception of noise in residential areas near airports. Some improvement, however, was noted insofar as particulate matter emissions have declined, the lead content of motor fuel has been reduced, PCBs are only permitted in closed systems, and chlorofluoro-carbon use in aerosols has been reduced 40 percent.

On the negative side of the ledger, significant increases in $NO_x$ emissions have occurred since 1978. Similarly, emissions of heavy metals, asbestos, and many halogenated hydrocarbons have increased, thereby increasing concern for the possible carcinogenic properties of such substances.

In commenting on the report, Interior Minister Gerhardt Rudolf Baum said that national air quality and noise control policies have been effective, but that more stringent regulation will be necessary to alleviate that air and noise pollution not yet responding to the measures provided under current law. The *Bundestag* will consider the report and its implications for amending the ECA in the fall of 1982, at which time the Interior Ministry will propose that carcinogens be brought under ECA authority. The report attributed most of the positive developments enumerated therein to the implementation of Federal laws and regulations, for example, limitations on the sulfur content of light heating oil and diesel fuel and the development of air quality plans and goals at the state level.

The Interior Ministry also plans to revise environmental standards and regulations in an effort to keep up-to-date with improvements in pollution control technology. Review of proposed revisions in technical guidelines for licensing industrial facilities should result in some changes as a result of the criticism the current guidelines received during public hearings. Nevertheless, the new guidelines will include ambient air quality standards for cadmium, lead, and thallium.

The Interior Ministry also is developing regulations for large heating plants that should result in drastic reductions in the emission of $SO_2$, $NO_x$ and halo-genated hydrocarbons, that is, fluoro- and chloro-carbons, in the hope that the acid rain problem affecting German forests can be solved domestically. Similarly, the Ministry intends to meet with the domestic automobile manufacturers during the summer of 1982 to press them to further reduce auto emissions, despite industrial agreements in the summer of 1981 to reduce auto emissions by 20 percent. In the EEC, the Interior Ministry will continue to lobby for a 50 percent reduction of auto emissions, as well as stringent limits on noise from motor vehicles.[195]

With regard to asbestos emissions, the Interior Ministry received a commitment in February from the cement industry to reduce the asbestos content of their building material by 30 to 50 percent over a three- to five-year period. In addition, the asbestos industry has agreed to work with the Ministry to develop substitute materials to reduce the consumption and, therefore, emissions of asbestos.[196]

## Hazardous Waste and Toxic Substances

Most of the other developments in environmental policy during early 1982 dealt in one way or another with the growing concern for both ambient levels of toxic substances and the emission and disposal of hazardous wastes. Inasmuch as three of the world's largest chemical producers are German companies (BASF, Hoechst, and Bayer),[197] hazardous waste disposal is a significant and growing pollution problem despite enactment of the Waste Disposal Act (WDA) a decade ago.

While almost 50,000 unregulated dump sites have been closed and 5,000 regulated sites opened since 1972, illegal dumping and the incineration of contaminated domestic waste to recover heat continue; seepage from old, closed dump sites threatens to contaminate large amounts of groundwater; significant amounts of hazardous waste continue to be shipped through the North Sea port of Antwerp (Belgium) for incineration or dumping at sea; and a three-fold increase in municipal sewage has resulted in a commensurate increase in contaminated sewage sludge. On the other hand, accomplishments include separation of most domestic waste from hazardous waste, regulation of the transport of hazardous wastes, and development of deterrents to illegal dumping. The Federal government concedes that the absolute magnitude of the problem increases.[198] There are limited dump sites, disposal at sea must ultimately be reduced, and regulations taking effect on April 1, 1983, may drastically reduce the use of contaminated sewage sludge as agricultural fertilizer.[199]

Notwithstanding the impending hazardous waste disposal crisis, Federal government-industrial agreements on recycling hazardous waste may alleviate the problem and defer the inevitable enactment of tougher hazardous waste disposal legislation. One such agreement seems to be working. In September 1980, German mercury battery producers and Interior Minister Baum agreed on a recycling scheme which resulted in the separation from domestic waste and recycling of 40 percent of the mercury batteries sold in 1981. About 40 million mercury batteries, used in cameras, calculators, watches, and hearing aids, are sold annually. If none were recycled, around 25 tons of mercury would be released into the atmosphere annually. Battery producers voluntarily pick up used batteries from retailers. German-produced batteries as well as those produced in the United States, United Kingdom, and Japan are marked with a unique symbol indicating their mercury content. Eventually, the symbol should appear on batteries produced in Hong Kong, Singapore, and Malaysia—all of which are negotiating with German officials. In Switzerland, a similar recycling program has resulted in the recycling of more than 80 percent of the mercury batteries sold.

The Interior Ministry anticipates similar recycling programs will be consummated with other industries during the next few years. If voluntary agreements are not worked out, the Interior Ministry will propose an amendment to the WDA authorizing the Ministry to mandate retrieval and recycling of specific, harmful products.[200]

In another hazardous waste-related matter, the *Bundesrat* in May approved state-administered regulations promulgated under the ECA and devised to reduce the risk to the environment resulting from hazardous waste accidents like that which occurred at a Seveso, Italy chemical plant in 1972. At Seveso, the nearby countryside was contaminated with toxic dioxin. The new regulations envision voluntary compliance by industry within the current state licensing process. However, the focus of plant licensing should now be primarily on environmental hazards from industrial plants and secondarily on the traditional concern for worker protection.

The regulations list the types of plants subject to the new licensing requirements. The applicability of the regulations to other plants is extended on a case-by-case basis by listing 139 individual chemicals and three general chemical classes that are also subject to the new licensing procedures. The 139 chemicals on the list were selected because of their acute toxicity, carcinogenicity and generally deleterious behavior when released to the environment; the three chemical classes, for their flammability or explosiveness.[201]

For new plants, the "safety analysis" required of the plant operator under the new regulations is part of the initial licensing process. For existing plants, the safety analysis must be submitted to state licensing authorities by September 1, 1982. The safety measures specified in the regulations are a prerequisite for continued industrial self-regulation. Continued reliance on voluntary compliance is contingent upon cooperation of industry. The regulations also require that information considered necessary for local authorities and citizens to effectively cope with an industrial accident be provided. In general, each industrial installation must provide information upon plant construction about pressures and temperatures typically experienced during plant operation and on the production process (or processes), especially storage of specified "dangerous" chemical substances.[202]

It should be noted that these regulations are consistent with the EEC's requirements for industrial accident prevention as provided in the Seveso Directive,[203] the final language of which was approved at the June 1982 EEC environment ministers meeting. The environment ministers also approved a directive on harmonizing monitoring and control procedures for titanium oxide waste; and therefore, the German titanium oxide industry may have to further curtail its dumping of wastes in the North Sea. Finally, a new directive on air quality standards for lead should not affect West Germany as its limits for lead in motor fuel are already more stringent than the EEC's.[204]

## Notes

1. John J. Putman, "West Germany: Continuing Miracle," *National Geographic*, August 1977, pp. 151 and 155.

2. U.S. Department of Commerce, *The Effects of Pollution Abatement on International Trade* (Washington, D.C.: U.S. Government Printing Office, 1973), p. A-34.

3. German Press and Information Office, *Protection of the Environment and Conservation of Nature* (Bonn: Federal Republic of Germany, 1972), p. 1.

4. U.S. Department of Commerce, *Effects of Pollution Abatement on Trade*, p. A-34.

5. Phillip W. Quigg, "Global Report," *World Environment Newsletter*, August 14, 1973, p. 1.

6. German Press, *Protection of the Environment*, p. 1.

7. Otto Kimminich, "The Law of Waste Disposal in the Federal Republic of Germany: The Role of Private Contractors," *Environmental Policy and Law*, Vol. 1, No. 1 (1975), p. 28.

8. Phillip W. Quigg, "Umweltfreundliche Deutschland," *World Environment Newsletter*, July 3, 1973, p. 3.

9. R. J. van Schaik, "The Impact of the Economic Situation on Environmental Policies," *The OECD Observer*, No. 79 (January-February 1976), p. 25.

10. Robert E. Mims, et al., "Special Report: The Slow Investment Economy," *Business Week*, October 17, 1977, pp. 62-63. For an explanation of the "Concerted Action" program, see Dr. Otto Schlecht's "The Role of the Concerted Action Program in the Economic Policy of the Federal Republic of Germany," in *The German Economic Review*, Vol. 8, No. 3 (1970), pp. 255-60.

11. U.S. Department of Commerce, *Effects of Pollution Abatement on Trade*, p. A-35.

12. *International Environment Reporter* (Washington, D.C.: Bureau of National Affairs, Inc., 1978), p. 241:0101.

13. Kimminich, "The Law of Waste Disposal," p. 30.

14. *International Environment Reporter*, pp. 241:0101 and 0102.

15. U.S. Department of Commerce, *Effects of Pollution Abatement on Trade*, p. A-35.

16. *International Environmental Guide—1975* (Washington, D.C.: Bureau of National Affairs, Inc., 1975), p. 61:1501.

17. German Press, *Protection of the Environment*, p. 2.

18. Henry J. Gumpel, *World Tax Series: Taxation in the Federal Republic of Germany*, 2d ed. (Chicago: Commerce Clearing House, Inc., 1969), p. 318.

19. Quigg, "Umweltfreundliche Deutschland," p. 4.

20. *International Environmental Guide—1975*, pp. 61:1501-1502.

21. *International Environment Reporter*, pp. 241:0102 and 0103.

22. *International Environmental Guide—1975*, pp. 61:1501-1502.

23. *International Environment Reporter*, pp. 241:0102 and 0103.

24. Max Streibl, "Environmental Policy in a Federal State," *Environmental Policy and Law*, Vol. 1, No. 3 (1975), p. 139.

25. *International Environment Reporter—Current Report*, April 10, 1978, pp. 111-12.

26. Quigg, "Umweltfreundliche Deutschland," p. 3.

27. Juergen Kilius and Ernest C. Stiefel, *Business Operations in West Germany*, Tax Management Portfolio No. 174-2nd (Washington, D.C.: Bureau of National Affairs, Inc., 1971), p. A-15.

28. *International Environmental Guide—1975*, p. 61:1503.

29. *International Environment Reporter*, p. 241:0105.

30. *International Environment Reporter—Current Report*, January 10, 1978, p. 7.

31. G. Hartkopf, "Representative View for the Others," *Environmental Policy and Law*, Vol. 2, No. 2 (1976), p. 91.

32. *International Environment Reporter—Current Report*, April 10, 1978, p. 111.

33. *International Environment Reporter*, p. 241:0105.

34. *International Environment Reporter—Current Report*, January 10, 1978, p. 21.

35. OECD Environment Directorate, *Economic and Policy Instruments for Water Management in Germany* (Paris: Organization for Economic Cooperation & Development, 1976), pp. 4 and 5.

36. *International Environmental Guide—1975*, pp. 61:1502 and 1503.

37. OECD Environment Directorate, *Water Management in Germany*, pp. 12 and 13.

38. Allen V. Kneese and Charles L. Schultze, *Pollution, Prices and Public Policy* (Washington, D.C.: Brookings Institute, 1975), p. 98.

39. Ibid., p. 109.

40. Putman, "West Germany: Continuing Miracle," p. 1.

41. Kneese and Schultze, *Pollution, Prices and Public Policy*, p. 109.

42. *International Environment Reporter*, pp. 241:2301 and 2302.

43. Kneese and Schultze, *Pollution, Prices and Public Policy*, pp. 90-91.

44. William A. Irwin, *Charges on Effluents in the United States and Europe* (Washington, D.C.: Environmental Law Institute, 1974), pp. 23-24.

45. Frederick R. Anderson et al., *Environmental Improvement Through Economic Incentives* (Baltimore: Johns Hopkins University for Resources for the Future, 1979), p. 62.

46. Irwin, *Charges on Effluents*, p. 24.

47. Allen V. Kneese and Blair R. Bower, *Managing Water Quality: Economics, Technology, Institutions* (Baltimore: Johns Hopkins University for Resources for the Future, 1968), p. 251.

48. Anderson, *Environmental Improvement Through Economic Incentives*, pp. 62-63.

49. Irwin, *Charges on Effluents*, p. 48.

50. Ibid., p. 55.

51. OECD Environment Directorate, *Water Management in Germany*, pp. 7, 12, and 16.

52. Federal Waste Water Charges Act, Articles 1, 9, and 18.

53. Ibid., Article 13.

54. Ibid., Article 2, Paragraph 1.

55. Ibid., Article 2, Paragraphs 2 and 3.

56. Ibid., Article 3, Paragraph 1.

57. Ibid., Article 3, Paragraphs 2, 3, and 4.

58. Ibid., Articles 7 and 8.

59. Ibid., Article 11, Paragraph 1.

60. Ibid., Article 9, Paragraph 3.

61. OECD Environment Directorate, *Water Management Policies and Instruments* (Paris: Organization for Economic Cooperation and Development, 1977), p. 92.

62. OECD Environment Directorate, *Water Management in Germany*, pp. 16, 18, and 20.

63. Federal Waste Water Charges Act, Article 9, Paragraphs 2 and 3.

64. Ibid., Article 10.

65. OECD Environment Directorate, *Water Management in Germany*, pp. 20 and 21.

66. Federal Waste Water Charges Act, Article 9, Paragraph 6.

67. Ibid., Article 4.

68. Ibid., Article 9, Paragraph 5.

69. Ibid., Article 15.

70. Ibid., Article 14.

71. *International Environment Reporter—Current Report*, January 10, 1978, pp. 10-11.

72. U.S. Department of Commerce, *Effects of Pollution Abatement on Trade*, pp. A-35 and A-36.

73. *International Environment Reporter*, p. 241:0104.

74. *International Environment Reporter—Current Report*, January 10, 1978, p. 7.

75. *International Environmental Guide—1975*, p. 61:1502.

76. *International Environment Reporter*, pp. 241:0103 and 0104.

77. *International Environmental Guide—1975*, p. 61:1502.

78. *International Environment Reporter—Current Report*, January 10, 1978, p. 11.

79. *International Environment Reporter*, pp. 241:0103 and 0104.

80. *International Environmental Guide—1975*, p. 61:1502.

81. *International Environment Reporter*, pp. 241:0103 and 0104.

82. *International Environmental Guide—1975*, p. 61:1502.

83. *International Environment Reporter*, p. 241:0104.

84. *International Environmental Guide—1975*, p. 61:1502.

85. *International Environment Reporter*, p. 241:0104.

86. Quigg, "Umweltfreundliche Deutschland," pp. 3 and 4.

87. *International Environment Reporter—Current Report*, June 10, 1978, p. 166.

88. Ibid., July 10, 1978, p. 204.

89. Ibid., June 10, 1978, p. 175.

90. Kimminich, "The Law of Waste Disposal," pp. 28-29.

91. OECD Environment Directorate, *Waste Management in OECD Countries* (Paris: Organization for Economic Cooperation and Development, 1976), p. 59.

92. Henner Bornemann and Heimo Emminger, "Environmental Protection in Germany: Recycling Going Strong," *Naturopa*, No. 21 (1974), p. 4.

93. *International Environmental Guide—1975*, p. 61:1503.

94. OECD Environment Directorate, *Pollution Charges*, p. 64.

95. *International Environmental Guide—1975*, p. 61:1503.

96. U.S. Department of Commerce, *Effects of Pollution Abatement on Trade*, p. A-36.

97. Kimminich, "The Law of Waste Disposal," pp. 29-31.

98. Section 3(3) of the Federal Waste Disposal Act of June 7, 1972.

99. Kimminich, "The Law of Waste Disposal," p. 31.

100. OECD Environment Directorate, *Waste Management in OECD Countries*, p. 48.

101. Kimminich, "The Law of Waste Disposal," pp. 31-34.

102. *International Environmental Guide — 1975*, p. 61:1503.

103. U.S. Department of Commerce, *Effects of Pollution Abatement on Trade*, p. A-36.

104. Bornemann and Emminger, "Recycling Going Strong," p. 4.

105. Richard Grace and Jonathan Fisher, *Beverage Containers: Re-Use or Recycling* (Paris: Organization for Economic Cooperation and Development, 1978), pp. 45, 135, 145, and 146.

106. *International Environment Reporter — Current Report*, January 10, 1978, p. 21.

107. Patricia K. DeJoie, "Wastes Around the World," *Environment*, Vol. 19, No. 7 (October 1977), p. 35.

108. *International Environment Reporter*, p. 241:0105.

109. OECD Environment Directorate, *Waste Management in OECD Countries*, p. 28.

110. Ibid., pp. 47, 48, 59, 70, and 71.

111. DeJoie, "Wastes Around the World," p. 35.

112. *International Environment Reporter — Current Report*, February 10, 1978, p. 30.

113. Ibid., July 10, 1978, p. 216.

114. Ibid., January 10, 1978, pp. 7-8.

115. Phillip W. Quigg, "EEC: The European Communities Constitute the One International Organization in the Environmental Field with (Baby) Teeth," *World Environment Newsletter*, November 7, 1972, p. 3.

116. OECD Environment Committee, *The Polluter Pays Principle* (Paris: Organization for Economic Cooperation and Development, 1975), pp. 74-75.

117. Kilius and Stiefel, *Business Operations in West Germany*, pp. A-30 to A-32.

118. Gumpel, *World Tax Series*, pp. 949-61.

119. Juergen Kilius, *Business Operations in West Germany*, T. M. 174-4th (Washington, D.C.: Tax Management, Inc., 1978), p. A-22.

120. Juergen Kilius and Jakob Strobl, "Trends — Recent Developments in International Taxation: Germany," *Tax Management International Journal*, December 1977, pp. 36-37.

121. Kilius, *Business Operations in West Germany*, p. A-22.

122. Kilius and Stiefel, *Business Operations in West Germany*, p. A-30.

123. Hermann Soell, *Beitrage Zur Umweltgestaltung: Depreciation Allowances or Subsidies* (Berlin: Verlag, 1975), pp. 24-45.

124. Gumpel, *World Tax Series*, pp. 424 and 486.

125. *Doing Business in Germany* (New York: Price Waterhouse & Co., 1975), p. 114.

126. OECD Environment Directorate, *Waste Management in OECD Countries*, p. 28.

127. U.S. Department of Commerce, *Effects of Pollution Abatement on Trade*, p. A-37.

128. Bornemann and Emminger, "Recycling Going Strong," p. 4.

129. Irwin, *Charges on Effluents*, p. 23.

130. OECD Environment Directorate, *Water Management in Germany*, pp. 6-12.

131. Streibl, "Environmental Policy in a Federal State," p. 140.

132. Kimminich, "The Law of Waste Disposal," p. 31.

133. Streibl, "Environmental Policy in a Federal State," p. 140.

134. OECD Environment Directorate, *Waste Management in OECD Countries*, p. 71.

135. OECD Environment Directorate, *Water Management in Germany*, pp. 10, 11, and 41-44.

136. *International Environment Reporter — Current Report*, January 10, 1978, pp. 10-11.

137. OECD Economic and Development Review Committee, *OECD Economic Survey: Germany* (Paris: Organization for Economic Cooperation and Development, 1975), pp. 37-40.

138. "Germany: Indicators Hint the Boom Is on Its Way," *World Business Weekly*, November 6, 1978, pp. 27-28.

139. Waste Water Charges Act, Article 13.

140. OECD Environment Directorate, *Water Management Policies and Instruments*, p. 96.
141. Waste Water Charges Act, Article 10.
142. U.S. Department of Commerce, *Effects of Pollution Abatement on Trade*, p. A-37.
143. Soell, *Depreciation Allowances or Subsidies*, pp. 24-45.
144. *International Environment Reporter—Current Report*, August 13, 1980, p. 318.
145. Ibid.
146. Ibid., February 13, 1980, pp. 48-49.
147. Ibid., January 9, 1980, pp. 13-14.
148. Ibid., March 12, 1980, p. 92.
149. Ibid., May 14, 1980, p. 185.
150. Ibid., February 13, 1980, p. 48.
151. Ibid., August 13, 1980, p. 318.
152. Ibid., June 10, 1981, p. 8.
153. Ibid., July 8, 1981, pp. 922-23.
154. Ibid., November 11, 1981, pp. 1078-80.
155. Ibid., September 12, 1979, pp. 851-52.
156. Ibid., May 14, 1980, p. 177, and April 9, 1980, pp. 131-132.
157. Ibid., April 11, 1979, pp. 624-25.
158. Ibid., August 13, 1980, p. 336.
159. Ibid.
160. Ibid., December 10, 1980, pp. 539-41.
161. Ibid., September 10, 1980, pp. pp. 411-12.
162. Ibid., December 10, 1980, pp. 539-41.
163. Ibid., March 11, 1981, pp. 701-2.
164. Ibid., December 10, 1980, p. 541.
165. Ibid., November 19, 1978, p. 386.
166. Ibid., September 12, 1979, pp. 851-52.
167. Ibid., December 10, 1980, p. 541.
168. Ibid., April 11, 1979, pp. 624-25.
169. Ibid., December 10, 1980, p. 541.
170. Ibid., July 8, 1981, p. 929.
171. Ibid.
172. Ibid., December 10, 1980, p. 541.
173. Ibid., December 9, 1981, p. 1190.
174. Ibid., August 13, 1980, p. 318.
175. Ibid., September 10, 1980, p. 414.
176. Ibid., July 8, 1981, pp. 923 and 929; and June 10, 1981, pp. 894-95.
177. Ibid., February 10, 1982, p. 78-79.
178. Ibid., October 14, 1981, pp. 1043-1045.
179. Ibid., September 9, 1981, p. 1034.
180. Ibid., July 8, 1981, p. 923.
181. Ibid., September 9, 1981, p. 1021.
182. Ibid., August 12, 1981, pp. 969-70.
183. Ibid., October 14, 1981, pp. 1067-68.
184. Ibid., August 13, 1980, pp. 340-41.
185. Ibid., October 14, 1981, p. 1068.
186. Ibid., August 13, 1980, pp. 341-42, and November 12, 1980, pp. 505-6.
187. John W. Wilson and Robert F. Ingersoll, "Growing German Economic Strains Threaten West," *Business Week*, November 24, 1980, pp. 64-72.
188. *International Environment Reporter—Current Report*, September 9, 1981, pp. 1012-13.
189. Ibid., February 10, 1979, pp. 509-10.
190. Ibid., September 9, 1981, pp. 1012-13.

191. Ibid., August 12, 1981, pp. 964-65.

192. Bradley Graham, "Schmidt's W. German Governing Coalition Avoids Collapse by Agreeing on Budget Outline for 1983," *Washington Post*, July 1, 1982. Unfortunately, the Free Democratic Party switched its support from the Social Democratic Party to the Christian Democratic Party on October 1, 1982, thereby allowing the election of Christian Democrat Helmut Kohl as the new Chancellor until the next national election in 1983. Thus, the new government's 1983 budget will be austere to reflect its more conservative political philosophy.

193. "Green at the Gills," *The Economist*, June 19, 1982, pp. 60-61.

194. "Flirting With Power in Germany," *The Economist*, July 10, 1982, pp. 40-41.

195. *International Environment Report—Current Reporter*, June 9, 1982, pp. 219-20.

196. Ibid., April 14, 1982, pp. 140-41.

197. "West Germany's Corporate Alchemists," *The Economist*, June 5, 1982, pp. 82-83.

198. Reinhard Spilker, "A Black Hole in the North Sea for Toxic Wastes," *Ambio*, Vol. 11, No. 1 (1982), p. 57.

199. *International Environment Report—Current Reporter*, June 9, 1982, pp. 220-21.

200. Ibid., May 12, 1982, p. 181.

201. Ibid., March 10, 1982, pp. 100-101.

202. Ibid., June 9, 1982, p. 274.

203. Ibid., March 10, 1982, pp. 100-101.

204. Ibid., July 14, 1982, p. 277; and "June in the EEC," *The Economist*, July 10, 1982, p. 46.

# 6

# Sweden

Probably more than any other country in Western Europe, Sweden can attribute its pollution problems to urbanization and industrialization.[1] Sweden's eight and one-quarter million inhabitants are very unevenly distributed over a country that is almost as large as France. Almost 60 percent of this Nordic country is covered by forests and 95,000 lakes.[2] More than 80 percent of the population is urban,[3] and over one-third lives in the three major metropolitan areas of Stockholm, Gotenberg, and Malmo. In addition, the Swedes have enjoyed Europe's highest standard of living for many years.

The social costs of continuously increasing Sweden's high standard of living can be attributed primarily to the expansion of industrial output, greater use of automobiles, increased use of packaging, the increasing demand for energy (primarily electric power and oil), and the construction of urban housing and urban transportation systems. For example, these activities generate considerable pollution of the airshed through the emission of particulate matter, corrosive and toxic gases, and unpleasant odors. In addition, climatic conditions in Northern Europe may effect the importation of as much as one-half of Sweden's air pollution from Germany, Denmark, the Benelux countries, and the United Kingdom, including the sulfur emissions that contribute to the acid rain phenomenon.

The largest single industrial source of water pollution is also one of Sweden's most important industries—the pulp and paper industry. While this industry accounts for most of the organic pollutants in Sweden's surface waters, the forest products and mining (iron ore) industries contribute most of the suspended solids to streams, rivers, and lakes. Another source of harmful effluent is the growing chemical industry. However, the largest source of water pollution is probably still the poorly treated and untreated wastes of the municipal sewage systems along with the wastes transported by urban run-off.[4]

The magnitude of the solid waste pollution problem has only been fully realized in the past few years,[5] even though Sweden has been a full-fledged member of the purchase, consume, and throwaway society for two decades.[6] Most of the solid waste results from the growing consumption of both packaged goods and short-lived products in a society with Sweden's high standard of living. Household, commercial, and office refuse generates over three-quarters of the volume of solid waste, with industry creating the balance.[7] In Stockholm, Sweden's largest urban area, two-thirds of the household refuse consists of paper and paper products.[8]

Environmental quality first became a substantive political issue in Sweden during the mid-1950s. During the 1960s, public concern for environmental protection grew steadily to become one of the country's leading political issues along with economic growth, social welfare,[9] energy (especially nuclear power), tax reform, and government decentralization. By 1969, the consensus on environmental protection had resulted in the Environmental Protection Act. The overwhelming public support for environmental protection in 1969 is apparent from the results of a public opinion survey taken that year[10] and reproduced in Table 6-1. Broad support for pollution control measures was easily sustained for several years thereafter, aided both by Sweden's hosting of the United Nations Conference on the Human Environment in June 1972 in Stockholm and by a unique, government-sponsored environmental education program wherein

> Under the auspices of the Ministry of Education, 250,000 Swedes were given instructions on the technical and legal aspects of pollution control. Of these, 10,000 went on to take two-week government-sponsored training courses. Finally, about 1,000 were chosen as pollution control officers and instructed to organize anti-pollution campaigns at grassroots level.[11]

## Pollution Abatement Programs

The Environmental Protection Act (EPA) of 1969 is the first environmental law that has had a significant impact on environmental quality in Sweden. Prior to the EPA, various other laws were enacted in an attempt to control specific types of environmental degradation, including the Water Act of 1918, Public Health Acts (since 1874), and the Nature Conservancy Act of 1964.[12] The EPA defines pollution indirectly through a very broad explanation of the law's applicability to "Whosoever engages in activities of a harmful nature shall adopt whatever protective measures that may be

### Table 6-1
### 1969 Swedish Public Opinion Poll
### on Environmental Issues
### (in percentages)

| Issue | For | Against | Don't Know | No Answer |
|---|---|---|---|---|
| Put charges on polluting industries | 84 | 8 | 8 | 0 |
| Prohibition against nonreturnable bottles | 75 | 9 | 12 | 4 |
| Higher local taxes to fight water pollution | 69 | 21 | 10 | 0 |
| Cut down GNP growth rate to save the environment | 54 | 17 | 25 | 4 |
| Prohibition against synthetic detergents | 45 | 29 | 23 | 4 |

*SOURCE:* Lennart Lundqvist, "Sweden's Environmental Policy," *Ambio*, Vol. 1, No. 3 (June 1972), p. 92.

required to prevent or remedy the situation,"[13] insofar as the activities originate from the continuous use of real property.[14]

The EPA contains comprehensive provisions for control of air, water, and noise pollution. The two government agencies responsible for administering the licensing provisions of the law are the National Environment Protection Board (NEPB) and the Concession Board for the Protection of the Environment, more commonly known as the National Franchise Board (NFB). These two agencies divide the authority to grant to governmental, industrial, commercial, cooperative, and municipal institutions licenses that allow these institutions to generate permissible levels of pollution.[15] Thus, the EPA covers the air, water, and solid waste pollution of a firm's industrial facilities, but the law does not apply to any damages that may result from its industrial products.[16]

The EPA applies to any continuous use of real property that might cause a disturbance, an interference, or a nuisance in the environment, including discharges of gas, wastewater, or solid material from buildings, facilities, or land into lakes, water courses, or other water areas, as well as emissions of harmful gases and noise into the airshed. Any natural or legal person (entity) engaging in such activities is required to adopt whatever protective measures are necessary to prevent damage or remedy injury. This duty is not absolute, but it is determined through an evaluation of the characteristics of the affected area, the severity of the effects of the nuisance, the technical feasibility of alternative remedies, and an environmental cost-economic benefit analysis of the economic activity under scrutiny.

The EPA requires an entity to apply for an operating license before it constructs or alters any facilities that are liable to generate environmental disturbances.[17] This licensing program applies to industrial plants and generally to any other facilities that discharge pollutants into public waters or municipal treatment facilities, as well as into the airshed and on public land. Both the NEPB and NFB issue operating licenses on a plant-by-plant basis. As has been true of most Swedish environmental legislation in recent years, the EPA designates very broadly the types of pollution to be controlled, but it does not establish rigid, compulsory standards for each type of pollutant.[18]

The construction of the environmental protection legislation is not unique since the recent trend has been towards greater use of general clauses in legislation enacted by the Swedish Parliament (*Riksdag*), especially for such technically complex fields as pollution control and energy conservation. Thus, the EPA consists of very broad general rules or principles on the scope, intent, and compliance requirements for national environmental protection.

General rules of law or legal principles enacted by Parliament are initially interpreted in greater detail through ordinances issued at the ministry level, such as the Environmental Protection Ordinance (EPO) promulgated by the Ministry of Agriculture shortly after the EPA became law in 1969. The clarifying interpretations of a Ministry are more precisely delineated through administrative guidelines or regulations developed by regulatory agencies like the NEPB through consultation with the interest groups most affected thereby. Periodically, an agency's regulations are reviewed in anticipation of amending them where they have been ineffective in attaining objectives or where technological developments should be reflected therein.

In the Swedish political system, there is a separation of policymaking ministries and policy-implementing (independent) agencies. Ministries (Agriculture) are relatively small, policymaking institutions that propose legislation to Parliament and issue clarifying ordinances like the Environmental Protection Ordinance. Even though agencies (NEPB) may have informal, informative discussions with ministerial representatives, the agencies generally formalize implementative interpretations of laws and statutes through boards of directors and advisory boards or committees, with heavy representation from "interest" groups (especially from industrial polluters).[19]

The NEPB, which was established in 1967 in anticipation of EPA, has been given very broad authority in environmental policy matters. However, several other national government agencies have specific responsibilities for environmental protection,[20] such as the National Board of Urban Planning and the Social Welfare Board. There is also an Environmental Advisory Committee made up of members of government and representatives of interest groups.[21]

NEPB's institutional structure includes a board of directors, a director-general to administer the daily operations of eight line and seven staff units (or departments), as well as the five advisory boards. Consensus and close cooperation with industry are achieved through industrial representation on NEPB's board of directors, its advisory boards, and its "working committees." Agricultural, labor, local government, and environmental representatives also serve on the board of directors and advisory boards. However, since their inception in anticipation of the EPA, the "working committees" have been dominated by industrial association and industrial firm representatives. The objective of the NEPB in establishing the "working committees" was to investigate the efficacy of implementing a system of emission or air pollution standards by cooperating with the polluters.

The NEPB policymakers rationalized their consensus tactics by arguing that the economic and technical information required to establish emission standards could most easily be acquired through cooperation with industrial polluters who could be assured that a reasonable set of standards would take effect upon enactment of the EPA. The NEPB representatives contributed information on existing foreign emission standards to the data and information on different pollution control technologies and the economic implications of such technologies that was provided by industrial representatives.[22]

The administration of and planning for most of the environmental protection measures explained below are the responsibility of the NEPB and the County (Regional) administration officials for the NEPB. As provided in the EPA, the NEPB and the NFB share the responsibility for issuing operating licenses to those governmental, industrial, commercial, cooperative, and municipal installations obligated to apply for a pollution permit before construction or material alteration of their facilities. The scope of the NEPB's authority in environmental protection matters has continually expanded since the agency's inception in 1967 so that it is now responsible for air pollution abatement, water pollution control, noise abatement, nature conservation, recreation, coordination of environmental research, control of hazardous products (shared with the Products Control Board and the National Board of Industrial Safety), and the monitoring of all discharges— emissions, effluent and solid and hazardous wastes—into the environment.[23] As explained below, local public health officials have traditionally been charged with supervising municipal sanitation through the provisions of the public health laws. However, NEPB's authority is much broader than that given to public health authorities.

Compulsory advance examination through NFB permits and NEPB exemptions and through notification of the County administration causes most questions as to the detrimental effects of significant polluting activities to be brought to the attention of the appropriate officials without their having to initiate investigative and supervisory action. The NEPB is responsible for supervision at the national level and it also coordinates local supervision through its control of the twenty-three County administrations that carry out such supervision.[24]

The EPO specifies procedural rules for how the NFB, NEPB and County (Regional) Administration share authority for implementing and enforcing the general requirements of the EPA.[25] The EPO enumerates the 38 types of industrial and commercial facilities, as well as wastewater treatment plants, that may not be erected or substantially altered without receiving an operating license or permit from either the NFB or the NEPB. Mines, steel works, pulp and paper mills, refineries, and power plants are included in the list.[26] In addition, the Ordinance specifies 25 other types of facilities that must apply to or notify the County Administration before construction or substantial alteration.[27]

### Licensing of Facilities for Environmental Protection

The NFB normally issues permits, but the EPA provides an alternative for industrial, commercial, and public facilities that are required to obtain an operating license because of their potential environmental effects. Owners of such proposed facilities may apply to the NEPB for an "exemption" from their duty to secure a permit from the NFB.

The NFB's procedural rules provide for an adversary process, which is similar to a court of law, in evaluating the permissibility of polluting activities. The NFB can inspect sites and hold public hearings as part of its deliberations. When the board grants a permit, it entitles the recipient to legal protection from the subsequent imposition of more stringent pollution standards by statutory or administrative lawmakers for a maximum of ten years, provided the polluter complies with the conditions established in the NFB's decision.[28]

The membership of the NFB includes a Chairperson, an environmental technologist, an industrial representative, and a person competent in conservation and environmental quality matters. The Chairperson is a lawyer with judicial experience. In cases involving municipal affairs, the industrial representative is replaced by a person with local government experience. In general, the issues brought before the Franchise Board are of much greater importance than those brought before the NEPB, which has been given the authority to grant an "exemption" to any entity required to apply for a permit under the EPA. The major industrial and municipal cases are usually decided by the NFB.[29]

About two-thirds of the environmental protection cases are resolved through the NEFB's "exemption" procedure. The Exemption Unit, which is one of the fifteen units and departments of the NEPB, issues "exemptions" from the duty to apply for a permit. The exemption is not a dispensation from the duty to take all precautionary measures that are "economically feasible" and "technically practicable" to prevent pollution. The exemption procedure is a negotiative process wherein the NEPB and the applicant agree on the conditions or precautionary measures that the applicant must maintain to continue or start operations. Unlike the NFB permit, the NEPB exemption is not legally binding if its conditions—specific precautionary measures—are met, but the exemption procedure is still a very attractive alternative for industrial and commercial operations because it is much less time-consuming and difficult to obtain than an application for a legally binding permit.[30]

Inasmuch as there is a high degree of consensus in both the legislative and the administrative branches of government in Sweden, current social issues, such as the environmental protection issue, are incorporated into the political process with great facility. Adversary processes in Parliament, in local government, and in ministries and agencies are infrequent.[31] The NFB's permit issuance activity is the only example of an adversary process within environmental protection administration wherein a public hearing provides for the participation of opposing interest groups in a proceeding that closely resembles a court of law. In contrast, the NEPB's exemption process is marked by the close cooperation of the regulator and the regulatee or polluter, especially in respect to what precautionary measures are "economically feasible" and "technically practicable."[32]

Facilities that existed when the emission standards and permit procedures authorized by the EPA were promulgated created special problems for the NFB, the NEPB, and its regional representatives in the twenty-three County administrations. In principle, establishments existing when the EPA became effective on July 1, 1969, were required to install pollution control devices and to take other precautions necessary to counteract water, air, and noise pollution. In practice, existing establishments have received special dispensation in that implementation of "instructions" on purification installations and operational "adjustments" have been spread over a reasonable period of time. The "instructions" and "adjustments" were limited by the costs that could be reasonably borne by the owners of the existing facilities, and there was a subsidy program. In spite of the installation of pollution control devices, a facility could still be closed down by the government (King's Council) if pollution therefrom was so significant that continued operation was deemed indefensible.

The interim regulations for the EPA provided the NFB with the absolute power to permit older facilities whose operations had begun before July 1, 1969, and had continued without increased or new pollution to operate indefinitely as long as the establishment or its operations were not materially altered. Although the NEPB and its regional administrators have the authority to intervene in the operations of both existing facilities and altered or planned establishments and even though other interested persons can file suit in the applicable real estate court, the right to request examination by the NFB (for permission to operate) effectively discouraged both unanticipated intervention by the NEPB and suits filed in a real estate court. Finally, where existing facilities were operating with permission granted by the Water Rights Court, the EPA does not apply.[33] However, with regard to all post-June 30, 1969, action deemed to be construction in water under provisions of the 1964 amendments to the Water Act, there will be some duplication of examination procedures. A facility that anticipates both construction in water—including water intakes, wastewater conduits, and docks—and the disposal of wastes will be judged first by the NFB or NEPB in accordance with the provisions of the EPA and other environmental protection legislation. Then the Water Rights Court will judge the facility in accordance with the provisions of the Water Act.[34]

## Enforcement of the Environmental Protection Law

Initially, the supervisory authorities must attempt to effect voluntary compliance with environmental protection law in respect to facilities built both before and after the EPA became effective on July 1, 1969.[35] Several means of coercion are available if a legal or natural person fails to comply with the law voluntarily. If a permit has not been granted, the NEPB can call upon the NFB to prohibit an activity or require precautionary measures. The County administration can issue injunctions and prohibitions for notification matters, but with exceptions for those situations where a permit for an activity has been issued. If a permit holder disregards the terms and conditions of its license, the County administration can direct that matters be corrected at the permit holder's expense, or direct him to correct them himself. A penalty can also be stipulated in connection with any injunction.[36] Finally, national and local environmental officials have the authority to gain access to and to carry out an investigation of polluting facilities and their surroundings. In some circumstances, the owner of the facilities may be obligated to give supervisory officials whatever information the officials may require for their investigation.

The penalty for failing to comply with the EPA requirement that there be an application for a permit or that there be notification can be imprisonment for a maximum of one year or a fine. The same penalty is applicable in cases where a person disregards the terms and conditions stipulated by the NFB, NEPB, or a County administration in a permit matter. Whenever the question of legal responsibility arises, the public prosecutor can request an ordinary court of law to resolve the issue. In addition to any penalties, the County administration can secure an injunction against or prohibit negligent activity until the situation is rectified.

As explained above, persons and groups affected by or interested in a decision on permissibility to be taken by the NFB, NEPB, or County administration have little opportunity to be heard because the Franchise Board conducts public hearings while the Environmental Protection Board and its regional representatives cooperate with polluting establishments. In addition, with regard to permit decisions and decisions involving prohibition or injunction on a notification matter, a party can appeal a government official's decision if the party feels that the authority assessing reasonableness failed to reflect his (or its) interests. The right of appeal does not exist in the case of a decision on an examination by the NEPB or County administration because such decisions lack the binding force of law. In cases where a permit for a polluting activity does not exist or where an examination is not relevant, an interested party can initiate legal action in special real estate courts to effect prohibition or control measures.

The EPA also provides compensation for damages resulting from a polluting establishment. Claims for compensation can be handled through the real estate courts[37] as follows:

> The main principle is that anybody who causes a nuisance by polluting activity shall provide compensation for such nuisance. Compensation is payable in the first instance if the nuisance is caused by negligence. If negligence cannot be proved, compensation is payable only if the nuisance is at all substantial and should not reasonably be tolerated in view of the circumstances in the locality or . . . its general occurrence in comparable circumstances.[38]

In addition to the provisions of the EPA which deal with the location of industrial or similar polluting activity, the Building Act has provided since January 1, 1973, for governmental review of any proposed location for such activity. Thus, when the location decision for any significant polluting activity is vitally important to the management of Sweden's land and water resources or of its energy and cellulose materials for forest industries, the central government will scrutinize the location choice.[39]

National Environmental Planning

Planning for and administration of the national environment is, therefore, also the responsibility of the National Board of Urban Planning (NBUP), as provided for in the Building Act. In principle, the NBUP has been primarily responsible for commencing large-scale planning for the use of water resources, whereas, land use planning has traditionally been the responsibility of local government. Municipalities, however, are limited in their planning authority by "activity" and "geographical" guidelines adopted by the central government as part of national physical planning. County administrations have actually been given the task of coordinating municipal planning objectives and the central government's guidelines in order to effect a balance that promotes favorable development from both local and national viewpoints. Even though the County administration is a regional branch of the central administration, its governing body (County council) includes both municipal experts and spokesmen for the local commercial and industrial community.[40]

Traditionally, the Swedish national government has engaged in a sectoral-type of national planning wherein a central government agency (and its twenty-three county offices) responsible for a specific sector of public administration performs short-term planning for that sector.[41] Those few central government agencies that are responsible for coordinating the plans in all, or most, sectors for from one- to five-year time periods are said to be engaging in cross-sectoral or "intersectoral" national planning. Thus, the preparation of the central government's budget by the Ministry of Finance is an example of such intersectoral planning. More recently, national physical (land and water use) planning has been expanded so that it rivals economic planning and regional development planning as one of the most significant aspects of intersectoral planning in Sweden.[42]

Though traditionally considered primarily a local matter, land and water use planning has increasingly become an issue of national import because conflicting claims for the utilization of land and water resources often occur in respect to unique natural assets, such as wetlands, coastal areas, forested areas, lakes, rivers, mountains, and wilderness areas. Those wishing to exploit Sweden's natural resources have too often prevailed over those parties interested either in keeping an area free of exploitation or in seeing an area exploited in a manner that is more compatible with the natural environment. In general, the conflict is between urban-industrial development groups and the recreational-conservationist interest groups.[43]

The crisis that accentuated the need for national physical planning occurred in the mid-1960s when the conflict caused by the location of a pulp and paper mill (industry) on a previously undeveloped section (wilderness) of the west coast of Sweden became a national issue.[44] Natural assets not only determine the physical environment of man, but they also supply necessary factors of production. Consequently, the long-term welfare of a society depends on optimal use of natural resources for both economic growth and environmental health. It follows then that the objective of national physical planning should be to create a pattern of economic development and growth that is compatible with environmental quality for both present and future generations. In other words, both regional socioeconomic development and environmental protection considerations are implicit in the national management of land and water resources.

The implementation of the national physical planning system was carried out in several stages and began in 1967 with preliminary investigations and methods studies. A second preparatory stage during 1969, 1970, and 1971 included (1) a comprehensive investigation and analysis of the probable future claims on available land and water resources by various interest groups; and (2) a proposal for the organization and legislation of a system of continuous national planning for land and water use. Most of the work during this second stage was carried out within the Ministry of Physical Planning and Local Government, which was given the responsibility for such matters in 1969, with participation by the County Administrative Boards and other regional bodies and in consultation with the NEPB and several groups with an interest in recreation and conservation.[45]

During early 1972, the proposal for national physical planning was disseminated to many central government authorities and boards, all municipalities, the labor unions, and many other types of organizations for their comment. After all interested government officials and interest groups had been given sufficient time for consideration and comment on the proposal, the central government introduced a revised version of its originally proposed legislation in November 1972. Parliament enacted the proposed amendments to the Building Act in December 1972, to take effect on January 1, 1973.[46]

Inasmuch as the "activity" and "geographical" guidelines for national physical planning enacted by Parliament in December 1972 amended the existing provisions for regional and local planning, the primary responsibility and authority for physical planning remains with the local levels of government where there is both comprehensive general planning (master plans) for municipalities and specific planning (district or subdivision plans) within the municipalities. Before the December 1972 amendments to the 1947 Building Act, only a few citizens had had the opportunity to vote on the adoption of a local master plan because most municipalities were reluctant to draft such comprehensive plans when there was no legal requirement for all local governments to commit themselves for the long term.[47] By the July 1974 deadline, 272 out of a total of 278 municipalities were committed to implementing the national guidelines in their drafting or revision of master plans,[48] even though the 1972 amendments do not mandate master planning by local government. The amendments do prescribe that the central government has the authority to require the development of a master plan for a specific area and to provide in that master plan for the "national interests" as prescribed in the guidelines for national physical planning.

Implementation of the guidelines laid down by Parliament has been proceeding in the context of municipal and regional planning in two stages. The first stage, which occurred during 1973 and early 1974, was a *programming* stage wherein each municipality drafted a program detailing the measures necessitated by the "activity" and "geographical" guidelines established for the national management of land and water resources. Thus, each municipality had to rationalize all its specific plans, its proposed plans, and any comprehensive plan with the national guidelines by July 1974. Both County Administrative Boards and central government administrative officials scrutinized and commented on these municipal programs.

In early 1975, the Swedish government[49] (cabinet or King's Council[50]) adopted a series of resolutions, one for each county and each based on programs drawn up by the municipalities within each county. These resolutions delineated a timetable for the second stage of implementing the guidelines, including a provision for local government to report to County Administrative Boards in the spring of 1977 as to the measures the local government had taken during the preceding two-year period. Hence, the second stage has been a *planning* stage in which the municipalities have been concerned with general physical (or master) planning so as to develop a comprehensive, but crude, outline of future land use within their jurisdictions. In addition, County Administrative Boards have been charged with the scheduling for implementation of protection for important natural areas, as well as for areas of historical interest.[51]

The "activity" guidelines which Parliament enacted for national land use planning in December 1972 describe how selected types of activity, such as farming and fishing, ought to be integrated with physical planning at the local and regional levels. The "geographical" guidelines delineate how selected natural resources, such as wilderness and water, are to be managed to preserve their utility for both economic welfare and nature conservation.[52] Thus, nature conservancy and outdoor recreation should be encouraged wherever "either unspoiled, untouched, unique, or easily accessible areas exist."[53] The location of heavily polluting industry in coastal areas is restricted to specified sections of the west, southeast, and northeast coast. In general, heavily polluting industry is restricted to locations that have already experienced some industrial development.

### National Environmental Planning and the Licensing of Facilities for Environmental Protection

The 1975 amendments to the Building Act established coordination between the EPA and the Building Act regarding the permissibility of a polluting facility's location. Thus, the government normally consults with the NFB before granting permission for a location. The NFB's participation is procedurally similar to its decision-making process in environmental protection matters, with both public hearings and site inspections being available for the board's use.

The central government's decision on the permissibility of a proposed industrial location is binding both on subsequent planning under the Building Act and on any deliberations requisite to the EPA. Thus, the NFB can neither sanction an activity where the government has not granted permission nor refuse to sanction an activity deemed permissible. However, responsibility for prescribing the conditions under which the polluting activity can be conducted remains with the NFB pursuant to the provisions of the EPA and EPO. Moreover, the government can grant permission for a proposed location contingent upon an activity's successful permissibility application to the NFB within a specified time period.

Pursuant·to the 1972 and 1975 amendments to the Building Act, the location of those industries of major importance to the management of land, water resources, forests, and energy is initially reviewed by the central government. Following its consultation with those agencies responsible for and groups interested in labor market policy, regional policy, industrial policy, environmental policy (NFB and NEPB), and national physical planning, the King's Council makes an overall assessment of the proposed industrial location before deciding to permit the construction, alteration, or expansion of an industrial facility. Finally, both the County Administrative Board and the municipality in which the heavy industry proposes to locate are consulted by the government, so that the municipality can formally approve or disapprove (veto) the proposed location. If the municipality disapproves, the government will not grant its permission for the industrial location.[54]

The central government's permission is mandatory only with regard to the location of the following new industrial facilities:

> iron and steelworks, non-ferrous metal works and ferro-alloy work; saw mills with an annual production capacity of at least 5,000 m$^3$ sawed timber products, pulp mills, paper mills and plants producing fibreboard, chipboard or plywood; plants for the manufacture of petrochemical products; oil refineries; nuclear power stations; plants for the reprocessing of nuclear fuels; fossil fuel power plants with a power input exceeding 500MW; plants for the production of fertilizer products; and, cement works.[55]

As a result of 1976 amendments to the 1918 Water Act, the government also has the power to make the final decision as to the permissibility of major hydroelectric power projects. Finally, the government must pass special resolutions in regard to the following establishments to reserve its right to assess the permissibility of their location pursuant to the provisions of the Building Act: (1) Material expansion or alteration of the aforementioned activities which if they were new establishments would automatically be subject to location assessment; and (2) establishment, material expansion, or significant alteration of any activity, which although not specifically subject to location assessment, are of principal importance to the management of land use, water resources, forests, and energy.[56]

Figure 6-1 outlines the national (Central), regional (County), and local government agencies with responsibility for one or more aspects of environmental administration and/or policy.

### Water Pollution Control

There is no system of effluent standards for stationary sources of water pollution because pre-EPA laws, such as the 1918 Water Code Preventing Water Pollution, the 1956 Law on Supervision of Watercourses, Lakes, and Other Bodies of Water, and the 1956 Act Concerning Measures Against Water Pollution from Vessels, have provided for a systematic examination of wastewater and water-polluting facilities at the local and regional levels of government. However, the EPA has had a substantial effect on water pollution control by providing that the terms and conditions for receiving a permit or exemption now depend on the most efficient technology, that is, "economically feasible" and "technically practicable" precautionary measures. This provision for a continual tightening of terms and conditions of permissible water pollution in concert with improvements in abatement technology, and the traditional concern for water pollution control have rendered source effluent standards much less of a necessity. Moreover, effluent guidelines are inherently less useful in controlling water pollution than they are in the abatement of air pollution because of fundamental differences in the assimilative capacity of water and air and because of the importance of purely local factors in coping with noxious effluent.[57]

### Air Pollution Control

The NEPB's Air Quality Department originally planned to use the information gained from the investigations of the "working committees" to set up an air pollution control system in which both ambient air quality and source emission standards would be adopted.[58] The proposed system would have combined "a 'best practicable means' approach, based on technical and economic premises and applied to the source of emissions, with an 'air resources management' approach, based on considerations of the effects of pollutants on public health."[59] The system actually adopted was based solely on source emission guidelines or standards. The original guidelines took effect in 1970 and have already been revised twice (in 1973 and 1977).[60]

When the source emission standards were recommended in June 1969, ambient air quality standards were not proposed at the same time because the investigations by "working committees" generated incomplete information on the effects of air pollution on the environment and especially on public health. The recommended emission standards were not finalized for several months to allow sufficient time for thoughtful commentary by interest groups. When the final guidelines were released in December 1969, about one-seventh of the standards had been modified in favor of industrial groups' complaints that certain proposed guidelines were so stringent that unreasonable economic costs would be incurred by firms forced to comply therewith. In the case of a few standards, the final guidelines were more stringent than those originally proposed.[61]

These source emission standards are compulsory only for those facilities with operating licenses either by permit from the NFB or by exemption from the NEPB. Those already operating on "exemption" status must comply with any subsequent revisions in emission standards within a specified period after the revisions take effect for new applications for both exemptions and permits. The air pollution standards in effect at the time a facility receives a permit from the NFB apply under all operating conditions and for the life of the installation. Standards are usually expressed in and compliance measured by determining a monthly arithmetical mean for total emissions.[62]

The original emission standards or guidelines have been through two major revisions since their December 1969 promulgation. The first major revision, which occurred in August 1973, included new guidelines for industries in which insufficient knowledge existed in 1969 to establish standards.[63] The 1973 revisions not only reflected new developments in the "best available technology," but for selected industries also continued different guidelines for new and old facilities. The most recent revisions occurred in the late 1970s.[64]

The source emission standards have been designed to provide guidelines for the NFB and NEPB in making decisions on permits and exemptions, respectively. Therefore, they reflect emission limits for both existing plants and new installations. Older facilities have benefited not only from less stringent guidelines, but also from a deferral of the effective date to July 1, 1974. As noted above, the standards stipulate discharge limits in terms of the average value (arithmetical mean) of total emissions during a prolonged period—usually one month. Therefore, maximum permissible emissions are an average of normal discharges with higher discharges experienced during abnormal operating circumstances.[65]

In addition to the more general source emission guidelines for air pollution from stationary sources described above, the severity of the sulfur dioxide ($SO_2$) pollution problem and the resulting acidification of Sweden's air, property, land, and water have necessitated specific standards for the sulfur content of fuel oil burned in industry, by utilities, and in homes. Thus, there is an implicit air quality standard for $SO_2$ because the allowable amount of sulfur in fuel oil has been reduced several times since the Royal Ordinance Concerning the Restriction of Sulphur Content in Fuel

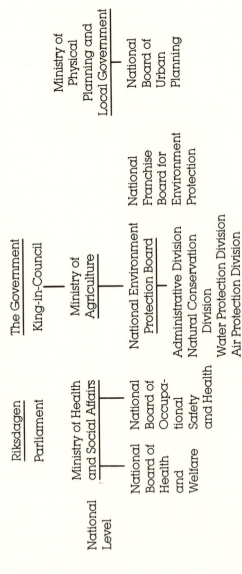

Figure 6-1. Environmental Administration in Sweden

National Level

Riksdagen
Parliament

The Government
King-in-Council

Ministry of Health
and Social Affairs

Ministry of
Agriculture

Ministry of
Physical
Planning and
Local Government

National
Board of
Health
and
Welfare

National
Board of
Occupa-
tional
Safety
and Health

National Environment
Protection Board

National
Franchise
Board for
Environment
Protection

National
Board of
Urban
Planning

Administrative Division
Natural Conservation
Division
Water Protection Division
Air Protection Division

Division for Environmental
Hygiene
Solid Waste Management
Unit
Products Control Division
Products Control
Committee
Research Committee
Research Laboratory

County Council/Administrative Board

Regional
Level        Nature Conservancy Section        County Architect

Municipal Board/Council

Local
Level     Local Public Health Committee    Local Housing Committee    Ad Hoc Health and Environment
                                                                     Committee/Large Cities

SOURCE: Swedish Institute, "Environmental Protection in Sweden," *Fact Sheets*, July 1977 and November 1973.

209

Oil, which was promulgated in 1968,[66] established a sulfur content limit of 2.5 percent by weight for July 1, 1969, and provided for the lowering of that limit when the government deemed more stringent standards necessary.

In the larger metropolitan areas (Stockholm, Gothenberg, and Malmo) and in areas of the country (eight of twenty-three counties) where land and water are most susceptible to acid precipitation, the sulfur content of fuel oil burned must be 1 percent or less by weight. By October 1, 1984, the 1 percent limit will be applicable to the entire country. As of October 1, 1977, the allowable sulfur content of fuel oil burned for home heating was 0.5 percent, and since October 1, 1979, the limit has been 0.3 percent. Finally, a central government committee established to study the sulfur dioxide pollution problem has recommended that Swedish industry be required to use sulfur-free fuel in the future in spite of the dwindling worldwide supply of sulfur-free fuel and the requisite investment in refining facilities designed to produce sulfur-free fuel.[67]

Regarding motor vehicle emissions, Sweden originally adopted standards similar to those established by the U.N. Economic Commission of Europe (ECE). However, inasmuch as ECE standards would not have led to an improvement in air quality, Sweden adopted the more stringent U.S. emission standards for model years 1976 and later. Consequently, for 1976 models, Swedish regulations on air pollution from motor vehicles approximated those applicable to 1973–74 models in the United States with an average reduction of hydrocarbon emissions of about 70 percent, of carbon monoxide emissions of about 65 percent, and of nitrogen oxide emissions of about 40 percent. These reductions are based on the emissions of untreated vehicles from a car fleet of a type corresponding to that in Sweden.[68]

## Solid Waste Pollution Control

The municipal government's virtual monopoly of solid waste disposal evolved because these wastes have historically been a critical problem for local public officials charged with implementing and enforcing all of the Public Health Acts since 1874. This traditional monopoly was legally sanctioned in 1970 by enactment of the Municipal Sanitation Act (MSA) which compels the municipal governments to provide household refuse collection and waste-disposal facilities as well as to clean up litter.[69] Pursuant to the mandate of the MSA, municipalities have been required by law since 1972 to collect, transport, and treat household refuse. Some municipalities, like Stockholm and several of its suburbs, have consummated agreements with private contractors who carry out one or more of the three steps of household refuse disposal: collection, transportation, and treatment. The final treatment of household refuse, over half of which consists of paper products, has usually been dumping or incineration with some composting and landfill operations. However, waste sorting and recycling will inevitably have to be adopted to reduce the residuals from final treatment; for example, households are being required to separate newspapers and magazines for recycling.[70]

About one-third of household refuse is incinerated, and most of the remaining two-thirds is finally disposed of in dumps and sanitary landfills.[71] Both methods of final treatment can result in violations of environmental law, such as the Water Act of 1918 (and the 1964 amendments thereto) and the EPA.[72] Furthermore, the Nature Conservancy Act of 1964 severely limits the disposal of paper and plastic products (packaging), glass products (nonreturnable bottles), and sheet metal products (nonreturnable cans, appliances, and motor vehicles)[73] where dumping occurs in nature reserves and national parks, on the shores of seas, lakes, and rivers, or on any site if such land use could effect significant changes in the landscape.[74]

In 1975, the Swedish Parliament authorized the government to issue directives providing for the separation by households of all recoverable wastepaper from other types of household refuse. By 1980, the government had fully implemented mandatory wastepaper collection wherever it was technically and economically feasible. However, the separation of other types of household refuse will be done at a later stage of waste disposal through mechanized processing.[75]

Even though the MSA requires municipalities to handle the household refuse within their jurisdictions, the law leaves them with unlimited discretion in dealing with other types of waste. A local public health ordinance is drafted and voted on at the municipal level.[76] Such local ordinances, which are necessitated because of the central government's residual authority as provided in Public Health Acts, typically include the following provisions:

Municipal wastes may not be treated by the person producing it (as manure, incineration, etc.); a waste treatment plant may not be set up without the prior authorization of the local public health authority (except when a license has been given according to the Environment Protection Act); the manner in which waste is to be prepared for collection; wastes other than "municipal" wastes which the municipal authority agrees to deal with; and responsibility for wastes which the municipal authority refuses to treat.[77]

Under the 1973 Ordinance Concerning Charges on Certain Beverage Containers, which is based on the central government's authority implicit in the 1964 Nature Conservancy Act, a small excise tax or charge is payable on all nonreturnable beverage containers. In addition to raising the price of beverages bottled in nonreturnable containers by the amount of the tax or charge, the ordinance has resulted in a switch from nonreturnable to returnable bottles. Nonreturnable cans have maintained their market share.

In summary, the price of all beverages has increased at a rate faster than that predictable by reference to the general rate of inflation.[78]

One of the detrimental effects of the Swedish variety of automania is the continual problem of abandoned junk automobiles and the related problem of poorly managed salvage yards.[79] With taxes totaling at least 100 percent of the "real" sales price of a new automobile, one wonders why Swedes continue to spend more of their income on automobiles than on housing when housing is in short supply and public transportation is abundant and relatively inexpensive.[80] The following taxes are imposed on the purchase of an automobile: an excise tax, the value added tax without credit, and annual registration.[81] The provisions of the 1975 Wrecked Automobile Act (WAA) should eventually eliminate the junk automobile problem.

The WAA established a system for licensing auto salvage operations authorized to issue a vehicle deregistration certificate entitling the automobile owner to a scrapping premium. The premium or rebate is financed through an excise tax levied on the producer (or importer) of automobiles. The excise tax revenue, which is accumulated in a trust fund, can also be used for payment of grants to municipalities for the removal of junk autos abandoned[82] on public property as authorized under the Municipal Sanitation Act.[83]

No legislation has been enacted primarily to regulate the disposal of harmful industrial waste. Much of the legislation enumerated above does deal to some extent with industrial waste. In addition, the Marine Dumping Prohibition Act of 1971[84] and the 1973 Law on Products Hazardous to Health and to the Environment[85] provide some implicit regulation of noxious industrial waste. Regulations under the Municipal Sanitation Act also allow municipal collection of harmful industrial and commercial wastes.

Two special enterprises,[86] known as SAKAB and INTERKAB, have been established through separate joint ventures between the State and the private sector. The central government and local governments are the majority owners of both enterprises. The objective of INTERKAB is to coordinate with municipal governments the development of an optimal, cooperative system of solid waste collection, transportation, treatment, and disposal (or recycling) for both household and industrial residual substances. SAKAB is primarily responsible for conducting research and coordinating activities related to the disposing, destroying, converting, or recycling of dangerous chemical substances.[87] One company (Reci) collects most of the waste oil. SAKAB and selected municipalities dispose of or recycle the oil.[88] Individual industrial firms can dispose of their own hazardous wastes if permission to operate has been granted by the NEPB or NFB. Otherwise, SAKAB arranges disposal of industrial chemical wastes.[89]

In principle, SAKAB has a monopoly for treating commercial and industrial wastes hazardous to the environment. This monopoly position should facilitate the development by 1980 of measures for collection and treatment of harmful wastes that are satisfactory from both an environmental and a public health point of view. In summary, one can presume that any entity receiving an operating permit from the NFB or an operating exemption from NEPB will have made arrangements to properly dispose of any noxious industrial or commercial wastes.[90]

The 1973 Law on Products Hazardous to Health and to the Environment (Hazardous Products Act) was enacted to regulate the use of hazardous products not explicitly covered in the EPA, EPO, MSA, Public Health Acts, Nature Conservancy Act, or the Marine Dumping Prohibition Act. Since all prior environmental protection legislation dealt with pollution caused by production of goods and services or the disposing of gaseous, liquid, or solid wastes, the 1973 Hazardous Products Act (HPA) "represents a substantial broadening and tightening-up of public control over (hazardous) products"[91] such as poisons and substances dangerous to man, pesticides, and polychlorinated biphenyls (PCBs) and products made therefrom, as well as other products hazardous to health and to the environment[92] because of their toxic chemicals or physio-chemical properties.

The burden of proof is on the manufacturer, seller, and importer under the HPA with even the suspicion of a hazard constituting sufficient grounds for central government intervention.[93] Anyone manufacturing, selling, handling, or importing hazardous products has a legal duty to take those steps and such precautions as are necessary to prevent or minimize damage to the environment and humans. There is a duty to investigate the composition and other properties of a hazardous product to determine its potential effect on public health and the environment. Moreover, information on a product's ill effects must be clearly marked on its packaging or container. Some hazardous products may only be manufactured or imported under a license issued by the Products Control Board, which was established by the HPA. Others may be banned.[94] Finally, the Products Control Board which can mandate that a hazardous product or waste derivative may be destroyed, converted, or otherwise dealt with only with special permission from that board.[95]

Any entities generating residuals hazardous to the environment are obligated to report annually on the nature, composition, quantity, and handling of such wastes.[96] These reports were initially made in 1976 to the NEPB which forwarded the information to the Environmental Information System (EIS). The EIS is a broad-based environmental information service managed by the Swedish Council for Environment Information, which was set up by Parliament in 1974. Other EIS projects on solid waste statistics have been designed to provide better information on the content, amounts, and distribution of domestic and commercial wastes so that sufficient data on all wastes will be available for implementation of an efficient recycling system.[97]

There is also a compulsory licensing program for enterprises that provide land transportation on a commercial scale for any hazardous residuals. Finally, an institution that intends to export harmful wastes can do so only if permission is granted by the NEPB.[98] Thus, harmful wastes generated in the production and consumption of hazardous products are monitored and controlled.

### Government Assistance

As a member of the OECD, Sweden has adopted the "polluter pays principle." Insofar as the Environmental Protection Act (EPA) of 1969 applied both to existing establishments and to new or altered facilities, the only equitable method for alleviating the inevitable injustice to pre-July 1969 establishments was to subsidize the required investments in pollution control facilities for a transitional period. In principle, to retroactively impose the more stringent pollution control standards authorized in the EPA without subsidization would effectively require older facilities to incur all the costs for complying with new, previously unpredictable environmental quality requirements.[99] Central government subsidization of municipal sewage facilities began many years before the EPA was enacted.[100] Since the EPA became law, the central government has also subsidized plants for treatment of municipal solid waste and of noxious industrial waste. In summary, the three types of pollution control facilities that the national government has subsidized are (1) antipollution investments of industrial establishments existing before July 1, 1969; (2) municipal sewage plants; and (3) municipal plants for treatment of solid waste.[101]

The subsidy for industrial pollution control facilities was originally planned for the five-year period from fiscal year 1969-70 to fiscal year 1973-74. The subsidy program authorized by the central government in June 1969 was originally planned to appropriate 50 million Swedish crowns annually — over $8 million.[102] The program was extended for more than a year, and additional subsidies were authorized during several recessionary periods in the early 1970s.

In practice, the subsidies were available only to industrial polluters with facilities operational on the date (July 1, 1969) the EPA became effective. The industrial polluter had to apply either for a permit to operate from the NFB or for an exemption from the NEPB and agree to the conditions stipulated in the permit or the exemption to be eligible for any subsidy. The normal procedure was for the industrial concern to apply for a permit or exemption and for a subsidy at approximately the same time. The NEPB was given responsibility for determining which firms should receive a grant for pollution control investments.[103] However, the central government can, in principle, decide where it is in the public interest to allow a grant for any industrial firm, old or new, which generates appreciable pollution of the air or water,[104] especially as an incentive for technological development.[105]

Financial assistance was normally granted for pollution control facilities designed for pre-EPA, pre-expansion production operations, although grants for pollution control measures designed for the expansion of production could be sanctioned by the central government under special circumstances, for example, where an industrial firm is the major employer in a depressed region. Grants were made only to firms with a high probability of surviving the foreseeable future. Finally, firms that could easily maintain profitability in spite of the unforeseen, increased cost of production attributable to complying with more stringent environmental quality requirements were not granted subsidies.

The financial assistance was intended for measures that effect pollution abatement, especially facilities for the purification of industrial effluent and emissions. Recycling facilities for wastewater were also eligible. Facilities that effectively connected an industrial establishment to a municipal sewage system were also treated as qualifying pollution abatement expenditures, but subsidies were not granted for the substantial reorganization of production for the purpose of reducing air and water pollution. In addition, the central government could decide to recommend financial assistance to a firm for treatment of its wastes in facilities separate from the firm's location.

The amount of financial assistance was limited to 25 percent of the cost of qualified pollution control investment and to 2.5 million Swedish crowns (over $400,000) for one and the same enterprise.[106] The 25 percent limit was raised to 75 percent once and to 50 percent twice during recessionary periods in the early 1970s when additional subsidies were allocated for industrial pollution control. Moreover, the five-year financial assistance program was extended for an additional fiscal year (1974-75) and into 1975-76 during which the NEPB primarily subsidized firms willing to attempt full-scale tests of innovative technical solutions for pollution abatement. Grants for technological development continue today.

The subsidization of water and sewage treatment facilities began in the 1920s. Originally, the subsidies were simply a means of creating jobs. Later, they were intended to equalize the costs for consumers of water in municipalities relying on different tax bases. In 1959, the system of subsidies was redesigned to aid in the financing of especially expensive sewage treatment facilities. More recently, in 1968 the grants for sewage purification facilities were modified to encourage municipalities to invest in those more sophisticated treatment processes capable of producing a very high degree of purification.

The most important condition that a sewage treatment plant must meet is that it be of public importance.[107] This test requires that the facility serve at least ten habitations. However, the plant need not be run by a municipality as long as the facility "is declared public in accordance with the legislation on public water and sewage works."[108] Grants will be awarded only for facilities designed to provide more than sludge separation treatment. Grant applications are made to the appropriate County administrator with regional authority for environmental protection who forwards his or her report, along with the application to NEPB for processing.

The grant base, that is, amount of the investment which qualifies for financial assistance, consists of the cost of new construction or the cost of additional construction (or reconstruction) of the treatment facility itself and the cost of outflow conduits from the facility to the recipient. Thus, the base does not include the conduits into the facility except in two exceptional situations: (1) where from a water conservation point of view it is desirable to locate a sewage treatment facility in the proximity of a recipient and at a longer distance from other users of the plant who must incur greater costs for intake conduits; and (2) where water conservation justifies a sewage treatment facility that serves several communities that must incur greater costs for transit conduits.

The subsidy limit is between 30 and 50 percent of the grant base, with the actual percentage being directly related to the degree of purification anticipated. The maximum amount of financial assistance for any one facility is 5 million Swedish crowns (over $800,000).[109] A subsidy of at least 30 percent is accorded to any sewage treatment facility designed for biological purification with the normal capacity to reduce biochemically oxygen-consuming substances, that is, biological oxygen demand, (BOD) by at least 60 percent. The 50 percent subsidy is available only to facilities designed for both biological and chemical purification which have the operational effectiveness to regularly reduce both the BOD content and the phosphorus content of effluent by at least 90 percent. Between the 30 and 50 percent limits, the percentage of subsidy is directly related to a schedule differentiated in 5 percent steps of, or improvements in, purification.

As in the case of the subsidy program for industrial pollution control facilities, the financial assistance provided for public sewage treatment facilities was augmented on three occasions during recessionary periods in 1971-72, 1972-73, and 1973. During these periods, the limits on government subsidies were increased 25 percent, and the percentage range became 55 to 75 percent rather than 30 to 50 percent. Table 6-2 summarizes the Swedish government's subsidization of industrial and municipal pollution control facilities during fiscal years 1968-69 to 1975-76. The direct subsidy to industry for pollution control facilities since 1969 has been at least 785 million Swedish crowns (about $130 million). In principle, no indirect subsidy is implicit in the 1,315 million Swedish crowns (about $220 million) granted to municipalities for the financing of sewage treatment facilities because the subsidies were only for dwelling units. In computing the grant base, a deduction was made therefrom to reflect the costs that can be properly allocated to expected industrial use of the sewage treatment facility.[110]

The third type of financial assistance for pollution abatement investments is actually two subsidy programs: one for facilities treating waste other than household refuse,[111] and the second for facilities designed for either recycling or other use of municipal waste or for specified conventional treatment.[112] Since July 1, 1972, financial assistance has been available for facilities treating nonhousehold waste where from the public standpoint the plant is urgently needed. Financial assistance has also been made available for the cost of storing chemical or other noxious wastes.[113] The grants can cover up to 50 percent of the cost of the facility where the investment is in the public interest, for example, the treatment of toxic wastes in regional plants.

State financial assistance for facilities designed to recycle, use, or treat municipal refuse began in 1975. Generally, grants are awarded for up to 50 percent of cost. The grant can cover up to 80 percent of the investment in certain cases.[114]

The central government decided upon enactment of the EPA in 1969 to provide transitional financial assistance for controlling pollution from pre-1969 industrial and commercial facilities in the form of direct grants over a five-year period. Accordingly, the Swedish system of taxation has not provided indirect subsidies in the form of tax reduction or deferral for investments in pollution control equipment. Although capital expenditures for pollution control equipment would seem to qualify for amortization under the Swedish investment reserve system, an explanation of this unique type of incentive to invest in plant and equipment when the national level of economic activity is low is beyond the scope of this study. In addition, this incentive is applicable to most capital expenditures by industrial and commercial concerns. Therefore, it would not, in principle, be a tax expenditure or indirect subsidy for investments in pollution control, although empirical evidence might support the notion that there had been a subsidy in practice because pollution abatement expenditures benefited more from the investment reserve system.

## Future Environmental Policy

When the Governmental Commission on the Revision of Environmental Legislation issued a report in December 1978, it proposed five major changes in the 1969 EPA:

(1)  More efficient procedures for licensing polluting activities.

(2)  More effective supervision and control of polluting plants and factories.

(3)  Tougher sanctions against polluters willfully or negligently violating pollution control requirements.

**Table 6-2**
**Government Subsidies for Emission Control Measures,**
**Fiscal Years 1968–1969 to 1975–1976**
**(Millions of Swedish Crowns)**

| Subsidized Sector | Type of Subsidy | Rate of Subsidization[a] | FY 68/69 | FY 69/70 | FY 70/71 | FY 71/72 | FY 72/73 | FY 73/74 | FY 74/75 | FY 75/76 | FY 68/69 FY 75/76 Totals |
|---|---|---|---|---|---|---|---|---|---|---|---|
| Municipal-ities | Basic | 30-50 | 40 | 50 | 60 | 80 | 130 | 130 | 130 | 130 | 750 |
| | Increased | 55-75 | | | | 365 | 100 | 100 | | | 565 |
| Industry (Plants in operation before July 1, 1969) | Basic | 25 | | 50 | 50 | 50 | 50 | 50 | 50 | 15 | 315 |
| | Increased | 50 or 75 | | | | 300 | 100 | 70 | | | 470 |
| TOTAL | | | 40 | 100 | 110 | 795 | 380 | 350 | 180 | 145 | 2100 |

[a]Based on percentage of capital expenditures.

*SOURCE:* Swedish Delegation, *Environmental Policy in Sweden* (Paris: OECD, 1977), p. 109.

(4) Greater opportunities for public participation in the licensing process.

(5) Widened rights of complaint against licensing decisions.[115]

The Commission concluded that the EPA was too weak in view of the stronger public support for environmental protection. Hence, the law must be strengthened to keep pace with current and future demands for environmental quality.[116] The Parliament was scheduled to enact most of the Commission's recommendations strengthening the EPA in 1981.[117]

Although "general" subsidies for private sector investments in pollution control ended in 1976, firms may still qualify for government grants where they invest in new and untested pollution control technology. There are also subsidies available for waste disposal as well as reception stations for waste from ships.[118]

## Improving Licensing Procedures

As explained above, the basic premise of the EPA is that environmentally hazardous activities should be licensed so that polluting operations can be regulated to avoid or minimize their harmful effect on environmental quality. The EPO, which was issued under the authority of the EPA, specifies which industries and activities must be licensed to expand, change, or begin operations. The NFB issues such licenses. Furthermore, for those activities specified as less hazardous in the EPO, NFB licensing is not necessary, but County environmental administration must be notified. Finally, the NEPB can grant "exemptions" from the licensing requirements of the EPO.

When the NEPB (or in some cases a County environmental administration) responds to an application for exemption, it dictates operating conditions similar to those that would be included in a license from the NFB. If the business or municipality accepts such conditions, the NEPB may grant exemption from the NFB licensing requirements. Thus, the "exemption" option is supposed to provide a quicker, less expensive means of accomplishing an environmentally sound solution to the pollution control problem. However, NFB licensing does guarantee immunity from most changes in environmental law for up to ten years.

During the early 1970s, the exemption system was probably necessary to expedite pollution control requirements as provided by the EPA, inasmuch as a large number of licensing cases would have been delayed by the NFB's time-consuming deliberations. In addition, the central government subsidized private investment in pollution control facilities and municipal government investment in sewage plants. It was, therefore, expeditious that the NEPB handle both the conditions of the exemption and the granting of subsidies.

In recent years, industry and municipalities have seldom applied for exemptions from the NEPB because of the uncertainty associated with the operating conditions therein when more stringent environmental regulations and legislation may be imposed at any time. Moreover, the bulk of the subsidies were exhausted by the mid-1970s. Thus, NFB licensing is now preferable to NEPB exemptions inasmuch as the NFB guarantees immunity from the imposition of all but a few of the requirements of subsequent changes in environmental law.[119]

The Commission recommended that the NEPB be freed of the tradeoff duties inherent in its responsibilities for granting exemptions. Consequently, it could give its undivided interest to its primary responsibilities, that is, being a spokesperson for the public interest in environmental quality. Thus, only the NFB and the County environmental administration would be involved in licensing industrial and municipal operations. The NFB should concentrate on licensing large and heavily polluting industrial operations and municipal sewage plants, while the County environmental administration's role would be expanded as it would be responsible for all other polluting activities for which the rule of compulsory notification would be applicable. The NFB permit decisions would continue to include all the pollution control requirements that have traditionally been deemed necessary to maintain environmental quality. Similarly, the County environmental administration's statement on a pollution activity, which must be issued well in advance of the inception of a polluting activity, should specify all the pollution control requirements deemed necessary by the administration to preserve environmental quality.[120]

Thus, criticism of the exemption procedure by the public and media should end when the exemption system is abolished. Although some of that criticism resulted from a misunderstanding that the NEPB "exempted" polluters from the EPA, there was a real conflict of interest inherent in the NEPB's responsibilities for exemptions and representation of the public interest in environmental affairs, especially in its appearances before the NFB. The NEPB should have more resources available to accelerate its efforts in developing guidelines for licensing polluting activities.

## Improving Monitoring and Control

The Commission concluded that, while the licensing and exemption procedures worked relatively well, the monitoring of compliance with the operating conditions specified in permits has been unsatisfactory. For example, division of the monitoring responsibilities between the NEPB and the County environmental administration is unclear. Thus, there has been a lack of coordination between the NEPB and the County administrators. Furthermore, the resources devoted to monitoring have been totally inadequate. County authorities have been making about 14,000 inspections annually, while the NEPB inspections have been running at between 4,000 and 5,000 per year, which the Commission considered adequate. Insofar as the NEPB would be relieved of its responsibilities for granting exemptions, it should use some of its freed resources to increase its monitoring and control efforts, especially with regard to the more serious sources of pollution and the study of the long-term effects of pollution on the environment.[121]

To improve monitoring and control, the Commission proposed that Public Health Committees (PHCs) be made responsible for all local control of polluters, while County environmental administrations would be formally responsible for controlling only those polluters with NFB permits. The local PHCs would be given a five-year transition period over which they could assume an increasing proportion of their new responsibility. To assist the local PHCs, the County environmental administrations should develop control programs for those polluters that are to be controlled by PHCs. Similarly, the NEPB should establish a Specialist Group on Supervision and Control to advise County environmental administrations and PHCs in complicated cases.[122]

As noted, the PHCs are to be given major responsibilities for inspection as well as the surveillance and monitoring of noise, odors, and other nuisances. In addition, PHCs should be given increased responsibilities for other pollutants primarily because local public health officials have more intimate knowledge of local conditions. In addition, public health officials already have inspection duties mandated by other legislation, for example, regulation of food hygiene and the work environment, and so a more efficient use of time would be achieved. County environmental administrations should also be given control of more complex types of industrial pollution, even though such polluters are required to apply for licenses from the NFB.

The Commission made no proposals for reorganizing the measuring and sampling of pollution. There are over 100 laboratories that make water pollution measurements. However, the Commission members from the Social Democratic party submitted an addendum that included a proposal for an "Environment Monitoring, Inc." that would be owned by the central government, with its operations modeled after the national motor vehicle inspection agency. The Social Democrats felt that such an organization would be necessary to ensure the effective functioning of the PHCs.[123]

Finally, the Commission proposed that polluters partially reimburse the local government for the cost of surveillance and monitoring. Thus, the introduction of a "control fee" to be paid by polluters to the community should provide some of the revenue needed to hire more public health officials[124] inasmuch as almost half of the local governments have only one public health official.

### Tougher Sanctions for Violations of the EPA

Prosecution for violations of the EPA have been minuscule. The Commission determined that public and media criticism of the failure to prosecute violators and the limited economic consequences for those few violators was well founded.[125] For example, during the 1970–78 period only about two polluters annually were found guilty of violations of the EPA or licensing conditions. Investigation and prosecution have been neglected so much that polluters assume they can get away with violations of the EPA and the operating conditions specified in their licenses or exemption agreements.

To correct the central government's neglect of the investigation and prosecution of violators, the Commission proposed that environmental agencies be legally required to investigate and prosecute all suspected violations of the EPA and licensing conditions. When violations are proved, the polluter should be assessed an "environmental protection fine," which is to be equal to the excess profits earned by violating the law. However, the maximum fine would be 25,000 Swedish crowns (about $4,000) per day for as long as the violation continues. Moreover, the statute of limitations for violations should be increased from two years to five years. Finally, a violator should be subject to imprisonment for up to two years,[126] rather than one year as provided under the current law, in cases of gross negligence or clear intent to violate the EPA or licensing conditions.

### Public Participation in Licensing

Currently, the EPA limits the opportunities of individuals to contest NFB licensing decisions to situations where an individual is directly and economically affected as the owner of property near a polluting facility. In addition, the NEPB can appeal NFB licensing decisions on general environmental protection grounds. Environmentalists have called for appeal rights for their own organizations as well as citizens' groups.[127]

To provide greater opportunity for meaningful participation by the public in the NFB licensing process, the Commission proposed two major changes. First, it recommended that each permit application contain information on the environmental consequences of a polluter's activities and on the control measures proposed by a polluter to mitigate the harmful effects of its activities. Second, the Commission proposed that all prospective polluters should be required to notify the County environmental administration at a very early stage in their planning for a polluting facility. Then the County environmental administration could determine whether the proposed facility warrants a preparatory public hearing. Any changes and modifications made pursuant to a required preparatory hearing must be included in the subsequent NFB permit application or notification of the

County environmental administration where a permit is not necessary. Inasmuch as the site inspection and public hearing preceding an NFB licensing decision will be continued, the public will have at least two opportunities to affect the outcome of the licensing process.

## Broadening Rights of Complaint

As noted above, the NEPB can appeal licensing decisions based on its general right of complaint in its role as the spokesperson for the public interest in environmental quality. Consequently, only those individuals who are "objectively concerned" (where direct economic effects to their property may ensue from pollution in their neighborhoods) have complaint rights under the original EPA. Thus, the Commission proposed that there should be greater opportunity for complaints against NFB licensing and statements by County environmental administrations upon notification by prospective polluters. Although general complaint rights should not be given to citizens, the Commission recommended that the complaint right be given to local government, concerned citizens (for example, environmentalists), and local trade unions.[128]

## Environmental Impact Statements

Swedish environmentalists have publicly called for a system of environmental impact statements like those in effect for a decade in the United States and recently proposed by the EEC. However, the Commission has not proposed any changes in the documentation required in an application for licensing a polluting activity. Such documentation is now restricted to technical and economic descriptions of plant, equipment, and other aspects of the operation of a prospective polluting facility.[129]

## Parliamentary Action on EPA Revisions

In early 1981, Parliament enacted legislation to revise and strengthen the EPA, which had served as Sweden's primary environmental protection law for over a decade. The amendments, which basically follow the recommendations made by the Governmental Commission on the Revision of Environmental Legislation in December 1978, became effective on July 1, 1981.[130]

## 1982 Addendum

Although the Social Democrats were unsuccessful in their efforts to include mandatory monitoring of water pollution by a government-chartered national monitoring corporation in the 1981 amendments to the EPA, environmental issues are not expected to be subject to much debate during the September 1982 national elections. The major political issues should be public spending cuts, tax reform, and the Social Democrats' proposal for trade-union buy-out of big business.[131] As mentioned above, a consensus on environmental issues has characterized Swedish politics for over a decade with the 1980 referendum on phasing out nuclear power being the only significant controversy on environmental policy since 1969. However, the Environment party, which was formed by a former Liberal party *Riksdag* member in late 1981, had by mid-1982 attracted enough support according to opinion polls so that it may pass the 4-percent-of-the-popular-vote threshold necessary to merit seats in the *Riksdag* (Swedish parliament).

If the new Environment party receives sufficient votes to win *Riksdag* seats and the Social Democrats do not elect a majority of the *Riksdag*, the Social Democratic party (SDP) or current Center-Liberal coalition might organize a coalition government that included the Environment party. As a member of either coalition government, the environmentalists could force the *Riksdag* to vote on some of their pet projects, including banning the use of herbicides in Swedish forests and raising sheep therein to control undergrowth while producing food and wool or taxing raw materials, energy and new technology in an effort to redirect economic policy from maximising output to improving the quality of life. Notwithstanding such radical policies, the Environment party may be able to attract sufficient votes from those protesting the policies of the other five parties to deny SDP or current Center-Liberal coalition from organizing a new government without the environmentalists' participation,[132] and, thus, the consensus on environmental issues will be tested.

## Acid Rain

The "acid rain" problem continues to be a major concern because of all the transboundary air pollution that gravitates to Sweden from Great Britain, West Germany, the Benelux, and Eastern Europe. At the end of a three-day June 1982 conference held in Stockholm, representatives from 21 European and North American countries, five international organizations, and several non-governmental organizations (NGOs) issued a communique supporting establishment of "concerted programs" for the reduction of $SO_2$ and $NO_x$ emissions as "a matter of urgency." These programs should be established and implemented within the framework of the 1979 UN Economic Commission for Europe Geneva Convention on Long-range Transboundary Air Pollution to which there are 35 signatories, including the United States and Canada. The Swedish government called the conference to focus international attention on acid rain, which affects Sweden more than it affects any other European nation.

The conference communique emphasized that the acid rain problem which is aggravated by transboundary air pollution renders the use of high stack chimneys as an unacceptable alternative to emission control technology. Therefore, the "best available technology," including fuel cleaning, should be used to control $SO_2$ and $NO_x$ emissions. Pollution control should be enhanced with increased energy conservation and improved monitoring programs.

When the ten EEC countries ratified the aforementioned Geneva Convention, it brought the number of countries that have ratified it to 22. The Convention is the first international legal agreement reflecting the "golden rule" principle agreed to at the United Nations-sponsored Conference on the Human Environment held in Stockholm ten years earlier. Insofar as ratification of the treaty is proceeding at a normal pace, it should take effect in 1983. Its implementation will not occur soon enough for Sweden and North America. Although $SO_2$ emissions have not increased and are not expected to increase in Europe and North America, the cumulative effect of acid rain has killed many lakes in Sweden and North America.[133] Sweden is also concerned about the impact of acid rain on their forests, and probably the greatest natural resource covering half the country, and yet Sweden may contribute as little as one-fourth of the $SO_2$ in acid rain.[134]

## Energy Policy and Air Pollution

Realizing that transboundary air pollution is not the only cause of acid rain and needing to reduce its dependence on expensive imported oil as an energy source, Sweden has continued to develop alternative, less polluting sources of energy and to encourage energy conservation by industry and households. Because a lengthy description of Swedish energy policy is not germane, the cursory review which follows integrates environmental and energy policy.

Almost 70 percent of total energy consumption is in the form of imported oil. About 20 percent is in the form of electricity with two-thirds produced by hydropower. However, half of Sweden's total hydropower potential may not be developed because it lies mainly in wilderness rivers (in northern Sweden) that are protected by stringent environmental laws. Nuclear power will be expanded until 1990, but the 1980 referendum on nuclear power requires a phasing out of that energy source early in the next century. Almost half of the imported oil is used to heat buildings, and yet under Swedish law the use of electricity to heat new buildings is forbidden in all but the most exceptional circumstances. In summary, Swedish energy planners searching for substitutes for imported oil find that a number of alternatives are precluded by law.

In May 1981, the *Riksdag* enacted an energy law that ambitiously projects an almost 50 percent reduction in imported oil by the early 1990s, even though Sweden like other developed countries has markedly decreased imports of crude oil since 1974. This objective will be achieved by developing alternative energy sources and through greater energy conservation. The government will subsidize energy research and development, oil substitution programs, and energy conservation investment. By increasing taxes on imported oil and oil products, revenue will be available for the aforementioned subsidies, and the higher price of oil will render alternative sources of energy more competitive with imported oil.

The major alternatives to imported oil will be coal, indigenous fuels (primarily peat and forest wastes), solar energy, alternative fuels for transportation, greater use of efficient electrical heat pumps, and more extensive use of district heating, especially in Stockholm where even waste heat from the Forsmark nuclear power plant may be used. Most of the coal will be imported, and it will either be low-sulfur or treated before combustion to remove much of the sulfur.[135] For example, a purified liquid coal known as "carbogel" could be used for electric power generation and district heating thereby reducing oil imports by one-third.[136] Environmentalists criticize the use of low-sulfur coal (or liquid coal) because its use inevitably increases $SO_2$ emissions and aggravates the acid rain problem. Similarly, environmentalists are concerned about any massive effort to "strip mine" peat deposits since peat bogs are found in over 10 percent of Sweden's land area. The concept of "energy plantations" to harvest fast-growing trees for fuel has been ignored by environmentalists, most probably because the species of trees are still experimental.

Since 1973-74 when the OPEC oil embargo precipitated the Energy Crisis, the central government has heavily subsidized energy conservation, especially through district heating, the proper maintenance of heating facilities, and building codes requiring high levels of insulation. These subsidies will be continued and increased. Moreover, subsidies and grants for energy research and development will be increased.[137]

Sweden's extensive development of district heating is surely unique among the developed countries, especially since the municipal government is responsible for energy planning and management. As explained above, Swedish municipalities are the fundamental public planning unit in respect both to land use and a variety of other public services, such as education, transportation, housing, electricity, heating, and other social services. Due to a concern for environmental quality (especially air pollution), several large municipalities introduced district heating systems in the late 1950s and early 1960s. Currently, over 50 municipalities of various sizes provide district heating services, while some larger cities provide virtually all the heating in their communities and over 40 percent of the national population either live or work in district-heating buildings. In addition, municipalities are responsible for over 20 percent of the nation's electrical production capacity and more than 40 percent of the total distribution of electricity.

Because about one-seventh of the total energy consumed in the country is distributed by municipalities, the *Riksdag* provided for mandatory municipal energy planning by local government in legislation that took effect July 1, 1977. As a consequence, municipalities must assess energy use within their jurisdiction, project electricity and heating demand for five to ten years in the future, and consider the alternative sources of energy available for meeting that future demand, especially satisfying the demand for heating through introduction or extension of district heating, the recovery of industrial waste heat, and incineration of municipal solid waste. One of the most important factors to be considered in evaluating these alternatives is the environmental impact of each energy source. Other factors considered include overall primary energy use, security of the supply of any source, and the financial resources available to the municipality.[138]

In practice, most district heating systems still rely primarily on oil for fuel. Many in the future will switch to low-sulfur coal, forest wastes, or peat as a primary fuel. Some larger cities, however, also rely on household solid waste for part of their fuel supply, for example, Stockholm incinerates most of its residents' solid waste, thereby conserving oil and reducing its need for sanitary landfill space.[139] In fact, the university city of Uppsala, which is 40 miles north of Stockholm, recycles virtually all waste. After most noncombustible wastes are separated from other household waste, the remaining trash and garbage are incinerated to supply 15 percent of the city's space heating. Then the noncombustible wastes, industrial wastes (such as scrap metal, plastics, and packaging refuse), and construction wastes (such as scrap building materials, metal, and discarded packaging) is taken to a one-man, automated recycling plant that separates recoverable materials (for example, scrap metal, plastics and refuse that can be used as fuel) from wastes that must be dumped.[140]

Since the early 1970s, many municipalities have worked out agreements with industrial plants within or contiguous with their jurisdiction to recover industrial waste heat for use in district heating systems. However, industrial waste heat could only have supplied about one percent of 1975 national energy supply if all the waste heat of major industrial plants in or near municipalities were recovered for use in district heating. There are, of course, many technical problems associated with recovering industrial waste heat. For example, plants do not operate 24 hours a day; they do not operate at a constant level during the day or year; and heat is produced all year round, especially during the warmer months when waste heat can be a pollution problem for Swedish water bodies. On the other hand, the equipment and materials required for optimizing the recovery of waste heat are already well developed.

Not only is progress being made on recovering waste heat from existing industrial plants, but many new industrial developments are being designed to provide their waste heat for district heating systems. Probably the largest waste heat recovery project in the world was completed in Gothenberg,

Sweden when *Svensk* Shell refinery on the Gota River began operating in late 1980 and supplying heat for Gothenberg's district heating system. Since 1974, the *Boliden Kemi AB* sulphuric acid plant has been supplying waste heat to the district heating system of the city of Helsingborg. Subsequent *Boliden* plant expansions completed through early 1982 have supplied sufficient waste heat to Helsingborg so that its municipal power utility has been able to curtail plans to construct co-generation facilities that also would have supplied heat for the district system. *Boliden* has also reduced its emission of hot water in the *Öresund* Straits.[141]

Even industrial plants that are miles away from a city may supply waste heat for district heating. Currently, a consortium, including two large Swedish corporations (ASEA and *Nynäs* Petroleum) and *Storstockholm Energi AB* (the municipal utility responsible for supplying Greater Stockholm with energy), is planning a huge coal gasification and synthetic fuel plant almost 40 miles southeast of the Swedish capital. The plant should supply about 10 percent of the current motor fuel consumption when completed at a cost of about $500 million in 1987 or 1988. Current plans call for piping waste heat to supply Stockholm, thereby reducing by 70 percent the amount of oil consumed by that city in producing heat.[142]

## Solid and Hazardous Wastes

With regard to solid and hazardous wastes, Sweden continues to be a pioneer in both recycling and disposal. Beginning in 1983, most used aluminum beverage containers will be recycled. An aluminum can compacter, which compresses the containers and determines the cash value thereof for the customer, will be located in around 30,000 retail stores throughout Sweden. A special deposit company will be established by PLM (the Malmö packaging company manufacturing the compacter), the breweries, and retailers. The 1985 goal of the deposit company and the government is to recycle 75 percent of all aluminum beverage containers and as a result aluminum imports should be eliminated.[143]

In May 1982, Sweden agreed to cooperate with the EEC on the latter's research and development program on recycling urban and industrial waste. The EEC's program, which was initiated on November 1, 1979, will continue until October 31, 1983. Sweden will contribute its expertise and funding to a program which has already included studies on the sorting of household waste, thermal treatment of waste, fermentation and hydrolysis, and recovery of rubber wastes. Thus, Sweden will be trading its expertise for the EEC's.[144]

As explained above, there has been cradle-to-grave monitoring and regulation of specified hazardous wastes in Sweden since 1975. Although such waste results from the use of hazardous materials by many industries and most households, the amount of hazardous waste produced probably amounts to less than one percent of the total waste produced in the country and about three-fourths of the waste is produced by a few plants. Therefore, hazardous waste has been relatively easy to control in Sweden and such waste results in minimal damage to the environment. The main source of hazardous waste is used oil and chemical solvents; however, many other types of wastes that are small in amount can be considerably more damaging when released in the environment.

All the major cities and many smaller municipalities (in total 60 out of 280 municipalities) have assumed responsibility for disposing of hazardous wastes. All other hazardous waste must be given final treatment by SAKAB, licensed companies (of which there are about 20), or the companies that produce such waste if such companies notify local authorities in advance. SAKAB operates several regional collection stations and two treatment plants. SAKAB is constructing a third treatment plant in central Sweden to be operational in 1984. The new plant will include facilities for incineration of hazardous waste, decanting of oil and extended storage.

Slightly less than half of the hazardous waste is treated by the companies that produce it. Very little of such waste is exported by SAKAB or the producers of the hazardous waste. Producers must receive permission from NEPB to export their hazardous waste. Some of the exported waste is incinerated at sea and some is reprocessed. The balance is disposed of in foreign countries.

Although Sweden has generally solved its hazardous disposal problem during the past decade, it was common practice during the 1960s and earlier to dispose of toxic substances and used chemicals in landfills along with household and non-hazardous industrial and construction wastes. Thus, older landfill and dump sites are a potentially serious environmental problem if they begin to release their hazardous wastes into groundwater, surface water, or the atmosphere à la Love Canal. When such hazardous wastes begin to leak, the government, and thus the taxpayer, will have to incur all the costs of clean up where responsibility for the waste dumping cannot be determined.[145]

## Notes

1. U.S. Department of Commerce, The *Effects of Pollution Abatement on International Trade* (Washington, D.C.: U.S. Government Printing Office, 1973), p. A-72.

2. Ake Flacker and Lennart Holm, *Urbanization and Planning in Sweden* (Stockholm: Royal Ministry of Agriculture, 1972), p. 7.

3. Swedish Delegation to the OECD Environment Committee, *Environmental Policy in Sweden* (Paris: OECD, 1977), p. 135.

4. U.S. Department of Commerce, *Effects of Pollution Abatement on Trade*, p. A-72.

5. Thorsten Sundstrom, "Environmental Policies of the City of Stockholm," *Studies in Comparative Local Government*, Vol. 6, No. 2 (1973), p. 29.

6. Fack Sweden, "Habitat: Zap Goes the Garbage Man," *Scandinavian Times*, Vol. 7,

No. 5 (November 1970), p. 20.

7. U.S. Department of Commerce, *Effects of Pollution Abatement on Trade*, p. A-72.

8. Sundstrom, "Environmental Policies of Stockholm," p. 29.

9. Lennart J. Lundqvist, "Shaking the Institutions in Sweden," *Environment*, Vol. 16, No. 8 (October 1974), p. 27.

10. Lennart J. Lundqvist, "Sweden's Environmental Policy," *Ambio*, Vol. 1, No. 3 (June 1972), p. 92.

11. U.S. Department of Commerce, *Effects of Pollution Abatement on Trade*, p. A-73.

12. Ibid., pp. A-73, A-80, and A-81.

13. Corinne Halberg, "Environment: Safeguarding Scandinavia's Assets," *Scandinavian Times*, Vol. 7, No. 1 (March 1970), p. 14.

14. Lundqvist, "Shaking the Institutions," p. 28.

15. U.S. Department of Commerce, *Effects of Pollution Abatement on Trade*, p. 73.

16. Halberg, "Environment: Safeguarding Scandinavia's Assets," p. 14.

17. Lundqvist, "Shaking the Institutions," pp. 28-29.

18. U.S. Department of Commerce, *Effects of Pollution Abatement on Trade*, p. A-73.

19. Lundqvist, "Shaking the Institutions," pp. 27-28.

20. Ibid., pp. 27-29.

21. *Sweden's Reply to the United Nations Inquiry in Connection with the United Nations' Conference on the Human Environment* (Stockholm: Royal Ministry for Foreign Affairs, 1970), pp. 31-33.

22. Lundqvist, "Shaking the Institutions," pp. 27-30.

23. Swedish Institute, "Environmental Protection in Sweden," *Fact Sheets on Sweden*, November 1973, p. 3.

24. Swedish Delegation, *Environmental Policy in Sweden*, p. 14.

25. EPO Section 1.

26. EPO Sections 2, 3, and 5.

27. EPO Sections 8 and 9.

28. Lundqvist, "Shaking the Institutions," p. 28.

29. Lennart Persson, *Environmental Protection Act and Marine Dumping Prohibition Act* (Stockholm: National Environmental Protection Board, 1972), pp. 17-21.

30. Lundqvist, "Shaking the Institutions," pp. 27-29.

31. M. Donald Hancock, *Sweden: The Politics of Post-Industrial Change* (Hinsdale, Ill: Dryden Press, 1972), pp. 154 and 199.

32. Lundqvist, "Shaking the Institutions," pp. 27-29.

33. Persson, *Environmental and Marine Dumping Act*, pp. 47-49.

34. Ibid., pp. 45-46.

35. Swedish Delegation, *Environmental Policy in Sweden*, p. 14.

36. Ibid., pp. 14 and 15.

37. Ibid., pp. 15 and 16.

38. Ibid., p. 16.

39. Ibid., pp. 16 and 132.

40. Swedish Institute, "Central Planning in Sweden," *Fact Sheets on Sweden*, August 1972, pp. 1-2.

41. Ministry of Physical Planning and Local Government, *Management of Land and Water Resources* (Stockholm: Royal Ministry of Foreign Affairs, 1971), p. 60.

42. Swedish Institute, "Central Planning," p. 1.

43. Ministry of Physical Planning, *Management of Land and Water*, pp. 7-10.

44. Swedish Delegation, *Environmental Policy in Sweden*, p. 75.

45. Ministry of Physical Planning, *Management of Land and Water*, pp. 8-11.

46. Swedish Delegation, *Environmental Policy in Sweden*, p. 76.

47. Swedish Institute, "Central Planning," p. 2.

48. Swedish Delegation, *Environmental Policy in Sweden*, p. 86.

49. Ibid., pp. 76, 77, 82, and 131.

50. Pierre Vinde, *Swedish Government Administration* (Stockholm: Swedish Institute, 1971), pp. 11-18.

51. Swedish Delegation, *Environmental Policy in Sweden*, p. 77.

52. Ibid., pp. 78 and 131.

53. Ibid., p. 131.

54. Ibid., pp. 84, 85, 131, and 132.

55. Ibid., p. 85.

56. Ibid., pp. 85 and 86.

57. Swedish Delegation, *Environmental Policy in Sweden*, pp. 18-20.

58. Lundqvist, "Shaking the Institutions," pp. 27-30.

59. Ibid., p. 30.

60. Swedish Delegation, *Environmental Policy in Sweden*, pp. 136-37.

61. Lundqvist, "Sweden's Environmental Policy," pp. 90-101.

62. *International Environmental Guide — 1975*, p. 61:2602.

63. Lundqvist, "Shaking the Institutions," p. 30.

64. Swedish Delegation, *Environmental Policy in Sweden*, pp. 16, 136, and 137.

65. Ibid., p. 18.

66. *International Environmental Guide — 1975*, p. 61:2602.

67. Swedish Delegation, *Environmental Policy in Sweden*, pp. 16, 17, 101, and 102.

68. Ibid., pp. 95-97, 133, and 134.

69. *International Environmental Guide — 1975*, p. 61:2602.

70. Sundstrom, "Environmental Policies of Stockholm," pp. 29-30.

71. Lars Emmelin, "Environmental Planning in Sweden: Energy Needs for Environment Protection," *Current Sweden*, ENV No. 73 (May 1976), p. 5.

72. Persson, *Environmental and Marine Dumping Act*, pp. 9-12, 45, and 46.

73. *International Environmental Guide — 1975*, pp. 61:2601 and 2602.

74. *Sweden's Reply to the UN Conference on the Human Environment*, p. 31.

75. Swedish Delegation, *Environmental Policy in Sweden*, pp. 102-3.

76. OECD Environment Directorate, *Waste Management in OECD Member Countries* (Paris: Organization for Economic Cooperation and Development, 1976), pp. 32-33.

77. Ibid., p. 32.

78. Ibid., p. 67.

79. Swedish Delegation, *Environmental Policy in Sweden*, p. 103.

80. "City Dweller's Demand: Off the Car!" *Scandinavian Times*, Vol. 7, No. 6 (January 1971), p. 15.

81. Martin Norr, Frank J. Duffy, and Harry Sterner, *World Tax Series: Taxation in Sweden* (Boston: Little, Brown & Co., 1959), pp. 144-47; Martin Norr, Claes Sandels, and Nils G. Hornhammar, *The Tax System in Sweden* (Stockholm: Skandinaviska Enskilda Banken, 1972), pp. 141-52; and Swedish Institute, "Taxes in Sweden," *Fact Sheet on Sweden*, March 1977, p. 2.

82. Swedish Delegation, *Environmental Policy in Sweden*, p. 103.

83. De Joie, "Wastes Around the World," pp. 36-37.

84. Persson, *Environmental and Marine Dumping Act*, pp. 81 and 82.

85. Ingemund Bengtsson, *The Act on Products Hazardous to Health and to the Environment* (Stockholm: Royal Ministry of Agriculture, 1973), pp. 7-23.

86. Swedish Delegation, *Environmental Policy in Sweden*, p. 102.

87. Tage Erlander, *Sweden's National Report to the United Nations on the Human Environment* (Stockholm: Royal Ministry of Foreign Affairs, 1971), p. 60.

88. OECD Environment Directorate, *Waste Management*, pp. 50-51.

89. De Joie, "Wastes Around the World," pp. 36 and 37.

90. Swedish Delegation, *Environmental Policy in Sweden*, pp. 102-3.

91. Bengtsson, *Act on Products Hazardous to Health and to the Environment*, p. 8.

92. Ibid., p. 17.

93. Ibid., pp. 10-11.

94. Ibid., pp. 13-16.

95. Ibid., p. 19.

96. Swedish Delegation, *Environmental Policy in Sweden*, p. 103.

97. Lars Emmelin, "Environment Planning in Sweden: The Swedish Environmental Information System," *Current Sweden*, ENV No. 78 (February 1977), pp. 1-3.

98. Swedish Delegation, *Environmental Policy in Sweden*, p. 103.

99. Lundqvist, "Shaking the Institutions," p. 30.

100. Erlander, *Sweden's National Report to the United Nations on Human Environment*, p. 60.

101. Swedish Delegation, *Environmental Policy in Sweden*, p. 21.

102. Persson, *Environmental and Marine Dumping Act*, p. 51.

103. Lundqvist, "Shaking the Institutions," p. 30.

104. Persson, *Environmental and Marine Dumping Act*, p. 51.

105. Swedish Delegation, *Environmental Policy in Sweden*, pp. 22 and 127.

106. Perrson, *Environmental and Marine Dumping Act*, p. 50.

107. Swedish Delegation, *Environmental Policy in Sweden*, pp. 22-25 and 127.

108. Persson, *Environmental and Marine Dumping Act*, p. 50.

109. Ibid., pp. 50-51.

110. Swedish Delegation, *Environmental Policy in Sweden*, pp. 24, 25, and 109.

111. Ibid., p. 21.

112. OECD Environment Directorate, *Waste Management*, p. 33.

113. Swedish Delegation, *Environmental Policy in Sweden*, p. 21.

114. OECD Environment Directorate, *Waste Management*, pp. 9, 33, 50, and 72.

115. Governmental Commission Report SOU 1978:80, *Battre Miljoskydd I* (Stockholm: Ministry of Agriculture, 1978), pp. 1-5.

116. Lennart J. Lundqvist, "Sweden Tightens Up Anti-Pollution Laws," *Ambio*, Vol. 8, No. 2/3 (1979), p. 123.

117. Swedish Institute, "Environmental Protection in Sweden," *Fact Sheets*, April 1981, p. 3.

118. Ibid., p. 4.

119. Lars Emmelin, "Environmental Protection Act Under Revision," *Human Environment in Sweden*, (July 1979) No. 12, pp. 2-3.

120. Lundqvist, "Sweden Tightens Up Anti-Pollution Laws," p. 123.

121. Emmelin, "Environmental Protection Act Under Revision," p. 3.

122. Lundqvist, "Sweden Tightens Anti-Pollution Laws," p. 123.

123. Emmelin, "Environmental Protection Act Under Revision," pp. 3-4. If the Social Democrats win sufficient seats in Parliament during the 1982 elections and they are able to form a new government, their addendum may be included in legislation that further revises and strengthens the EPA.

124. Lundqvist, "Sweden Tightens Up Anti-Pollution Laws," p. 123.

125. Emmelin, "Environmental Protection Act Under Revision," p. 4.

126. Lundqvist, "Sweden Tightens Up Anti-Pollution Laws," p. 123.

127. Emmelin, "Environmental Protection Act Under Revision," p. 5.

128. Lundqvist, "Sweden Tightens Up Anti-Pollution Laws," p. 123.

129. Emmelin, "Environmental Protection Act Under Revision," p. 5.

130. *Svensk författningssamling*, Law No. 1981:420, June 10, 1981.

131. "An Election Loser," *The Economist*, May 1, 1982, p.

132. "Back to the 1680s in Sweden," *The Economist*, July 10, 1982, p. 41. As expected, the Social Democrats prevailed in the September 1982 national election. Their receipt of 45.6 percent of the popular vote along with the 5.6 percent support for the Communists means the nonsocialist parties polled less than 50 percent of the vote. The Green party polled less than 4 percent.

133. *International Environment Reporter-Current Report*, July 14, 1982, pp. 281-282; and "Criticisms Rain Down on Cross-border Polluters," *The Economist*, July 10, 1982, pp. 79-80.

134. Bertil Hägerhäll, "Acidification of the Environment," *Current Sweden*, No. 281 (February 1982), pp. 3-4.

135. Bill Dampier, "Sweden's Energy Plan: 50 Percent Less Oil Within 10 Years," *Ambio*, Vol. 10, No. 5 (1981), pp. 216-18.

136. Don Henrichsen, "Shaking Off the Sheiks," *Sweden Now*, Vol. 15, No. 1 (1981), p. 36.

137. Dampier, "Sweden's Energy Plan," *Ambio*, pp. 216-18.

138. Stephen Tyler "Municipal Energy Planning—The Swedish Experience," *Human Environment in Sweden*, No. 13 (September 1979), pp. 1-3.

139. Robert Skole, "Waste Not, Want Not," *Sweden Now*, Vol. 14, No. 1 (1980), p. 38.

140. Don Henrichsen, "Uppsala Recycles Itself," *Sweden Now*, Vol. 15, No. 6(1981), p. 35.

141. Skole, "Waste Not, Want Not," *Sweden Now*, pp. 39-40.

142. Don Henrichsen, "New Plant Will Beat Energy Costs," *Sweden Now*, Vol. 15, No. 5 (1981), p. 81.

143. "Aluminum Recovery? Can Do," *Sweden Now*, Vol. 16, No. 2 (1982), p. 36.

144. *International Environment Reporter-Current Report*, July 14, 1982, p. 298.

145. Olov von Heidenstam, "Managing Hazardous Waste," *Ambio*, Vol. 11, No. 1 (1982), p. 66.

# 7

# The United Kingdom

Inasmuch as the Industrial Revolution began in nineteenth-century Britain, one should not be surprised that the concern for pollution problems dates back over a century in the United Kingdom. Climatic conditions, topography, industrialization, and population movements have contributed to a long history of air, water, and solid waste pollution. These pollution problems have been most serious in the heavily populated industrial districts of Megalopolis England, in the coastlands and valleys of south Wales, in the urban areas of central Scotland, and in the urban area surrounding Belfast, Northern Ireland.[1] During the period 1875-1970, Parliament enacted no less than fifty-five separate laws to regulate pollution and to protect the natural environment.[2]

The United Kingdom is a small nation with a population of over 56 million and with most of the settlement concentrated in the London-English Midlands corridor, central Scotland, and south Wales. It is comparable in area, population, and economic development to that urban complex of the northeastern United States which includes the Boston to Washington corridor and the heavily industrialized extension of that corridor into western New York, western Pennsylvania, and northeastern Ohio.[3]

Traditionally, the most damaging form of air pollution has been smog or smoke from industrial and domestic coal-burning fires required to produce goods and to heat homes, respectively. This problem has been reduced significantly through restrictions on the burning and sale of smoke-producing fuels. Now there are the problems of particulate matter and sulfur dioxide to contend with as a result of increased industrial production, power generation, and conversion to central heating.[4] Finally, a threefold increase in the use of motor vehicles since 1960 has resulted in a commensurate increase in auto emissions—hydrocarbons, carbon monoxide, and nitrogen oxides.

The close proximity of the cities in Megalopolis England aggravates urban air and water pollution problems. Often heavily polluted air is blown by prevailing winds from one city to another downwind. Downstream cities use river water for domestic and industrial purposes after industrial effluent and municipal sewage have been discharged upstream.[5] Thus, water pollution problems are worst in the industrialized, urban areas of Megalopolis England and in south Wales. About one-half of industrial effluent is discharged into public sewer systems with the balance ending up in rivers, lakes, and the sea. Moreover, many of the sewer systems are obsolete and overloaded, resulting in sewage treatment of low quality and sewage discharge that is still heavily polluted.

Since the United Kingdom is such a small, densely populated country, solid waste disposal in land dumps is a relatively more serious problem there than on the Continent with the obvious exception of the Benelux countries. Moreover, the legacy of the nineteenth- and early twentieth-century industrialization and extraction of minerals is found[6] on the 91,000 acres of land in Wales and the English Midlands that have been so seriously damaged that extensive reclamation will be necessary before it can be used again.[7] In addition to the large quantity of domestic and commercial refuse collected by local government every year, even more solid waste is generated through industrial production, in the generation of electric power, and during the extraction of coal, clay, gravel, and other minerals. Finally, there is a trend toward the production of more complex, more dangerous wastes, such as heavy metals, chemical residues, and radioactive materials.

Traditionally, the British have coped with their limited number of waste-disposal sites and with their water pollution problems by dumping wastes, effluent, and sewage in the sea—the North Sea as well as the Irish Sea, the Atlantic Ocean, and the English Channel.[8] Although such pollution could cause damage to marine life and to human life where contaminated seafood is consumed, studies of the North Sea by international organizations have shown no evidence of dangerous pollution so far, much to the chagrin of those who have attributed the decline of commercial fishing to increased marine pollution. Since oil and gas production began in the North Sea during the early 1970s, there has been a new pollution threat. There have been no major oil spills, although accidents have occurred on several occasions,[9] including the heavily publicized blowout of a Phillips Petroleum oil well in April 1977. While the sea has a great capacity for assimilating wastes, a few estuaries have become overloaded because the water flowing in those tidal rivers does not circulate and mix freely enough. Continual discharges into rivers, estuaries, and coastal waters will eventually exceed the environmental assimilative capacity of the sea if no steps are taken to reduce water pollution and the dumping of wastes offshore.[10]

Since the United Kingdom's pollution problems have resulted primarily from too many people and too much industry concentrated in a limited area of the small island nation, the British began their first serious program for physical planning shortly after World War II.[11] As a result of thirty years of land use planning and pollution control measures to reduce emissions and effluent, the British environment has improved overall. Dramatic success has been achieved through the remarkable reduction of air pollution in London, Manchester, and other major metropolitan areas and through the significant improvement in water quality, as in the Thames River. Nonetheless, increased urbanization, industrialization, mechanization of agriculture, and motor vehicle use, along with increases in the British standard of living and consumption, continue to place severe pressure on the environment at a time when the Queen's subjects are demanding more of an overall improvement in the quality of that limited environment.[12] Finally, Britain's relatively low per capita income when compared to that of its European neighbors and its newfound oil and gas wealth have accelerated economic growth and thereby accentuated the pollution problems related thereto.

## Pollution Abatement Programs

The earliest pollution control legislation in the United Kingdom was enacted in the mid-nineteenth century.[13] It established a tradition and principle of environmental law that is very much adhered to today. In principle, the primary responsibility for resolving pollution problems should lie with local or regional government. Therefore, the proper role of the central government should be to provide the statutory framework which delimits the controls on individual and institutional behavior. Implementation and enforcement of the statutory framework are generally relinquished to local authorities or regional water authorities. For determining the permissible levels of pollution in their areas, these governmental bodies have been given broad discretionary powers that enable them to integrate local resources and priorities, local land use planning considerations, and the environmental assimilative capacity of the locale in the standard-setting process.

In delineating a statutory framework, the central government judiciously retains full responsibility for matters of national concern, issuing guidance, advice, and requests in the form of directives to or regulations for the proper local or regional government. In addition, the central government may in a limited number of cases set national standards, such as for exposure to radioactive materials or for products and operations that pollute in more than one region or locale. The central government may also exercise control through a national agency. Thus, Britain's pollution control policy differs from that commonly found in other industralized countries: other developed nations usually establish uniform standards controlling specific pollutants and ambient quality through detailed regulations, whereas the United Kingdom adheres to the principle that standards should be "reasonably practicable."

This principle requires that the use and maintenance of equipment, as well as the supervision and operation of processes, be carried out in a manner to ensure the abatement of pollution as far as is practicable in view of the local environmental assimilative capacity, the economic implications of marginal increases in abatement, and the current state of medical and technical knowledge. Consequently, it is possible to establish standards for pollutants from each factory and to make the standards more stringent as technological advances and environmental demands occur. This system gives greater flexibility than the system of uniform statutory standards.[14] In summary, the British have adhered to three fundamental antipollution principles: (1) application of emission and effluent standards on a local or regional basis; (2) application of the "reasonably practicable" or "best practicable means" standard, for example, enforce what is technically and economically feasible for each firm; and (3) enforcement of standards by persuasion (or moral suasion) in lieu of prosecution.[15]

As one of the first nations to industralize, the United Kingdom was also one of the first to experience large-scale pollution. The early strategy for effectively alleviating pollution problems was based on decentralized regulation. Thus, even though there has otherwise been an erosion of local control in the United Kingdom, local and regional governments continue to be deeply involved in implementation and enforcement of national environmental law.

Since its inception in November 1970, the Department of the Environment (DOE) has had responsibility for alleviating most pollution problems in England.[16] As many as eleven other ministries may have concurrent responsibility for any one environmental issue. However, the Secretary of State for the Environment has been assigned the general responsibility for coordinating all government activity related to pollution and the environment. In Wales, Scotland, and Northern Ireland similar responsibilities have been given to the Secretaries of State for their respective countries.[17]

In Wales, there is a Welsh Office of DOE, whereas the Scottish Development Department (SDD) and the Northern Ireland Office of the DOE exercise functions similar to those of the DOE within their respective countries. Most environmental quality legislation applies to all of the United Kingdom, but there is some variation to reflect different circumstances in each country. Except as otherwise explained, one can assume that the pollution abatement program and government assistance discussed below are equally applicable to all the countries making up the United Kingdom.[18] The most recent development in environmental quality legislation occurred in the summer of 1978 when Parliament enacted the Pollution Control and Local Government Order. This law explicitly extends the law (Pollution Control Act) to Northern Ireland insofar as it covers water, air, solid waste, and noise pollution.[19]

The DOE was created by combining three former ministries—Transport, Housing and Local Government, and Public Building and Works. As a result, responsibility for controlling pollution and managing the environment became more integrated. Several departments retained their own unique sectoral functions because it was neither logical nor feasible to combine all departments with some responsibility for environmental matters. Inasmuch as the Secretary of State for the Environment coordinates all the pollution control activities of the central government, there is at least a structural safeguard against conflicting government actions.[20]

Other central government agencies along with their pollution control responsibilities are (1) the Mnistry of Agriculture and Fisheries which controls the use of agricultural chemicals, the disposal of farm wastes, the monitoring of contaminants in foodstuffs, and the protection of sea and fresh water fishing; (2) the Department of Trade and Industry which controls oil pollution at sea and pollution around airports;[21] (3) the Health and Safety Commission which controls workplace pollution and specified air emissions; (4) the Department of Health and Social Security and the Scottish Home and Health Department which ensure the dissemination of advice on the public health implications of pollutants to the appropriate departments;[22] and (5) the Department of Education and Science which ensures that the research councils supervise and generate adequate scientific knowledge on environmental problems and issues. Thus, the major pollution problems—air pollution, inland water pollution, and solid and radioactive waste disposal—are the responsibility of DOE and its offices in Wales and Northern Ireland.[23]

The three departments that merged into the DOE have continued their responsibilities as ministries under the coordination of the Secretary of State for the Environment. (1) The Ministry for Local Government and Development supervises local government, land use planning, transportation planning (road building and traffic management), recreational use and conservation of the countryside, and the disposal of refuse, sewage, and wastewater; (2) the Ministry for Housing and Construction supervises the development and financing of housing programs, the construction industry (for example, building codes), the construction and maintenance of government property, the protection of ancient monuments and historic buildings and the development of new towns; and (3) the Ministry of Transport Industries supervises the nationalized transportation industries (such as railroads, ports, and waterways), including the operational management, financing, and resource allocation thereof.[24]

The other major subdivision within the DOE is the Deputy Secretariat for Environmental Protection which consists of five offices: (1) the Central Unit on Environmental Pollution; and (2) the four Undersecretariats (or Directorates) for water, road safety, vehicle engineering and inspection, and noise, clean air, and wastes. The Central Unit consists of a team of administrators and scientists that coordinate government programs on air pollution, water pollution, and noise, including research.[25]

Traditionally, commissions, committees, and councils have been established periodically to advise the government on public problems.[26] For example, air pollution has been analyzed by the Beaver Commission and the Clean Air Council; and water pollution, by the National Water Council. Since 1970, there has also been a permanent Royal Commission on Environmental Pollution (RCEP). The RCEP is an independent body that advises on national and international issues related to environmental quality, especially research needs and future problems.[27]

National Environmental Planning

As suggested above, pollution abatement policy in the United Kingdom is characterized by its two-tier framework of planning (national) and enforcement (local). The central government, subject to the advice and consent of Parliament, devises national strategy and establishes the necessary national standards to be carried out primarily by the local governmental authorities.[28] Local government, which consists of over sixty county (or regional in Scotland) councils and about 430 district councils, derives its power to provide specified services from Parliamentary Acts. The power is vested in the elected county (or regional) and district councils which act through their respective professional administrations. The county (or regional) councils are generally responsible for major planning, such as land use and transportation; district councils are responsible for other services, including environmental or public health.[29]

The Town and Country Planning Acts for England and Wales (1971) and for Scotland (1972) are both derivatives of the modern-day basis for the British system of land use planning, the Town and Country Planning Act of 1947. The laws provide for a nationwide planning system under the Secretaries of State for the Environment, Scotland and Wales. Compulsory planning duties are imposed on all local planning authorities—counties and districts. Each local planning authority is required to prepare a development plan for its area. Then the proposals and objectives delineated in the "structure plan" are reviewed by the appropriate central government ministry. When the appropriate ministry approves the plans, more detailed proposals, including plans for action in areas where comprehensive development or improvement is anticipated, are prepared. While these more detailed plans are seldom submitted for ministerial review, each "structure plan" is subjected to review continuously. It may be amended either when the broad policies for use and development of land should be changed or when new "action areas" develop.

Most forms of construction, natural resource extraction, and any significant change in the use of either land or existing buildings can occur only if approved by the appropriate local planning authority. There are also special provisions for controlling the location of industry and office buildings. The broader problem of industrial development is the joint responsibility of several central government departments, although each industrial development plan must normally receive local planning authority approval.

Regional economic planning by the central government relies on the advice of local planning authorities and regional economic planning councils. The local authorities and regional councils confer at length before "structure plans" are prepared so that the economic planning and development of each region can be integrated. In some cases, several local authorities in a region will submit a joint plan for ministerial approval.

The central government's regional development policy is based on studies of the socioeconomic resources of each region, and its plans focus on those selected areas where land and infrastructure are deemed favorable for development. In addition, there are provisions for development of new towns, expansion of existing towns near large cities (town development), urban renewal, the preservation of historic buildings and monuments, the establishment of open space (or "greenbelts") in and around urban areas, protection of coastlines, and development of national parks, forest parks, and nature reserves.

Public and interest group participation in the planning process is encouraged at the earliest stages of the drafting of both "structure plans" and the more detailed "local plans." Each local planning authority must adequately publicize the objectives and proposals in "structure plans" and "local plans," those proposals for development that differ materially from the approved objectives of a plan, applications for permission to undertake specified types of development, and any other schemes that could be expected to affect a substantial number of the local population. After the publicity has been generated, members of the public and interest group representatives are given an opportunity to comment on the plans or development affecting their respective areas.[30] Since the "structure plans" are simply general policy statements that indicate trends and broad patterns of development in the future, the appropriate minister seldom allows public or interest group objections at an inquiry. Therefore, virtually all public inquiries are held at the discretion of local planning authorities, and commentary is limited to the detailed proposals and "action area" plans included in a "local plan."

While the local plan system relieves the central government of much of the detailed evaluation work for which it would otherwise be responsible, the minister maintains residual powers for intervening in difficult or controversial cases. The local plan must be forwarded to the appropriate minister for his (or her) information. He not only has the right to intervene, but he can also stipulate that a plan shall not become effective until he approves it. A minister retains the power to recall a plan for reevaluation at which time he can approve, amend, or reject at his discretion.

Conversely, any domestic, commercial, or industrial development, or material change in the use, of one's land or building can occur only with the permission of local planning authorities. If a natural or legal person is refused planning permission or is granted conditional permission, there is a right of appeal to the appropriate minister, who may transfer the appellant's case to an inspector. A public inquiry will then be held by an inspector before he makes a decision on the appealed application for development. Thus, "obtaining planning permission" is purely an administrative law procedure with no right to appeal for judicial review except where there is a disputed point of law or defect in the administrative procedure. A favorable decision by the court does not usually reverse the administrative decision on the appellant's proposed development. The development is broadly defined as the execution of operations (for example, engineering, construction, or mineral extraction) on, over, in, or under land, or materially changing the use of buildings or land. In summary, if a person wants to conduct operations that constitute development under the land use planning law, he does not have an automatic right to do so, but rather he must "obtain planning permission" from the appropriate local planning authority.[31]

Thus, in matters of land use planning, the relationship of the central government and the local government is very much like the working partnership attributed to British pollution control policy. The Central government, upon the requisite consent of Parliament, initiates national strategy and establishes goals and guidelines, whereas local government (districts with a population of 75,000 to 100,000 and counties of 500,000 or more) develops "structure plans" and implements detailed plans that have been designed to reflect local circumstances and that often are adapted in response to local public opinion.[32]

In England and Wales, the two levels of local government split the responsibility. There are over fifty County councils and about 370 district councils. The County councils are responsible for strategic planning for land use, waste disposal, transportation, highways, and public safety (police and fire protection). The district councils are responsible for environmental health,[33] solid waste collection, public housing, building regulation, and local planning and development. Both levels of local government have concurrent powers of eminent domain, that is, the power to acquire and dispose of land for planning purposes, including development, redevelopment (urban renewal and land restoration), parks, open space, and coastal protection.[34] Education and other personal social services are also provided at the County level except in the densely populated, metropolitan counties of West Midlands (Birmingham and Coventry), Greater Manchester, Merseyside (Liverpool), West Yorkshire (York, Leeds and Bradford), South Yorkshire (Sheffield), and Tyneside (Newcastle). In these urban counties, metropolitan district councils are responsible for education and other social services, in addition to local planning and development, housing, building regulations, solid waste collection, and various other environmental health functions.[35]

The local government in metropolitan London was extensively reorganized almost a decade before that in the rest of Great Britain because the environmental problems of the capital city were much more acute. Since 1965, Greater London has been governed by a two-tier system of local government with one metropolitan-wide government called the Greater London Council (GLC) and thirty-three subdivisions called boroughs. The GLC, like the metro counties, concentrates on strategic services, such as police and fire protection, broad planning, transportation, and solid waste disposal. The boroughs are responsible for local planning and development, education, and other personal social services, such as welfare, solid waste collection, and various environmental health functions. The GLC and its boroughs share responsibility for housing: the GLC oversees large-scale slum clearance and rehabilitation projects to improve public housing, and the boroughs build new public housing.[36]

Scotland has had a separate land use planning statutory and administrative framework since 1947, but the Scottish Town and Country Planning Act of 1972 essentially parallels its namesake in England and Wales.[37] One major difference between the two laws accrues from a somewhat different apportionment of planning responsibilities between the two levels of local government—the nine regional councils, fifty-three district councils, and three island area councils.[38]

## Water Pollution Control

Great Britain has suffered extensively from the industrial and urban pollution of its surface waters. Thus, early attempts to control water pollution, such as the Harbors Act of 1814, Public Health Acts, and the Rivers Pollution Prevention Act of 1874, provided the enabling legislation for local and regional regulation of urban and industrial effluent.[39] The United Kingdom has fewer untapped water resources, both surface and ground water, than any other Western European country except Belgium, and its demand for water is expected to almost double by the year 2000 with a concomitant increase in wastewater. Understandably, then, public policymakers have increasingly focused on improved water quality management at a broader regional, river-basin level. The major objectives of water quality management in the United Kingdom are (1) to ensure an adequate supply of clean water for human and industrial use; (2) to provide ample sewerage and sewage treatment facilities for processing human, industrial, and agricultural effluent; (3) to improve the water quality of all major rivers by the early 1980s, in response to the need to take an increasing percentage of the country's drinking water out of surface waters that receive industrial effluent and sewage and as a result of the trend toward viewing rivers and streams as environmental assets or amenities instead of open sewers and drainage ditches; and (4) to expand the use of the country's surface waters beyond their traditional functions of water supply, sewerage, and drainage by developing them as recreational and aesthetic assets supporting wildlife and water sports.

In order to achieve these broad policy objectives, the management of the British hydrological system was reorganized during the mid-1970s.[40] England and Wales had almost 14,000 institutions (twenty-nine river authorities, 160 water suppliers, and 12,000 wastewater treatment institutions) responsible for water resource management before their reorganization into ten regional water authorities on April 1, 1974. On that same date, the new system of local government in England and Wales became operational.[41] The reorganization of both local government and water resource management in Scotland also took effect simultaneously, on May 16, 1975.[42]

The nine regional water authorities in England and the National Water Development Authority for Wales have overall responsibility for water resources. Specifically, they oversee water supply, wastewater treatment, sewage disposal, pollution control, land drainage, flood prevention, water recreation, fisheries, and, to some extent, navigation. The Secretary of State for the Environment appoints the Chairmen of the boards of trustees for the ten water authorities. The majority of the board members are appointed by County and district councils. The National Water Council, which is made up of the Chairmen of the water authorities and a few other water experts, was established by the Water Act of 1973, the same law that created the water authorities. This Council serves in an advisory capacity for the central government on national water policy.

Pursuant to the Local Government Act of 1973, Scotland's water supply has become the responsibility of nine regional and three island councils, along with the Water Development Board of Central Scotland. The Board supplies water to regional councils in central Scotland. The duties and powers of these councils are provided for in the Scottish Water Acts (of 1946, 1949, and 1967) and the Scottish Local Government Acts (of 1973 and 1975). There are seven river purification boards and three island councils, and they are responsible for water pollution control within Scotland as stipulated in the Rivers Prevention of Pollution Acts (of 1951 and 1965).

Both the ten regional authorities in England and Wales and the Scottish regional councils, river purification authorities, and island councils have been given broad jurisdiction. However, the effective date of the Pollution Control Act (PCA) of 1974, the omnibus legislation enacted to enable the governmental authorities to operationalize water quality management,[43] has been deferred because the United Kingdom has recently experienced a period of severe stagflation.[44] The PCA provides for the central government's comprehensive regulation of the pollution of almost all inland and coastal waters, as well as underground water. Until the water pollution provisions of the omnibus PCA are implemented, the Public Health Acts and the Water Acts will continue to provide the local and regional government in England, Wales, and Scotland with piecemeal, overlapping powers to manage the water resources of Great Britain.[45]

The Public Health Act of 1936 is a general nuisance and hygiene law covering both air and water pollution at the local level and granting local government the right to prosecute.[46] The law not only establishes responsibility for disposal of sewage, but also specifies that sewage discharge be allowed only where it will not create a nuisance. The 1936 act provides for fines in cases of contamination of drains or public sewers and for polluting rivers, streams, or other bodies of water. In addition, this Public Health Act gives local authorities responsibility for inspections to determine nuisances, for issuance of abatement orders, for imposition of fines or court action (prosecution) where more appropriate, and for recovery from the polluter of costs incurred to eradicate a nuisance. Finally, the Public Health Acts of 1937 and 1961 stipulate that, where possible, the effluent of trades should be discharged into the public sewers provided by local authorities. Inasmuch as the consent of the local authority is required before trade effluent discharges are allowed, local control through provision of conditions and levying of service charges is effected.

There are three principal types of water polluters: individuals, communities, and industries. All three must pay to abate their pollution. Most individuals pay for household sewage disposal through property tax exactions. In rural Britain, individuals either install their own septic, providing a satisfactory effluent, or build a cesspit that must be cleaned out regularly at their own expense except in Scotland where such services are provided without charge.

A community, through its local wastewater treatment authority, is responsible for ensuring that its sewage does not pollute underground aquifers and surface waters and that the effluent from its sewage treatment facilities satisfies the standards of the regional water authority. Local wastewater treatment authorities are financed through both (real) property tax revenue and charges for commercial and industrial wastewater treatment.

Commercial and industrial establishments discharge their effluent either directly into surface waters or into a community sewage system. In respect to surface water discharges, the establishment at its own expense must satisfy the Regional Water Authority's effluent standards. With regard to sewage system discharges, each establishment must pay to have its effluent treated by the local authority. On occasion, partial treatment by the polluter may be necessary before it can discharge into the community (or municipal) sewage system.[47]

The most common charge system used by local authorities is based on the cost of biological oxygen demand (BOD) treatment; the cost of sludge treatment; the cost of primary treatment; and the cost of effluent conveyance. In principle, charges for industrial and commercial effluent are intended to recover the incremental cost of investment in and/or operation of local sewage and wastewater treatment facilities.[48]

In England and Wales, the Water Act of 1945 and the 1948 amendments provide for the establishment of local by-laws so that statutory water suppliers can protect the water within their jurisdiction.[49] The Water Act also empowers statutory water suppliers to require the construction of water treatment facilities with funds provided through service charges for water use and to condemn and acquire land necessary for protection of underground water. In 1963, Parliament enacted the Water Resources Act. This law established and delineated the functions—including establishment of effluent standards for discharge consents—of the original river authorities, which were reorganized into ten Regional Water Authorities by provisions of a more recently (1973) enacted Water Act. The 1963 Act and 1968 amendments required the development of a national water policy and established a Water Resources Board to advise river authorities responsible for conserving, redistributing, and augmenting water resources.[50] The Water Resources Act provides the authorities with the power to prevent underground water pollution, with emergency power to clean up accidental river pollution, and with condemnation power to acquire land necessary for the protection of reservoirs (surface water supply) and underground water supply.

In Scotland, the duties and powers of the water supply authorities are delineated in several laws, including the Scottish Water Acts of 1946, 1949, and 1967 and the Scottish Local Government Acts of 1973 and 1975. The Scottish Prevention of Pollution of Rivers Acts of 1951 and 1965 gives the seven river purification boards and the three island councils responsibility for controlling water pollution in inland waters throughout Scotland.[51] These Pollution of Rivers Acts require approval or consent from river purification boards before effluent can be discharged into Scottish rivers and estuaries. Consent orders may stipulate special conditions reflecting downstream water use and the assimilative capacity of the natural river.[52]

The domestic use of water is also paid for through the property tax levy. Nondomestic users may pay for their water by volume used or in relation to the value of their property, whichever is greater. In England and Wales, commercial properties pay by meter or by property value, whichever is larger, while industrial users pay either at a progressive rate with a minimum charge or at a flat rate per 1,000 gallons. Similarly, in Scotland, many nondomestic users pay either at a progressive rate or a flat rate. In some circumstances, commercial and industrial users pay according to the value of their property.[53]

## Water Quality and the Pollution Control Act

When fully implemented, the 1974 Pollution Control Act (PCA) will extend protection from water pollution to virtually all ground, surface, and coastal waters and will allow for more stringent controls to ensure protection from the possibility of accidental water pollution. The act, which is applicable to the United Kingdom, extends the general prohibition on polluting nontidal inland waters to the tidal stretches of inland waters, specified underground water, and coastal waters. Consent of the appropriate Regional Water Authority (river purification board in Scotland) will be necessary before discharges of trade effluent or sewage can be made into a body of water,[54] both directly through pipelines and injection and indirectly from working mines and land disposal. The PCA not only stipulates procedural rules for granting, withdrawal, revocation, and appeal of discharge consents, but it also provides for transitional arrangements and conditional variances. Finally, the public disclosure requirements of the PCA include the advertising of each application for consent to discharge and the release of information in applications for consent and in consents issued upon a proper request from the public. The Secretary of State may grant specific exemptions from these public disclosure requirements[55] in cases where proprietary information could be compromised.

By the end of 1979, the water quality management provisions of the 1974 Pollution Control Act should have been fully effective. Regional Water Authorities and Scottish river purification boards are in the process of developing river quality objectives, as well as reviewing existing consents for industrial and other discharges. The authorities and boards inherited an irregular system of discharge consents in which the permissible effluent levels were often not logically related to water quality objectives.

Some polluters were able to acquire much less stringent consents than those of similarly situated polluters on the same body of water. Effluent standards on some heavily polluted rivers are less stringent than those on relatively less polluted bodies. Moreover, many existing consents do not properly reflect the capabilities of current wastewater treatment facilities, nor do they appropriately reflect the funds available to the public and private sector for water pollution abatement.

In implementing the water quality management provisions of the PCA during 1978 and 1979, the central government endorsed the recommendations developed by the National Water Council (NWC) during their two-year study (1975-77). The recommendations were prepared after consultation with water authorities and experts and interest groups from the scientific, environmental, and industrial communities. The NWC recommendations are generally supported by the water authorities which will have to implement them. If the NWC recommendations are carried out, the water authorities are going to determine quality objectives for bodies of water within their jurisdiction within one year; to delineate long-run goals where water is already of adequate quality and short-run goals where water quality must be upgraded; to complete a formal review of existing consents within two years; and to determine as accurately as possible the maximum level of pollutants that can be allowed in effluent discharges (assuming existing and anticipated sources for five to ten years) without jeopardizing the achievement of quality objectives.

In determining water quality objectives, environmental and other interest groups will have the opportunity to comment on and influence priorities for water quality improvement. These objectives will be based on parameters that can be attained by 95 percent of the samples made.[56] A proposed river quality classification scheme would "provide a common basis for general description of corresponding limits on water quality."[57] When such a scheme is adopted, it would be adapted to local conditions as necessary and subject to amendment as technical knowledge improves. As a result of the review of existing consents, a public register of discharge consents and actual effluent discharges will be available for comparison with river quality objectives, but any changes in existing consents will have to be approved nationally.

Having determined the maximum loads of permissible pollutants in effluent discharges for a given river basin, fixed figures will be used in the approval of conditions for discharge consents with regard to both existing and new sources of effluent. The water authorities are not going to be relaxing their operations in existing wastewater treatment plants. Nor will they be deferring their construction of new plants except where it would be cost-effective to reallocate resources to improve river quality elsewhere without significant harm.[58]

## National Air Pollution Control

Great Britain's first pollution control legislation was the Alkali and Works Regulation Act of 1863, which became the forerunner of the currently effective air pollution law—the Alkali and Works Regulation Act of 1906.[59] Air pollution from selected industrial sources is regulated by the central government under the authority specified in the Alkali and Works Regulation Act of 1906, the Health and Safety at Work Act of 1974, and the Pollution Control Act of 1974. Other immobile sources of air pollution are regulated by local government as authorized by both the Public Health Acts and the Clean Air Acts.

The emissions from over sixty industrial processes, or "scheduled processes," are regulated at the national level by the HM Alkali and Clean

Air Inspectorate or HM Industrial Pollution Inspectorate in Scotland. Among the regulated processes are the thermal generation of electricity, the production of ammonia, metal manufacturing, cement and brick works, ceramic and related industries, and chemical and fuel plants. The Alkali Inspectorate or Industrial Pollution Inspectorate is responsible for the external environment, while the Factory Inspectorate and Nuclear Installations Inspectorate are responsible for the working environment. Both are part of the Health and Safety Executive established by provisions of the Health and Safety at Work Act.

The Secretary of State for the Environment (or Secretary of State for Scotland) may expand or contract from the list of scheduled processes subject to central government review. All scheduled processes must acquire annually a certificate of registration to operate. In applying for the first certificate of registration, a facility must be equipped with the "best practicable means" for abating atmospheric emissions either by preventing discharges of smoke, dust, grit, and offensive noxious gases or by rendering these discharges inoffensive and harmless where of necessity discharged. The pollution control devices and facilities must be maintained for efficient, effective, and continuous operation. Moreover, the pollution abatement equipment and plant of "registered works" are subject to regular inspection. In fact, the proprietors of registered works must provide facilities for inspectors so that the latter can enter the premises to observe and carry out tests. The owners must also disclose proprietary information about their processes upon the request of inspectors who are required to maintain such information in strictest confidence.

The more than sixty scheduled processes[60] generate about "60% of the most noxious of offensive emissions to air from industrial processes."[61] The Alkali Act specifies emission limits for only four of these processes; for the others, there is a requirement that the "best practicable means" be implemented and maintained. The term "best practicable means" has been interpreted as: the provision, efficient maintenance, and effective use of equipment for abating discharges—such as smoke, dust, grit, and noxious $(SO_2/NO_x)$ gas; but encompassing the effect of such abatement on the operation of and cost to production processes with the objective of optimizing (cost-benefit ratio of 1) between the degree of harm or nuisance effected and the amount of cost incurred. The obligation to use the "best practicable means" often requires alteration of plant and equipment, as well as production method, as improvements in technology occur.

The Alkali Act requires that certain tests be performed periodically. Such analyses may be carried out by the proprietor, by a qualified industrial testing laboratory, by the government chemist, or by any other appropriate government department. In addition to this periodic monitoring of emissions at the source, the central government conducts a National Survey of Smoke and Sulfur Dioxide and other surveys of vehicle emissions, metals, aerosols, acid particles, and dust from 1,200 monitoring sites.

Failure to register annually or to use the "best practicable means" to abate emissions can result in legal proceedings by the Inspectorate's Chief Alkali Inspector (or Procurator Fiscal in Scotland). Upon the request of local authorities, the Inspectorate can advise on and inspect processes and operations that are not otherwise within the scope of the Alkali Act and the Health and Safety at Work Act. However, enforcement remains with local governmental authorities.

### Local Air Pollution Control

As provided in the Clean Air Acts and the Public Health Acts, emissions from stationary sources not otherwise within the scope of the central government's authority are regulated by local government.[62] The Clean Air Acts provide for the regulation of smoke from the combustion of coal, oil, and gas to generate power, operate industrial plants, and heat homes. The law authorizes local government to establish chimney height standards and "smoke control areas," as well as to require prior approval for furnaces in new plants. The acts also provide for application of the "best practicable means" standard where central government intervention is necessary because local government does not effectively exercise all its authority, and for penalties if unauthorized fuel is used in "smoke control areas."

All of the provisions of the Clean Air Acts were effective for new facilities by November 1971 and for existing facilities after January 1978. In addition, the 1956 act established Clean Air Councils for England, Wales, and Scotland.[63] These Councils are authorized to review the effectiveness of legislation for the control and abatement of air pollution and to advise the appropriate Secretary of State on air pollution problems.[64]

The major impetus for the Clean Air Act of 1956 was the London Smog of 1952 in which some 4,000 people were alleged to have died from the effects of a coal-smoke and fog mixture that enveloped the Greater London atmosphere. The Clean Air Act of 1968 strengthened the 1956 law by adding smoke generated by oil and gas combustion to that generated by burning coal.[65] Thus, polluters could no longer avoid the law by switching from coal to oil or gas. The Clean Air Acts prohibit emissions of "dark smoke," which is defined as smoke as dark as or darker than No. 2 shade on the Ringelmann Chart, both from the chimney of any building (except for limited periods, such as during ignition and stoking), and from any trade or industrial premises (except through an approved chimney or in other specified situations, such as the burning of timber on demolition sites).

As provided in the Clean Air Acts, all new furnaces, except domestic boilers with limited heating capacity, must be capable "as far as practicable" of being operated continuously without smoke emissions if burning the type of fuel for which they were designed. All new furnaces, except those exempted (such as those of limited size) by the Secretary of State or local government, must be equipped with arrestment devices that are properly operated and maintained so that "smokeless" combustion occurs.

Local government is authorized to issue "smoke control orders." These orders for establishing "smokeless zones" are subject to confirmation by the Secretary of State for the Environment. In such zones, smoke emissions from any buildings in part (or all) of a local government's jurisdiction are prohibited.

Both where a new chimney is to be built to serve a new or existing furnace, and where a furnace with an existing chimney is enlarged, the chimney height related thereto must usually be approved by the local government. Approval should be forthcoming when the local authorities are satisfied that the chimney will be high enough to ensure that the smoke, dust, grit, and gases therefrom will not affect public health or become a nuisance. Upon the request of a local government official, the central government will provide assistance in determining the appropriate height for a chimney under specified circumstances.[66]

The nuisance provisions of the Public Health Acts were originally enacted to regulate emissions odors from the wastes of humans and animals. The provisions authorize local government to control smoke emissions from sources other than industrial premises and chimneys—for example, burning dumps and bonfires. These "smoke" nuisance provisions have been extended through the Clean Air Acts to encompass selected emissions from industrial processes.[67] However, local government does not have the power to require prior approval of facilities except where combustion is involved. Generally, local governments can act only after a nuisance from emissions has occurred, even though they have a general right of entry as authorized by the Public Health Acts.

There are often numerous inspections by local authorities in problem areas, for example, "smokeless zones," but most local inspections occur on an irregular basis. The Pollution Control Act of 1974 provides local authorities with expanded powers to obtain information on the emissions from industrial establishments.[68] Although these new discretionary powers became effective in January 1976, most local authorities have forgone their use. Many are awaiting reports from local advisory committees as to the emissions for which data should be requested. The DOE predicted that by the end of 1978 all large urban authorities were collecting information on emissions from firms within their jurisdiction. The aggregation of this data is the basis for DOE statistics on air pollution control that are explained in the 1979-82 update below.

The Pollution Control Act authorizes the collection of information on smoke, dust, grit, sulfur dioxide, and various other air pollutants emitted by industrial and trade firms. The law requires that the information collected be published in the public register. The public disclosure of a firm's emissions would provide more information for local and central government authorities, as well as environmental groups. Therefore, such disclosure will inevitably result in more regulation of emissions by industry.[69] For example, the Pollution Control Act provides for the central government's regulation of the sulfur content of fuel oil with actual limits varying for different areas of the United Kingdom. Enforcement is the responsibility of local government except in respect to "registered establishments" regulated by the central government as provided in the Alkali Acts. Without implementation of local government's discretionary authority to request emission data from firms, local enforcement of the limit on the sulfur content of fuel oil cannot be effected.[70]

### National Versus Local Air Pollution Control

In comparing the regulation of emissions from stationary sources by authorities at the national and local levels, at least two generalizations are possible. First, the central government (Alkali Inspectorate) relies primarily on the concept of "best practical means" in controlling air pollution from "scheduled processes," while local authorities do not uniformly rely on such a concept for regulating emissions. Second, the central government's regulatory methods are generally more sophisticated and broader in scope than those of local government—for example, the Inspectorate's enforcement through prior approval of facilities and application of the "best practicable means" concept.

Although local governments do not have broad authority to approve new and expanded industrial facilities before operations commence, they do have "prior approval" powers over those processes wherein combustion of fossil fuels occurs and over the chimney heights related thereto. In contrast, new industrial establishments that utilize one or more of the "scheduled processes" must acquire prior approval of their pollution abatement equipment from the Alkali Inspectorate before operations may begin. As explained above, the law does not specify the type of equipment to be used, but rather it delineates the performance level necessary for achieving the "best practicable means" requirements, including plant maintenance and supervisory procedures.

The "best practicable means" notion generally does not rely on statutory ambient-air-quality or source-emission standards. Therefore, industries that employ one or more of the "scheduled processes" are always required to utilize the best available abatement to minimize emissions or at least to render those unavoidable airborne pollutants inoffensive and harmless to public health. In essence, plant design and maintenance requirements, as well as those few emission limits that have been stipulated, have been established to effect acceptable ambient air quality. At the same time, they allow for consideration of local conditions, the economic limits of each industry, and the present and future state of pollution abatement technology.[71] The Alkali Inspectorate is responsible for determining "the level of pollution, the best practicable means of pollution control, and the pollution deemed harmless and inoffensive. The chief inspector makes the final decision on the requirements for each industry, but always in close consultation with industry and other interested parties."[72] Finally, at least twice a year officials from the Inspectorate conduct unannounced audits of "scheduled processes" wherein emission tests are made and other data on the operation of the process(es) are recorded in strictest confidence.

In late 1977, the HM Alkali and Clean Air Inspectorate proposed that several industrial processes now under the control of local government be subject to central government air pollution control regulations. Even though the actual type and number of processes affected will not be known until the revision process is complete,[73] the principal recommendations of the Alkali Inspectorate will be

> To shift some processes now registered under the inspectorate to local authorities while bringing others controlled by local bodies back under the central Government; to add some asbestos works to the processes subject to central Government controls. Currently none is under its control; and to broaden application to the type of iron and steel cupola which now falls under central Government supervision.[74]

"Some processes" would probably include several chemical and metal processes.

If no controversy results, the revision process will take about six months to complete once the Alkali Inspectorate actually makes its proposal. The process includes a three-month period for public comment and a three-stage review within the central government. During the drafting stage, the affected industries and other interested parties are consulted through informal discussions with representatives of the Alkali Inspectorate to minimize future controversy. If the industries affected by proposed revisions do not concur with the emission limits therein, they can appeal to the Secretary of State for the Environment. The industries affected by the new requirements are usually given three to five years to adapt their older facilities, but new plants must comply immediately.

By the end of 1977, the Alkali Inspectorate regulated the emissions of sixty-two different industrial processes at approximately 2,200 industrial locations. Local government was responsible for approximately 50,000 other industrial establishments. In principle, the sixty-two scheduled processes are regulated by the Alkali Inspectorate because they have the greatest potential for pollution. Insofar as the Inspectorate generally has maintained a technical edge over local authorities, the Inspectorate has traditionally expanded its regulatory responsibility to those industrial processes where technical problems have arisen. At the same time, it relinquishes responsibility to local government where technical problems have been resolved.[75]

### Abatement of Motor Vehicle Emissions

Emissions from motor vehicles are the other major source of air pollution in the United Kingdom. These sources have never been considered as great a threat to public health as the domestic and industrial sources of smoke and other emissions. Nonetheless the Motor Vehicles (Construction and Use) Regulations were enacted in 1973 based on a projected doubling of motor vehicles by the year 2000[76] and the requirement that the United Kingdom enact environmental legislation to conform with EEC directives on pollution abatement.[77] Pursuant to the provisions of the Road Traffic Acts of 1960 and 1972, these regulations require that the design engineering and maintenance of motor vehicles be carried out so that no avoidable smoke and other visible vapor are emitted. Motor vehicles with gasoline engines must be designed to or be fitted with equipment that recirculates crankcase emissions. Diesel-powered motor vehicles must conform with special emission limits—especially severe on black smoke—established by the central government's British Standards Institution,[78] but these special limits apply only to those vehicles manufactured after October 1972.

The Motor Vehicles Regulations provide for the control of emissions of CO, unburnt hydrocarbons, and $NO_x$. The limits for hydrocarbons became more stringent for motor vehicles after April 1977. New models manufactured after October 1977 have been required to comply with standards for emissions of $NO_x$. In addition, the 1974 Pollution Control Act authorized the central government to regulate the content of fuels for motor vehicles. Consequently, the lead content of gasoline has been limited to 0.45 grams per liter since January 1, 1978.[79]

### Solid Waste Pollution Control

The omnibus legislation enacted in 1974 and popularly referred to as the Pollution Control Act (PCA) also provides for the comprehensive regulation of waste disposal on land. When the applicable provisions of the law have been fully implemented, local government will be responsible not only for establishing extensive plans for the disposal of household, commercial, and industrial wastes but also for ensuring the adequate disposal of such wastes.[80] Under the PCA, a site-licensing system has recently been established, thereby creating a disposal permission mechanism.

Waste disposal may only be conducted according to the provisions of a site license issued by the appropriate local authority for each disposal site. The license specifies both the operating conditions and the type of wastes for each landfill site.[81] Local authorities have been given the enforcement powers necessary to make certain that specified operating conditions are heeded. Persons disposing of wastes without a license are subject to legal sanctions. Under provisions not yet fully implemented, producers of dangerous wastes will be regulated.[82] The Deposit of Poisonous Wastes Act of 1972 provides only for controls on the ultimate disposal of wastes which are poisonous, noxious, or polluting. Under the law, accounting for toxic wastes will be made easier by monitoring production, use, and disposal.

Even though local government officials have had the authority to control solid waste since enactment of the Public Health Act in 1936, the waste-disposal problem was virtually ignored until the enactment of the 1947 Town and Country Planning Act. This law gave local government the authority to regulate all new land development, including landfill disposal sites and surface mines. Land dereliction occurring prior to 1947 was not affected. However, land restoration has been financially supported by the central government[83] through public works programs like "Operation Eyesore," which was designed both to beautify the countryside and to relieve the high unemployment during the 1972-73 recessionary period.[84]

Beginning in the mid-1960s with the reorganization of the government in Greater London and ending in the mid-1970s with the consolidation of local government authorities in England, Scotland, and Wales, a new, more streamlined pattern for local government authority came into being. The purpose of the reorganization was to strengthen participatory democracy through the establishment of fewer local jurisdictions with the economies of scale required to carry out their parliamentary-designated functions, including environmental planning and solid disposal waste.

The lower level of local government (district or borough) is responsible for solid waste collection, and the higher level (County or region) is charged with disposing of solid wastes. Even so, waste management in the United Kingdom remains a goal rather than a reality because the amount of waste continues to grow at a rapid rate while very little recycling or incineration to produce electric power or heating has occurred. In a small, densely populated area like the United Kingdom such trends must be reversed. In 1974, the Department of Industry established an experimental Waste Materials Exchange which excluded household refuse, scrap metals, second-hand equipment, and other recyclable materials for which markets already exist. The central government has also established a Waste Management Advisory Council made up of experts from industry, local government, labor unions, and public interest groups representing the consumer, the environment and neighborhoods. Working Groups within the Council have made detailed studies of the recovery of ferrous metals, nonferrous metals, and packaging. In addition, a Working Group has studied the economics of waste recovery, especially the reuse of wastepaper because the British paper industry is so dependent thereon.

Before the enactment of PCA, several laws were available for the public control of wastes which often create special problems—waste oil, used tires, abandoned motor vehicles, discarded household appliances, mining and quarry wastes, and litter. Until the PCA is fully implemented, oil waste will be regulated as provided in the Deposit of Poisonous Wastes Act; tires, vehicles, and appliances, as provided in the Civic Amenities Act; extractive industry waste, as provided in the Mines and Quarries (Tips) Act; and litter, as provided in the Litter Act and Dangerous Litter Act.[85]

Pollution Taxes

Apparently, no pollution taxes are imposed in the United Kingdom, even though the local government's authority to charge for waste-disposal services and the Regional Water Authorities' power to establish charges for water use and wastewater treatment within their jurisdiction provide the institutional mechanism for implementing effluent taxes and waste-disposal taxes. In general, the Parliamentary Acts governing local authorities provide for the financing of their functions through the levying of taxes on real property (30 to 35 percent) and through central government grants (45 percent). The other major source of local government revenue is rents and service charges.[86] Local authorities may charge for the pickup of commercial waste and must charge for the industrial waste they collect, but household waste collection is financed with property tax revenue.[87]

## Government Assistance

The "polluter pays principle" is adhered to in the United Kingdom as it is in the other member-countries of the OECD. While in principle pollution control costs are incurred by the polluter, there are two major exceptions in the British Isles. Where it is not technically and economically feasible for a private firm to comply with specified standards, there may be a variance from or waiver of pollution abatement standards by application of the "best practicable means" standard. In addition, the nationalized industries can receive long-term financing from the central government to assist in capital investment programs, as well as some assistance for operating costs. Since the subsidy programs for the nationalized industries are broadly applied, capital investments in pollution abatement equipment and current costs of pollution control are often covered.

A subsidy program also exists for partial reimbursement of the costs incurred by individual households that must convert to appliances suitable for smokeless use. The subsidy is from the local officials. In summary, the conversion costs can be borne as follows: homeowner or renter—30 percent; local authority—30 percent; and central government—40 percent.[88]

Capital investment in water supply, wastewater treatment, and sewage disposal facilities are not subsidized by the central government. There are some exceptions, however. In England and Wales, there were grants from the central government under the Rural Water Supplies and Sewage Acts, 1944-65 and the Local Employment Acts, 1960-72. The Secretary of the State for the Environment may subsidize the operating expenses of local (water board, government, or water company) or Regional Water Authorities to provide or improve the water supply in a rural area, as well as to adequately provide for the wastewater treatment or sewage disposal facilities in a rural locale. County councils are also authorized to subsidize the operating expenses of a Regional Water Authority. In regions designated development areas by the central government because of high unemployment, population changes, migration, or regional economic policy, subsidies from the central government are available for the improvement of water supply, wastewater treatment, and sewage disposal.[89] Under provisions of the Industry Act of 1972, the central government may also subsidize the processing of scrap and other waste materials.[90]

In Scotland, the Secretary of State may make direct grants to regional and island councils in their capacity as water suppliers and operators of wastewater treatment facilities to improve the water supply, wastewater treatment, and sewage disposal in rural areas and to assist the development of industry. Indirect grants are made in the form of subsidies to support a specified level of water and sewage rates.[91]

The British income tax law does not provide for any special treatment of investments in pollution control equipment and facilities. Rather, pollution abatement investments are treated just like any other capital expenditure incurred by a trade or business.[92] A first-year capital cost allowance (depreciation) of up to 100 percent of a taxpayer's investment in industrial equipment and machinery for use in a trade or business can be deducted in determining taxable income. A 25 percent declining balance depreciation is allowed for any portion of the taxpayer's investment that is not deducted in the first year. The maximum capital cost allowance applicable to investments in industrial buildings is 50 percent in the first year and then 4 percent declining balance for the investment that is not depreciated in that initial year. Capital investments made in conducting scientific research related to a taxpayer's trade or business, such as in the development of pollution control technology, are also fully deductible in the year the expenditures are incurred.[93]

British tax policy as reflected in the income tax, value added tax (VAT), and excise taxes imposes a relatively large burden from the taxpayers who own automobiles. The income tax law provides for a first-year capital cost allowance of up to 100 percent for only those motor cars for hire or carriage of the public.[94] All other cars used in a trade or business qualify for the 25 percent of declining balance method of capital cost recovery. The 8 (now 15) percent VAT introduced when the United Kingdom became part of the Common Market applied to motor vehicles, along with a car tax at 10 percent of wholesale value, which is the wholesale price of a car net of taxes.[95] VAT fell on the retail price of a car, including the 10 percent car tax. When the VAT rate was increased to 15 percent, the car tax was abolished.[96] There is also an annual motor vehicle tax of £25 (approximately $40) on private cars. Motor vehicles used to transport goods are taxed according to their unladen weight, while buses and taxis are taxed by seating capacity. The exactions for "goods" vehicles range between £30 ($50) and £567 ($1,000) annually.

Even though energy sources, namely, coal, gas, electricity, oil, and gasoline (or petrol), are generally not taxed under VAT,[97] gasoline for private automobiles is taxed at a special 12.5 percent rate. Public transportation is exempt from the VAT.[98] In addition, there is a levy or customs duty on imported oil which varies in relation to final use of the crude as gasoline, diesel fuel, fuel oil, kerosene, or lubricating oil. The exaction on gasoline and diesel fuel is 22.5 pence (40¢) per Imperial gallon, whereas the duty on other derivatives of crude oil is 1 pence (1.75¢) per Imperial gallon. Similar excise taxes are imposed on indigenous crude oil products, for example North Sea oil and gas. Exemption from these exactions on crude oil derivatives is applicable to their use in fishing vessels and lifeboats, as refinery fuel, as part of industrial processes, and as a raw material for gas-making and chemical synthesis. Moreover, the impost on gasoline and diesel fuels is reduced to that applied to fuel oil when they are burned in approved furnaces.[99] Finally, waste oils that are recycled for lubrication and for use in heating are exempt from the excise tax on crude oil derivatives.[100]

## Environmental Policy Developments, 1979-1981

When Prime Minister Margaret Thatcher's Conservative party won a decided majority in Parliament's House of Commons in the spring (May) of 1979, it marked a significant turning point in the governing of postwar Britain. With the exception of a brief period in the early 1970s, the Labour party has been in power continuously since the end of World War II. During its more than thirty years' tenure as the majority party, it pioneered and advanced the political economy of Lord Keynes in building a welfare State of massive proportions with a huge bureaucracy and many State-owned basic industries, including steel, coal, oil, natural gas, motor vehicle and aircraft manufacturing, and air transport. Whenever a problem was perceived, it seemed to be the duty of the government to provide a solution through regulation or government assistance.

Thatcher's Conservative party was elected on a platform that rejected the policies of the previous thirty-five years. Consequently, her government has not only reduced the public sector's influence on social and economic affairs, but it has also significantly lowered income taxes. Even though valued added taxes on consumption were increased, the overall rate of tax revenue growth has been reduced, thereby requiring reductions in government expenditures—much to the delight of the constituency of the Conservative party.[101] As is explained below, those government agencies with responsibilities for pollution control and environmental quality have suffered budget cuts just like other agencies. While the budget cuts have affected the enforcement and research activities of agencies with environmental protection responsibilities,[102] the implementation and development of environmental law, regulation, and monitoring continued during the 1979-81 period. The most significant of these developments are described below.

### State of the Environment

The Conservative government's DOE has changed the emphasis in environmental protection from accumulating data to cope with immediate issues to using data to monitor progress in attaining environmental quality objectives. When the DOE published its annual pollution reports for 1979 in early 1980, a digest of environmental pollution statistics was included for the second time. By including both types of information, along with an assessment of the costs of both the preventive or remedial measures and the environmental damage in the DOE's annual state of the environment publication, it is anticipated that better decisions can be made on establishing priorities for pollution control actions that bring steady progress in improving the environment.

Based on the 1979 publication, water quality in rivers continues to improve with detergents no longer a major pollution problem. However, the discovery of extremely low concentrations of a variety of organic micropollutants has emerged as a potential health problem, and levels of lead, cadmium, mercury, and selected pesticides are being monitored in rivers where such discharges may be occurring. About 10 percent of the households in Great Britain (except Northern Ireland) use drinking water with lead levels above the international public health standard. Where public health problems persist, the water authorities will take remedial action, such as water treatment, source replacement, or pipe replacement.

Air pollution from home heating and power generation has decreased significantly over the past two decades, whereas emissions from motor vehicles have grown because of the increased number of vehicles in service. More specifically, smoke emissions have decreased 80 percent since 1960. $SO_2$ emissions, which were fairly constant during the 1960s, have fallen about 16 percent since 1970, although little decrease therein has occurred in recent years. There has been a more dramatic 50 percent decrease in ambient levels of $SO_2$ in urban areas since the early 1960s. Conversely, motor vehicle emissions of $NO_x$, CO, and hydrocarbons are now a major contributor to air pollution in the central city sections of most urban areas. Increased emissions from gasoline-powered vehicles (nearly 20 percent for 1970-77) and diesel-powered vehicles (10 percent for 1970-77) should eventually be brought under control insofar as the United Kingdom has adopted the emission standards established by the Economic Commission of Europe for new motor vehicles.

Progress is being made on the reclamation of urban land that was formerly used for dumping hazardous wastes, as well as sites of gas works and sewage farms. In fact, many of the contaminated sites may be used for development after remedial measures to mitigate environmental problems are completed. The reclaimed land will be used for housing, recreation, community services, industry, or farming.

About 85 percent of household and other nonhazardous urban waste (except that disposed of by private contractors) is being disposed of in licensed sanitary landfills without prior treatment. Because of the increased reliance on incineration, the amount of waste disposed of in landfills without prior treatment increased by only about 30 percent during the 1965-75 decade, although the quantity of waste to be disposed of increased much more than that. Research on the disposal of modest amounts of hazardous wastes with urban wastes in landfills demonstrates that such disposal can be done safely.

Inasmuch as noise complaints to environmental health officers increased at a fairly constant rate during the 1972-78 period, noise from motor vehicles and aircraft has emerged as a very serious problem near highways and airports. Its seriousness is accentuated by the fact that most complaints are made to highway authorities and airport operators. In August 1980, the Transport Ministry announced tougher noise pollution standards for motor vehicles produced for 1983 and thereafter.[103]

In its 1980 annual publication on environmental pollution and statistics, the DOE reported little overall improvement in air and noise pollution, while progress on waste-disposal and wastewater treatment continued but at increased costs. As noted above, downward trends in smoke and $SO_2$ emissions and ambient levels had indeed leveled off during 1979, while emissions from motor vehicles continued to grow at a steady rate. Similarly, noise complaints to environmental health officers continued to show a steady rise.[104]

In the National Water Council's study for the 1976-80 period, a substantial decrease in water pollution in the rivers and estuaries in England and Wales for the 1970s was reported, but prospects for continued overall improvement were not promising because of economic stagnation and significant budget cuts. In general, as a result of significant investments in pollution control during the early 1970s, the sources of many pollutants are under control, and fish have returned to many rivers that had been lifeless. Continued and changing sources of pollution, along with the deferral of some pollution control investment during the late 1970s, has resulted in the deterioration of water quality in some rivers that were previously not considered serious problems. However, such deterioration has seldom affected users of these rivers, and overall water quality has improved.[105]

In Scotland, over two-thirds of the rivers are in compliance with the EEC's requirements for fresh water quality to support marine life. The cleanliness of Scottish rivers reflects both the importance of the fishing industry and its lack of the heavily polluting industry found in England and Wales. However, the budget cuts in London and a stagnant economy have forced the water authorities in Scotland to concentrate their water pollution control investments on noncomplying rivers rather than to try to improve water quality in all rivers.[106]

## Air Pollution Control

While the responsibility for water pollution control is primarily left to the Regional Water Authorities, the central government retains responsibility for controlling air pollution. As mentioned earlier, industrial, commercial, and household air pollution was brought under control during the past two decades, although improvements have leveled off. Air pollution complaints from the inspection of licensed and unlicensed plants were down in 1979 as compared to the peak year of 1978. The Alkali and Clean Air Inspectorate has published, and is developing for publication, guidelines on the best practicable means for controlling emissions in a wide range of industries using scheduled processes. A set of guidelines for sampling and analyzing emissions is also going to be developed as a complement to the best practicable means guidelines.

In addition to a shortage of available funds for investment in better pollution control equipment in a stagnant economy, there are, of course, many recurring industrial air pollution problems. Licensed chemical plants are still guilty of too many accidental releases of gases and fumes. Although some British Steel Corporation plants that violate the best practicable means standard have been closed, others have been unable to reduce emissions because of cutbacks in central government expenditures. Because of the international oil supply situation in 1979, some heavier crude oil had to be used in British refineries, thereby increasing $SO_2$ emissions. Coal-fired power plants continue to experience problems with plant start-up, intermittent operational difficulties, and adverse meteorological conditions. Infractions by mineral producers did not decline in 1979, although most problems were quickly corrected once they were pointed out. Metal recovery operations also continued to cause many emission problems in 1979.

Finally, infractions by cement plants were still too high in 1979, although the complaints are typically associated with low-level emissions. Hopefully, the establishment in 1979 of more stringent limits on particulate emissions for new cement kilns, along with the requirement that continuous particulate emission monitors be attached to the new kilns, will solve the problem with low-level emissions.[107] In summary, the improper use and maintenance of air pollution control equipment results in most of the problems of recurring emissions.[108]

### Hazardous Waste Disposal

The disposal of urban household and commercial wastes is the exclusive responsibility of local government, which seems to be handling the problem well on its own or through licensed disposal sites. However, the PCA of 1974 authorizes the central government to regulate the disposal of hazardous wastes, although local government is generally responsible for monitoring their disposal too. Nonhazardous industrial wastes may also have to be regulated by local government when industry is unable to properly handle its own residuals at the plant site.

In February 1981, the Environment Ministry implemented its controversial new regulations for the transport and disposal of hazardous wastes to be effective in mid-March 1981. As provided in the PCA, these Control of Pollution Special Waste Regulations (CPSWR) supersede the 1972 Deposit of Poisonous Wastes Act. Although the CPSWR adds for the first time a set of regulations that control the transport of hazardous wastes to a disposal site, their controversial nature resulted from a change in prenotification requirements applicable to waste producers and transporters upon transfer of hazardous waste to a licensed disposal site. In order to overcome opposition from Members of Parliament, local governments, and environmentalists, the DOE agreed to review the CPSWR after one year to alleviate fears that the new regulations would increase the environmentally unsound disposal of hazardous wastes.

In practice, the new prenotification system provides that waste producers determine which of their wastes are "dangerous to life," that is, special wastes according to general criteria specified in the regulations, rather than a threat to water supplies or the environment generally. Although industry had hoped for specific criteria, for example, an inclusive list of substances classified in terms of form, quantity, and/or concentration, the regulations include groups of substances rather than specific substances and there are no references therein to form, quantity, and/or concentration. "Dangerous to life" is based on the threat to the life of a hypothetical 20-kilogram child who is four and one-half years old. In summary, about one-third of the hazardous wastes requiring prenotification under the 1972 Deposit of Poisonous Wastes Act will now be required to prenotify waste disposal. The reduction in prenotification activity should allow local waste-disposal authorities to devote more resources to monitoring the operations of licensed disposal waste sites and along with industry to investigate the toxic effects of its wastes.[109]

Toxic Substances

The United Kingdom, like its fellow EEC members France and West Germany, had until September 18, 1981, to implement the requirements of the Sixth Amendment to the EEC Directive on Classification, Packaging, and Labeling of Dangerous Substances in respect to notification by producers and importers of toxic substances. The regulations implementing the Sixth Amendment were the joint effort of the DOE and the Health and Safety Commission (Executive) under the authority of the 1974 Health and Safety at Work Act and the European Communities Act.[110] The regulations were not proposed for comment until mid-February 1981.[111] Consequently, interim regulations had to be issued in early September 1981 in order to meet the EEC's September 18 deadline.[112] Later, in early December 1981, final regulations implementing the Sixth Amendment were issued by the Health and Safety Commission (HSC).[113]

In 1978, about 800 dangerous chemicals were included in the Packaging and Labeling of Dangerous Substances Regulations. Another 121 dangerous chemicals were added to the list in August 1981 in a revision that was part of the implementation of the Sixth Amendment. In 1978, producers of dangerous chemicals were only required to put a simple label on the container warning the user of its dangers, explaining its use and handling procedures and how to package the substances in containers of sound construction, but the August 1981 revisions required that the label include risk warnings for handling twenty-four of the substances on the original list of 800.[114] Finally, the December 1981 regulations, referred to above, completely revised the 1978 regulations, required more risk warnings, and increased the number of chemicals listed to over 1,000. The main features of the 1978 regulations were integrated with the new regulations.[115]

The land transport and importation by ship of dangerous substances are also regulated. In August 1981, the HSC issued its Dangerous Substances Conveyance by Road in Road Tankers and Tank Containers Regulations.[116] Later, in December 1981, the HSC published a Code of Practice explaining how dangerous substances are to be classified and labeled for road transport purposes.[117] Similarly, in August 1981, the Shipping Ministry issued the Merchant Shipping (Tankers) Regulations applicable to tankers carrying oil, gas, and chemicals in bulk into and out of British ports.[118] With regard to ships carrying dangerous chemicals in nonbulk form, such as canisters, barrels, or other containers, the Department of Trade Marine Division issued a special Notice to Mariners to regulate the inadvertent loss of containers of dangerous chemicals in and near British ports until the 1973-78 Marine Pollution Convention takes effect in 1983.[119] However, cradle-to-grave regulation of most toxic substances is not yet a reality.

### Future Environmental Policy

Because of the Conservative party's overwhelming majority in the House of Commons, there will most likely be no national election until the spring of 1984, which marks the expiration of the five-year term of each Member of Parliament. Therefore, Prime Minister Thatcher will govern the United Kingdom according to the Conservative party's platform until early 1984. Her political security assured, Thatcher will continue to dismantle the bureaucracy, including government agencies with environmental protection responsibilities. As already noted, agencies with responsibilities for maintaining and improving environmental quality have indeed not been spared from the Thatcher government's general budget-cutting policy.

Budget cuts reflect both the Conservatives' economic philosophy of reducing income taxes to stimulate investment and long-run economic growth, and their political philosophy of deregulation where the costs of government control are greater than the benefits. Deregulation is not intended to be no regulation, but reflects a desire to make the central government more cost-effective by either transferring some regulatory responsibilities to local government or relying on voluntary compliance by institutions and individuals. If environmental quality as measured by pollution indices decreases, one would assume that the deregulation trend would be reversed in respect to those aspects of pollution control linked with the deterioration in environmental quality. Unfortunately, the stagnant United Kingdom economy may cause industry and local government to forego investments in pollution control. On the other hand, the closing of unprofitable industrial installations often reduces pollution.

What follows is a summary of the more significant impending environmental problems, as well as anticipated environmental policy developments. More specifically, water pollution, air pollution, solid waste, hazardous waste, noise, and toxic substances will continue to be problems, although progress has been made in controlling the most serious aspects of these problems. Furthermore, several changes in or additions to environmental law and environmental protection administration have already been announced or proposed. Such environmental policy developments are explained below with the pollution problem to which they are applicable.

### Water Pollution

As discussed earlier, river water quality improved dramatically during the 1970s when public water and river authorities, as well as industry, made massive investments in wastewater treatment. Now that much of the gross water pollution is under control, more subtle sources and types of water pollution have emerged. The stagnant economy has already jeopardized improvements in water quality because investments to replace the older sewage treatment facilities have been deferred. Such deferral results in the overloading of the aging facilities as urban population grows, and industrial and commercial polluters comply with effluent regulations. Such plants are not designed to deal with the greater complexity of many industrial and commercial effluents, thereby increasing the level of exotic contaminants in surface water.

The agricultural pollution of rivers has apparently caused the deterioration of rivers previously classified as unpolluted.[120] Because of urban encroachment on some of the best agricultural land, farming is being conducted more intensively with the increased use of pesticides and synthetic fertilizers, as well as livestock feedlots, generating much of the pollution from agricultural run-off. Generally, agricultural activity is exempt from land use planning controls, and there are only voluntary controls on the use of agricultural chemicals.

The Thatcher government has accepted in principle the recommendation from the Royal Commission on Environmental Pollution (RCEP) that the Agricultural Ministry take more responsibility for control of pollution from farming activities. However, the government is reluctant to include agriculture in land use planning at a time when it is attempting to streamline such development controls. Of the eighty recommendations made by the RCEP in its study of agriculture and pollution, Thatcher's Environment Minister Tom King agrees that research on the environmental impact of farming should be increased notwithstanding overall reductions in the budget for research. However, limited fiscal resources will require that research priorities be established. Similarly, priorities will have to be established for the many other recommendations made by the RCEP.[121]

The National Water Council (NWC) has revised its river classification scheme for evaluating water quality in developing its 1980 report on water quality. The new scheme takes into account the potential use of rivers in such an evaluation. Furthermore, a new taxonomy is applicable to estuaries, excluding not only nontidal rivers and canals, but also some sections of surface water previously considered tidal rivers. As a result, most sections of tidal rivers and, for the first time, many of the major estuaries are included in a new "estuaries" classification, while all other surface water is included in the "rivers" classification.[122] Table 7-1 summarizes the state of surface water quality as described in the 1980 NWC report.

In the past, the monitoring of pollutants in estuarine, coastal, and marine waters focused on dangerous substances that were known to be toxic, persistent, and bio-accumulative, for example, heavy metals, pesticides, and PCBs. Recently, a system was established in England and Wales to

## Table 7-1
## 1980 Surface Water Quality

| Quality | Rivers[a] | Estuaries[a] |
|---------|-----------|--------------|
| Good[b] | 36.2/33.1 | 67.3 |
| Fair[c] | 21.6 | 23.5 |
| Poor[d] | 7.6 | 5.2 |
| Bad[e] | 1.5 | 4.0 |

SOURCE: *International Environment Reports—Current Report*, January 13, 1982, p. 9.

[a]Percentages of total length of rivers and estuaries surveyed.

[b]For rivers, there are two levels of "good quality" rivers. Class 1A is defined as water of high quality suitable for potable supply abstraction, game, or other high-class fisheries, and high amenity value. Class 1B is defined as water of less quality than 1A but usable for substantially the same purposes. For estuaries, the definition of "good quality" is the same as that for Class 1A rivers.

[c]"Fair quality" is defined as waters suitable for potable supply after advanced treatment and supporting reasonably good coarse fisheries with moderate amenity value.

[d]"Poor quality" is defined as waters that are polluted to such an extent that fish are absent or only sporadically present. Such river water can be used for low-grade industrial abstraction purposes, and it has considerable potential for further use if cleaned up.

[e]"Bad quality" is defined as waters that are grossly polluted and are likely to cause nuisance.

measure discharges of the type of pollutants specified in the EEC directive of dangerous substances.

Once the relevant provisions of the 1974 PCA are fully implemented, discharge regulations will be applicable to tidal and coastal waters. The expanded monitoring of pollutants in such waters will generate sufficient data to develop the PCA discharge controls. Research on the limited monitoring data indicates that serious damage to marine life has not occurred. The objective of the monitoring has been to discern the public health hazards of marine pollution with regard to fish and shellfish consumed by humans. However, industrialized areas on estuaries have caused some damage to marine life. Therefore, monitoring of the largest and potentially most polluting discharges, along with periodic review of the most important pollutants and related indicators, will be necessary. Notwithstanding the limited damage from estuary pollution, all industrial and municipal discharges will be subject to regulation, and the discharge controls will be reviewed regularly to ensure that the regulation is effective.[123]

Because the United Kingdom is an island nation, most of its substantial international trade enters and leaves the country through its seaports. Moreover, the North Sea development of oil and gas should continue for several years. As a consequence, the British are concerned about marine pollution from shipping and natural resource development. Generally, the central government has preferred international treaties to unilateral national control of marine pollution. However, environmentalists and politicians representing coastal areas have leveled accusations that reliance on lengthy, complicated international negotiations is just an excuse for inaction. Although negotiations should be continued under the auspices of the United Nations, IMCO (Intergovernmental Maritime Consultative Organization), and MARPOL (Marine Pollution Convention), national legislation should be enacted as an interim solution. Extension of territorial waters from 3 to 12 miles would expand the United Kingdom's jurisdiction enough to render national legislation effective for most sources of marine pollution affecting the country's coastal areas.[124]

With respect to the regulation of shipping in European ports, EEC members, Norway, and Sweden have been working on a treaty providing for stringent ship inspection requirements in each nation's ports. However, the EEC is resisting the treaty arrangement, which would be based primarily on existing international agreements that have been difficult to enforce. Thus, the EEC is threatening to establish even more stringent inspection standards based on a proposed directive developed by an EEC working group if its members consummate the treaty.[125]

In October 1981, the Royal Commission on Environmental Pollution (RCEP) released its report entitled "Oil Pollution of the Sea" in which it concluded that oil pollution does not pose a serious long-term threat to the marine environment, whereas heavy metals and radioactive waste may. The RCEP's report, which was prompted by the *Amoco Cadiz* oil tanker disaster in the English Channel, includes over forty recommendations for improving marine pollution prevention, containment, and treatment, as well as a significant reorganization of the United Kingdom's antipollution capability.

The RCEP did not discount the serious local damage associated with oil spills. It did, however, conclude that approximately 60 percent of oil pollution results from land discharges, that is, oil in effluents discharged into rivers as a result of the current exemption from standards that refineries enjoy because their installations are aged, while 20 percent each results from tanker accidents and deliberate discharges and from natural seepages and discharges of general shipping operations. The expansion of North Sea offshore oil and gas development increases the chances of a major blowout releasing large quantities of oil into the sea. The RCEP was especially concerned with blowouts near the coast, protection of pipelines, and the use of large tankers for storing oil in the North Sea near the shore.

Although the RCEP did not recommend regional action on the control of oil pollution as long as there is hope for early adoption of international agreements thereon, a European agreement on marine pollution controls may be necessary if the prospect of consummating international agreements within a reasonable period of time fades. Thus, both the extension of territorial waters from 3 to 12 miles and a European (the EEC and other European nations) agreement on enforcing international (IMCO, MARPOL, United Nations Law of the Sea, and so on) standards for ships that enter European ports may be necessary. Currently, over 100 nations define their territorial waters on the basis of a 12-mile limit.

Although sewage sludge often contains traces of heavy metals, about 28 percent of the United Kingdom's sewage sludge will continue to be dumped at sea, while the rest of the sludge is disposed of on agricultural land (40 percent), on land (22 percent), and by incinerating it (10 percent). Because the sludge is dumped at sea where strong tidal currents sweep the waste away from the coastline, dumping at sea is often the most economical and environmentally safe alternative for disposing of sewage sludge. However, the ocean's capacity to assimilate and disperse wastes is not unlimited. Consequently, dumping sites must be carefully selected, and the dumping controlled as well as monitored to ensure that marine life is not adversely affected.[126]

The EEC is developing an amendment to the 1976 Directive on Discharges of Dangerous Substances into the Aquatic Environment to regulate the disposal of sewage sludge at sea. Thus, the United Kingdom's policy of dumping sewage sludge in the sea may eventually be constrained by the EEC, even though the alternatives for disposal may be more expensive.[127]

The Thatcher government has introduced legislation that would regulate deep sea mining temporarily until the United Nations Law of the Sea Treaty is agreed upon. The proposed law, which is known as the Deep Sea Mining Act, would exempt deep sea mining from the 1974 Dumping at Sea Act and establish a licensing system and regulations for controlling the activities of mining companies hunting mineral-rich nodules on the seabed. Similar laws have been enacted in the United States and Germany and are being planned in other European countries, including France.[128]

## Air Pollution

As noted above, $SO_2$ and particulate matter emissions from industry, commerce, and households have decreased significantly over the past decades, while motor vehicle emissions have increased primarily because of the growth of automobile ownership. $SO_2$ emissions are expected to decrease until 1985, when the substitution of coal (and nuclear power) for oil begins to overcome general enhancement of all pollution abatement equipment. Although motor vehicle emissions of CO and hydrocarbons increased during the 1970s, dramatic reductions in the same emissions from households seem to have offset such increases. While lead emissions from motor vehicles increased during the early 1970s, no increases occurred during the 1975-79 period. Finally, the increase in $NO_x$ emissions showed a steady rise through 1979 inasmuch as there was no countervailing decrease to mitigate the increased burning of motor fuels. Eventually, the Economic Commission of Europe's (ECE's) pollution control regulations applicable to new motor vehicles should effect a reduction in CO and hydrocarbon emissions. In February, 1980, the EEC agreed that the ECE's automobile emission standards should be written into EEC environmental law.

With $SO_2$ and particulate matter emissions currently under control, the DOE has turned its attention to emissions of other significant air pollutants, including CO, hydrocarbons, $NO_x$, lead, mercury, cadmium, asbestos, $CO_2$, ozone, and chlorofluorcarbons. In fact, the levels of at least fifteen air pollutants have been monitored at twenty urban sites for the past five years to ascertain trends in ambient air quality for lead, cadmium, and other dangerous air pollutants. Similar monitoring activities have been ongoing in rural areas for nine years. In summary, the levels of dangerous air pollutants seem to remain in ambient concentrations well below those that could cause public health problems, and some pollutants evidence downtrends in concentrations.[129]

By far the most significant development in air pollution control relates to the future capability of the HSC to enforce its regulations and to determine potential atmospheric pollution problems inasmuch as the Thatcher government has continually reduced its inspectorate and research budgets.[130] The United Kingdom has a century-old tradition of health and safety inspectorates. By 1880, there were a dozen different health and safety inspectorates. By then, environmental inspectorates for air pollution could declare a furnace a nuisance unless its smoke emissions were abated "as far as practicable." Similarly, they could demand that water polluters use the "best practicable and reasonably available means" to make their discharges harmless. In general, the inspectorates are responsible for enforcing laws and regulations, relying on cooperation and persuasion rather than strict enforcement. On the other hand, the inspector effectively decides whether the inspectorate should prosecute a violator. As a consequence, an inspector's recommendations essentially have the force of law, especially insofar as an inspection will normally result in the discovery of one or more violations.[131]

The HSC has coped with its budget cuts by reducing its factory inspections at a time when maintenance of air pollution abatement equipment needs improvement; preparing more industry guidance material for achieving voluntary compliance with air pollution control law and regulations; cutting expenditures for research while retaining its testing and laboratory services; cutting back on the nationwide monitoring of $SO_2$ and particulate matter to allow for increased monitoring of other pollutants; and periodic transferring of inspection responsibilities to local authorities where complex technical problems of pollution control have been resolved.[132]

For several years, lead in the environment has been a controversial issue with regard to motor vehicle emissions in the United Kingdom.[133] As mentioned earlier, ambient levels of lead have been stable for the past few years. In deference to those who argue that inner city lead concentrations are too high and constitute a public health problem, especially for children, the Environment Minister Tom King announced that the maximum level of lead in gasoline will be reduced to 0.15 grams/liter from the current standard of 0.40 g/l by 1985. In adopting the German standard for leaded gasoline, the proposed EEC ambient air quality standard of 2 micrograms or less per cubic meter should be attainable in all metropolitan areas, including London. The Lawther report (named for the toxicologist P. J. Lawther who chaired the committee of experts that studied the effects of lead on the environment) was issued in April 1980. The report also recommended the EEC standard but not a reduction in the lead in gasoline because the experts could not conclusively link lead in gasoline with high levels of lead in the blood.

King also announced other measures to control lead hazards in water, food, and paint. Inasmuch as current controls on industrial emissions of lead are adequate to meet the 2-microgram standard, no measures to further restrict such emissions were recommended. Hazards from old, lead-based paint are to be dealt with by providing information and advice to local authorities and by increased emphasis on health education for the public. With respect to lead-contaminated water, it is anticipated that government subsidies for home improvements will be available to alter plumbing systems with lead-lined tanks. Moreover, local government, water authorities, and public health officials will warn that segment of the population affected by lead contaminated water supplies. Although lead residue in food was subjected to more stringent standards in 1980, the Food Additives and Contaminants Committee has been instructed to review the effect of using lead-based solder in cans.[134]

## Land Use Planning and Waste Disposal

Insofar as the United Kingdom has already developed an extensive system of local and regional land use planning, it would only support the EEC's proposed directive on environmental impact assessment where the EIA directive can be amended to allow the integration thereof with the existing land use planning and site licensing system. The disposal or incineration of wastes, especially hazardous wastes, is an important element of land use planning in the United Kingdom. Thus, the location of any economic activity, whether industrial, commercial, extractive, or agricultural, that generates wastes, especially the hazardous type, should be contingent upon the availability of environmentally sound waste-disposal methods and sites. As noted above, comprehensive regulations were issued in December 1981 to control the disposal and transporting of hazardous wastes. In summary, the disposal of wastes generated in the future should be under control if land use planning is conscientiously conducted in the context of current national regulations for waste transport and disposal. Nevertheless, land use planning must continue to play an increasingly important role in improving and maintaining environmental quality in the United Kingdom.

Problems from poor disposal of hazardous waste do exist. However, no Love Canals are apparent at this time.[135] Conversely, strip-mining, which is an important and growing source of the coal necessary for the conversion from imported oil, continues to be a problem because such coal-mining pollutes surface water and squanders scarce land where it is not restored to its original use as farmland or forest, or some alternative use, such as parkland.[136]

## 1982 Addendum

In February 1982, Environment Under-Secretary Giles Shaw announced the long-awaited implementation of the water pollution control provisions of the 1974 Pollution Control Act (PCA). Implementation, which began in July 1982, should be virtually complete by July 1986 with the most important provisions effective by July 1984. Thus, the power of water authorities in England, Wales, and Scotland to control pollution of inland, underground, coastal, and tidal waters will be strengthened and extended. However, industry will not be forced to pay more for its effluent (treated or untreated) until some future date when its profitability improves.

As provided in the PCA, "consent provisions" were introduced first in July 1982 so that discharge consents can be issued by the water authorities before the penalty provisions of the act take effect in July 1983. Nevertheless, many existing discharges to previously unregulated water bodies will not be required to obtain discharge consents until the Environment Secretary issues the necessary order to place those waters under full control. This delay should enable planning for the continued enhancement of water quality consistent with water-use objectives and the best use of the monetary resources of industry, commerce, and water authorities. For example, tidal waters are only subject to control in specific areas, including those main estuaries for which special powers were granted.

In addition, the PCA provides for public advertisement of new discharge proposals. Such public registers will include information on effluent discharges and water quality. Although the public register provisions will not be implemented until 1984 in an effort to minimize administrative costs, the regulations necessary for implementation of the provisions will be developed before then so that the water authorities can plan for the collection of pertinent data.[137]

As a result of substantial public investment in sewage systems and treatment facilities as well as more stringent controls on industrial effluent, the quality of water in nontidal rivers and canals has improved significantly over the past two decades. Over 90 percent of the pollution in effluent passing through public treatment facilities is removed before discharge to nontidal rivers, while over 50 percent is removed before discharge to tidal rivers. In respect to some coastal waters, treated effluent is often discharged too close to shore. Full implementation of the PCA should result in even better pollution control and improved water quality.[138]

Notwithstanding the desirability of implementing the water pollution control provisions of the PCA, the licensing of commercial fishing on the Thames in April 1982 marked another step in the rehabilitation of that famous river's once lifeless waters. The Thames River Authority also pleased sport fishermen by predicting a significant increase in the salmon "running" the river in 1982.[139] Although other less-publicized, equally dramatic examples of the improvement of water quality exist, the PCA's full implementation will be necessary to ensure quality water for greater recreational use and increased human consumption, insofar as groundwater supplies are being depleted and contaminated.

Inasmuch as the water pollution control provisions of the PCA generally reinforce current environmental policy, they are consistent with the requirements of the EEC's directive on protection of groundwater—the major source of drinking water in Great Britain. The directive envisages control of specified substances so that they do not contaminate groundwater directly or indirectly. Such control should be implemented through a consent procedure. Where necessary, appropriate measures should be taken to ban the release of "blacklisted" or toxic substances into aquifer and to regulate the entry of less harmful, "greylisted" substances into groundwater. Water authorities responsible for water quality and local authorities responsible for waste disposal and mineral extraction planning and regulation are the "competent authorities" to whom the central government has delegated the power necessary to effect compliance with the EEC's groundwater protection directive.[140]

Air Pollution

In monitoring air pollution from registered (licensed) industrial plants and processes, the Chief Inspectorate of Clean Air indicated in its 1980 annual report issued in March 1982 that there were 1969 registered plants involving 2862 scheduled processes. Although there were fewer public complaints on air pollution and over 85 percent of the complaints concerned registered plants, the Clean Air Inspectorate "audited" the registered plants at only a slightly increased rate when compared to 1979, that is, 14,120 inspections in 1980 to 14,099 in 1979. All ten prosecutions going to court were successful and eleven more were initiated. Only four plants received improvement notices in 1980.

In explaining the reduced number of public complaints, the Chief Inspector Jim Beighton concluded that reduced production levels linked to the economic recession in Britain was the best explanation. Ironically, the recession also forces industry to defer capital investment, as well as to close the marginally profitable, more polluting plants. Thus, short-term amelioration of industrial air pollution may soon be overcome by deterioration of emission control standards when the operations of aging plants are increased during future economic expansion.

One controversial smokeless-fuels plant in South Wales received 14 improvement notices in 1980 and several resulted from public complaints. Generally the plant failed to satisfy the requirement to use the best practical means available to control emissions of smoke, raw coal gas, dust, and grit. As a result of the efforts of the local inspector, the plant's owners agreed to invest £4 million (about $7 million) to make the improvements necessary to control its emissions. Of course, the inspector is monitoring the progress of the plant on installing the "best practical means" technology.

Following the monitoring of the level of lead in the blood of children living near one lead-smelting plant, the plant was forced to close in 1980. Lead-laden wastes were also removed from the plant site. Lead-smelting infractions at five other plants were also reported in 1980, but improvements were made at all five and the Clean Air Inspectorate will closely monitor the plants to ensure that further planned improvements are also consummated[141]

The controversy over lead pollution from auto emissions in urban areas continued during 1982. In February, it was revealed that the Chief Medical Officer of the Health Department, Sir Henry Yellowlees, had sent a letter in March, 1981, to the Ministeries of Environment, and Education and Science, as well as to the Home Office, explaining that, based on his analysis of the research results available, the link between lead in gasoline and brain damage to children is strong, notwithstanding the conclusions of the 1980 Lawther Report. Coincidentally, the central government announced in May 1981, that the maximum allowable level of lead in gasoline would be reduced from 0.4 to 0.15 gram per liter by 1985.[142]

Then it was revealed in March that British Petroleum (BP) had privately urged the government in 1981 to prohibit lead in gasoline rather than further reduce the allowable limit to 0.15 gram per liter, that is, the lowest limit provided in the EEC's directive on lead in gasoline. BP's reservations on lead-free gasoline included concern about the cost of redesigning engines to burn lead-free gasoline, the possible loss of auto exports if the EEC does not decide to adopt a directive prohibiting lead in gasoline, and the cost of building or adapting refineries to produce lead-free gasoline. A poll also taken in March indicated 9 out of 10 Britons supported a ban on lead in gasoline even if it resulted in higher prices for gasoline and automobiles.

Although the *British Medical Journal* attacked lead-free gasoline in one of its February 1982 issues,[143] the British Medical Association's (BMA's) Council unanimously approved a report at its May meeting in which the BMA's science and education board concluded that it is scientifically impossible to establish a level at which lead is harmful to humans. The report, which has been submitted to the Royal Commission on Environmental Pollution (RCEP), also concluded that it should be obvious to everyone what should be done when the opportunity arises to reduce lead concentrations in the environment, especially because lead is bio-accumulative and the body can only rid itself of the element over a long period.

The National Executive Committee of the opposition Labor Party also announced in May its commitment to ban lead in gasoline, to require that all new cars sold in the United Kingdom be manufactured with engines that use lead-free fuel, and to impose higher taxes on leaded gasoline. The opposition Liberal and Social Democratic parties also are expected to adopt similar policy positions on lead-free gasoline.[144]

In response to the public concern for lead pollution, the RCEP felt compelled to undertake a major study of the environmental effects of the substance. Conflicting expert opinion on the public health effects of low-level lead concentrations in humans and the role low-lead gasoline plays in effecting such concentrations also prompted the RCEP study. Even if this new study reaffirms the 1980 Lawther Report,[145] Parliament may eventually take action to ban leaded fuel. For example, a private bill that was introduced in Parliament in June would require all filling stations to sell unleaded fuel and all new automobiles to be designed to operate on lead-free fuel. However, private bills are seldom enacted.[146] Although the United States currently stands alone in requiring that all new cars operate on lead-free gasoline, Japan and Australia are in the process of enforcing similar requirements, and there is some support for an EEC directive requiring that only lead-free fueled automobiles be sold in member countries.[147]

## Hazardous Waste

In spite of the extensive body of environmental protection law in the United Kingdom, the problem of hazardous waste disposal is far from being solved. Unfortunately, local government is ill-equipped to handle the hazardous waste disposal responsibilities that now accompany its traditional responsibility for disposal of household, commercial, and industrial waste. As mentioned above, the Control of Pollution Special Waste Regulations (CPSWR) promulgated in 1980 under authority provided in the PCA replaced the 1972 Deposit of Poisonous Wastes Act as of March 16, 1981. The CPSWR, deemed necessary to implement the 1978 EEC directive on toxic and dangerous substances, actually reduced by about two-thirds the number of wastes for which local government disposal authorities must be pre-notified. Needless to say, environmentalists are upset by what they perceive to be a relaxation of controls on hazardous waste disposal.

Environmentalists are even more concerned about how effectively local government has been in implementing the CPSWR (and its predecessor the Deposit of Poisonous Wastes Act). When the House of Lords Select Committee on Science and Technology held hearings during late July 1981, the testimony of Department of Environment (DOE) officials revealed that central government is generally unaware of the quantity and type of hazardous waste produced and disposed of in the United Kingdom and the DOE does not coordinate efforts of local government to dispose of hazardous waste. The Select Committee's survey of 165 local waste disposal authorities also revealed that of the 140 local authorities responding to the Committee's questionnaire only 12 had waste disposal plans or data on waste in spite of the seven year period since enactment of the PCA provision that requires such data and planning.

Even if the local plans were available, the DOE officials indicated that they were not prepared to analyze them to determine patterns of hazardous waste disposal so that the burden of disposal could be shared throughout the country. Under the CPSWR, disposal sites are to be licensed by local authorities. They are also responsible for enforcing the operating terms and conditions of the license, controlling the transportation and disposal of hazardous wastes, and ensuring that permanent records of the location of all hazardous wastes so disposed are maintained. Nevertheless, few local waste disposal authorities know how much or what type of hazardous waste is generated or disposed of in their jurisdiction.

The Select Committee concluded that about half of all industrial waste is disposed on the site where it is generated. Although such sites must be licensed, notification of disposal is not required, and very little is known about the amount and type of this waste, or of how it is disposed. The rest of industrial waste is disposed of by private contractors primarily in licensed landfills, but it is also disposed of by dumping in the sea, through incineration, by biological, physical or chemical treatment, or even in underground storage facilities. The private contractors tend to frequent those large landfills which will handle large amounts of hazardous waste with a minimum amount of supervision. Thus, less than ten percent of the landfill sites handle about two-thirds of the hazardous waste.

Notwithstanding the concentration of sites for disposal of hazardous wastes, most local authorities are ill-equipped to monitor and enforce the terms and conditions under which they have granted licenses insofar as they are still responsible for collecting and disposing of household and commercial waste at a time when reductions in government spending preclude the hiring of additional inspectors and scientific staff. Although no Love Canal incidents have occurred, it may be only a matter of time before one or more do occur not only because of current laxity in regulating most licensed dumps but also because many dumps were not regulated before enactment of the 1972 Deposit of Poisonous Wastes Act. The DPWA was primarily a reaction to the revelation that sodium cyanide drums were dumped indiscriminately in the west Midlands at Nuneaton.

Although a more extensive analysis of the Select Committee's report and its recommendations would be interesting, it also would be repetitive and beyond the scope of this study. However, the Conservative government agreed to review the CPSWR after March 1982 and to report about September 1982. Moreover, the Select Committee's report was debated in November 1981 by the House of Lords. The Conservative government did declare during the debate that it intended to develop the ways and means to implement much of what the Committee recommended, but it embraced only the recommendation calling for the establishment of a Hazardous Waste Inspectorate to strengthen the control over and guidance of local waste disposal authorities. The government was also interested in the recommendations that waste producers be registered, a system of quarterly reports from hazardous waste producers be instituted, and a licensing system for hazardous waste handlers be established.

On the other hand, the government rejected outright some specific recommendations. For example, replacement of 165 local waste disposal authorities with eight regional blocs of disposal authorities was rejected because it would create another costly level of bureaucracy. Other specific recommendations made by the Select Committee include: (1) the burden of proof for determining whether a waste is "special" should be shifted from local authorities to the waste producer who has the greater expertise for such a determination; (2) the local authorities should license all professional handlers of hazardous wasted; (3) the waste producers should be responsible for all the costs of disposal, thereby relieving the local ratepayer of the additional cost for hazardous waste disposal; and (4) public reporting of the results of investigations at monitored sites should be instituted to improve public confidence in the safety of dumpsites. Finally, there was no commitment by the government to produce a white paper in anticipation of a major legislative proposal, although it admitted that some legislative proposals may be necessary.[148]

# Notes

1. U.S. Department of Commerce, *The Effects of Pollution Abatement on International Trade* (Washington, D.C.: U.S. Government Printing Office, 1978), p. A-84.

2. Marion Clawson and Peter Hall, *Planning and Urban Growth* (Baltimore: Johns Hopkins University Press, 1973), p. 254.

3. Ibid., pp. 62-65.

4. U.S. Department of Commerce, *Effects of Pollution Abatement on Trade*, p. A-84.

5. Clawson and Hall, *Planning and Urban Growth*, pp. 237 and 240.

6. U.S. Department of Commerce, *Effects of Pollution Abatement on Trade*, p. A-84.

7. Department of Environment, *The Human Environment: The British View* (London: Her Majesty's Stationery Office, 1972), p. 10.

8. Ibid., pp. 33-34.

9. Philip W. Quigg, "How Much Pollution in the North Sea?" *World Environment Newsletter*, August 14, 1973, p. 3.

10. DOE, *Human Environment*, p. 34.

11. *Britain, 1974: An Official Handbook* (London: Her Majesty's Stationery Office, 1974), p. 174.

12. Eric T. Lummis, "Environmental Protection in the U.K.," *Environmental Policy and Law*, Vol. 1976, No. 2 (1976), p. 88.

13. U.S. Department of Commerce, *Effects of Pollution Abatement on Trade*, pp. A-86, A-90, and A-91.

14. Lummis, "Environmental Protection in the U.K.," p. 88.

15. U.S. Department of Commerce, *Effects of Pollution Abatement on Trade*, p. A-85.

16. DOE, *Human Environment*, p. 14.

17. Lummis, "Environmental Protection in the U.K.," p. 87.

18. *International Environment Reporter*, p. 291:0101.

19. *International Environment Reporter—Current Report*, July 10, 1978, p. 228.

20. DOE, *Human Environment*, p. 14.

21. *International Environmental Guide—1975*, p. 61:3001.

22. *International Environment Reporter*, p. 291:0101.

23. DOE, *Human Environment*, p. 14.

24. *International Environment Guide—1975*, p. 61:3001; and *International Environment Reporter*, p. 291:0101.

25. *International Environmental Guide—1975*, p. 61:3001.

26. U.S. Department of Commerce, *Effects of Pollution Abatement on Trade*, p. A-85.

27. *International Environment Reporter*, p. 291:0101.

28. *International Environmental Guide—1975*, p. 61:3001.

29. *International Environment Reporter*, pp. 291:0101 and 291:0102.

30. *Britain, 1974: An Official Handbook*, pp. 174-80.

31. Clawson and Hall, *Planning and Urban Growth*, pp. 160-66.

32. DOE, *Human Environment*, p. 18.

33. *International Environmental Guide—1975*, p. 61:3002.

34. *Britain, 1974: An Official Handbook*, p. 174.

35. British Information Services, "Local Government Revamped," *British Record*, May 2, 1974.

36. "If Only Other Cities Were Like London," *Business Week*, May 30, 1970, pp. 64-67.

37. Clawson and Hall, *Planning and Urban Growth*, p. 160.

38. *International Environmental Guide—1975*, p. 61:3002.

39. U.S. Department of Commerce, *Effects of Pollution Abatement on Trade*, pp. A-86.

40. DOE, *Human Environment*, pp. 31-32.

41. British Information Services, "Britain's Cleaner Rivers," *British Record*, May 2, 1973, pp. 1 and 2.

42. OECD Environment Directorate, *Economic and Policy Instruments for Water Management in the United Kingdom* (Paris: OECD, 1976), pp. 1, 10 and 11.

43. *International Environmental Guide — 1975*, pp. 61:3001, 61:3003, and 61:3004.

44. R. J. Van Schaik, "The Impact of the Economic Situation on Environmental Policies," *OECD Observer*, No. 79 (February 1976), p. 25.

45. *International Environment Reporter*, pp. 291:0104 and 291:0105.

46. U.S. Department of Commerce, *Effects of Pollution Abatement on Trade*, p. A-90.

47. OECD Environment Directorate, *Water Management in the U.K.*, pp. 15-18.

48. OECD Environment Directorate, *Pollution Charges: An Assessment* (Paris: Organization for Economic Cooperation and Development, 1976), pp. 18, 19, and 21.

49. *International Environment Guide — 1975*, p. 61:3004.

50. U.S. Department of Commerce, *Effects of Pollution Abatement on Trade*, pp. A-91 and 92.

51. *International Environmental Guide-1975*, p. 61:3004.

52. U.S. Department of Commerce, *Effects of Pollution Abatement on Trade*, p. A-92.

53. OECD Environment Directorate, *Water Management in the U.K.*, pp. 15-18.

54. *International Environmental Guide — 1975*, p. 61:3004.

55. *International Environmental Reporter*, pp. 291:0104 and 291:0105.

56. *International Environment Reporter-Current Report*, May 10, 1978, pp. 134-35.

57. Ibid., p. 134.

58. Ibid., pp. 134-35.

59. U.S. Department of Commerce, *Effects of Pollution Abatement on Trade*, p. A-86.

60. *International Environment Reporter*, p. 291:0102.

61. Lummis, "Environmental Protection in the U.K.," p. 88.

62. *International Environment Reporter*, pp. 291:0102 and 291:0103.

63. U.S. Department of Commerce, *Effects of Pollution Abatement on Trade*, pp. A-90 and A-91.

64. *International Environment Reporter*, p. 291:0101.

65. U.S. Department of Commerce, *Effects of Pollution Abatement on Trade*, pp. A-85, A-90, and A-91.

66. *International Environmental Guide-1975*, p. 61:3003.

67. *International Environment Reporter*, p. 291:0103.

68. *International Environment Reporter — Current Report*, January 10, 1978, p. 10.

69. Ibid., February 10, 1978, p. 35.

70. *International Environmental Guide — 1975*, p. 61:3003.

71. *International Environment Report — Current Report*, January 10, 1978, pp. 9 and 10.

72. Ibid., p. 10.

73. Ibid., January 10, 1978, pp. 9 and 10.

74. Ibid., p. 10.

75. *International Environment Report — Current Report*, January 10, 1978, pp. 9 and 10.

76. DOE, *Human Environment*, p. 31.

77. *International Environment Reporter*, p. 291:0101.

78. *International Environmental Guide — 1975*, p. 61:3003.

79. *International Environment Reporter*, p. 291:0103.

80. Ibid., p. 291:0104.

81. *International Environment Reporter — Current Report*, June 10, 1978, p. 167.

82. *International Environment Reporter*, p. 291:0104.

83. U.S. Department of Commerce, *Effects of Pollution Abatement on Trade*, pp. A-86, A-87, A-90, and A-91.

84. Anthony Wolff, "World Progress Report: Operation Eyesore," *Saturday Review/World*, January 26, 1974, p. 8.

85. OECD Environment Directorate, *Waste Management in OECD Member Countries* (Paris: Organization for Economic Cooperation and Development, 1976), pp. 62, 63, 68, and 75.

86. Office of Information, *New British System of Taxation*, pp. 44-45.

87. OECD Environment Directorate, *Waste Management in OECD Countries*, pp. 34 and 35.

88. U.S. Department of Commerce, *Effects of Pollution Abatement on Trade*, p. A-87.

89. OECD Environment Directorate, *Water Management in the U.K.*, pp. 32-33.

90. OECD Environment Directorate, *Waste Management in OECD Countries*, p. 52.

91. OECD Environment Directorate, *Water Management in the U.K.*, pp. 32-33.

92. Malcolm J. Forster, "Taxation As an Instrument of Pollution Control," *Environmental Policy and Law*, Vol. 2, No. 2 (June 1976), p. 92.

93. M. R. Ratledge "Tax Planning for Business," *Taxation*, October 16, 1976, pp. 41 and 42; M. Rawlinson, "Trends-Recent Developments in International Taxation: United Kingdom," *Tax Management International Journal*, December 1977, pp. 41-43; and Eugene L. Gomeche, 68-5th T. M., *Business Operations in the United Kingdom* (Washington, D.C.: Bureau of National Affairs, Inc., 1973), pp. A61-A65.

94. Ratledge, "Tax Planning for Business," p. 41.

95. British Central Office of Information, *The New British System of Taxation*, (London: Her Majesty's Stationery Office, 1973), pp. 33-34.

96. *Britain, 1974: An Official Handbook*, p. 216; and David Wilson and Patrick D. Daniels, 68-7th T.M., *Business Operations in the United Kingdom* (Washington, D.C.: Bureau of National Affairs, Inc., 1981), p. A111.

97. Office of Information, *New British System of Taxation*, pp. 33 and 41.

98. Linda S. Avelar (ed.), *Business Study: United Kingdom* (New York: Touche Ross International, 1978), p. 131.

99. Office of Information, *New British System of Taxation*, p. 31.

100. OECD Environment Directorate, *Pollution Charges*, p. 64.

101. Geoffrey Smith, "Britain: A Managing Woman," *Atlantic*, March 1980, pp. 4-12.

102. *International Environment Reporter—Current Report*, September 12, 1979, pp. 886-88; December 12, 1979, pp. 987-88; September 10, 1980, pp. 412-13; July 8, 1981, p. 937; and November 11, 1981, pp. 1089-90.

103. Ibid., March 12, 1980, pp. 95-97, and September 10, 1980, p. 410.

104. Ibid., February 11, 1981, pp. 635-636.

105. Ibid., January 13, 1982, p. 9.

106. Ibid., September 10, 1980, pp. 418-419.

107. Ibid., May 13, 1981, pp. 843-44.

108. Ibid., May 14, 1980, pp. 186-87.

109. Ibid., November 12, 1980, pp. 503-4; February 11, 1981, p. 628; and March 11, 1981, pp. 684-85.

110. Ibid., January 9, 1980, p. 4.

111. Ibid., March 11, 1981, p. 681.

112. Ibid., December 9, 1981, p. 1118.

113. Ibid., January 13, 1982, p. 3.

114. Ibid., September 9, 1980, pp. 406-7, and August 12, 1981, pp. 968-69.

115. Ibid., January 13, 1982, p. 3.

116. Ibid., September 9, 1981, p. 1010.

117. Ibid., January 13, 1982, p. 3.

118. Ibid., September 9, 1981, p. 1014.

119. Ibid., March 11, 1981, p. 696.

120. Ibid., January 13, 1982, p. 9.

121. Ibid., March 12, 1980, pp. 105-7.

122. Ibid., January 13, 1982, p. 9.

123. Ibid., March 12, 1980, p. 96.

124. Ibid., January 9, 1980, pp. 6-7 and 18-20; July 9, 1980, p. 289; and October 8, 1980, p. 468.

125. Ibid., August 12, 1981, p. 963.

126. Ibid., November 11, 1981, p. 1087-89.

127. Ibid., December 9, 1981, p. 1119.

128. Ibid., June 10, 1981, pp. 881-82.

129. Ibid., March 12, 1980, pp. 95-96, February 11, 1981, pp. 635-36, and January 14, 1981, p. 583.

130. Ibid., September 10, 1980, pp. 412-13; July 8, 1981, p. 937; and November 11, 1981, pp. 1089-90.

131. See Gerald Rhodes, *Inspectorates in British Government* (Winchester, Mass.: Allen & Unwin, 1981).

132. *International Environment Reporter—Current Reports*, February 13, 1980, pp. 37-38; May 14, 1980, pp. 186-87; September 10, 1980, pp. 412-13; May 13, 1981, pp. 834-35 and 843-44; July 8, 1981, p. 937; and November 11, 1981, p. 1089.

133. Ibid., January 10, 1979, pp. 471-73; February 10, 1979, pp. 511-12; May 9, 1979, pp. 688-89; June 13, 1979, p. 727; October 10, 1979, pp. 894-95; November 14, 1979, pp. 931-32; June 11, 1980, p. 260; December 10, 1980, pp. 544-45; April 8, 1981, p. 793; November 11, 1981, p. 1986; and February 10, 1982, p. 85.

134. Ibid., May 14, 1980, pp. 185-86; May 13, 1981, p. 838; and June 10, 1981, pp. 882-83.

135. Ibid., January 9, 1980, pp. 21-22; November 12, 1980, p. 512; and December 9, 1981, p. 1118.

136. Ibid., June 11, 1980, p. 253.

137. Ibid., March 10, 1982, pp. 111-12.

138. Ibid., April 14, 1982, pp. 150-51.

139. Ibid., May 12, 1982, p. 203.

140. Ibid., April 14, 1982, p. 150.

141. Ibid., pp. 151-52.

142. Ibid., March 10, 1982, pp. 121-22.

143. Ibid., April 14, 1982, pp. 138-40.

144. Ibid., June 9, 1982 pp.

145. Ibid., May 12, 1982, pp. 180-81.

146. Ibid., June 14, 1982, p. 293.

147. Ibid., April 14, 1982, p. 138.

148. Wendy Barnaby, "Relying on Luck to Avert Disaster," *Ambio*, Vol. 11, No. 1 (1982), pp. 53-56.

# 8

# The United States

The United States suffers from a full assortment of pollution problems. Almost any serious pollution problem found in Canada, France, West Germany, Sweden, and the United Kingdom is also a material pollution problem somewhere in the United States, whereas the pollution problems of the desert Southwest are unique. More specifically, with urban development, industry, agriculture, and resource extraction widespread throughout the country, water pollution is a problem in virtually every U.S. water basin system. On the positive side, it should be noted that there are several examples of significant improvement in water quality, including the Willamette, the Hudson, the Manongahela, and the Ohio rivers. Similarly, there are cases of significant improvement in ambient air quality. Nonetheless, air pollution remains a serious problem in most major metropolitan areas because of increased industrial activity and motor vehicle use. Coincidently, the apparent stabilization and isolated reduction in the level of both waste water discharges and emissions correlate with the continued growth of solid and hazardous wastes from households and industry, respectively. Finally, inasmuch as the increased incidence of cancer in the general population has been linked to increased levels of carcinogenic and toxic substances in the human environment, there has been a preoccupation with the control of the production, use, and disposal of toxic chemical substances in recent years.[1]

## Pollution Abatement Programs

Although piecemeal Federal legislation to control air, water, and solid waste pollution had been enacted by 1965,[2] congressional approval of the

National Environmental Policy Act of 1969 (NEPA) became the first comprehensive statement of Federal environmental policy. The act remains the paramount declaration of national environmental policy. NEPA not only required all Federal agencies henceforth to consider the ecological implications of their actions, but also created the Council on Environmental Quality (CEQ),[3] which is responsible for evaluating the effect of Federal activities and programs on the environment; advising the President on policies that should improve environmental quality; monitoring the condition of and trends in national environmental quality; and appraising the interrelationship of environmental policy with other Federal policy objectives.

In addition, the CEQ supervises the preparation of environmental impact statements (EIS) by Federal agencies. NEPA requires Federal agencies to consider the environmental impact of their decisions, including the preparation of an EIS where an agency's action significantly affects the quality of the human environment. The courts have generally construed NEPA as requiring a comprehensive environmental analysis of a wide range of Federal government actions, including the allocation of Federal funds and the issuing of permits and licenses.[4]

The 1970 Presidential Reorganization Plan No. 3 created the Environmental Protection Agency (EPA) by consolidating the Federal government's environmental programs in one agency. This consolidation gave the EPA responsibility for administration of Federal programs to control air, water, solid waste, pesticide, and noise pollution; establishment and enforcement of emission, effluent, and disposal standards; assistance to State and local governments in their pollution control efforts; monitoring and evaluation of environmental quality; and conducting of research and demonstrations related to improving or maintaining environmental quality.[5]

Since 1970, the EPA has been given other environmental duties: review and comment on the environmental impact of the action(s) of any other Federal agency affecting EPA responsibilities; regulation of the transportation of materials to be dumped and the issuing of permits for the dumping thereof in the ocean; and regulation of the manufacture, distribution, use, and disposal of chemical substances that may be hazardous to humans or the environment. In addition, the EPA shares responsibilities with other Federal agencies for control of radiation (Nuclear Regulatory Commission), of coastal water pollution by oil and other noxious substances (U.S. Coast Guard and the Federal Maritime Commission), of noise (Federal Aviation Administration and Federal Highway Administration), and of ocean dumping (U.S. Army Corps of Engineers).

Several other Federal government agencies have relatively exclusive responsibilities for selected aspects of the environment. The Department of Agriculture is responsible for soil conservation; the Commerce Department's National Oceanic and Atmospheric Administration, for meteorological services, including weather modification, and coastal zone management which includes fishery conservation; the U.S. Army Corps of Engineers, for control of navigable waters; the Department of the Interior, for management of the outer Continental Shelf, Federal lands, and strip-mining operations; the Department of Housing and Urban Development, for community development which includes water supply and sewage; and the recently organized Department of Energy, for energy-related environmental issues.[6]

## Water Pollution Control

Federal interest in water pollution control goes back to the Refuse Act of 1899, which was designed to prohibit impediments to navigation of interstate rivers and lakes, the Public Health Service Act of 1912, and the Oil Pollution Act of 1924. Water pollution was not recognized as a national problem, however, until enactment of the Water Pollution Control Act of 1948, which was extended and strengthened through amendments in 1952, 1956, 1961, 1966, and 1970. The Federal Water Pollution Control Act (FWPCA) and its amendments established the principle that the water quality problem is best dealt with at the local level. It provides for (1) Federal government assistance through low-interest loans and grants for municipal abatement facilities that comply with Federal effluent standards and (2) Federally supported research of water pollution abatement techniques and strategies. The 1961 amendments extended abatement enforcement to intra-State navigable rivers and lakes, and a 1970 Executive Order established a permit program for industrial effluent discharge into navigable waters under the authority provided in the Refuse Act of 1899. However, the State and local governments retained virtually all responsibility and authority for water pollution control until 1972 when the FWPCA amendments were enacted. The FWPCA superseded all previous amendments to the Water Pollution Control Act of 1956 which had permanently extended the 1948 law.[7]

In principle, the States continue to have primary responsibility for the national water pollution program. However, the EPA is required to usurp a State's right to ensure the attainment of national standards where the State does not meet its responsibilities under the act. In practice, the FWPCA established two programs for water pollution control. One program relies on national effluent guidelines on an industry-by-industry basis, while the other utilizes water quality standards on a specific body of water basis.[8]

The 1972 amendments established both a national goal of complete elimination of effluent discharges into navigable waters by 1985 (amended to July 1, 1987) and an interim goal of water quality suitable for water recreation (swimming) and for the propagation and protection of marine and wildlife wherever possible by 1983 (amended to July 1, 1984). Title I of the 1972 act authorizes the EPA both to prepare comprehensive programs for preventing and eliminating pollution in navigable and ground waters, and to appropriate funds to assist States in administering such programs. Under Title II, the EPA is authorized to make grants to State and local governments for up to 75 percent of the construction costs of those treatment plants owned by the public.

In Title III, the act requires the EPA to establish specific effluent limitations. All nonpublic stationary sources were required to employ the "best practicable control technology currently available" by July 1, 1977. By July 1, 1983 (amended to July 1, 1984), an additional effluent limitation for the above nonpublic sources will require that they employ "the best available technology economically achievable." For all publicly owned treatment facilities, the best practicable control technology currently available as of July 1, 1977, had to be operational by June 30, 1978. By July 1, 1983, a stricter effluent limitation for all publicly owned treatment works will require that they employ the "best practicable waste treatment technology over the life of the works." In addition, the EPA was required to issue new source effluent standards for twenty-one designated industries for the more common water pollutants and toxic pollutants.

Title IV requires that a permit be issued to discharge any pollutant into navigable waters from a stationary source. This title replaces the permit program established in 1970 by Executive Order under the Refuse Act of 1899. Finally, the civil penalties provided in Title III are limited to $10,000 per day, whereas the criminal penalties are from $2,500 to $25,000 per day or up to $50,000 per day.[9]

In addition to delaying the deadlines for control of unconventional, nontoxic industrial discharges, the Clean Water Act of 1977 tightens controls on toxic pollutants and extends the Federal grant program for municipal sewage treatment plants. The 1977 act amends the timetable established in the FWPCA of 1972 by requiring nonpublic stationary sources to control conventional pollutants (suspended solids and effluent that increases biological oxygen demand) by July 1, 1984. Nonconventional pollutants, which is a new designation applicable to selected nontoxic organic, chemical, or thermal effluent, will have to be controlled by July 1, 1987. The principal objective of the 1977 amendments was to require the EPA to establish effluent standards for toxic substances. These standards, which must be met by July 1, 1984, were prompted by a concern for the safety of drinking water taken from surface waters.[10]

In respect to the effluent standards for conventional and nonconventional pollutants, the 1977 act provides waiver procedures for those effluent dischargers who cannot comply with the extended deadlines. There are no provisions for a waiver of the effluent standards established for toxic substances. For those polluters who cannot comply with the effluent standards for toxic, conventional, and/or unconventional pollutants by the 1984 and 1987 deadlines, there will be a penalty in the amount of the cost of complying with standards.

Industrial dischargers that utilize publicly owned sewage treatment facilities are subject to a different set of effluent standards than other nonpublic dischargers are. This separate set of requirements for use of publicly owned facilities establishes pre-treatment standards for permitted discharges into public systems, prohibits specified hazardous or toxic discharges, and establishes a tax or user charge system to partially finance operations and maintenance of public facilities.

With regard to the national industrial effluent guideline system, the EPA had established standards for more than fifty industrial categories by early 1978, including several types of food processors, textiles, cement manufacturing, feedlots, electroplating, inorganic chemicals, several types of organic chemicals and products derived therefrom, fertilizer and phosphate manufacturing, iron and steel manufacturing, ferroalloy manufacturing, nonferrous metals manufacturing, conventional electric power generation, leather tanning and finishing, petroleum refineries, glass and asbestos manufacturing, rubber processing, several types of wood products and chemicals manufacturing, coal and other minerals mining and processing, offshore oil and gas extraction, pharmaceutical manufacturing, paving and roofing materials, paint and ink formulating, pesticides and explosives manufacturing, photographic processing, and hospitals. For "new stationary sources," or facilities constructed after publication of effluent guidelines and standards, there are separate and stricter requirements that mandate the achievement of no discharge where practicable, but at least of the best demonstrated control technology, operating methods, or processes.

For the water pollution control program that utilizes water quality standards on a specific body of water basis, nonpublic effluent dischargers must meet stricter standards for a specific river or lake. The principal mechanism for enforcing both the standards for a specific body of water and the national effluent limitations is the National Pollutant Discharge Elimination System (NPDES) wherein either a State or the Federal government issues a NPDES permit for each stationary source of effluent into navigable waters in the United States. The FWPCA provides for either optional State legislation establishing a permit program subject to EPA approval or Federal government enforcement through the EPA's permit program.[11] By the end of 1977, twenty-nine States and the Virgin Islands had established EPA-approved permit programs, including California, Delaware, Illinois, Indiana, Michigan, Minnesota, Missouri, New York, Ohio, Tennessee, and Wisconsin. However, several other major industrial States, including Kentucky, Louisiana, New Jersey, Massachusetts, Pennsylvania, and Texas, had not established EPA-approved programs for administering NPDES.[12]

The FWPCA imposes penalties for failure to report discharge of oil and other "hazardous" pollutants into navigable water, and makes the polluters liable for clean-up costs. The EPA has issued regulations applicable to discharges of oil, to oil pollution prevention, to the limited liability for small onshore oil-storage facilities, and to the civil penalties for violations of oil pollution prevention requirements. In addition, there are Federal Maritime Commission regulations on the financial responsibility of ship owners for oil pollution clean-up, as well as U.S. Coast Guard regulations both on notification of oil spills and on payments into and disbursements from an oil pollution clean-up fund.[13]

### Air Pollution Control

The Clean Air Act of 1963 as amended has led to increasing Federal control over air pollution regulation. Originally, the Air Pollution Act of 1955 established the policy of Federal guidelines and State and local government responsibility for control of air pollution. Since 1963, the Federal government has effectively preempted State and local authority by (1) setting standards for motor vehicle emissions; (2) establishing procedures for issuance of required air quality criteria; (3) establishing atmospheric areas and air quality control regions; (4) prescribing national ambient air quality standards for each air pollutant with the potential to adversely affect public health and/or public welfare; (5) requiring States to submit plans for implementing and enforcing primary and secondary ambient air quality standards and requiring implementation of such plans approved by the EPA within a specified period of time thereafter; (6) authorizing the EPA to prescribe implementation and enforcement plans for States not so providing or for States with plans unapproved by the EPA; (7) requiring the EPA to recommend ways State and local governments can achieve national ambient air quality standards; (8) requiring the EPA to adopt standards for stationary sources of pollution, as well as for hazardous air pollutants, where ambient air quality standards would be ineffective in controlling such pollution; (9) mandating emissions standards for new motor vehicles relating to carbon monoxides and hydrocarbons for model year 1975 and nitrogen oxides for model year 1976; and (10) providing for economic sanctions and/or incarceration for those violating air pollution control statutes and regulations.[14]

The EPA has determined that the following seven pollutants meet the dual requirements of having an adverse effect on health or welfare and of being "public," that is, being caused by numerous or diverse sources: hydrocarbons, carbon monoxide, particulate emissions, sulfur oxides, nitrogen dioxide, photochemical oxidants, and lead. Moreover, the Clean Air Act requires the EPA to add to this list any air pollutant which it determines to be harmful to public health and/or public welfare. With regard to any pollutant on the EPA's list of those it finds harmful and diverse, the Clean Air Act requires the EPA to establish both primary and secondary national ambient air quality standards. The primary standards are deemed to be those necessary to protect public health; and the secondary, those necessary to protect public welfare.

As mentioned above, the EPA has also been required to publish air quality criteria and control techniques for air pollutants it finds harmful and diverse. The air quality criteria have included information on the public health and welfare effects of the EPA's list of pollutants in the ambient air. The EPA's control techniques have delineated available air pollution control technology, alternative means of preventing and controlling air pollution, and alternative methods of operating, processing, and powering that eliminate or minimize emissions.

The Clean Air Act has established three approaches for controlling EPA-specified air pollutants and for making sure that once ambient air quality standards are attained, they are maintained or improved: required State plans for implementation and enforcement of national ambient air quality standards; regulation of new stationary sources of pollution; and emission standards for new motor vehicles.[15] Although State implementation plans are the principal mechanism for achieving national ambient air quality standards, several areas are currently in violation of EPA-approved standards.

The 1977 Clean Air Act Amendments provided for EPA approval of all State implementation plans by July 1, 1979. Where plans were not submitted and approved by the July 1 deadline, the EPA was authorized to block construction or modification of any major sources of emissions in areas not complying with national ambient air quality standards. States must fully disclose in their implementation plans those areas wherein primary standards are not being met. With regard to six of the emissions (all but lead) for which the EPA has established ambient air quality standards, the primary standards, which were formerly to be achieved under the 1970 amendments by 1975-76, must now be met by December 31, 1982. The 1977 amendments provide for additional deferral of carbon monoxide and photochemical oxidants until December 31, 1987, under extenuating circumstances. The 1977 amendments also required the EPA to review all existing ambient standards by December 31, 1980, and every five years thereafter.[16] In addition, the EPA's controversial "significant-deterioration-prevention" policy was "codified" in the Clean Air Act Amendments of 1977. This EPA policy, which was originally promulgated by administrative regulation, has been designed to ensure that stricter ambient air quality standards than national levels would apply to "clean air" areas. Finally, the 1977 amendments mandated uniform reporting of ambient air quality by requiring all Air Quality Control Regions to adopt the Pollutants Standards Index (PSI) by August 1978.[17]

The 1977 amendments required the installation of the best available control technology as defined by the EPA for those new stationary sources of air pollution that the EPA is required to regulate. By the end of 1977, the EPA had promulgated standards for over twenty-five different major sources of emissions, including incinerators, cement plants, various types of conventional power plants, refineries, nonferrous metal production plants and smelters, iron and steel plants, ferroalloy production facilities, sewage treatment plants, and coal preparation plants. In addition, the EPA has issued emission standards for both new and existing sources of asbestos, beryllium, mercury, and vinyl chloride,[18] all of which are considered hazardous air pollutants.

The most publicized aspect of the 1977 amendments to the Clean Air Act concerned the relaxation of motor vehicle emission standards enacted in 1970. The 1977 legislation deferred the effective date of the 1977 emission standards for two more years. It also established stricter standards for hydrocarbon and carbon monoxide emissions in 1980 and for carbon monoxide and nitrogen oxides in 1981. The EPA was authorized to waive the stricter carbon monoxide standards for 1981 where a further deferral was both in the national interest and compatible with public health objectives. The EPA may also allow smaller motor vehicle manufacturers to delay compliance with the nitrogen oxides standard.[19] Lead emissions are controlled through limits on the lead in gasoline for vehicles equipped with engines that consume leaded gasoline. Moreover, the emission control systems that rely on a catalytic converter cannot use leaded gasoline without damaging the converter and the engine. However, the national ambient air quality standard for lead is still in the proposed stage because of difficulty in assessing the impact of lead emissions and the ambient levels thereof on human health when air is not the only major source of exposure.[20]

## Solid and Hazardous Waste Pollution Control

Waste disposal and its related land use problems are the other major pollution concern faced by our industrial, consumer-oriented society. Moreover, amelioration of air and water pollution problems often results in more solid waste and more hazardous and toxic wastes. Federal legislation enacted in 1976 may go a long way in solving the problem of safely and economically disposing of waste from mineral extraction, industrial production, and the "final" consumption of goods and services. The Resource Conservation and Recovery Act (RCRA) of 1976 provided, for the first time, for direct Federal government involvement in solid waste control and established, in principle, the systematic government control of hazardous waste from generation to ultimate disposal.

The amount of hazardous waste to be reckoned with is most probably less than 100 million tons annually. Although hazardous wastes are potentially more harmful to the environment and more threatening to public health, the sheer magnitude, diversity, and uneven distribution of over 5 billion tons of solid waste have created serious pollution problems throughout the country. In the past, State and local governments have exercised little control over land disposal sites wherein annually much of the over 145 million tons of municipal waste (trash and garbage) and some of the 340 million tons of industrial waste are deposited. In 1976, less than half of approximately 16,000 municipal disposal sites complied with existing State regulations. In addition, sludge from both 23,000 municipal wastewater abatement plants and 50,000 industrial abatement facilities was unregulated.

The RCRA prohibits the establishment of new open dumping sites, requires the EPA to establish criteria for sanitary landfills, and mandates that all open dumps be either closed or upgraded to sanitary landfills by 1983. The act also requires a national inventory of all "open dumps," with the States being given the responsibility for identifying and controlling such land disposal sites. The RCRA also requires the EPA to publish guidelines to assist each State both in developing its solid waste management plan and in identifying regions for solid waste planning. The EPA is required to establish Resource Conservation and Recovery Panels, which are teams of professionals from industry and government with expertise in technical, marketing, financial, and institutional matters, in order to provide technical assistance to State and local governments. Finally, the EPA is authorized to provide direct financial assistance for the establishment of comprehensive solid waste planning wherein environmentally sound disposal methods and programs for resource recovery and conservation are optimally combined.

Municipal and industrial wastes are dwarfed by agricultural and mining wastes, which account for over 2.3 and 1.7 billion tons, respectively. As already noted, the agricultural waste program is shared with the Agriculture Department in the case of soil erosion and other agricultural run-off.[21] Although no Federal legislation currently addresses the problem of mining wastes comprehensively, the Surface Mining Control and Reclamation Act (SMCRA) of 1977 purports to solve the strip-mining waste problem. The SMCRA, which is the exclusive responsibility of the Interior Department, requires strip-mine operators to restore the mined land to its "original" contour in as many cases as possible. The act also provides protection for prime agricultural land and river valleys. Enforcement authority for the SMCRA will be turned over to those States whose reclamation plans meet the minimum Federal standards.[22]

There seems to be some confusion as to where the hazardous waste provisions of RCRA end and those of the Toxic Substances Control Act (TSCA) of 1976 begin. The TSCA empowers the Federal government both to control and, where necessary, to stop use and production of chemical substances that present an unreasonable risk of injury to public health or the environment. The TSCA also requires chemical manufacturers to give notice of plans to produce a new substance or market an old substance for a new use, to test selected substances, to keep extensive records on both the properties and production of substances, and to disclose the significant health effects of dangerous substances.

Although the RCRA mandates a Federal or State regulatory, inspection, and enforcement program in every State, the act anticipates State acceptance of responsibility for hazardous and solid waste control, with Federal guidance, technical assistance, and financial aid. The RCRA provides for identification and listing of hazardous wastes, along with performance standards for the generation, transportation, treatment, storage, and disposal thereof. Hazardous waste includes toxic chemicals, pesticides, acids, caustics, flammables, and explosives. Although such wastes are generated by farms, mining, Federal government installations, laboratories, and hospitals, the primary source is industry. Many States have responded positively to the RCRA by passing hazardous waste legislation. Several States—California, Illinois, Kansas, Maryland, Minnesota, New Mexico, Oklahoma, Oregon, Texas, and Washington—had already enacted and implemented legislation designed to control hazardous wastes. Other States are considering hazardous waste legislation.[23]

## The Emergence of Pollution Taxation

Environmental policy in the United States is implemented primarily through the use of regulation wherein compliance with regulatory rules, administrative procedures, performance standards, and prohibitions is effected by one or more Federal and State agencies with responsibility and authority for interpreting, administering, and enforcing environmental law. However, the use of regulatory taxation (effluent, emission, and excise taxes on water, air, and solid waste pollutants, respectively) has emerged in the United States during the recent concern for the excessive cost and somewhat limited welfare of an increasingly more stringent and pervasive regulatory system.

The Clean Air Act Amendments of 1977 include two types of noncompliance penalties which some political economists would consider an emission tax or charge. One penalty provides for an exaction equal to the cost of compliance in situations where a firm fails to meet national emission standards or performance standards. With regard to heavy-duty vehicles, such as trucks and buses, that fail to meet performance standards, firms may be penalized in various ways, including by pollutant, by vehicle, or by engine.[24] In addition, the 1978 Energy Tax Act provides for an excise tax on motor vehicles manufactured for the 1980 model year and thereafter depending upon their fuel efficiency, that is, a "gas guzzler" tax.[25]

The 1977 amendments to the FWPCA also provide for noncompliance penalties which more closely resemble effluent taxes or charges than arbitrary impositions. In general, penalties for noncompliance with the deadlines for effluent limitations are set at the cost of complying with such standards.[26]

The Resource Conservation and Recovery Act of 1976 required the EPA to establish a committee to study the efficacy of excise taxes on solid waste.[27] The Resource Conservation Committee issued a report in late 1978 recommending imposition of a solid waste-disposal charge based on the material content of products that become part of municipal solid waste.[28] Beverage container deposit charges are imposed in six States and at Federal facilities. Finally, the SMCRA provides for a tax on all coal, the proceeds from which are to be used to reclaim abandoned strip-mined land.[29]

## Integration of Federal and State Environmental Policy

Federal and State environmental policy on air, water, and solid and hazardous waste pollution is characterized by Federal prohibitions, regulatory rules, guidelines, standards, technical assistance, and direct financial aid with State implementation, monitoring, and enforcement. In addition, State and local governments have enacted various other laws on environmental protection to deal with their unique or most critical pollution, conservation, or environmental quality problems inasmuch as there is no prohibition against State or local laws which are more stringent than Federal environmental legislation. Therefore, several States have constitutional language and/or State laws that provide for more stringent and/or more comprehensive laws on environmental protection. New York has been known for its progressive sanitation laws and its statewide environmental administration, whereas Michigan has been considered a leader in conservation legislation.[30] However, California is now generally considered to be the leader among the States in dealing with environmental protection and growth, especially air pollution.

A few States have reacted somewhat sooner to an environmental protection issue. For example, the Oregon "bottle bill" is designed to control the litter from throwaway beverage containers, and the Florida legislation on comprehensive land use is designed to protect especially unique and/or fragile areas. But California has unquestionably been the pacesetter in enacting comprehensive legislation to deal with the problems of environmental preservation and growth since 1970. Consequently, one should analyze California's environmental protection legislation and administrative machinery to better understand State responses both to Federal mandates and to the demands of its populace.[31] Like Federal environmental policy, California and other State legislatures enact environmental protection laws establishing a mechanism for statewide regulation, but allow local or regional implementation and/or enforcement.[32]

### Government Assistance

As a leading member of the Organization for Economic Cooperation and Development, the United States is committed to the "polluter pays principle" adopted by the twenty-three members of the OECD in May 1972. Therefore, financial assistance in the form of direct grants to polluters has traditionally not been utilized by the Federal government to reimburse the private sector for the costs of pollution control, although voluntary, indirect subsidies exist. Beginning with enactment of the original Water Pollution Control Act of 1948, Federal financial assistance (in the form of low-interest loans) for municipal wastewater treatment facilities has been an integral part of the effort to maintain and improve water quality.[33] To the extent that industrial and commercial enterprises are charged less for their use of local publicly owned facilities than they would have to pay a private sector producer of such abatement services, the Federal and local governments have subsidized industry and commerce. Over the past thirty years, the subsidy for local publicly owned wastewater treatment facilities has evolved from a low-interest loan program to a direct grant program. Moreover, the annual amount of Federal appropriations has increased from about $20 million to several billion.[34]

The 1977 Clean Water Act Amendments (FWPCA) also included a provision that authorized a financial aid program for small business to help them control air and water pollution in compliance with EPA emission and effluent standards.[35] The 1976 Small Business Act Amendments provided both for loans to farmers attempting to control water pollution from agricultural activities[36] and for Small Business Administration guarantees of loan contracts between small businesses and public authorities. These loans are for the installation and operation of air and water pollution abatement equipment wherein a local or State authority issues tax-exempt, industrial development bonds to finance the pollution control investment. To be eligible, the small business must not only be under an order to comply with Federal (or State) air or water standards, but it must also be sponsored by a bank or other financial organization, which will certify that other long-term financing is either unavailable or too costly in spite of the small company's creditworthiness.[37] Even though this new financial aid program could result in a direct grant from the Federal government to a small business in apparent violation of the OECD's "polluter pays principle," both the social effect of supporting small business, and the limited effect of the subsidy program on the price of goods and services exported from the United States, fall within the exceptions provided by the OECD, that is, each member can allow for its own unique national circumstances in implementing the "polluter pays principle."

The WPCA of 1948 authorized a fund of $22.5 million annually for five years to be used for low-interest loans to local, publicly owned wastewater treatment plants. However, the loans were limited to the lesser of one-third of capital expenditures or $250,000. The loan program was extended in 1952, 1956, and 1961. Then in 1966 the WPCA was amended with passage of the Clean Water Restoration Act wherein $3.9 billion was authorized primarily for grants to construct municipal wastewater treatment plants, but also for research and development projects. The $250,000 or one-third of construction cost limitation was abolished by the 1966 amendments. The 1972 FWPCA authorized the EPA to make grants of up to 75 percent of the capital costs of State and/or locally owned wastewater abatement facilities,[38] and a total of $18 billion was authorized over a five-year period.[39] Most recently, the 1977 amendments to the 1972 FWPCA authorized the expenditure of $24.5 billion over a five-year period.[40]

## Federal and State Tax Policy

The major source of Federal tax revenue is the income tax, whereas most States rely on both the income tax and the sales tax. To the extent that provisions of Federal, State, and local tax law allow investments in pollution control, special treatment through more rapid amortization, investment tax credit, tax exemption, and so on, tax policy is being used to complement environmental policy through implicit subsidies or tax expenditures. In other words, to the extent that an investment in pollution control is either not taxed or triggers a lower tax than a comparable amount of investment in other property, the Federal, State, or local government has subsidized the capital expenditure for pollution control.

The diametric use of tax policy to control pollution is more in the nature of a penalty for polluting as opposed to the rewards or subsidies explained above. If the penalty for polluting is an exaction per unit of pollutant, such exactions are referred to as pollution taxes, and they are essentially regulatory taxes, for example, the environmental excise tax explained in Chapter 10.

The more conventional use of tax policy to control pollution is, of course, the tax expenditure or indirect subsidy. Since comprehensive Federal and State environmental policy did not develop until the late 1960s and early 1970s, one should not be surprised that tax policy relating to it developed almost simultaneously. At the Federal level, three sections of the Internal Revenue Code of 1954 provide preferential treatment for investments in pollution control as follows:

(1) Section 169 Amortization of Pollution Control Facilities.

(2) Section 38 Investment (in Certain Depreciable Property) tax credit as limited in Section 46 (c) (5) for pollution control investments.

(3) Section 103 Interest (on Certain Governmental Obligations) excludable from gross income as limited by the Section 103 (b) provision in respect to Industrial

Development Bonds, the proceeds of which are used to finance pollution control investment.

Chapter 9 provides a detailed explanation of the taxation of investments in pollution control, including Sections 169 and 38.

Interest earned on obligations of a State or local government is generally excluded from gross income as provided in Section 103 (a). A major exception was enacted in 1968 to effectively include in gross income interest earned on industrial development bond issues of more than $1 million after April 30, 1968. In addition, there is an exception for certain types of facilities financed with industrial development bonds, including in Section 103 (b) (4) (E) and (F) air or water pollution control facilities and sewage or solid waste-disposal facilities in respect to which there is no limit on the amount of tax-exempt industrial development bonds that can be issued.

Industrial development bonds are defined in Section 103 (b) (2) as any obligation issued as part of a bond issue all or a major portion of the proceeds are used directly or indirectly in any trade or business carried on by any person who is not a tax-exempt person. Moreover, the payment of the principal or interest on the bonds (under the terms of such an obligation or any underlying arrangement) must be wholly or primarily secured by an interest in property used (or to be used) in a trade or business or in payments in respect to such property. Alternatively, the payment of the principal and interest may be derived from payments in respect to property, or borrowed money, used (or to be used) in a trade or business. The term "exempt person" means (1) a governmental unit, or (2) an organization described in Section 501 (c) (3), that is, religious, charitable, scientific, literary or educational organizations, and exempt from tax under Section 501 (a), but only with respect to a trade or business carried on by an organization that is not an unrelated trade or business determined by applying Section 513 (a).

The rising cost of effective pollution control facilities and sewage or solid waste disposal facilities has led many businesses to use tax-exempt industrial development revenue bonds (IDBs) as a means to finance pollution or waste disposal costs. Industrial development bonds are a debt obligation issued under the name of a state or local government for the benefit of a private corporation. As noted above, interest earned on obligations of a State or local government is generally excluded from gross income as provided in Section 103(a). After the bonds are sold by the government agency, the proceeds are made available to the business at the same low interest rate.

Treasury Regulation Section 1.103-8 sets forth the requirements as to the definition of air and water pollution control facilities, sewage disposal facilities and solid-waste disposal facilities. The term "pollution control facility" (water or air) means property used, in whole or in part, to abate or control water or atmospheric pollution or contamination by removing, altering, disposing, or storing pollutants, contaminants, wastes, or heat. This property includes the necessary sewers, pumping, power, and other equipment. The facility must be designed to meet or exceed applicable Federal, State, and local requirements for the control of atmospheric or water pollutants or contaminants. In addition, the facility must *not* have a significant purpose other than the control of pollution.

The regulations define the term "sewage disposal facility" as any property used for the collection, storage, treatment, utilization, processing, or final disposal of sewage. "Solid waste disposal facilities" are similarly defined. The term solid waste has the same meaning as in the Solid Waste Disposal Act; material may not be solid waste, however, if at the time of the issuance of the bonds the facility has a market value or any value for which any person would be willing to purchase it. The Crude Oil Windfall Profit Tax Act of 1980 extended Section 103 to include tax-exempt 103 financing to certain property used primarily to convert fuel derived from solid waste into steam and to convert solid waste into alcohol. In summary, the key requirement to financing air pollution control, solid waste, or sewage facilities with industrial development revenue bonds is that "substantially all" of the proceeds must be used to finance the facilities.

The Economic Recovery Act of 1981, which was enacted on August 13, 1981, included a new tax credit designed to encourage the research and experimentation or development (R&D) activity necessary for American industry to maintain its technological edge in the highly competitive world economy. The Federal tax law has traditionally allowed a business to either expense or capitalize and amortize its R&D expenditures as provided in Section 174. This expense or capitalize option is still applicable, but as of July 1, 1981, a 25 percent tax credit is also allowable on the "incremental amount" of R&D expenditures, that is, the increase in such expenditures as compared to the base period which is the average R&D expenditures for the prior three-year period. R&D expenditures incurred to develop pollution control technology may qualify for this Section 44F credit.

Insofar as State and local governments rely on income, sales, and property taxes to finance their activities, a variety of tax incentives for pollution control investments exist at the State and local levels. There are no tax incentives for pollution control in eight States, and in several States property used in production may be exempt from tax. Table 8-1 is a summary of the exemptions, deductions, and credits available for investments in pollution abatement facilities classified by State and tax base.

## Future Environmental Policy

The election of Ronald Reagan as the fortieth President of the United States on November 4, 1980, signaled the beginning of a new era in

## Table 8-1
## State and Local Tax Incentives
## of Air and Water Pollution Control Investments

| State | Income Tax | Property Tax | Sales & Use Tax |
|---|---|---|---|
| Alabama | D | E | E |
| Arizona | D | — | — |
| Arkansas | — | — | E |
| California | D | — | — |
| Colorado | C | — | — |
| Connecticut | C | E | E |
| Florida | — | a | — |
| Georgia | — | E | E |
| Hawaii | D | E[b] | E[b] |
| Idaho | — | E | E |
| Illinois | — | c | E |
| Indiana | — | E | — |
| Iowa | — | E | — |
| Kentucky[i] | d | e | E |
| Maine | — | E | E |
| Maryland | — | — | E |
| Massachusetts | D or C | E | — |
| Michigan | — | E | E |
| Minnesota[n] | C[g] | E | — |
| Mississippi[i] | D | — | — |
| Missouri | — | — | E |
| Montana | — | h | NA |
| Nebraska | — | — | R |
| Nevada | NA | E | — |
| New Hampshire | — | E | NA |
| New Jersey | — | E | — |
| New York | D | E | — |
| North Carolina | D | E | — |
| Ohio | E | E[i] | E[i] |
| Oklahoma | C | — | — |
| Oregon | C | E | NA |
| Pennsylvania | — | E[j] | E |
| Rhode Island | D | E | E |
| South Carolina | — | E[k] | E[o] |
| Tennessee | — | E | f |
| Utah | — | — | E[l] |
| Vermont | — | E | — |
| Virginia | — | E | E |
| Washington | — | — | E |
| West Virginia | D | m | — |
| Wisconsin | D | E | E |
| Wyoming[n] | NA | E | — |

Note: Alaska, Delaware, Kansas, Louisiana, New Mexico, North Dakota, South Dakota, and Texas have not enacted incentives.

SOURCE: *1982 CCH State Tax Guide* (Chicago: Commerce Clearing House, 1982), pp. 664-65.

Key to Symbols: D = deduction allowed for purchase or investment in air or water pollution equipment; E = exemption; R = refund; and C = credit. NA = not applicable. NA indicates tax not imposed, and, therefore, there is no need for a tax incentive in the form of deduction, exemption, refund, and/or credit.

[a]Florida: Valued at not greater than salvage value.

[b]Hawaii: Air pollution control equipment only.

[c]Illinois: Special assessment applies.

[d]Kentucky: Excluded from property factor of apportionment formula.

[e]Kentucky: Subject to State taxation only.

[f]Tennessee: Taxes at lower rate.

[g]Minnesota: Expired December 31, 1980.

[h]Montana: Lower assessment applies to air pollution control equipment.

[i]Kentucky, Mississippi, Ohio: Special tax treatment applies to water, air, and noise pollution control equipment.

[j]Pennsylvania: Exempt from capital stock tax.

[k]South Carolina: Applies only to machinery used in mining, quarrying, compounding, processing, or manufacturing tangible personal property when installed and operated for compliance with an order of an agency of the United States or South Carolina.

[l]Utah: Exemption expires for tax years beginning on or after February 1, 1980.

[m]Pennsylvania: Exempt from capital stock tax.

[n]Minnesota, Wyoming: Special tax treatment applies to water, air, and land pollution control equipment.

[o]South Carolina: The exemption applies only to machinery used in mining, quarrying, compounding, processing, or manufacturing tangible personal property when installed and operated for compliance with an order of an agency of the U.S. or of this State to prevent or abate pollution of the air or water.

## Federalism.

Reagan campaigned heavily on the issue of deregulation, and he promised to "get the Federal government off the backs of the American people." More diplomatically, one might characterize his commitment as the promise to decontrol virtually every facet of Federal regulation, including any environmental regulations that cost more than they benefit American society. The battle cry of the Reagan Administration during its first year was deregulation!

Developments in Federal environmental policy during 1981 do indeed reflect a successful initial campaign against government regulation.

(1) Controversial appointments of allegedly anti-environment individuals to the Secretary of the Interior and EPA Administrator posts.

(2) Establishment of a cabinet-level Task Force on Regulatory Relief under the chairmanship of Vice-President George Bush along with Office of Management and Budget review of regulations.

(3) Significant reduction of both the Interior Department's and the EPA's budget and staff, as well as that of the Council on Environmental Quality.

In addition, 1982 promises two other major changes in Federal environmental policy — extension of the Clean Air and Federal Water Pollution Control Acts; and the return of much of the implementation and enforcement of environmental regulations to the States. Furthermore, the proposed Federal budget for fiscal year 1983 will include additional cuts in the EPA budget and staff. Finally, two major environmental laws were enacted during the "lame-duck" session of Congress following the November 4, 1980, elections in anticipation of a more conservative 97th Congress. The Superfund, establishing an environmental excise tax system (explained in detail in Chapter 10), for the production of hazardous substances, the proceeds of which are to be used to clean up hazardous waste dumpsites, was certainly a surprise. Likewise, the final compromise on the Alaskan land bill was probably the best that environmentalists could have realistically expected.

### Controversial Appointments at Interior and the EPA

Although the appointment of James G. Watt as Secretary of the Interior was probably more controversial than that of Anne M. Gorsuch as EPA Administrator, they equally arouse the ire of environmentally sensitive Americans because of their actions since they took office. As the overseer of nearly 3 billion acres of Federally owned property (including over 2 billion acres of outer Continental Shelf ocean bottom), Secretary Watt has considerable power over the natural resources of the United States. Many of his proposals to open public lands for private development, especially for the extraction of resources, have had to be tempered or delayed because of public pressures. One would expect environmentally activist groups such as the Sierra Club, National Audubon Society, Natural Resources Defense Counsel, Environmental Defense Fund, and Wilderness Society to be against him, but even the conservative National Wildlife Federation has announced its opposition to Watt's continued residence at the Interior Department. Reagan has stood by his appointment, and the "Sagebrush Rebellion" of the western States, where the bulk of Federally owned lands are located, seems generally pleased with Watt's position on the increased private development of public lands.

Given the fiscal conservatism of our times, Watt's desire to spend funds on repairing and restoring national park facilities rather than acquiring new parkland may be prudent. It is not an easy task to maintain a balance between the enjoyment of natural resources by our generation and the preservation of them for future generations. Watt also demonstrated that he could stand up to powerful development interests in November 1981, when he refused to approve a dam at the confluence of the Verde and Salt rivers that would have inundated the Yavapai tribal lands in Arizona. Another example of his will occurred in November when Watt testified before the House Interior Committee against a bill that would give him the power to grant rights-of-way across private property for coal slurry pipelines to transport coal mined in the western States. Although the Reagan Administration acknowledged that such Federal power would expedite development of pipelines to compete with the railroad monopoly in the transport of western coal, Watt emphasized that the legislation would be inconsistent with the administration's "New Federalism" policy of returning power to the States. Thus, Watt pleased the railroads at the cost of offending the coal mining industry, electric utilities (which are under a Federal mandate to convert from oil and gas to coal), and western States with coal resources.

Watt's participation in the Reagan Administration's deregulation program is most apparent in his reorganization of the enforcement activities related to the Surface Mining Control and Reclamation Act (SMCRA) of 1977, which established the Federal authority for regulating strip-mining. The Interior Department issued revised regulations in the April 17, 1981, *Federal Register*, relaxing requirements related to bond-posting and the closing of mines for violations of the SMCRA. Watt also reorganized staffing for the Office of Surface Mining (OSM), which is responsible for implementing strip-mining policy. The staff of the five regional offices of the OSM will be replaced with two technical centers and fourteen State "liaison" offices for the thirty-one coal-producing states. In addition, the inspection and enforcement divisions have been reclassified as subdivisions and their staff reduced by two-thirds.[41]

Since Anne M. Gorsuch's appointment as EPA Administrator followed Watt's takeover at Interior by a few months, she was not initially as controversial because of the vengeance with which he attacked his job. However, her subsequent administrative actions were no less criticized by the Environmental Lobby, State officials, and even selected industry representatives. At the outset, Gorsuch's appointment was disturbing to environmentalists because, like Watt, she lacked experience with the management of environmental affairs. Even Watt's experience as president of an anti-environmental regulation industry group (Mountain States Legal Foundation) and as former member of the Federal Power Commission better qualified him for a top-level management position when compared to her limited involvement in enacting environmental legislation as a member of the Colorado State Legislature.

The new EPA Administrator, like the Interior Secretary, has surrounded herself with top-level aides who were formerly employed by or were representatives of industry. In following the Reagan Administration's policy of deregulation, she cooperated with the budget-cutting tactics of the director of the Office of Management and Budget (OMB), David A. Stockman, for fiscal years 1981 and 1982. However, Gorsuch recently experienced a "change of heart" with regard to the EPA budget for fiscal 1983, when Stockman tried to slash her budget even more than she had cut it.

Apparently, Gorsuch did not consult any of the senior EPA career staff or her predecessors at the EPA, such as, Russell Train, William Ruckelshaus, and Douglas Costle, when she arrived in Washington in anticipation of her confirmation by the Senate. Upon her confirmation in April 1981, she began working with her immediate staff to reorganize the EPA without the advice and counsel of senior EPA career staff. As a result of the budgetary pressures and deregulation policy established by the Reagan Administration, Gorsuch decided to promote voluntary compliance by industry with environmental regulation. Therefore, she eliminated the EPA's Enforcement Division by assigning enforcement lawyers to other EPA divisions which have responsibility for developing pollution abatement programs for water, air, solid waste, and pesticides and toxic substances, for research and development, and for administration. These EPA divisions have traditionally worked in a cooperative and nonthreatening environment with industry and with State and local governments in developing pollution abatement programs.

As a result of the elimination of the Enforcement Division, the enforcement actions referred to the Department of Justice, as well as the initiation of new enforcement actions, declined significantly. This result is consistent with both the deregulation goals of the Reagan Administration and its New Federalism policy of giving enforcement responsibilities to the States.[42]

Although the "foot-dragging" on promulgating regulations for the TSCA of 1976 began in the Carter Administration, not much progress has been made during Gorsuch's short tenure at EPA.[43] Some would argue that the EPA's policy of voluntary compliance by industry is just going to encourage noncompliance, but even the chemical industry realizes that no regulations under TSCA means that producers of potentially toxic chemicals are faced with uncertainty and the possibility that the Federal courts may inconsistently establish standards for toxic substances. In fact, top-level EPA officials held unprecedented *secret* meetings with chemical industry representatives in late 1981 to develop acceptable standards for the use of some chemicals suspected of being carcinogens.[44] Thus, cynicism abounds in Congress, as well as among environmentalists, as to the intentions of the Reagan EPA in regulating toxic substances.

Another EPA policy disturbing to environmentalists and Congress was the announcement in August 1981 that Gorsuch was forming internal task forces in order to review regulations already promulgated on various environmental laws for which the agency has enforcement responsibilities.[45] For example, the EPA announced in late November 1981 that it was considering a major relaxation of auto emission standards as a method of settling eight lawsuits filed by the auto industry on EPA regulations, ranging from emission standards for trucks to warranty requirements for antipollution components for motor vehicles.[46]

Deregulation

Although deregulation began in the Carter Administration with the establishment of the Regulatory Analysis Review Group,[47] the current impetus for regulatory reform reflects the support Ronald Reagan received during his presidential election campaign for his promise to "get the government off the back of the American people." President Reagan wasted no time in taking action on his promise: only nine days after he took office, he ordered all Federal agencies to freeze regulations issued during the last days of the Carter Administration. His order also placed a sixty-day moratorium on the issuance of new regulations, except regulations necessitated by emergencies or serious economic hardship.

As a focal point for his Administration's deregulation efforts, Reagan issued Executive Order 12291 on February 18, 1981, which created a new cabinet-level Task Force on Regulatory Relief (hereinafter the Task Force) under the Chairmanship of Vice-President Bush. The order also authorized David Stockman's OMB to review and guide the deregulation efforts of Federal agencies, such as the EPA, the Interior Department's OSM, and the Labor Department's Occupational Safety and Health Administration (OSHA). In principle, the order does not empower the Task Force or OMB to assume the EPA's (or any other agency's) authority and responsibility for promulgating regulations. However, the OMB's persuasive powers are significant because it can cut an agency's regulatory and enforcement budget. One need only review the budget cuts for the revised fiscal year (FY) 1981 budget, the FY 1982 budget, and the proposed FY 1983 budget to appreciate how the OMB can implicitly control the deregulation process by cutting funds and staff for the regulatory and enforcement activities of a Federal agency, such as the EPA, Interior's OSM, and Labor's OSHA.

Furthermore, the OMB has been given the authority under the Paperwork Reduction Act of 1980 to control the amount of forms, reports, and the like, that the Federal government requires individuals and institutions to provide for executive branch agencies. Consequently, the OMB can refuse to approve the forms, reports, or other paperwork requirements of an agency unless it cooperates with the regulatory reforms preferred by the budget agency. However, President Reagan's OMB has probably not needed to exercise such authority for the EPA or Interior Department because of the zeal with which Gorsuch and Watt have proposed cuts in their budgets.[48]

Indeed, the regulatory explosion has been real. For example, the text of the *Federal Register* grew from about 20,000 pages in 1970 to over 70,000 by 1979. However, estimates of the costs of regulation vary considerably, and its benefits are even more difficult to determine. Clearly, however, the trend is being reversed by the Reagan Administration.

Business often cites the EPA's regulations as the most costly of all those promulgated by Federal agencies during the 1970s. Thus, it is no surprise that EPA regulations headed the list of most costly Federal compliance rules when the Business Roundtable published its 1979 study of regulation.[49] More recently, the Commerce Department conducted a survey to determine which Federal regulations are the most burdensome and costly for business and industry. The top twenty, ranked according to number of responses and estimated cost, include five for which the EPA is responsible — hazardous waste management rules, standards for the national pollution discharge elimination permit system, the Clean Air Act's pre-treatment standards, notification and testing requirements under the TSCA, and proposed rules for "cradle to grave" records of chemicals that pose possible health or environmental risks. Four out of the first five on the list were EPA regulatory responsibilities. Vice-President Bush is using this list and others compiled by the Task Force as a basis for the Reagan Administration's deregulation efforts.[50] A more detailed analysis of EPA deregulation actions is provided below.

Thus far, deregulation is being accomplished by budget cuts, agency reorganizations, and specific policy directives from either the OMB or the Task Force. However, Congress has not overlooked public and business support for deregulation. There is apparently support in Congress for statutory authority to implement a "legislative veto" process. Although Congress has specific veto power for at least 190 laws, a general "legislative veto" statute is being developed as part of the Senate's omnibus regulatory reform legislation. Such legislation is contingent upon both House cooperation and judicial review to determine constitutionality because of the "separation of powers" doctrine.[51]

### Budget Cuts

In order to balance the Federal budget during his Administration as he promised during the 1980 campaign, President Reagan apparently felt he needed to begin with the fiscal year 1982 budget which his predecessor, former President Jimmy Carter, had submitted to Congress. Thus, Reagan's Budget Director David A. Stockman developed revised FY 1982 budget figures which Congress approved with few changes, thanks to a coalition of conservative Southern Democrats and Republican loyalists in the House. Similarly, Congress passed several budget rescissions proposed by the OMB as part of the budget reconciliation process for the FY 1981 budget, including cuts in the EPA's budget. Furthermore, Gorsuch did not spend all of the funds appropriated for her agency by Congress.[52] Ironically, the EPA's operating budget for FY 1982 is larger than its actual FY 1981 budget although the Reagan Administration reduced Carter's proposed EPA operating budget.[53]

For FY 1982, the EPA operating budget was increased to $1.35 billion.[54] In spite of the small budget increase between FY 1981 and FY 1982, there is to be a reduction in EPA personnel from 11,800 in FY 1981 to 11,400 in FY 1982,[55] with most of the loss to be borne by EPA's enforcement staff. As noted, Gorsuch abolished the EPA Office of Enforcement, temporarily reassigning attorneys in that office to other divisions of the agency. As a result, the EPA referred less than fifty cases to the Justice Department in 1981 after averaging 200 cases per year over the past decade.[56]

Dramatic decreases in the FY 1983 EPA budget and personnel as proposed by Gorsuch were apparently not sufficient for the OMB which slashed EPA's budget and personnel[57] to $780 million and 7,340, respectively. Fortunately, President Reagan supported Gorsuch and overruled the OMB's cuts in the EPA budget and staff, so that for FY 1983 operating budget appropriations would be $975 million and personnel would be 8,500.[58] This budget still represents a 40 percent drop from the FY 1982 budget appropriations for the EPA. Hence, the Environmental Lobby and the environmentally conscious members of the Senate and House may force the Congress to increase EPA's budget and staff to FY 1982 levels.

Interior's OSM is another agency with environmental responsibilities that has suffered from budget and personnel cuts in the FY 1982 budget. Ironically, however, Labor's OSHA received a slight budget increase for FY 1982, although their staff was cut 20 percent.[59]

Apparently, the Council on Environmental Quality (CEQ), which is responsible for monitoring EISs required by the National Environmental Protection Act (NEPA) for any major Federal agency actions with environmental impact, has suffered the most drastic reductions in budget and staff as a result of the budget cuts. Thus, the CEQ's budget for FY 1982 was reduced to about $1.5 million.[60] These cuts mean that the CEQ will be fortunate to have one staff member who can devote full time to the review of the 700 EISs submitted annually. (It devoted six or seven people to such tasks during the Carter Administration.[61]) If the CEQ does not satisfactorily handle the EIS process, Congress may allocate more funds and staff to the council with specific instructions to carry out its NEPA responsibilities at the same level as maintained during the Carter Administration, which had successfully streamlined the EIS process.[62] Another casualty of the budget and staff cuts is the CEQ's annual report on the state of the environment—*Environmental Quality*, which was published from 1970 through 1980.[63]

To mitigate the effect of drastic cuts in Reagan's proposed FY 1983 budget, the President announced in his January 26, 1982, State of the Union Message to Congress that he plans to implement his New Federalism policy by turning over many Federal social and environmental programs to the States. Although some Federal monies will also be given to the States to help finance their expanded role in Federal programs, State governments will have to pay a significant portion of the cost of their new responsibilities.[64] Therefore, the States would have either to raise taxes or to limit their efforts to the level that can be funded from promised Federal aid. Of course, Congress can choose to restore Federal funding for such social and environmental programs to pre-FY 1983 levels in response to demands by the States that they cannot raise taxes to finance what they consider to be Federal responsibilities. In fact, congressional support in the form of legislation may be necessary before President Reagan can implement his New Federalism. Consequently, Congress may decide to approve Reagan's plan and authorize more funds to help States finance their responsibilities, or it may reject his New Federalism and authorize more funds for Federal agencies with responsibility for social and environmental programs. Many States and environmental groups may lobby for the latter.

## The New Federalism

The following list indicates the Federal environmental programs for which the States already have some responsibility:

National Environmental Policy Act of 1969

Clean Air Act of 1970

Federal Water Pollution Control Act of 1972

Coastal Zone Management Act of 1972

Safe Drinking Water Act of 1974

Hazardous Materials Transportation Act of 1974

Resource Conservation and Recovery Act of 1976

Toxic Substances Control Act of 1976

Surface Mining Control and Reclamation Act of 1977

Federal financial aid for many of these State-operated programs has been woefully inadequate. Therefore, those States that have decided not to supplement Federal funding are probably not carrying out program objectives as well as those States that do supplement Federal financial assistance.

In summary, the New Federalism may exacerbate the differences that already exist between States in the implementation of Federally mandated social and environmental programs. These differentials may be especially acute in the case of environmental programs because both the Clean Air and Water Pollution Control Acts must be renewed in 1982 and because the implementation of the Resource Conservation and Recovery Act and the Toxic Substances Control Act is entering a critical period of development. It would be very unfortunate if certain States were to allow their environmental quality to deteriorate either because of funding shortages or because of local or regional opposition to environmental regulation.

Clean Air Act

Although budget cuts and the New Federalism will occupy much of the Congress's attention in 1982, renewal of the Clean Air Act (CAA) will probably be the single most important environmental policy issue that must be resolved before the end of the 97th Congress. In theory, the Clean Air Act was supposed to have been renewed by September 30, 1981. Both the Senate Environment and Public Works Committee and the House Subcommittee on Health and Environment have held hearings on the act, but action has been delayed because the Reagan Administration decided in August 1981 not to submit legislation, but rather to establish broad objectives for revising the act.

Before analyzing proposals for revising the CAA, a brief summary of its provisions is provided as a basis for understanding such revisions. In the early 1960s, Federal air pollution control legislation was limited in scope, with the Federal role consisting of financial assistance to the States and research and development. When the public and Congress learned that air pollution problems were worse than originally perceived, Federal air pollution control legislation was amended, culminating in the enactment of the Clean Air Act of 1970. Now the CAA must be renewed to extend Federal authority in establishing and enforcing air pollution control standards.

The CAA requires that the EPA set national ambient air quality standards designed to protect human health within a margin of safety. Standards have been set for various types of emissions for mobile (automotive) and stationary (industrial) sources, including ozone, carbon monoxide, sulfur dioxide, hydrocarbons, lead, nitrogen dioxide, and various hazardous pollutants, such as asbestos, mercury, and vinyl chloride.

Other CAA provisions require States to design a State implementation plan for achieving Federal air quality standards if the States have not already done so. It also provides for the prevention of significant deterioration (PSD) of certain areas that actually have cleaner air than the standards require. This PSD provision prevents industries from moving out of developed areas to less developed areas in order to avoid pollution control requirements; and it keeps U.S. national parks and wilderness areas from deteriorating and losing the very attributes that make them attractive.

The new source performance standards limit emissions from major new industrial sources. In this way, as new industry gradually replaces old, the degradation of air quality will lessen and eventually improve. Similarly, if a new industry is located in an area that does not meet Federal air quality standards, that industry must insure that a previously existing industry reduces its pollution by more than the amount the new source will emit. This regulation has the net effect of reducing emissions while allowing industrial development to occur in areas where air pollution is already a serious problem. This arrangement has been characterized as the "offset policy."

A related development introduced by the EPA in 1979 is known as the "bubble concept." This approach to air pollution control allows a polluter to treat a multisource industrial complex as a single source of a specific air pollutant, that is, the entire plant complex is treated as a "bubble." Consequently, the polluter may abate its overall emission of that pollutant in a cost-effective manner by successfully utilizing the least costly technology until the standard for the polluting substance is satisfied. The bubble concept eliminates the necessity both for source-by-source optimization within a plant complex and for establishment of government-specified pollution controls for each source. By the end of 1981, at least seventy "bubble" arrangements were being developed and another thirty had been submitted to the EPA.

If the bubble concept were combined with the offset policy, several companies and their plants could be included in a multiple-company "bubble." In other words, a market in pollution rights could be created whereby one company's pollution abatement would become an exchangeable commodity to be sold to another firm (new or expanding) in the same air pollution control region. Thus, a decrease in pollution from one company's plant could be offset by an equal or lesser increase in emissions from another firm's industrial complex. Examples of such arrangements have already occurred at a General Motors' facility in Oklahoma City and the Volkswagen plant in New Stanton, Pennsylvania.[65] Such arrangements could be authorized in the new Clean Air Act, but the EPA apparently has the authority to establish regulations to formalize combined "bubble concept-offset policy" arrangements.

The Clean Air Act has probably been the most controversial environmental legislation enacted by the Congress, although both RCRA and TSCA may eventually become more controversial. As a consequence, ideas from industry and business, as well as environmentalists, for the "reform" of the CAA are plentiful. Overall, the reformers support somewhat competing interests—industrial expansion, energy development, and cleaner air. More specifically, debate on renewal of the CAA will probably focus on scheduled deadlines for achieving ambient standards by those areas currently not complying therewith; the PSD program; the standards-setting process and the scientific basis for standards; the process for approving and revising State implementation plans; motor vehicle emission standards for both gasoline- and diesel-powered engines; and the inter-State and international transport of pollutants, especially sulfur dioxide which is the major ingredient in acid rain. The Canadians are lobbying for a significant reduction in the burning of high-sulfur coal by utilities and industry in the Midwest and the northeastern United States, although Federal energy policy has mandated a switch from burning fuel oil and natural gas to coal-burning.[66]

The Reagan Administration's position on renewal of the CAA was announced in early August, 1981 by EPA Administrator Gorsuch. The Administration's statement of basic principles, which are based on the EPA's proposed legislation for the renewal of the CAA, includes the following recommendations for major revisions:

(1) Secondary air quality standards, that is, those that protect nonhealth qualities such as visibility in national parks, should remain a Federal responsibility, but the PSD program should be limited to national parks and wilderness areas.

(2) National ambient air quality standards should be based on a "risk analysis" method that will set primary air quality standards to protect humans from "real health risks."

(3) Outside of national parks and wilderness areas, new sources of air pollution should be controlled by requiring uniform pollution control technology.

(4) There should be a deferral of deadlines prescribed by the CAA for meeting national ambient air quality standards in respect to those large urban areas that cannot possibly attain standards by the prescribed deadlines.

(5) A more effective hazardous emissions program should be established.

(6) Research on the effects and pervasiveness of acid rain should be accelerated.

(7) Motor vehicle emission standards should be relaxed.

(8) States should be given more authority, as well as more flexibility, in designing State implementation plans with the EPA's role being one of review and monitoring.[67]

In spite of lengthy debate, no legislation reenacting the CAA was passed in 1981, but just before the House and Senate completed the first session of the 97th Congress in December 1981, Congressman Thomas A. Luken (D-Ohio) introduced a bill that makes moderate revisions in the CAA. The legislation is co-sponsored by Congressman John D. Dingell (D-Mich.), who is Chairman of the House Energy and Commerce Committee which has jurisdiction over air pollution control legislation, and Congressman James T. Broyhill (R-N.C.), who is the ranking minority member on the House Energy and Commerce Committee. It was a Dingell-Broyhill coalition that helped enact air pollution control legislation in 1977. In general, industry (but not the chemical industry or the utilities) seems to accept the bill as a necessary compromise because it is probably the best it can do.[68]

The Reagan Administration formally endorsed Luken's bill in mid-January 1982. Speaking for the Administration, EPA Administrator Gorsuch said that she was very pleased with the bill and that Vice-President Bush would help lobby for the revisions insofar as the changes are consistent with the Reagan Administration's policies of deregulation and New Federalism. The following provisions of the Luken bill would seem to be consistent with the general principles announced by the Administration in August 1981:

(1) The rules for locating new pollution sources in areas not complying with national air quality standards would be relaxed by replacing the "lowest achievable emissions rate" with a requirement that the new source use the "best available control technology."

(2) The EPA would be allowed to defer to 1993 the 1982 and 1987 deadlines for any State's compliance with national air quality standards if the State guarantees compliance by 1993.

(3) Auto emissions for new motor vehicles would be doubled because new cars would still run cleaner than the old cars they are replacing, thereby reducing total pollution from motor vehicles.

(4) Penalties for States that fail to submit implementation plans for compliance with the CAA would be eliminated, and the EPA's review and approval process for changing State implementation plans would be expedited.[69]

Only time will tell how well the Luken bill will fare. It will most probably be passed in a somewhat amended form because it is supported by industry and the Reagan Administration, and it is co-sponsored by two of the most influential Congressmen on matters concerning the environment.

### Other Environmental Legislation

In addition to the Clean Air Act, water pollution control legislation must be reauthorized so that the sewage treatment construction program can be continued. The Endangered Species Act of 1973 must also be reenacted or it will expire in 1982. As always, water resources project authorizations[70] will be controversial with environmentalists in combat with local developers and politicians. Reagan's budget-cutting objectives will probably result in significant reductions in funds available for both the sewage treatment construction program and water projects.

### Environmental Administration at EPA, CEQ, and OSM

Budget and staff cuts at the EPA, CEQ, and OSM have apparently already affected enforcement efforts. The reorganization of both the EPA and OSM has resulted in a deemphasis of enforcement activities, with the anticipation that industry and business will voluntarily comply, especially if environmental regulations are simplified. Moreover, State enforcement efforts are supposed to be expanded to compensate for less Federal enforcement. Similarly, the CEQ will have little staff to review EISs. Therefore, it must concentrate on the most significant Federal agency actions with environmental impact.

Of course, environmental groups perceive such budget and staff cuts as the dismantling of the pollution control regulatory system that they worked so hard to develop. Conversely, business and industry welcome the friendlier atmosphere in Federal agencies with environmental regulation responsibilities. Business and industry would probably characterize the atmosphere as one of cooperation, while during the previous decade the atmosphere was one of confrontation by antagonistic bureaucrats. Judicial and congressional review of the actions of the Reagan Administration's environmental regulators will most certainly limit any intentions they have to dismantle the pollution control regulatory system developed over the past decade.

While the 1970s were years of development for Federal environmental regulatory authority, the 1980s should be a period of implementation and "fine-tuning" for the regulatory system necessary for ensuring environmental quality. Most public opinion polls indicate that support for pollution control legislation is strong and stable, although recent economic problems may erode some support. Thus, the EPA in particular has arrived at a critical point in its evolution as the preeminent pollution control agency.[71] Its operating budget may be slashed from almost $1.35 billion in FY 1982 to less than $1 billion in FY 1983, and its staff cut to about 8,500 from the approximately 10,500 to be employed in FY 1982.[72] Although there was only a small cut in the EPA operating budget between FY 1981 and FY 1982, the full-time staff was cut by about 1,000 through attrition.[73]

An outline of the EPA agenda for the 1980s should provide sufficient support for the conclusion that the agency has reached the crucial point in its history. Therefore, the budget and personnel cuts it faces may hamper its ability to carry out its responsibilities. Of course, less Federal environmental regulation can be compensated for by more activity at the State and local levels of government. The uniformity of standards will suffer, however, because there will be considerable variation in how the States implement and enforce Federal pollution control mandates. Whereas the EPA maintains that achievement of environmental quality goals is an intergovernmental responsibility and that States know best their unique environmental problems and their optimal solution, environmentalists argue that the States will not allocate sufficient funds or staff to environmental regulation and that there will be inconsistencies between States, providing business and industry with the opportunity to operate in States with less rigorous standards and enforcement. The following is a cursory outline of the EPA's agenda for the 1980s:

(1) Siting of hazardous waste-disposal facilities as required by RCRA.[74]

(2) Implementation of the provisions of TSCA for regulating over 50,000 old substances as well as new chemicals, which is woefully behind schedule because of "foot-dragging" during the Carter Administration and a change in policy under the Reagan Administration.

(3) Rewriting of approximately 10,000 water discharge permits to encompass toxic waste controls.

(4) Negotiation of over 40,000 new water discharge permits to regulate and limit the amount of toxic wastes in sewage effluent.

(5) Implementation of the provisions of the Safe Drinking Water Act of 1974 for over 35,000 water systems.

(6) Application of the new Superfund law to clean up the potentially dangerous Love Canal-type hazardous waste dumpsites.[75]

The above agenda, while abbreviated, nevertheless suggests the Herculean task facing the EPA and the States in effecting environmental regulation as mandated by Federal legislation. Both Congress and the Federal courts will have considerable impact on the timetable for and rigor with which the EPA and States carry out their responsibilities for pollution control. If public support for environmental regulation continues at its current high level, many of the deregulation and New Federalism objectives of the Reagan Administration may be overruled by congressional and judicial mandates. In respect to the environment, these are indeed interesting times.

### 1982 Addendum

During the first six months of 1982, the Reagan Administration's environmental policy makers maintained a low profile. Ironically, opinion polls indicate that public support for environmental protection regulation continues at a high level, in spite of a consensus that less government and deregulation are desirable national priorities.[76] Thus, the Administration has toned down its rhetoric on environmental policy issues. However, the Administration has not abandoned its cooperative effort with industry to materially revise the major environmental acts that are to be renewed or reviewed during 1982, that is, the Clean Air Act, Water Pollution Control Act, Resource Conservation and Recovery Act, and the Endangered Species Act.[77]

As part of its public relations effort, the Reagan Administration's Council on Environmental Quality (CEQ) released a July 1982 report on environmental quality similar to those annual reports released by the CEQ in the ten years prior to Reagan's taking office. In the report, which stressed the significant progress made in reducing air pollution since the early 1970s, Reagan's CEQ took issue with an environmental report card published a month earlier by the Conservation Foundation.[78] The Foundation asserted that the Reagan Administration had terminated a decade of bipartisan consensus on environmental protection policy by advocating "deregulation, defederalization and defunding." The Conservation Foundation concluded that the Administration's policy has caused a polarization in relations among Federal environmental agencies, the Congress and environmental groups, discouraged the communication required to facilitate environmental programs, distracted the nation's attention from a new set of problems that could potentially harm public health and the environment, and terminated progress on the development of the information and analysis concerning natural resources and the environment by severely cutting the budget for scientific research thereon.

The Conservation Foundation's report, like the CEQ's report, did enumerate the significant progress the country has made in dealing with some traditional environmental problems. In the report, it concluded that: (1) emissions of most major air pollutants have continued to decline; (2) progress toward water quality goals has not been comparable, although there is evidence that some of the worst pollution problems may be easing; and (3) more land has become available for outdoor recreation and more sensitive lands are being protected. On the other hand, many traditional problems remain. They include soil erosion, pollution of national parks, abandoned hazardous waste dumpsites, and pressure to develop sensitive ecological areas such as wetlands and barrier islands. Moreover, one-seventh of the population lives in urban areas that will not be able to satisfy ambient air quality standards for protection of human health from ozone by 1987, even if present motor vehicle emission standards are satisfied. With regard to the new set of problems, the report cited acid rain, groundwater contamination and depletion, hazardous waste disposal, indoor air pollution, and the accumulation of $CO_2$ and chlorofluorocarbons in the atmosphere.

The Conservation Foundation's report did include some suggestions for improving environmental programs. It recommended reorganization of environmental agencies (in which programs generally are compartmentalized) to more comprehensively deal with environmental problems. The report also advocated increased reliance on both market mechanisms and incentives, as well as increased integration of both statutory authorities and organizational arrangements.[79]

Insofar as the Conservation Foundation is a traditional environmental group like the National Wildlife Federation, its critique of environmental policy traditionally has been moderate and objective in its tone. Thus, its environmental report card received widespread publicity because of its credibility. In contrast, when ten of the more activist environmental groups issued a report in late March 1982 indicting the Reagan Administration's environmental policy, it did not receive as much publicity, although the EPA and Interior Department reacted negatively as one might expect. However, the White House refused to comment on the report,[80] although most of the final (but unpublished) environmental report of the Carter Administration's CEQ was available as a basis for an updated report on progress in solving environmental problems.[81]

Another objective report on environmental policy was released in the spring of 1982. This report, which was issued by the Government Accounting Office (GAO), concluded that the Federal government must evaluate the effectiveness of: Federal and State programs to reduce the dangers posed by hazardous and solid wastes; water pollution control programs; environmental regulatory strategies; and environmental impact statements. Furthermore, the Federal government should review: (1) the fiscal and administrative integrity of the sewage treatment construction grants program; (2) the Clean Air Act and what effect changes to it will have; (3) improvement in regulation of dangerous pesticides and chemicals; and (4) the safety of drinking water. With respect to emerging environmental problems, the GAO recommended that more attention be given to global environmental issues, such as acid rain, the "greenhouse effect" from $CO_2$ accumulation in the atmosphere, and the depletion of ozone in the atmosphere.[82]

Notwithstanding unyielding public support for environmental programs and well-publicized reports on continuing and emerging pollution problems, the Reagan Administration has to contend with a united group of environmental and conservation organizations—the Environmental Lobby. Ironically, these organizations are financially healthy as a result of the Administration's own rhetoric. The Administration also must cope with key Congressmen and Senators, that is, the leadership of committees with responsibility for environmental affairs, who generally hold a moderate position on environmental issues. As a consequence, it is increasingly likely that the Administration and industry lobbyists will be unable to effect deregulation and defederalization through significant revisions of the environmental laws due to be reenacted during the 1982 congressional session.[83]

### The EPA Budget

The spending targets established by Congress in approving its first 1983 concurrent budget resolution at the end of June are indicative of congressional moderation on environmental issues. As a result of a House-Senate conference committee's compromise, the Senate's proposal to maintain FY 1983 environmental spending at the level appropriated for FY 1982 was adopted.[84] Thus, the EPA's operating budget would remain at about $1.1 billion for FY 1983 rather than drop 12 percent to $961 million as President Reagan had requested. Similarly, Reagan's request for a 12 percent decrease in EPA staff from about 9,820 to almost 8,650 would not occur. In addition, the EPA's sewage treatment construction grants program and its research and development program would continue at FY 1982 budget levels, while appropriations for the "Superfund" would probably increase as proposed in the Reagan 1983 budget.[85]

Although the first budget resolution establishes spending goals, the House and Senate appropriations committees actually determine how much each government agency will be allowed to spend. These two committees should complete much of their appropriations task before the congressional summer recess begins in mid or late August at which time the actual FY 1983 EPA budget should be delineated. However, the actual EPA budget will have to survive the second congressional budget resolution in the fall, and the President's approval. The second resolution establishes spending ceilings in concert with estimates of Federal revenue,[86] thereby establishing an acceptable budget deficit. Unfortunately, the possibility of continued economic stagnation or only a mild economic recovery could mean a larger than anticipated budget deficit. Consequently, both additional tax revenue and across-the-board budget cuts would be necessary before a continuing budget resolution could be approved.[87] Finally, the President's approval and signature will be necessary.

Notwithstanding all the threats to continued funding of the EPA FY 1983 budget at the FY 1982 level, congressional dissatisfaction with the EPA's performance and previous cuts by the Reagan Administration in the EPA budget would seem to render additional cuts in that budget unlikely. Support for this conclusion is based on bipartisan concern expressed during EPA Administrator Gorsuch's appearances before Senate Environment and Public Works Committee in late February, as well as on the House Democrats' threat earlier that month to make EPA budget cuts a major campaign issue in the November 1982 election.[88]

Evidence of the Democratic threat to use the EPA budget cuts and the Reagan environmental record as a campaign issue occurred in late July at an unusual joint hearing of four House subcommittees. During the meeting, EPA Administrator Gorsuch was harassed for three hours in a packed hearing room by the Democratic members of the subcommittees. The Democrats grilled Gorsuch on the Reagan Administration's environmental record, especially its budget cuts, its implementation of environmental programs, and its proposed revisions of environmental laws. Of course, she defended the Administration's environmental record as being consistent with the President's policies of deregulation and defederalization based on the comparative advantage of States in delineating and solving environmental problems. In expressing her own dissatisfaction, Gorsuch pointed out that Congress had been unwilling to enact the Administration's legislative proposal for revising the Clean Air Act.[89]

## Clean Air Act and Air Pollution Control

The EPA could pursue aggressively sanctions against the 32 States that did not fulfill by July 1, 1982 the Clean Air Act's requirement to submit State implementation plans for making progress toward attaining national ambient air quality standards. The mandatory sanctions include the withholding of sewage treatment construction grants and grants to States to aid their air pollution control program, as well as the delay of any construction projects that would be a major source of air pollution if completed. Gorsuch has admitted on several occasions that vigorously enforcing the sanctions could pressure Congress to accelerate enactment of revisions to the Clean Air Act (CAA). However, she has been careful to follow a "due process" approach in enforcing the sanctions so that Congress will have the time to deliberately revise the CAA, especially sanctions associated with the original deadlines for state plans and for attaining ambient air quality standards.[90]

In early June, 32 States were notified that their failure to submit a State implementation plan by July 1 could result in a loss of Federal funds and in injunctions to halt construction of major industrial projects. All 32 States, including California and much of the industrial northeast, failed to meet the July 1 deadline. Thus, the EPA has decided to rely on lengthy notice-and-comment procedures to extend the review process for State implementation plans. These procedures should effectively delay action on sanctions until mid-1983 at which time Congress should have amended the CAA to allow the EPA more discretion in imposing sanctions. Notwithstanding the EPA's procedural postponement of mandatory sanctions, industry and the States should feel more pressure and they in turn should force Congress to enact revisions of the CAA quickly.

With regard to the 18 States that already have implementation plans approved by the EPA, there is a December 31, 1982 deadline for meeting national ambient air quality standards. Any of the 18 States, which include Florida, Maine, and Minnesota, not meeting the deadline would be subject to the same sanctions applicable to the 32 States not submitting implementation plans. Assuming amendments to the CAA are not enacted in 1982, the EPA would probably adopt the same time-consuming type of process for delaying imposition of sanctions so that revisions of the law could be enacted to delay the deadline for satisfying air quality standards.[91]

Although there has been some resistance in both the House and Senate to the extension of deadlines, relaxation of ambient air quality standards and the weakening of certain emission standards,[92] the major roadblock to reenacting the CAA in 1982 may be the addition of provisions in the Senate version of the revised law to mitigate the problem of acid rain. Although the Senate proposal would mandate a significant reduction of $SO_2$ emissions in the 31-State region adjacent to and east of the Mississippi River, consumers of the production from coal-fired plants (primarily electric utilities) would not begin to incur the costs of emission controls until the late 1980s inasmuch as the emission reduction would be phased in between then and 1995.[93] Notwithstanding the concern of the northeastern states and Canadian complaints about U.S. stalling on acid rain negotiations,[94] the White House is opposed to the Senate bill, as are the House and lobbyists for utilities and industry.[95]

The EPA did make one controversial announcement with regard to air pollution control early in 1982. In mid-February, it asked for public comment on the phasedown of the allowable lead content of gasoline and on its proposal to indefinitely suspend the October 1982 deadline for small refiners (up to 50,000 barrels per day of crude oil) to meet the limit on the lead content of gasoline that is already applicable to large refiners, that is, 0.5 gram per gallon. The EPA indicated that it is considering several options: continuation of lead phasedown, continuation of the lead phasedown in selected urban areas only, creation of "market rights" in annual lead content, relaxation of the allowable lead content of gasoline, or rescinding the lead limits altogether. The latter two options were a response to a mid-1981 request from Vice-President Bush's Task Force on Regulatory Reform that the EPA reconsider the need for lead phasedown regulations when at least half of the motor vehicles sold in the United States operate on lead-free gasoline.

Since 1975, most automobiles sold in the United States have been required to use lead-free gasoline because new autos must be equipped with catalytic converters that only control $CO_2$, $NO_x$, and hydrocarbon emissions if lead-free gasoline is used in operating the vehicle. Thus, as new automobiles burning lead-free gasoline replace the older autos burning leaded gasoline, there is a commensurate increase in the consumption of lead-free gasoline[96] and a decrease in ambient levels of lead in the atmosphere.[97] As a consequence, the issue arises as to whether lead phasedown regulations are still necessary to ensure that the ambient level of lead in the atmosphere is acceptable from a public health viewpoint when ultimately all automobiles will operate on lead-free gasoline and most other transportation vehicles will burn lead-free diesel fuel.

In March, the Center for Disease Control issued a report that indicated there had already been a 37 percent drop in mean blood lead levels between 1976 and 1980.[98] Nevertheless the EPA privately doubts the estimates from air pollution monitors of the ambient levels of lead in urban areas, though monitoring may have underestimated lead levels because monitors were not properly located, some modification of the lead phasedown regulations is not probable until after the November election.[99]

In April 1982, the EPA officially announced its long-anticipated expansion of the "bubble concept" to create a "market rights" approach to controlling industrial emissions. Although this announcement pleased industry, it received minimal criticism from environmentalists. The bubble concept will now be integrated with the "offset policy," which has been used in areas that do not meet national ambient air quality standards. As explained above, the bubble concept has been used to treat emissions from a plant complex (or geographic area) as if the complex (or area) were a "bubble" so that total emissions rather than emissions from a particular source (for example, a smokestack) would be the basis for satisfying emission standards. As a consequence, plant expansions or new plants were allowed in areas already in compliance with national air quality standards when average emissions for a specific pollutant within a plant complex satisfied emission standards. Conversely, the offset policy was used to prevent plant expansion or new plants except where emissions from existing plants are reduced to offset emissions from the plant expansion or new plant.

Under the EPA's new bubble concept policy, markets for buying, selling, and trading rights to pollute the air are also being established in those areas not meeting national ambient air quality standards. Not only did the EPA issue guidelines for the banking and selling of emission-reduction credits, but it also transferred to the States much of the authority for approving bubble arrangements. In principle, the expanded bubble concept should allow a company or area to reduce costs of pollution control because of increased flexibility in the selection of the most cost-effective pollution control technology for reducing overall standards. Moreover, expansion of the bubble concept should reduce air pollution inasmuch as industry should have both the financial incentive to abate pollution below required levels and the market incentive to develop improved pollution control technology. Based on EPA estimates, the 18 existing bubbles have saved industry $50 million to date. Similarly, approval of the 90 pending bubbles could result in the savings to industry of more than $1 billion according to the EPA.

Environmentalists criticize the EPA's new bubble concept policy insofar as it would allow areas not complying with national ambient air quality standards to delay their clean-up efforts. With regard to the shifting responsibility to the States, environmentalists worry that economic stagnation and the traditional unwillingness of the States to vigorously enforce Federal regulations could result in the approval of inadequate bubble plans, thereby aggravating air pollution.[100] Finally, questions have been raised by politicians and environmentalists about EPA's authority to both expand the bubble concept and shift administrative responsibility to the States, but Congress is expected to authorize the EPA to use its innovative market approach when the CAA is extended.

### Water Pollution Control

The Water Pollution Control Act (WPCA) which is to be extended in 1982 will probably suffer the same fate as the Clean Air Act did in 1981 when it was supposed to be reenacted, that is, extension of the WPCA will not occur until 1983. Both industry, and State and local government are lobbying for revisions in the Federal Water Pollution Control Act.

The Reagan Administration submitted 15 proposed amendments to the WPCA in late May. These proposed amendments would give State and local governments more flexibility in conducting their water pollution control programs. For example, the WPCA would no longer require the EPA to specify what water pollution control methods should be used by municipalities to abate water pollution so long as the local government promises that its program could achieve acceptable results. Needless to say, industry is lobbying for similar amendments so that the EPA could substitute effluent discharge and water body standards for the current technology-based approach to controlling industrial water pollution.[101]

With regard to the U.S. performance in meeting its commitments under the 1978 Great Lakes Water Quality Agreement, two reports by the GAO verify Canadian complaints about U.S. "foot-dragging" on abating its pollution of the Great Lakes. In its May 1982 report, the GAO concluded that the United States would not meet the agreement's goal of achieving secondary treatment of all municipal sewage discharged into the Great Lakes by December 31, 1982. While 99 percent of the Canadian municipalities had achieved the secondary treatment goal, less than two-thirds of the U.S. municipalities on the Great Lakes were using secondary treatment. Similarly, over 40 major municipal treatment plants in the United States were probably in violation of the phosphorus discharge standard agreed to in 1978. Thus, the EPA Great Lakes National Program Office must be given the resources and authority necessary to meet U.S. commitments. The GAO also concluded that the agreement should be revised to include other water pollution problems, for example, non-point source and toxic pollution.

In its late June report, the GAO stressed that the United States has neglected its role as the principal advisor for water quality problems of U.S.-Canadian waters as delineated by the International Joint Commission (IJC). The GAO's examples of this neglect include: (1) the failure of the United States to respond to the recommendations of the IJC's multi-million dollar 1980 study of pollution from land-use activities; (2) the formal response of the United States to only three of 16 reports issued by the IJC since 1972; and (3) the failure of several key Federal and State agencies involved in water quality activities to respond to requests from the IJC and its boards and committees for consultation, advice and information. In conclusion, the GAO reported that the IJC can not effectively carry out its water quality activities for the Great Lakes and other boundary waters without greater support and cooperation from the U.S. government.[102]

With half of the U.S. population depending on groundwater for drinking purposes, contamination and depletion of aquifers is a major threat to water resources in the lower 48 states. No less than six Federal laws deal more or less peripherally with the protection of groundwater. Some State laws also exist. The Safe Drinking Water Act (SDWA) of 1974 is most directly applicable to the contamination of groundwater, but its provisions have only been partially implemented by the EPA. Furthermore, the Reagan Administration has proposed amendments to the SDWA that would narrow the scope of the law and make it less useful as the statutory authority for developing a comprehensive groundwater conservation policy. Insofar as the SDWA does not expire until 1984 and over two-thirds of the States have reported serious groundwater contamination problems,[103] one should not expect enactment of the Administration's proposed amendments until 1983 or 1984, if ever.

### Hazardous Wastes and Toxic Substances

Two other Federal laws that could be used as statutory authority for developing a comprehensive program for the protection of groundwater are the Resource Conservation and Recovery Act (RCRA) and the Toxic Substances Control Act (TSCA). In fact, the continuing implementation of both the RCRA and the TSCA has been the impetus for most EPA regulatory activity during early 1982. Implementation of these two laws should continue into 1983. The scope of these two laws is much broader than simply the protection of groundwater. As noted above, the RCRA was enacted to allow the EPA and States to better regulate the disposal of municipal solid waste and to control the generation, transportation, treatment, storage, and disposal of hazardous waste, while TSCA allows the EPA to regulate toxic substances beginning with their production, throughout their use and finally upon their disposal.[104]

EPA actions on these two laws have brought controversy to EPA Administrator Gorsuch.[105] One should expect continued controversy from EPA actions on both laws through the balance of 1982 and into 1983. Concern in Congress about recent and anticipated EPA regulations on hazardous waste has prompted a negative reaction to the industry-supported amendments proposed by the Reagan Administration for extension of the RCRA. By mid-summer, Congress had authorized the expenditure of more money than the EPA had requested in the FY 1983 budget for carrying out the regulatory provisions of the RCRA. Furthermore, the House had passed a tougher bill than either the Administration or industry wanted. The bill includes several revisions of the hazardous waste provisions including a broadened definition of regulated waste generators. Finally, the Senate is not expected to soften the impact of the House bill, although industry could exert severe lobbying pressure.[106]

Congressional concern was increased dramatically in late February as a result of the EPA's 90-day suspension of a three-month-old ban on disposing of liquid hazardous wastes in landfills and its proposal to allow up to 25 percent of a landfill to include liquid hazardous waste. Because of complaints by members of Congress and environmental groups (some of which initiated lawsuits) that the EPA's action could result in more Love Canal incidents, the EPA reinstated the ban less than three weeks after the suspension thereof. In addition, the EPA approved an interim final regulation allowing only *de minimus* amounts of liquid hazardous waste to be dumped in landfills as a replacement for its three-week-old proposal to allow up to 25 percent of liquid hazardous waste in a landfill.

In defense of the EPA, it was also being sued by industry, that is, chemical producers and waste disposal firms, as to the statutory authority for the EPA's original ban on the disposal of liquid hazardous waste in landfills, especially in view of the slow pace at which the EPA has approved hazardous waste dump sites. As noted, the House bill extending the RCRA expands the EPA's authority to regulate hazardous waste so that industry suits would be moot. For example, the bill requires the EPA to promulgate regulations for minimizing the disposal of liquid hazardous waste in land-fills and to report to Congress annually on its progress in determining whether specific hazardous wastes should or should not be disposed of on land. The bill would also prohibit almost all underground injection of hazardous waste into or above aquifers. Oil and gas production wastes would be exempt from this prohibition.[107]

In mid-July, the EPA issued its long-deferred RCRA regulations for storage and disposal of hazardous wastes in Federally licensed sites. These regulations, which were developed over a six-year period, are generally acceptable to environmentalists and represent a sharp contrast to the deregulation philosophy of the Reagan Administration insofar as the EPA has estimated they will cost industry an additional billion dollars annually. Currently, only about 20 percent of the 150 million pounds of hazardous waste produced annually is disposed of in landfills or hazardous waste treatment facilities, while an estimated 80 percent of such wastes are retained at the same site where they are generated. In total, about three-fourths of hazardous wastes are now being disposed of in landfills. However, when these new hazardous waste regulations take effect at the beginning of 1983, many plant-site, commercial, and public landfill sites will not satisfy the tougher guidelines for licensing hazardous waste treatment, storage and disposal facilities.

The regulations require land disposal facilities, above-ground disposal sites, treatment facilities, and storage sites to obtain permits, that is, to be licensed. Permits are issued to each site or facility complying with specific requirements for monitoring its operation, controlling the flow of liquids on and off it, sealing the site or facility before closing it, obtaining liability insurance for it, and providing financially for over 25 years of monitoring after closing if hazardous wastes remain. All new land disposal facilities, that is, landfills and impoundments, must meet design and construction standards. Most importantly, new facilities must install an impermeable liner system that prevents contaminants from leaking to the soil or into the groundwater. However, old land disposal sites do not have to install an impermeable liner system if costs are deemed prohibitive. Needless to say, environmentalists are not pleased with the possibility that many old hazard-ous waste dumpsites will not have to install protective liner systems.

Notwithstanding industry's surprise at how tough the regulations are, environmentalists and many members of Congress are concerned about two aspects of the EPA's regulation of hazardous waste storage and disposal—

the monitoring of an estimated 10,000 licensed sites (of which 2,000 will be for permanent disposal) by the EPA to ensure that each is properly secured from contaminating either the soil or groundwater, and the exemption for generators of 1,000 kilograms (2,200 pounds) or less of hazardous wastes monthly.[108] As noted, Congress is authorizing the expenditure of more money than the EPA requested for hazardous waste regulation in the FY 1983 budget it submitted to Congress. Furthermore, the House bill extending the RCRA would prohibit the EPA from exempting generators of more than 100 kilograms (220 pounds) monthly from the hazardous waste disposal regulation.[109] In an effort to reassure Congress, environmentalists and local groups living near hazardous waste sites, the EPA has announced the following goals for its FY 1983 hazardous waste regulatory program: (1) 800 permits should be issued by EPA regional offices and States for hazardous waste tanks, containers and piles; and (2) inspections should be made by EPA regional offices and States of 2,126 groundwater monitoring facilities, 1963 waste treatment, storage and disposal facilities, and 2,734 waste generators and transporters.[110]

With respect to the continued implementation of TSCA, much of the EPA's activity has been acceptable to environmentalists. The addition of new substances to the TSCA chemicals-in-commerce inventory or the addition of chemicals from that inventory to the list of toxic substances for which exposure, use and production data must be reported to the EPA are generally noncontroversial regulatory actions, although ultimately some of the chemicals for which mandatory reporting is not required may be questioned when further research or experience demonstrates their health or environmental risk. Similarly, few questions about the EPA's research activity, that is, investigation of toxicity or carcinogenity of substances, have been raised. On the other hand, the EPA's informal agreements with many chemical producers to allow the latter a certain amount of discretion in testing their chemicals, that is, "voluntary" testing, may be challenged by environmentalists in litigation, insofar as TSCA provisions generally mandate government test procedures for any of the over 50,000 "old" chemicals within one year of their being recommended for testing by the Interagency Testing Committee.[111]

With regard to the testing of "new" chemicals, the EPA in late July issued controversial proposed regulations for exempting low-risk and low-volume chemicals from pre-manufacture testing by the Federal government. Although TSCA authorizes the exemption of certain chemical substances from pre-manufacture notification and testing, environmentalists and concerned members of Congress dispute the EPA's standards for determining low risk to public health or environment and its exemption of up to 10,000 kilograms (22,000 pounds) of production annually, that is, low-volume chemicals. Questions have also been raised about the reliability of tests conducted by chemical producers or "qualified" testing laboratories in determining that the new chemical satisfies the government's low-risk standards. Inasmuch as the new exemption regulations could apply to almost one-half of the 1,000 new chemicals developed each year, EPA Administrator Gorsuch argues that the Office of Pesticides and Toxic Substances will have more resources to concentrate on testing new chemicals that pose the greatest threat to public health and the environment, while reducing the regulatory burden on the chemical industry. In defense of the exemption regulations, they also provide for increased EPA monitoring and testing of certain chemicals subsequent to their having survived premanufacturing review and testing.[112]

Once a substance has been characterized as being hazardous to public health or the environment under TSCA, it most certainly is subject to hazardous waste regulation. In addition, other substances are considered hazardous wastes for purposes of the EPA regulation of their production, use, storage, and disposal under provisions of the RCRA, but what about all those hazardous substances produced and disposed of before TSCA and RCRA were enacted and implemented? Because most states did not regulate hazardous waste disposal before the RCRA was enacted to provide for Federal regulation of hazardous waste, there are a myriad of poorly documented, and often unsealed, disposal sites containing hazardous wastes. Their existence typically becomes publicly known when the hazardous waste disposed in one of these closed sites contaminates the soil or groundwater, for example, the Love Canal incident.

## The Superfund Program and Stripmining Regulations

As noted above, Congress enacted the 1980 "Superfund" law (CERCLA), which is another law applicable to groundwater protection, to finance part or all of the clean-up of closed and abandoned disposal sites that leak hazardous wastes. Ultimately, the Superfund program may be necessary to finance the clean-up of licensed dumpsites where their impermeable liner systems have failed to contain all the hazardous waste in the dumpsite. Because of their significantly lower cost, land disposal of hazardous wastes will continue to be preferred to costly alternatives such as chemical neutralization, deep-well injection, and resource recovery. Thus, there will be many more land disposal sites (landfills, impoundments and so forth) with the potential for contaminating groundwater and soil.

Reimbursement for the clean-up of most leaking dumpsites (old, current and future) should be forthcoming from the waste producers or waste dump operators. As provided for in CERCLA, clean-up responsibility for a leaking dumpsite vests in both the waste producer or waste dump operator, but interim financing will often be provided by the EPA's Superfund program to expedite clean-up. The possibility of treble damages should encourage waste producers to clean up leaking dumpsites in which their wastes were disposed.

As of mid-1982, the Superfund balance was $300 million. As will be explained in the next chapter, the Superfund balance is financed from environmental excise tax revenue and Federal appropriations. Only 115 hazardous waste dumpsites are scheduled to be cleaned up in 1982 (70 by the EPA at a cost of $52 million and 45 by States and liable parties)[113] because of delays in finalizing the revised National Contingency Plan required by CERCLA for determining priorities and enforcement standards for cleaning up hazardous substance spills and leaking hazardous waste dumpsites. In fact, the OMB's intransigence on several iterations of EPA proposals for implementing the Superfund-mandated clean-up finally ended when a public interest environmental law firm (Environmental Defense Fund) convinced a Federal district court in mid-February to order the EPA to issue its revised National Contingency Plan as well as enforcement standards for Superfund clean-up.

Within a month after the court order, the EPA issued its proposed plan which after some minor modifications in response to public comment became final in late spring. The revised plan, which formerly dealt only with the clean-up of oil and hazardous substance spills in navigable waters as authorized by the Water Pollution Control Act, now includes guidelines for coordinating Federal and State government actions in cleaning up spills of hazardous substances and leaking hazardous waste dumpsites, methods the EPA uses to identify and investigate abandoned dumpsites, criteria to be used in determining the clean-up priority of dumpsites and spills, broad standards for establishing an acceptable level of clean-up for each site and spill, and guidelines for determining when the Federal government can use Superfund trust money for cleaning up a hazardous waste dumpsite or a hazardous substance spill. Under the plan, the EPA can react in one of two ways to hazardous substance spills and hazardous dumpsites—an immediate clean-up because of an emergency situation, such as, a spill or flooded dumpsite, or a planned, methodical clean-up in response to a chronic, slowly developing hazardous waste problem at a dumpsite.

Although environmentalists and concerned members of Congress have criticized the EPA's revised plan, the chemical industry found it acceptable. In defense of the EPA, its Administrator Anne Gorsuch has requested $230 million for the FY 1983 Superfund program, that is, an increase in funding of $40 million from the FY 1982 budget for the hazardous substance spill and dumpsite clean-up program.[114] In addition, chemical producers have agreed to pay a total of $80 million to voluntarily clean up abandoned hazardous waste dumpsites. Finally, the EPA has doubled the number of cases it referred to the Justice Department for prosecution during the first five months of 1982 as compared to the same period in 1981 when Gorsuch was reorganizing the agency.[115]

Although there may be considerable disagreement with the conclusion that Gorsuch's early 1982 activities at the EPA were noncontroversial, one would be hard-pressed to argue for the characterization of Interior Secretary Watt's activities during the first half of 1982 as being controversial when compared to his tumultuous first year. In fact, Watt was almost silent until July 1982 when he announced a controversial program for leasing virtually all of the Federal government's offshore property for private sector oil and gas exploration. For example, Watt made no public comment after he agreed to an out-of-court settlement of a lawsuit challenging his proposed deregulation of strip-mining controls (originally developed by the Carter Administration) implementing the 1977 Surface Mining Control and Reclamation Act (SMCRA).

Under the mid-April settlement, most of Watt's proposed changes in strip-mining controls were deferred until the Department of Interior (DOI) could prepare a comprehensive environmental impact statement on the relaxation of 33 strip-mining regulations after holding two public hearings and allowing a 60-day public comment period. However, the DOI was allowed to proceed with 23 of the 33 proposed changes in the regulations after conducting limited environmental impact statements for each. Although the DOI agreed not to relax the regulation that allows citizens to accompany State strip-mine regulators on mine inspections, it refused to compromise on the proposal to allow States more freedom in establishing their own strip-mining controls under the SMCRA.

Another environmental group has sued the DOI to reinstate mandatory Federal standards for States that have established or plan to establish their own strip-mining controls. Traditionally, mining regulators in many States have been unable or unwilling to satisfactorily control the hazardous waste from and reclamation of surface mines; and therefore, Congress enacted SMCRA.[116] Thus, the mining regulations of several States may not survive the scrutiny of the Federal courts, even if Watt's proposed rule is sustained in court.

Watt did confront Congress on a procedural matter peripherally related to environmental regulation in late February. Specifically, Watt refused to comply completely with a House subcommittee's subpoena for documents related to Canadian firms' investments in U.S. oil and gas leases under instructions from the White House, which claimed executive privilege. Ultimately, the White House saved Watt from being held in contempt of Congress by working out a compromise on the review of the documents which the House Energy and Commerce Committee subpoenaed as part of its investigation of Canada's discriminatory energy policies.[117] Watt's problem with Congress was somewhat overshadowed by his coincident announcement to impose a moratorium on opening wilderness areas to natural resource development, thereby disarming environmentalist, conservationist, and Congressional critics of prior policy to allow drilling and mining on such public lands, but development continues on other Federal lands.[118]

As noted, the Reagan Administration's environmental record is going to be an issue in the fall 1982 election campaign. As a consequence, the Administration's chief environmental officers—EPA Administrator Gorsuch and Interior Secretary Watt—will be controversial because of their past efforts in carrying out Reagan's policies of deregulation and New Federalism. Moreover, if Gorsuch and Watt take further actions to deregulate or to transfer environmental regulatory responsibilities to the States, their criticism by Congressional and State candidates will be focused on their most current decisions. For example, EPA Administrator Gorsuch will probably be criticized for those hazardous waste dumpsites she does not order cleaned up immediately. Similarly, Secretary Watt will receive criticism in those coastal States that fear environmental damage from offshore oil and gas development in Federal areas off their coasts, for example, off the California and Alaska coasts. Finally, if the Republicans lose control of the Senate and lose a large number of House seats, the Reagan Administration's deregulation and defederalization of environmental policy could be stalled indefinitely by a Democratically controlled Congress.

## Notes

1. U.S. Council on Environmental Quality (CEQ), *Environmental Quality: The Ninth Annual Report* (Washington, D.C.: U.S. Government Printing Office, 1974), pp. 1-33, 90-130, and 159-84.

2. U.S. Department of Commerce, *The Effects of Pollution Abatement on International Trade* (Washington, D.C.: U.S. Government Printing Office, 1978), p. 7.

3. Norman J. Landau and Paul D. Rheingold, *The Environmental Law Handbook*, (New York: Friends of the Earth/Ballantine Books, 1971), p. 21.

4. *International Environment Reporter*, p. 61:0101.

5. U.S. Department of Commerce, *Effects of Pollution Abatement on Trade*, p. 13.

6. *International Environment Reporter*, pp. 61:0101-0102.

7. U.S. Department of Commerce, *Effects of Pollution Abatement on Trade*, pp. 72 and 73.

8. *International Environment Reporter*, p. 61:0102.

9. U.S. Department of Commerce, *Effects of Pollution Abatement on Trade*, pp. 14-15; and *International Environment Reporter*, pp. 61:0102-0103.

10. *International Environment Reporter—Current Report*, January 10, 1978, p. 13.

11. *International Environment Reporter*, pp. 61:0102-0104.

12. CEQ, *Environmental Quality: Ninth Annual Report*, pp. 105-8.

13. *International Environment Reporter*, p. 61:0101.

14. U.S. Department of Commerce, *Effects of Pollution Abatement on Trade*, pp. 8-9.

15. *International Environment Reporter*, p. 61:0101.

16. Eugene P. Seskin, "Clean Air Amendments of 1977," *Resources*, January-March 1978, pp. 11-12.

17. CEQ, *Environmental Quality: Ninth Annual Report*, pp. 1-4 and 61-71.

18. *International Environment Reports*, pp. 61:0101-0102.

19. Seskin, "Clean Air Amendments of 1977," pp. 11-12.

20. CEQ, *Environmental Quality: Ninth Annual Report*, pp. 4, 86, and 87.

21. Ibid., pp. 5, 45-52, and 90-92.

22. *International Environment Reporter*, p. 61:0104.

23. CEQ, *Environmental Quality: Ninth Annual Report*, pp. 45-52 and 90-92.

24. Seskin, "Clean Air Amendments of 1977," pp. 11-12.

25. Internal Revenue Code Section 4064.

26. *International Environment Reporter*, p. 61:0102.

27. CEQ, *Environmental Quality: Eighth Annual Report*, p. 49.

28. CEQ, *Environmental Quality: Ninth Annual Report*, pp. 172-74.

29. *International Environment Reporter*, p. 61:0104.

30. Landau and Rheingold, Environmental Law, pp. 19-25 and 144-46.

31. Roger M. Williams, "Anticipating America," *Saturday Review*, November 11, 1978, pp. 33-37.

32. See Carol S. Weissert, *State Mandating of Local Expenditures*. Washington, D.C.: Advisory Commission on Intergovernmental Relations, 1978.

33. U.S. Department of Commerce, *Effects of Pollution Abatement on Trade*, p. 12.

34. CEQ, *Environment Quality: Eighth Annual Report*, pp. 27-28.

35. *International Environment Reporter*, p. 61:0102.

36. CEQ, *Environmental Quality: Eighth Annual Report*, p. 42.

37. Byron Klapper, "Low-Cost Funds for Small Firms to Fight Pollution Stated in Trial Backed SBA," *The Wall Street Journal*, March 21, 1977, p. 26.

38. U.S. Department of Commerce, *Effects of Pollution Abatement on Trade*, pp. 12-14.

39. CEQ, *Environmental Quality: Eighth Annual Report*, p. 27.

40. Seskin, "Clean Air Amendments of 1977," p. 12.

41. Barbara Ferkiss, "Spectrum," *Environment*, October 1981, pp. 21-22.

42. Robert Cahn, "EPA Under Reagan," *Audubon*, January 1982, pp. 14 and 16.

43. Jacqueline M. Warren and Ross Sandler, "EPA's Failure to Regulate Toxic Chemicals," *Environment*, December 1981, pp. 2-4.

44. Cahn, "EPA Under Reagan," p. 16.

45. *International Environment Reporter—Current Report*, September 9, 1981, p. 1020.

46. Caroline E. Mayer, "EPA Studies Relaxation of Rules," *Washington Post*, November 26, 1981.

47. Burt Schorr, Albert R. Karr, and Stan Crock, "Federal Agencies Ease, Lift Some Regulations That Burden Business," *The Wall Street Journal*, September 4, 1979, pp. 1 and 16.

48. Peter Behr, "Deregulating the Regulators," *Environment*, April 1981, pp. 4, 5, and 41.

49. Victor E. Millar and Michael E. Simon, "The Measure of Regulation," *Arthur Andersen Chronicle*, Vol. 39, No. 1 (1979), pp. 13-16.

50. Associated Press, Washington, D.C., August 4, 1981.

51. "Congress vs. the Regulators," *Business Week*, January 18, 1982, pp. 115-16.

52. *International Environment Reporter—Current Report*, June 10, 1981, pp. 889-90.

53. "Regulation and the 1982 Budget," *Regulation*, July-August 1981, pp. 9-10.

54. *International Environment Reporter—Current Report*, January 13, 1982, pp. 10-11.

55. "Regulation and the 1982 Budget," *Regulation*, p. 10.

56. Caroline E. Mayer, "Reagan Administration Cutting Regulatory Reins," *Washington Post*, November 22, 1981.

57. Cahn, "EPA Under Reagan," p. 16.

58. *International Environment Reporter—Current Report*, January 13, 1982, pp. 10-11.

59. "Regulation and the 1982 Budget," *Regulation*, p. 10.

60. *International Environment Reporter—Current Report*, August 12, 1981, p. 975.

61. Peter Behr, "An Interview with A. Alan Hill," *Environment*, July/August 1981, pp. 38-39.

62. Robert H. Nelson, "Rethinking the Role of EISs," *Technology Review*, January 1982, pp. 8-9 and 86.

63. Behr, "Interview with Hill," p. 38. Late in 1982, the 1981 report was published.

64. Rich Jaroslovsky and Timothy D. Schellhardt, "Reagan Rejects Tax Rises to Curb Deficits, Offers Welfare Program Swap With States," *The Wall Street Journal*, January 27, 1982, pp. 3, 16, and 22.

65. Steven J. Marcus, "A Wall Street for Pollution," *Technology Review*, November-December 1981, pp. 11 and 87.

66. Lois Ember, "The Clean Air Act," *Environment*, April 1981, pp. 2-4.

67. Brian Haggerty, "Administration Philosophizes on Clean Air Act Revisions," *Engineering Times*, September 1981, pp. 1-2.

68. "A Clean Air Bill That Industry Will Buy," *Business Week*, January 18, 1982, p. 34.

69. Joanne Ornang, "Rewriting in the Sky," *Washington Post*, January 21, 1982.

70. Jim Ford, "Congress and the Environment," *Environment*, January-February 1981, pp. 33-37. Congress did extend the Endangered Species Act in late 1982.

71. Cahn, "EPA Under Reagan," pp. 14 and 16.

72. *International Environment Reporter—Current Report*, January 13, 1982, pp. 10-11.

73. Cahn, "EPA Under Reagan," p. 14.

74. Steven J. Marcus, "The New EPA: Who's Watching the Environment," *Technology Review*, January 1982, pp. 8-9 and 86.

75. Cahn, "EPA Under Reagan," p. 14.

76. Richard Anthony, "Trends in Public Opinion on the Environment," *Environment*, Vol. 24, No. 4 (May 1982), pp. 14-20, 33, 34.

77. Herbert W. Cheshire, "Washington Outlook: The White House Cleans Up Its Environment Act," *Business Week*, June 7, 1982, p. 131.

78. "Air Pollution Lower, Federal Report Shows," Associated Press, Washington, D.C., July 21, 1982.

79. *International Environment Reporter—Current Report*, July 14, 1982, pp. 301-2.

80. Katherine Barton, "Spectrum: Environmentalists Indict Reagan," *Environment*, Vol. 24, No. 4 (May 1982), p. 21.

81. Behr, "Interview with Hill," pp. 38-39.

82. *International Environment Reporter—Current Report*, June 9, 1982, p. 230.

83. Cheshire, "Washington Outlook," *Business Week*.

84. *International Environment Reporter—Current Report*, July 14, 1982, pp. 296-97.

85. Ibid., March 10, 1982, pp. 101-3.

86. Ibid., July 14, 1982, p. 297.

87. Lee Walczak, "Washington Outlook: A Deficit Time Bomb is Ticking," *Business Week*, August 1, 1982, p. 81.

88. *International Environment Reporter—Current Report*, March 10, 1982, pp. 101-2.

89. Andy Pasztor, "House Democrats Attack EPA's Record, Will Try to Make It Campaign Issue in Fall," *Wall Street Journal*, July 23, 1982.

90. Ibid.; and *International Environment Reporter—Current Report*, July 14, 1982, p. 297.

91. "The EPA's Hazy Stand on Clean-air Sanctions," *Business Week*, July 19, 1982, pp. 60, 62.

92. *International Environment Reporter—Current Report*, March 10, 1982, pp. 110-11.

93. "Senate Panel Clears Acid-Rain Curbs, Defies White House," *Wall Street Journal*, July 23, 1982; and *International Environment Reporter—Current Report*, July 14, 1982, pp. 287-88.

94. *International Environment Reporter—Current Report*, April 14, 1982, pp. 146-47, and July 14, 1982, pp. 248-49.

95. "Senate Panel Clears Acid-Rain Curbs, Defies White House," *Wall Street Journal*, July 23, 1982.

96. *International Environment Reporter—Current Report*, March 10, 1982, pp. 115-16.

97. Sandra Sugawara, "Lead Tests Too Lax, EPA Memo Says," *Washington Post*, July 11, 1982.

98. *International Environment Reporter—Current Report*, April 14, 1982, p. 162.

99. Sugawara, "Lead Tests Too Lax," *Washington Post*. In a dramatic reversal of policy, the EPA announced in late August 1982 that stricter regulation of the lead content in gasoline would take effect on November 1, 1982.

100. Katherine Barton, "Spectrum: EPA Expands Bubble Concept," *Environment*, Vol. 24, No. 4 (May 1982), p. 24. *See also* 47 *Fed. Reg.* 15076-86, April 7, 1982.

101. "Moves That May Dilute the Clean Water Act," *Business Week*, June 21, 1982, p. 48.

102. *International Environment Reporter—Current Report*, June 9, 1982, p. 231, and July 14, 1982, p. 286.

103. Jacqueline M. Warren, "Law: Environmental Statutes Under Attack," and Amy Horne, "Groundwater Policy: A Patchwork of Protection," *Environment*, Vol. 24, No. 3 (April 1982), pp. 2-4, and 6-11.

104. Horne, "Groundwater Policy," *Environment*, p. 9.

105. Pasztor, "House Democrats Attack EPA's Record," *Wall Street Journal*.

106. *International Environment Reporter—Current Report*, June 9, 1982, p. 230, and April 14, 1982, pp. 156-57; and Lee Walczak, "Washington Outlook: Capital Wrapup—Environment," *Business Week*, August 2, 1982, p. 81.

107. *International Environment Reporter—Current Report*, June 9, 1982, p. 230; April 14, 1982, p. 162; and March 10, 1982, pp. 118-20.

108. Philip Shabecoff, "EPA Releases Rules on Hazardous Waste," *New York Times*, July 14, 1982; and Carol E. Curtis and Alyssa A. Lappen, "Waste Makes Haste," *Forbes*, August 2, 1982, pp. 53 and 56.

109. *International Environment Reporter—Current Report*, June 9, 1982, p. 230.

110. Ibid., July 14, 1982, p. 304.

111. Ibid., pp. 288-290.

112. Ibid., May 12, 1982, p. 172, and March 10, 1982, p. 116; and "EPA Proposes Rules Easing Requirements for Federal Analysis of New Chemicals," *Wall Street Journal*, July 29, 1982.

113. Curtis and Lappen, "Waste Makes Haste," *Forbes*.

114. *International Environment Reporter—Current Report*, April 14, 1982, p. 148; and March 10, 1982, pp. 103, 120, 124 and 124.

115. Pasztor, "House Democrats Attack EPA's Record," *Wall Street Journal*.

116. Katherine Barton, "Spectrum: Strip-Mining Deregulation," *Environment*, Vol. 24, No. 4 (May 1982), p. 24.

117. "Watt Secures Approval to Moratorium Proposal," United Press International, Washington, D.C., February 23, 1982; "House Panel Votes to Cite Watt for Contempt," United Press International, Washington, D.C., February 26, 1982; and "Watt to Make Documents Available to House Panel," *Wall Street Journal*, March 17, 1982. In mid-December 1982, the EPA Administrator was held in contempt of Congress for refusing (on presidential orders) to turn over documents on the EPA's handling of the hazardous waste clean-up program.

118. Andy Pasztor, "Watt Drops Plan to Allow Firms in Wilderness," *Wall Street Journal*, February 22, 1982; and Andy Pasztor, "Watt Softens His Line, But Image as Extremist Cuts His Effectiveness," *Wall Street Journal*, June 2, 1982. In August 1982, both the House and the Senate voted overwhelmingly to ban oil and gas drilling in wilderness areas until the end of 1982. Later in the "lame duck" session, the House voted to extend that ban.

# 9

# U.S. Taxation of Pollution Control Investments

Almost any business involved in manufacturing, production, extraction, or furnishing of transportation, communication, or electrical energy services requires at least some investment in pollution control technology. Accordingly, for many businesses, the tax implications of owning or leasing pollution control facilities are significant. The tax problems implicit in owning or leasing pollution control facilities are similar to the problems associated with interests in most other fixed assets. Typical problems include financing, installation, maintenance, and replacement. However, the owner, lessor, or lessee of pollution control facilities should be aware that unique, and sometimes complex, issues of taxation are related to their use. Consequently, the analysis below includes the definitional, compliance, and planning aspects of pollution control technology under the Federal tax law. To make the analysis more comprehensible, the pertinent provisions of the Economic Recovery Tax Act of 1981 (but not the 1982 tax law)[1] are covered in somewhat more detail than might be necessary if the law were not so recent.

A taxpayer can either purchase or lease pollution control facilities for use in a trade or business. The tax treatment of buying or leasing pollution control equipment does not necessarily differ from that associated with the acquisition or rental of other tangible personal property. Nevertheless, a review of the general tax rules as applied to such equipment and its special tax problems is warranted. The following text therefore focuses on the pertinent capital cost recovery provisions, namely, depreciation or rapid amortization and the investment credit, for owners of pollution control facilities regardless of whether they are users or lessors. The tax treatment of the profits of a pollution control facility is also explained. In addition, the tax implications of leasing pollution control facilities are explained from both the lessor's and the lessee's point of view. Finally, there is an analysis of the tax treatment of Industrial Development Bonds, the proceeds of which are used to finance pollution control investment, and a comprehensive example is provided.

## Depreciation

The Economic Recovery Tax Act of 1981 (ERTA) made significant changes in the depreciation provisions of the Internal Revenue Code with its creation of the Accelerated Cost Recovery System (ACRS). New Section 168 typically allows greater deductions for most assets than were formerly available under either CLADR (Class Life-Asset Depreciation Range) or "facts and circumstances" Section 167 depreciation. Section 168 is mandatory,[2] while Section 167 allows a taxpayer to elect between CLADR or "facts and circumstances" depreciation. Although Section 168 ACRS deductions essentially replace Section 167 depreciation deductions, Section 169 rapid amortization is still theoretically available for pollution control facilities that are added to a plant in existence before 1976.

## Accelerated Cost Recovery System

Where pollution control facilities are an integral part of other capital equipment (or activity) that is included in CLADR classes of economic activity other than Land Improvements or Industrial Steam and Electric Generation and/or Distribution Systems, the property is treated as "three-year," "five-year," "ten-year," "fifteen-year real," or "fifteen-year public utility" property, depending upon the "present class life" (CLADR class) of the capital equipment. Section 168(b) specifies annual recovery percentages for each of the categories of recovery property. The phrase "integral part of other capital equipment" reflects the requirement under the CLADR asset guideline class system that property must be included in the class in which the property is primarily used.[3]

If pollution control facilities do not fall into the Land Improvements or Industrial Steam and Electric Generation and/or Distribution Systems—two of the general business assets classes 00.11 to 00.4—then they must fall into the applicable industrial category in classes 01.1 to 80.0.

### Table 9-1
### Three-Year Cost Recovery Percentages

| Recovery Year | Year Property Placed in Service | | |
|---|---|---|---|
| | *1981–1984* | *1985* | *After 1985* |
| 1 | 25 | 29 | 33 |
| 2 | 38 | 47 | 45 |
| 3 | 37 | 24 | 22 |

Thus, Table 9-1 would be used to determine the ACRS deduction only if the pollution control facility were part of "three-year property." Alternatively, Table 9-2 would be used to determine the ACRS deduction if a pollution control facility were part of "five-year property."

### Table 9-2
### Five-Year Cost Recovery Percentages

| Recovery Year | Year Property Placed in Service | | |
|---|---|---|---|
| | *1981–1984* | *1985* | *After 1985* |
| 1 | 15 | 18 | 20 |
| 2 | 22 | 33 | 32 |
| 3 | 21 | 25 | 24 |
| 4 | 21 | 16 | 16 |
| 5 | 21 | 8 | 8 |

Similarly, Table 9-3 would be applicable to pollution control facilities that were part of "ten-year property."

### Table 9-3
### Ten-Year Cost Recovery Percentages

| Recovery Year | Year Property Placed in Service | | |
|---|---|---|---|
| | *1981–1984* | *1985* | *After 1985* |
| 1 | 8 | 9 | 10 |
| 2 | 14 | 19 | 18 |
| 3 | 12 | 16 | 16 |
| 4 | 10 | 14 | 14 |
| 5 | 10 | 12 | 12 |
| 6 | 10 | 10 | 10 |
| 7 | 9 | 8 | 8 |
| 8 | 9 | 6 | 6 |
| 9 | 9 | 4 | 4 |
| 10 | 9 | 2 | 2 |

If the pollution control facility were part of "fifteen-year public utility property," Table 9-4 would be used to determine the applicable ACRS deduction.

Where the pollution control facilities consist of both tangible personal property (equipment) and depreciable real property (a building), the capital invested in the real property portion would not be recovered through use of one of the three above tables (or Table 9-4) because those tables are applicable primarily to tangible personal property. Rather,

**Table 9-4**
**Fifteen-Year Public Utility**
**Cost Recovery Percentages**

| Recovery Year | Year Property Placed in Service | | |
|---|---|---|---|
| | *1981–1984* | *1985* | *After 1985* |
| 1 | 5 | 6 | 7 |
| 2 | 10 | 12 | 12 |
| 3 | 9 | 12 | 12 |
| 4 | 8 | 11 | 11 |
| 5 | 7 | 10 | 10 |
| 6 | 7 | 9 | 9 |
| 7 | 6 | 8 | 8 |
| 8 | 6 | 7 | 7 |
| 9 | 6 | 6 | 6 |
| 10 | 6 | 5 | 5 |
| 11 | 6 | 4 | 4 |
| 12 | 6 | 4 | 3 |
| 13 | 6 | 3 | 3 |
| 14 | 6 | 2 | 2 |
| 15 | 6 | 1 | 1 |

Table 9-5 would be applicable to the depreciable real property portion of a pollution control facility, notwithstanding a three-year, five-year, ten-year, or fifteen-year public utility property classification for the tangible personal property portion of the facility that is an integral part of other capital equipment.

With respect to all classes of ACRS property (except fifteen-year real property), the half-year convention is applicable in determining the amount of the cost recovery deduction in the year the equipment is placed in service. Accordingly, the Section 168(b) schedules reflect a half year of cost recovery for the first year an asset is placed in service.[4] For example, the 15 percent 1981–84 rate for five-year property is one-half of the effective 150 percent declining-balance rate, that is, $150\% \times \frac{1}{5} \times \frac{1}{2}$. Furthermore, the definition of unadjusted basis in Section 168(d)(1) does not require an adjustment of capital cost to reflect the salvage value of the equipment.

ACRS permits a delay, at the taxpayer's option, in recognizing depreciation charges. Section 168(b)(3) provides for the use of straight-line depreciation for one or more classes of recovery property. This election to use straight-line also includes a corresponding election for the choice of recovery periods. Table 9-6 is a summary of the recovery period options available for each class of recovery property.[5]

Several special operating rules are applicable to the straight-line election

**Table 9-5**
**Fifteen-Year Real Property**
**Cost Recovery Percentages[a]**

| If the Recovery Year Is: | The Applicable Percentage is: (Use the column for the month in the first year the property is placed in service) | | | | | | | | | | | |
|---|---|---|---|---|---|---|---|---|---|---|---|---|
| | 1 | 2 | 3 | 4 | 5 | 6 | 7 | 8 | 9 | 10 | 11 | 12 |
| 1 | 12 | 11 | 10 | 9 | 8 | 7 | 6 | 5 | 4 | 3 | 2 | 1 |
| 2 | 10 | 10 | 11 | 11 | 11 | 11 | 11 | 11 | 11 | 11 | 11 | 12 |
| 3 | 9 | 9 | 9 | 9 | 10 | 10 | 10 | 10 | 10 | 10 | 10 | 10 |
| 4 | 8 | 8 | 8 | 8 | 8 | 8 | 9 | 9 | 9 | 9 | 9 | 9 |
| 5 | 7 | 7 | 7 | 7 | 7 | 7 | 8 | 8 | 8 | 8 | 8 | 8 |
| 6 | 6 | 6 | 6 | 6 | 7 | 7 | 7 | 7 | 7 | 7 | 7 | 7 |
| 7 | 6 | 6 | 6 | 6 | 6 | 6 | 6 | 6 | 6 | 6 | 6 | 6 |
| 8 | 6 | 6 | 6 | 6 | 6 | 6 | 5 | 6 | 6 | 6 | 6 | 6 |
| 9 | 6 | 6 | 6 | 6 | 5 | 6 | 5 | 5 | 5 | 6 | 6 | 6 |
| 10 | 5 | 6 | 5 | 6 | 5 | 5 | 5 | 5 | 5 | 5 | 6 | 5 |
| 11 | 5 | 5 | 5 | 5 | 5 | 5 | 5 | 5 | 5 | 5 | 5 | 5 |
| 12 | 5 | 5 | 5 | 5 | 5 | 5 | 5 | 5 | 5 | 5 | 5 | 5 |
| 13 | 5 | 5 | 5 | 5 | 5 | 5 | 5 | 5 | 5 | 5 | 5 | 5 |
| 14 | 5 | 5 | 5 | 5 | 5 | 5 | 5 | 5 | 5 | 5 | 5 | 5 |
| 15 | 5 | 5 | 5 | 5 | 5 | 5 | 5 | 5 | 5 | 5 | 5 | 5 |
| 16 | — | — | 1 | 1 | 2 | 2 | 3 | 3 | 4 | 4 | 4 | 5 |

[a]This table does not apply for short taxable years of less than 12 months.

*SOURCE:* Information Release, U.S. Department of Treasury, September 10, 1981.

**Table 9-6**
**Optional Straight-Line Recovery Periods**

| Recovery Period for ACRS | Elections for Straight-Line Recovery Period |
|---|---|
| 3-year property | 3, 5, or 12 years |
| 5-year property | 5, 12, or 25 years |
| 10-year property | 10, 25, or 35 years |
| 15-year public utility property | 15, 35, or 45 years |
| 15-year real property | 15, 35, or 45 years |

in Section 168(b)(3)(B), but most of them are beyond the scope of this analysis. One rule that should be noted is that the straight-line election applies to *all* the property placed in service in the same year in a recovery property class.

ACRS is equally applicable to both new and used pollution control facilities, but property placed in service by the taxpayer before January 1, 1981, may not be recovery property.[6] In addition, there are "anti-churning" rules to discourage trafficking in used property among related persons and to deny ACRS treatment to used property acquired in many nonrecognition transactions.

## Expensing and Additional First-Year Depreciation

The Section 179 additional first-year depreciation has been repealed for 1981 and replaced with the new Section 179 election to expense specified depreciable trade or business assets. Table 9-7 summarizes the amount of Section 179 property that can be expensed beginning in 1982.[7]

**Table 9-7**
**Section 179 Current Expense Election**

| Taxable Year Beginning | Amount Deductible |
| --- | --- |
| In 1982 or 1983 | $ 5,000 |
| In 1984 or 1985 | $ 7,500 |
| After 1985 | $10,000 |

As was true with additional first-year depreciation, the basis of the recovery property upon which a Section 179 deduction has been taken must be reduced by the amount ($5,000, $7,500, or $10,000) of the capital expenditure currently expensed. This adjustment is now required for both ACRS and investment tax credit (ITC) purposes.[8] The election is available only for pollution control facilities that are both Section 38 ITC property and acquired by purchase for use in a trade or business.[9]

Married taxpayers filing separately must share the annual amount of Section 179 deduction. Similarly, partnerships and controlled corporate groups are limited to one Section 179 deduction each year. Certain noncorporate lessors are also precluded from taking a Section 179 deduction on leased property that otherwise qualifies.[10]

It is noteworthy that fiscal year 1980 taxpayers, especially corporations, may find both Section 167 and the new Section 168 applicable to investments in fixed assets made during a taxable year that straddles calendar years 1980 and 1981. Similarly, old Section 179 additional first-year depreciation is applicable to acquisitions before 1981. However, acquisitions in 1981 are ineligible for both the old and new Section 179 provisions.

## Investment Tax Credit

The ITC is another tax benefit associated with the acquisition or use of most pollution control facilities. Accordingly, the acquisition of pollution control technology for use either in a trade or business can be expected to generate an investment tax credit.

Most pollution control equipment comes under the CLADR designation of assets that places the property in a recovery period of five years or more. Hence, there should be a Section 38 ITC equal to 10 percent of the pollution control equipment cost. Only in limited circumstances would the ITC for pollution control facilities be 6 percent of cost, for example, where the equipment is an integral part of an asset that qualifies for the three-year ACRS recovery period.[11] Examples would include pollution control equipment used with light transportation equipment, for research and experimentation, and as part of some industrial equipment.

The amount of ITC may be increased through the end of 1985 where the pollution control equipment is an integral part of property that qualifies for the additional energy credit.[12] Moreover, if the taxpayer is a corporation with an employee stock option plan that qualifies under Section 409A, the ITC percentage may be increased by as much as $1\frac{1}{2}$ percent through the end of 1982.[13]

A critical issue in the computation of the ITC is the determination of what constitutes the investment qualifying as Section 38 property. Generally, Section 38 property is tangible personal property. Qualifying tangible personal property must also be either Section 168 recovery property or any other property that is depreciable (or amortizable) and has a useful life of three years or more. Thus, the use of pollution control equipment as part of a pollution control facility qualifies the equipment for the investment tax credit.

As long as the pollution control facility does not include tangible real property or intangible personal property, the entire amount of the investment therein, that is, the pollution control equipment including peripheral equipment related thereto, would normally qualify for the investment credit.[14] Certain improvements in tangible real property that are part of the installation and operation of pollution control equipment may also qualify for the investment credit if the improvements are essentially an accessory of such equipment and not a structural component of a building. Moreover, structural components may also qualify for the investment credit if they and the pollution control equipment are a unified system.

On the other hand, the equipment may be an integral part of tangible real property, such as a structural component of a building. Thus, it is logical to deny the Section 38 property designation to otherwise tangible personal property that is an integral part of a system that is tangible real property. For example, an environmental control and alarm system that uses electronic sensors to relay data to a computer, which then initiates safety measures through other electronic devices, may be denied Section 38 property status by the IRS because the environmental control and alarm system is an integral part of the building's operations. The environmental control and alarm system would be considered a structural component of tangible real property.

After determining all the assets that qualify for Section 38 status, the new assets must be distinguished from those that are used, although "new" in use. The investment credit for used property placed in service during 1981-84 is presently limited to a maximum of $125,000. Beginning in 1985, the limit will be increased to $150,000.[15]

The new versus used distinction could be problematic for pollution control equipment because old equipment may be cannibalized for its salvageable parts usable in newer models. Nevertheless, it is probably unnecessary for all materials used in the construction of a new asset to be "new" in use; property may not be treated as reconditioned or rebuilt, even though it contains used parts. For example, when previously used circuit boards are reworked to perform a specific function and reassembled in a new computer, the full cost of that new computer will qualify as new property.[16]

If there are no Section 46(a)(3) carryover problems, the investment tax credit is generally taken in the year that the Section 38 property is "placed in service."[17] In the case of qualified property that is being constructed by or for a taxpayer and that has a normal construction period of two years or more, the taxpayer can elect to treat progress expenditures on such construction as qualified investment.[18] It should be noted, too, that the time a pollution control facility is placed in service also controls the year depreciation begins under ACRS.

With regard to pollution control equipment that requires testing before it can be used for business purposes, there is a question of when the asset is "placed in service." The tax court ruled in *LTV Corporation*[19] that a computer was placed in service only after there was mutual agreement between the customer and supplier that the equipment was operational and ready to perform the functions for which it was intended. Moreover, the IRS has indicated that it will not consider this critical standard specified in *LTV Corporation* to have been met if significant tests with appropriate materials are yet to be performed. The fact that some subsequent tests may remain to be performed to establish optimal pollution control capacity, however, is not sufficient to delay placed-in-service status if the other conditions are met and the asset is otherwise operational.[20]

### Other Expenses

The other expenses incurred in acquiring and using pollution control equipment are generally treated like expenses attributable to the acquisition and use of any other tangible personal property. For example, interest expenses incurred in financing the purchase of pollution control facilities would be deductible as Section 162 trade or business expenses, or Section 212 production of income expenses. Furthermore, Section 163 explicitly authorizes deduction of interest. Similarly, repairs and maintenance expenses are currently deductible. However, where such expenses prolong the life of the asset, increase its value, or change its use, the cost must be capitalized and amortized over the life of the asset.[21] There are no specified repair allowance percentages under the Section 168 capital cost recovery system, although annual asset guideline repair allowance percentages are still applicable to assets acquired before 1981 and depreciated under the CLADR system.[22]

A purchaser of pollution control equipment normally incurs a sales tax upon acquisition of the asset.[23] An annual property tax as the owner of tangible personal property is payable to some States and most local governments. Real property taxes to the extent a pollution control system includes or is included in building improvements may also be payable. In addition, the building materials would be subject to sales taxation. Both State and local taxes are deductible as trade or business expenses.

Furthermore, Section 164(a) specifically authorizes the deduction of State and local taxes on pollution control equipment sales and pollution control property. Finally, Section 266 allows the taxpayer to elect to capitalize *or* deduct taxes and other carrying charges (interest, and so on, during construction) incurred in acquiring pollution control systems, even though for accounting purposes such expenses *must* be capitalized.

## Leasing

Although much of the above analysis is applicable to the lessor of pollution control equipment, there are several tax provisions that have been designed to encourage leasing through a sharing of economic benefits among the lessor, lessee, and capital supplier. While the direct ownership of pollution control equipment guarantees all the tax advantages explained above, the end-user of a pollution control system may not be the owner. Consequently, the incidence of the tax benefits attributable to acquisition of pollution control equipment may vary somewhat depending on the leasing arrangement with the capital supplier, which may be a third party or the lessor.

An economic analysis of leasing arrangements suggests several explanations for this popularity. First, some users do not have sufficient credit to acquire pollution control equipment. Users may also be concerned about technological obsolescence or imposition of higher standards. They would therefore rather lease a pollution control system in an arrangement that allows them to trade in their equipment for a new model when it is available. Finally, there are many end-users who cannot take advantage of all, or even most, of the tax advantages of equipment ownership, so they trade tax benefits for a lower cost of pollution control services. This last reason led to several provisions in ERTA that make the tax advantages of leveraged leasing more attractive than ever before. In summary, then, leasing is advantageous because pollution control equipment users can conserve capital, mitigate the risk of technological obsolescence, and trade on the tax benefits.

In most cases, the lessor would be entitled to ACRS deductions (or depreciation prior to 1981), interest deductions, and the investment credit. In some leasing arrangements, however, the ITC can be passed on to the lessee.[24] The lessee is also usually entitled to a current deduction of all lease payments.

For years, the IRS has had established guidelines[25] that must be met before it will issue an advance ruling on whether a transaction would be treated as a lease (rental) rather than a conditional sale or financing arrangement. If a transaction did not meet the "safe harbor" tests established by the IRS, then the lessor could lose the ITC and depreciation deductions, while the lessee would not be guaranteed those tax benefits even if they could be used.

For lease agreements consummated after August 12, 1981, by corporate lessors (in respect to property placed in service after 1980), the more liberal "safe harbor" rules of Section 168(f)(8) apply. The new rules entitle qualifying lessors and lessees to share the ACRS and ITC tax benefits between them. A lessee may acquire pollution control equipment, for instance, and then sell it to the lessor in what is essentially a leveraged financing arrangement providing a leaseback. By the structure of their agreement, the lessor and lessee can stipulate who is entitled to the cost recovery deductions and the ITC on the pollution control equipment. As a result, these rules permit lessors and lessees of pollution control equipment to trade or effectively "sell" the tax benefits.

The following summary of the new "safe harbor" rules integrates the requirements of the old rules established by the IRS by specifying the factors that *no longer need to be taken into account* in determining if a transaction is a lease.

(1) Whether the lessor or lessee must take the tax benefits into account in order to make a profit from the transaction.

(2) The fact that the lessee is the nominal owner of the property for State and local law purposes (for example, has title to the property) and retains the burdens, benefits, and incidents of ownership (such as payment of taxes and maintenance charges with respect to the property).

(3) Whether a person other than the lessee may be able to use the property after the lease term.

(4)  The fact that the property may (or must) be bought or sold at the end of the lease term at a fixed or determinable price that is more or less than its fair market value at that time.

(5)  The fact that the lessee or related party has provided financing or has guaranteed financing for the transaction (other than for the lessor's minimum 10-percent investment).

(6)  Whether the obligation of any person is subject to any contingency or offset agreement.[26]

To state the new safe harbor rules positively, a corporate lessor must have "at risk" only 10 percent of the pollution control equipment cost at all times during the lease period. Tax benefits may be included by the corporate lessor in evaluating the profitability of the equipment lease arrangement.

These new safe harbor rules guarantee that a pollution control equipment transaction will be treated as a lease for both Section 168 ACRS deductions and the Section 38 ITC. Both the lessor and lessee must elect to treat the lessor—which must be a corporation (other than a Subchapter S corporation or personal holding company), corporate partners, or a grantor trust of corporations—as the owner of the leased property. Moreover, the term of the lease, including any extensions, may not exceed the greater of either 90 percent of the useful life of the leased property for purposes of Section 167 or 150 percent of the present class life (not recovery period) of such pollution control equipment.[27]

For leased pollution control equipment to qualify for the Section 168(f)(8) election, it must be recovery property (as defined in Section 168(c)) that is new Section 38 property leased within three months after such property was placed in service. When the lessor is the first party to acquire the equipment, it must qualify as new Section 38 property as if it had been acquired by the lessee. For pollution control equipment first acquired by the lessee, such property will qualify for the Section 168(f)(8) election only if its adjusted basis in the hands of the lessor does not exceed its adjusted basis in the hands of the lessee at the time of the lease. For equipment acquired by the lessee after December 31, 1980, but before August 13, 1981, the three-month period begins with the date of enactment—August 13, 1981.[28] Thus, lessees who acquired equipment during this period had until November 13, 1981, to arrange a sale and leaseback transaction that qualified for the Section 168(f)(8) election.

Insofar as the Section 168(f)(8) election does not apply to noncorporate lessors (including Subchapter S corporations and personal holding companies), the "old" safe harbor rules are still effective for such tax entities. Under the "old" safe harbor rules, a transaction is treated as a lease if:

(1)  the lessor's minimum at-risk investment in the property throughout the lease term is 20 percent of cost;

(2)  the lessor has a positive cash flow and a profit from the lease independent of tax benefits;

(3)  the lessee does not have a right to purchase the property at less than fair market value;

(4)  the lessee does not have an investment in the lease and does not lend any of the purchase costs of the owner; and

(5)  use of the property at the end of the lease term by a person other than the lessee must be commercially feasible.[29]

Accordingly, the noncorporate lessor meeting these conditions would be entitled to the Section 168 capital cost recovery deductions for leased pollution control equipment, assuming such a taxpayer is also "at risk" for the full cost of the equipment.

To the extent noncorporate lessors are "at risk" on leased pollution control equipment that satisfies the "old" safe harbor rules, Section 46(e)(3) provides additional limitations. A noncorporate lessor may claim the investment credit only under either of the following situations:

(1)  Where the lessor manufactured the leased pollution control equipment in the ordinary course of business.

(2)  If the life of the lease covers less than one-half of the present class life (not recovery period) of the equipment, and the lessor's Section 162 business expenses assigned to the equipment exceed 15 percent of the rental income produced during the first twelve months of the lease.[30]

As indicated above, the lessor may elect to pass through the ITC to the lessee of pollution control equipment while retaining the ACRS deductions except for sale and leaseback.[31] This election is available only for new Section 38 property and for conventional (non-safe harbor) leases. The lessor's rental receipts should also increase to reflect the economic benefit which would accrue to the lessee by using the ITC.[32]

Perhaps the major drawback of ACRS for both corporate and noncorporate lessors is that Sections 56 and 57(a)(12) could potentially subject the recovery deductions to the 15 percent add-on minimum tax on tax preferences. For most leased pollution control equipment, the ACRS tax preference amount equals the difference between the ACRS deduction (based on its recovery property class) and straight-line amortization of the equipment (based on its CLADR useful life). The total for each year, when combined with other tax preference items, is reduced by the exemption of $10,000 plus a corporate lessor's income tax or one-half a noncorporate lessor's income tax, both of which are after credits. The balance is taxed at 15 percent on the lessor's return, and the resulting tax may not be offset by most credits.

This unfortunate consequence is an often overlooked disadvantage of ACRS for lessors. Of course, by electing longer lives and straight-line recovery deductions, taxpayers can mitigate the effect of the add-on minimum tax—but with a corresponding reduction in ACRS deductions.

## Dispositions

The tax consequences attributable to the disposition of pollution control equipment depend upon whether the equipment is sold or exchanged. If the equipment is sold, then any resulting gain would usually generate "ordinary income," although some capital gain (Section 1231 gain) may result. If the equipment is sold at a loss, Section 1231 provides for "ordinary loss" treatment. Furthermore, the sale of pollution control equipment will sometimes result in the recapture of some or all of the investment tax credit taken thereon. Finally, the exchange of "old" pollution control equipment for "new" equipment will normally defer recognition of any gain or loss realized thereon.

### Depreciation Recapture

Although ERTA changes somewhat the tax treatment of dispositions of tangible personal property, the recapture of ACRS deductions as ordinary income is essentially the same as the recapture of depreciation under the pre-1981 law.[33] Unlike the CLADR system, however, ACRS does not provide for deferral of recognition of gains and losses on the otherwise taxable disposition of assets.[34] The definition of Section 1245 property for assets acquired before 1981 remains virtually unchanged.[35]

Insofar as most pollution control equipment is either Section 1245 property or Section 1245 recovery property, the sale, taxable exchange, or taxable involuntary conversion of such property held for more than one year will result in "ordinary income" gain to the extent that the amount realized exceeds the adjusted basis, but not the recomputed basis, that is, basis for depreciation of ACRS.[36] Any portion of the amount realized that exceeds the recomputed basis is treated as Section 1231 gain. Conversely, where the amount realized is less than the adjusted basis upon the disposition of the equipment held for more than one year, the loss thereon is treated as a Section 1231 loss.

In determining the adjusted basis of recovery property upon its disposition, Section 168(d)(2)(B) specifies that no ACRS deduction is allowed in the year of an asset's disposition. Section 168(d)(2)(B) also applies to the retirement of an asset,[37] but it is not applicable to nontaxable dispositions.

ITC Recapture

If ACRS property ceases to be Section 38 property before it has been held for the required five- or three-year period, certain recapture percentages become applicable.[38] See Table 9-8.

## Table 9-8
## ITC Recapture Percentage

| Holding Period | Recapture Percentage 5-, 10-, or 15- Year Property | ITC Percentage Allowed | Recapture Percentage for 3-Year Property | ITC Percentage Allowed |
|---|---|---|---|---|
| One year or less | 100 | 0 | 100 | 0 |
| More than one year, but not more than two | 80 | 2 | 66 | 2 |
| More than two years, but not more than three | 60 | 4 | 33 | 4 |
| More than three years, but not more than four | 40 | 6 | 0 | 6 |
| More than four years, but not more than five | 20 | 8 | 0 | 6 |
| More than five years | 0 | 10 | 0 | 6 |

In the event a lessee acquires qualified lease property from a lessor who had been treated as the owner of such recovery property, special rules under Sections 1245(a)(6) and 47(a)(5) operate to place the lessee in the recapture position of the lessor. Consequently, a lessee who purchases leased pollution control equipment from the lessor must be aware, when subsequently disposing of it, that all potential ACRS and ITC recapture (to the extent not previously recaptured by the lessor) can be triggered.

Nontaxable Dispositions

Where a disposition is the result of a nontaxable like-kind exchange (Section 1031) or involuntary conversion (Section 1033), the adjusted basis of the exchanged or involuntarily converted property is substituted for the cost basis of the new property. Nontaxable pollution control equipment transactions include trade-ins of used equipment and forced dispositions resulting from a casualty or theft. Section 168(f)(7) requires the Secretary of the Treasury to promulgate necessary regulations for determining the

ACRS deduction for recovery property during the taxable year in which such property is acquired or disposed of in a nontaxable transaction.

## Amortization of Pollution Control Facilities

As an alternative to Section 168 ACRS deductions, Section 169, which was originally enacted as a temporary measure as part of the Tax Reform Act (TRA) of 1969, provides for amortization over a sixty-month period for certified pollution control facilities that are added to or used in connection with a plant in existence before 1969 or before 1976. Section 169(d) defines a certified pollution control facility as a new identifiable treatment facility which is used in connection with a plant in operation before January 1976 (January 1969 for pre-1969 plants). The election for rapid amortization is available only for that portion of the property's basis attributable to the first fifteen years of its useful life. For example, if a facility has a twenty-year useful life, 75 percent (15/20) of its cost would be amortizable over sixty months, and 25 percent used to be subject to a regular Section 167 facts-and-circumstances depreciation over a twenty-year useful life. However, Section 168 would now seem applicable to 25 percent.

The useful life of a capital expenditure determines both its Section 167 depreciation period and its useful life for ITC purposes. In general, the useful life for facts-and-circumstances depreciation is the period during which the asset may reasonably be expected to be useful in a taxpayer's trade or business. Regulation 1.167(a)-1(b) provides that useful life is not the expected or maximum physical life, but rather it depends on

(1) Wear, tear, decay, or decline from natural causes.

(2) Technological changes, inventions, and current developments in a taxpayer's business.

(3) Geographical location, climate, and so on.

(4) Repair, replacement, and retirement policies used.

The courts have delineated other factors that are relevant in determining useful lives. In summary, the taxpayer may use his business judgment in determining an asset's useful life, although the IRS prefers statistical studies, for example, the asset guideline lives delineated in the CLADR system. It should be noted that the CLADR system of depreciation is also available when Section 169 amortization is elected and the useful life of the pollution control facilities is more than fifteen years.

Originally, Section 169 required the waste treatment facility to have been constructed, reconstructed, erected, or first placed in service by the taxpayer after December 31, 1968, and before January 1, 1975, for plants in operation before January 1, 1969. Presently, for plants beginning operations before January 1, 1969 or 1976, Section 169(d)(4) provides for an indefinite period after December 31, 1968 or 1975, during which investments can be made in tangible personal (and certain real) property that is certified by the appropriate Federal (EPA) or State certifying authority as a pollution control facility. The facility must be used to abate or control water or atmospheric pollution or contamination by removing, altering, disposing of, storing, or preventing the creation or emission of pollutants, contaminants, wastes, or heat. If the facility includes a building, the building must be devoted *exclusively* to pollution control to qualify for rapid amortization rather than be subject to Section 168, that is, ACRS deductions.

In addition, the facility must not significantly increase the output or capacity, extend the useful life, or reduce the total operating costs of a plant or production unit therein. Nor may it alter the nature of the manufacturing or production process. However, Regulation 1.169-2(a)(3) provides that where a facility serves a function in addition to the abatement of pollution, the appropriate Federal certifying authority (EPA) determines what percentage of a given facility can be allocated to the pollution control function.

Section 169(e) denies favorable tax treatment to taxpayers if the cost of their facilities will be recovered from profits made through the recovery of waste. If an abatement facility recovers marketable wastes and the estimated profits from the wastes are not sufficient to recover the entire cost of the facility, the amortizable basis of the facility will be reduced in accordance with Regulation Section 1.169-2(c). Generally, the amortizable basis is adjusted for cost recovery, where the latter amounts to revenue less selling expenses. Estimated profits do not include any savings to the taxpayer by reason of his or her reuse or recycling of wastes or other items recovered in connection with the operations of the plant or other property served by the treatment facility.

## Certified Pollution Control Facilities

As defined in Section 169(d)(7), a "certified pollution control facility" is a new identifiable treatment facility used, in connection with a plant or other property in operation before January 1, 1969, or 1976, to abate or control water or atmospheric pollution or contamination by removing, altering, disposing, storing, or preventing the creation or emission of pollutants, contaminants, wastes, or heat. Moreover, the Federal and State pollution control authorities must certify that the facility is in conformity with applicable Federal and State regulations. Certification procedures are explained below.

Although the EPA determines whether a facility is a treatment facility, that is, one used to abate or control pollution or contamination, it relies on a State certifying authority to determine compliance with various State environmental standards. In practice, the EPA Regional Administrator should consider the following factors (where applicable) in its determination:

(1) Whether the applicant is in compliance with all the regulations of federal agencies applicable to the use of the facility, including conditions specified in any permit issued to the applicant by the Army Corps of Engineers under Section 13 of the Rivers and Harbors Act of 1899, as amended.

(2) All applicable water quality standards, including water quality criteria and plans of implementation and enforcement established pursuant to Section 10(c) of the Federal Water Pollution Control Act or State laws or regulations.

(3) Plans for the implementation, maintenance, and enforcement of ambient air quality standards adopted or promulgated pursuant to Section 110 of the Clean Air Act.

(4) Recommendations issued pursuant to Section 10(e) and (f) of the Federal Water Pollution Control Act or Sections 103(e) and 155 of the Clean Air Act.

(5) Water pollution control programs established pursuant to Sections 3 or 7 of the Federal Water Pollution Control Act.

(6) Local government requirements for control of air pollution, including emission standards.

(7) Standards promulgated by the Administrator of the Environmental Protection Agency pursuant to the Clean Air Act.[39]

In addition to the guidance implicit in these factors, the EPA has published guidelines for use by its regional offices and taxpayers in determining which facilities are qualified. Thus, the following devices are considered air pollution control facilities: inertial separators (cyclones, and so forth); wet collection devices (scrubbers); electrostatic precipitators; cloth filter collectors (baghouses); direct fired afterburners; catalytic afterburners; gas absorption equipment; gas adsorption equipment; vapor condensers; vapor recovery systems; floating roofs for storage tanks; afterburners, secondary combustion chambers, or particle collectors used in connection with incinerators; and a contact sulfuric acid plant in a flash copper smelting furnace.[40] Similarly, the EPA considers the following types of equipment to be water pollution control facilities:

(1) Pre-treatment or treatment facilities that neutralize or stabilize industrial or sanitary waste for disposal in a municipal waste treatment facility.

(2) Skimmers or similar devices for removal of greases, oils, and fat-like materials from an effluent stream.

(3) Facilities that concentrate and recover vaporous by-products from a process stream for reuse as raw feedstock.

(4) A facility to concentrate and/or recover tars or polymerized tar-like materials from the waste effluent previously discharged in the plant effluents.

(5) A device used to extract or remove a soluble constituent from a solid or liquid by use of a selective solvent.[41]

In recent years, control of toxic waste—liquid and solid—has become the preeminent environmental issue. Such wastes must be more or less permanently stored either on the site of their production (or use) or at a licensed disposal site. However, facilities installed in an effort to comply with the Toxic Substances Control Act and/or the Resource Conservation and Recovery Act should generally be certified by State environmental authorities and/or the EPA Regional Administrator as qualifying for Section 169 rapid amortization where the facility is designed to abate or control water or atmospheric pollution or contamination by removing, altering, disposing of, storing, or preventing the creation or emission of pollutants, contaminants, or wastes.

In addition, a preventive facility will qualify as a pollution control facility if it does not "significantly" increase the output or capacity, extend the useful life, or reduce the total operating costs of a pre-1969 or pre-1976 plant to which the facility is attached. Such preventive facilities will not qualify if they alter the nature of the manufacturing or production process, or pollution control facility.[42] "Significantly" means a change of more than 5 percent, where determination thereof as it affects output, capacity, costs, or useful life is based on the operating unit most directly associated with the pollution control facility. Examples of qualified preventive facilities include equipment at a plant site that prevents pollution by removing sulfur from fuel before it is burned at the plant, and a recovery boiler that removes pollutants from materials at some point in a production process that is otherwise unchanged. However, when electrolytic processing equipment is added to a plant that previously employed heat to process a material, the replacement of heat with electrolysis does not constitute a qualified pollution control facility, although the electrolysis equipment may prevent the creation or emission of pollutants.[43]

The EPA has also delineated several devices that do not qualify as treatment facilities.

(1) Modification of boilers to accommodate "cleaner" fuels, for example, the removal of stokers from a coal-fired boiler and the addition of gas or oil burners.

(2) Replacement of a heavily polluting iron cupola furnace with a minimally polluting electric induction furnace in a cast iron plant.

(3) Any device that is a part of a disposal system for subsurface injection of inadequately treated industrial or sanitary wastes or other contaminants.

(4) In-plant process changes that may prevent the production of pollutants but that in themselves do not remove or dispose of wastes.

(5) Devices that simply disperse the pollutants, for example, a high-stack chimney except for those devices (a cooling tower) that dissipate heat and prevent increase in the temperature of the receiving stream.[44]

Although the term "facility" is used in Code Section 169 and the Treasury regulations, it is not defined therein. Similarly, EPA regulations do not define "facility," but they do provide that facilities are usually systems consisting of several parts.[45] A treatment facility system would include all auxiliary equipment used to operate and control the system and any equipment used to handle, store, transport, or dispose of the collected pollutants along with the devices that actually collect or extract pollutants.[46]

In the case of water pollution control, a facility does not necessarily begin at the point where effluent leaves the last unit of the production process or end at the point of discharge, but most water pollution control facilities would probably exhibit such characteristics, that is, they are self-contained units.[47] Thus, such facilities would include lagoons, ponds, and similar storage facilities.[48] Similarly, an air pollution treatment facility would include fans, blowers, ductwork, valves, dampers, and electrical equipment, used to operate and control an air pollution control system.[49] Undoubtedly, the definition of a facility has been the source of many disputes between taxpayers and the Federal government—IRS and/or EPA. It would seem that the EPA should have resolved such disputes through the certification process, which is explained in detail below.

### New Identifiable Facility

A new identifiable facility is defined as a treatment facility (as explained above) that must be:

(1) A new facility, that is, a facility constructed or acquired after December 31, 1975 (or December 31, 1968, for pre-1969 plants).[50]

(2) Of the character subject to depreciation under Section 167.[51]

(3) Tangible personal property or tangible property other than a building or its structural components except where such a building or structural components are special purpose structures used exclusively for the treatment of pollutants.[52]

Only the first of these three requirements is explained below in detail. The other two are described briefly, for a detailed explanation is beyond the scope of this chapter.

As noted earlier, Congress enacted the Section 169 rapid amortization election in 1969 to apply only to facilities acquired or completed after December 31, 1968,[53] and used in connection with plants in operation before January 1, 1969.[54] In essence, Congress seems to be willing to encourage voluntary compliance with pollution control standards by subsidizing the cost of modifying an old plant, inasmuch as such costs generally exceed the cost of incorporating pollution control facilities in a new plant. However, Congress did not anticipate that rapid amortization would have any substantial effect on the number of pollution control facilities installed. Therefore, Section 169 has been justified as an equity provision necessary to share the cost of pollution control between industrial polluters and the general public, since old plants were designed in compliance with pollution control standards effective at the time of construction. It was hoped that the rapid amortization incentive would tip the scales in favor of installation of antipollution devices as opposed to deferral of compliance through challenges to Federal pollution control law and regulations.[55] The Tax Reform Act of 1976 amended Section 169 by making the rapid amortization election, which was due to expire on December 31, 1975, permanent for plants or other property in operation after 1968, but before January 1, 1976, as well as pre-1969 plants and property.[56]

In order for a plant or other property to qualify as being in operation before 1976 (or 1969), the plant or other property must actually have been performing the function for which it was constructed or acquired before January 1, 1976 (or 1969).[57] However, a property is not disqualified just because it was being used either at partial capacity or as a standby facility before January 1, 1976 (or 1969).[58]

A plant or other property is defined as any tangible property, whether such property is used in a trade or business or held for the production of income. Examples include a refinery, a motor vehicle, or a furnace in an apartment.[59]

In some situations, it may be necessary to distinguish a plant from the various properties of which it is comprised. If a refinery, for example, were partially destroyed by a tornado and then repaired, would a new refinery have been created? What if a small, pre-1976 plant were significantly enlarged after January 1, 1976, would the expanded plant be classified as a pre-1976 plant? The following rule provides the answer to these two questions:

> In addition, if the total replacements of equipment in any single taxable year beginning after December 31, 1975, represent the replacement of a substantial portion of a manufacturing plant which had been in operation before such date, such replacement shall be considered to result in a new plant which was not in operation before such date. Thus, if a substantial portion of a plant which was in existence before January 1, 1976, is subsequently destroyed by fire and such substantial portion is replaced in a taxable year beginning after that date, such replacement property shall not be considered to

have been in operation before January 1, 1976. The replacement of a substantial portion of a plant or other property shall be deemed to have occurred if, during a single taxable year, the taxpayer replaces manufacturing or production facilities or equipment, which comprises such plant or without regard to the adjustments provided in Section 1016(a)(2) or (3) in excess of 20 percent of the adjusted basis (so determined) of such plant or other property determined as of the first day of such taxable year.[60]

This rule can be adapted for pre-1969 plants by substituting 1968 and 1969 for 1975 and 1976, respectively.

In summary, this rule distinguishes between the replacement of "a substantial portion" of a plant and the replacement of less than a substantial portion where "less than a substantial portion" would amount to the replacement of 20 percent or less of the adjusted basis (book value for income tax purposes) of such plant or other property as of the beginning of the taxable year. Thus, the replacement of a substantial portion creates a "new plant." However, replacements over a number of years would not result in a new plant insofar as the cost of replacement in any one year does not exceed the 20 percent level.

What if the replacement of equipment in a pre-1976 plant results in an increase in the capacity of the old plant? The following rule applies to this question:

A piece of machinery which replaces one which was in operation prior to January 1, 1976, and which was a part of the manufacturing operation carried on by the plant but which does not substantially increase the capacity of the plant will be considered to be in operation prior to January 1, 1976. However, an additional machine that is added to a plant which was in operation before January 1, 1976, and which represents a substantial increase in the plant's capacity will not be considered to have been in operation before such date. There shall be deemed to be a substantial increase in the capacity of a plant or other property as of the time its capacity exceeds by more than 20 percent its capacity on December 31, 1975.[61]

The rule for pre-1969 plants is identical except that 1969 and 1968 should be substituted for 1976 and 1975, respectively. Unfortunately, the "20 percent test" for a substantial increase in capacity is a cumulative, rather than annual, test. It also should be noted that both the substantial portion and substantial increase rules apply, either to "a plant" or "other property," that is, specific equipment and machinery.

To be considered a "new" facility, it must have been completed or acquired after December 31, 1975 (or 1968 for pre-1969 plants).[62] Moreover, such property will only qualify if the "original use" begins with the taxpayer and such use begins after December 31, 1975 (or 1968).[63] If a facility were completed after December 31, 1975 (or 1968), but begun before that date, Section 169 would only apply to that portion of a taxpayer's basis attributable to post-1975 (or 1968) construction, reconstruction, or erection.[64]

The term "original use" means the first use to which the property is put regardless of whether this use is the same as that of the taxpayer considering a Section 169 election. In other words, rebuilt equipment acquired by the taxpayer for pollution control is not considered as put to original use by the acquiring taxpayer, notwithstanding the fact that its prior owner used the equipment for pollution control or other purposes.[65] Thus, the Section 169 meaning of original use is consistent with its meaning for depreciation[66] and the ITC.[67]

The determination of whether a facility has been "acquired" after December 31, 1975 (or 1968), is the same as the determination thereof for depreciation and for the ITC.[67] Similarly, the determination of whether a facility has been "completed" after December 31, 1975 (or 1968), is based on the tax rules applicable to depreciation and the ITC.[68] The following is a summary of the rules for determining when a facility is acquired and completed:

(1) Property is considered as constructed, reconstructed, or erected by the taxpayer if the work is done for him in accordance with his specifications.[69]

(2) It is not necessary that the construction materials be new in use or acquired after 1975 (or 1968). If construction began after December 31, 1975 (or 1968), the entire cost may be taken into account in determining the adjusted basis of the facility, regardless of the time of purchase of the materials. Construction "begins" when physical work on the facility is started.[70]

(3) Property shall be deemed to be acquired when reduced to physical possession or control.[71]

The second requirement that a treatment facility be of the character subject to depreciation under Section 167 is intended to ensure that Section 169 does not broaden the definition of depreciable property. For this reason and also because of space limitations, a detailed explanation of "property of the character subject to depreciation under Section 167" is not provided here.[72]

The third requirement that a treatment facility be tangible personal property, but not buildings or structural components except special purpose tangible real property used exclusively for the treatment of pollutants, is consistent with the requirements property must satisfy to qualify for the

ITC.[73] A "building" is generally defined as any structure or edifice enclosing a space within its walls and having a roof. This definition encompasses factory buildings, office buildings, warehouses, barns, garages, railway stations, bus stations, store buildings, and apartment buildings.[74] "Structural components" include chimneys and other components related to the operation or maintenance of a building.[75]

Buildings and their structural components whose exclusive function is control or abatement of air or water pollution do, however, qualify for Section 169 rapid amortization if it can be shown that their *only* function is the treatment of pollutants or other wastes. Thus, the taxpayer must demonstrate that the only purpose for the facility (building and structural components) is to comply with Federal or State pollution control requirements. Fortunately, the provision of space in the building for workers or management responsible for the operation of the facility does not violate "exclusive function" requirement. Conversely, where a portion of the building is used for manufacturing or production, the entire structure loses its Section 169 character.[76]

Where a building and its structural components is essentially part of machinery or equipment or is an enclosure so closely combined with the machinery or equipment it supports that the useful life of the machinery determines the useful life of the building and its structural components, then the building with its structural components is a special purpose structure and is not treated as a building. Examples of such structures include storage tanks for oil and gas, grain storage bins, silos, fractioning towers, blast furnaces, coke ovens, brick kilns, and coal tipples.[77] In some cases, it would seem that chimneys that are an integral part of an air pollution control system would qualify as a special purpose structure.

### "In Connection With"

As explained earlier, the pollution control facility must be used "in connection with" a plant or other property in operation before January 1, 1976 (or 1969), but, neither Section 169 nor the regulations pertinent thereto define the phrase "in connection with." Based on the examples provided in the regulations, however, one would conclude that a treatment facility need not be physically connected to a plant or other property to be used in connection therewith. Consequently, it is only necessary that the facility be used to abate or control pollution that would otherwise be caused by the plant or other property.

What if a facility is used in connection with more than one plant where at least one of the plants was not in operation prior to January 1, 1976 (or 1969)? Section 169 does not require that a facility must be used *only* with a pre-1976 (or 1969) plant.[78] In such cases, Regulation Section 1.169-2(a)(3)(ii) provides that the basis of the facility must be allocated by the EPA among the various plants with Section 169 rapid amortization available only for that portion of the basis allocated to pre-1976 (or 1969) plants. In practice, the EPA Regional Administrator either certifies a taxpayer's allocation or makes a new allocation[79] inasmuch as the EPA Guidelines (see Appendix IV) require a taxpayer to make a reasonable allocation based on the taxpayer's detailed explanation in the certification application.[80] Thus, a facility that treats wastes discharged by more than one plant (or other property) is treated as two or three separate facilities, that is, one of which is used in connection with pre-1976 (or 1969) plants and one of which is used in connection with nonqualifying plants.

According to the EPA Guidelines, the allocation method should be based on a comparison of the effluent capacity of pre-1976 (or 1969) plants with the treatment capacity of the pollution control facility. For example, if the pre-1976 (or 1969) plant has a capacity of seventy-five units of effluent with an average *daily* output of sixty units, the new plant has a capacity of forty units with an average *daily* output of twenty units, and the treatment facility has a capacity of 150 units of effluent, then 50 percent (75/150) of the cost could be allocated to the old (pre-1976/1969) plant.[81] However, a taxpayer could present some different method of allocation in his (her or its) certification application, for example, allocation based on *annual* effluent flows. But the EPA Regional Administrator must certify such an alternative methodology.

### Profit-making Facilities

Because the recovery of waste products reduces the amount of pollutants that would otherwise be released, any equipment that recycles industrial by-products could be characterized as a pollution control facility. However, many firms install such recycling equipment for commercial, and not pollution control, purposes. Thus, Congress enacted Section 169(e) to limit the rapid amortization deduction for profit-making pollution control facilities.

Section 169(e) provides that the amortizable basis of a pollution control facility must be reduced by an amount reflecting the estimated profits derived from recycling over its useful life. Where it is apparent that a portion of the costs of a treatment facility will be recovered through recycling, the EPA Regional Administrator is responsible for identifying such facilities based on his/her evaluation of a taxpayer's application for certification thereof.[82] Notwithstanding the EPA's responsibility for determining those facilities that most probably would recover a portion of their costs and the nature of the wastes to be recovered, Regulation Section 1.169-2(d)(2) prescribes the method to be used by a taxpayer in estimating the actual profits from recovery which are then to be used in determining the amortizable basis for the Section 169 deduction.

More specifically, a taxpayer must initially compute the total estimated gross receipts that should result from the *sale* of recovered wastes over the facility's "actual useful life," that is, the shortest possible life as provided in the regulations for Section 167.[83] Then estimated profits are computed by reducing total estimated gross receipts by the sum of estimated average annual maintenance and operating expenses, which include utilities and labor, allocable to that portion of the facility the EPA has certified and estimated selling expenses. For facilities that provide functions in addition to waste treatment, only those expenses allocable to the treatment facility may be deducted from gross receipts in determining profits. Nevertheless, the amount of estimated profits is determined only at the time of the Section 169 election. Consequently, the computation of profits is not subject to revision on the basis of unanticipated subsequent events, for example, upon the change in the market price of recovered wastes or as a result of increases or decreases in operating expenses.

Although the EPA defers the general responsibility for defining "estimated profits" to the Treasury Department, the EPA regulations provide that such profits do not arise from the taxpayer's *use* or *reuse* of recovered waste.[84] Likewise, Regulation Section 1.169-2(d)(2) provides as follows:

> (2) *Estimated profits.* For the purpose of this paragraph, the term "estimated profits" means the estimated gross receipts from the sale of recovered wastes reduced by the sum of the (i) estimated average annual maintenance and operating expenses, including utilities and labor, allocable to that portion of the facility pursuant to paragraph (a)(1)(i) of this section which produces the recovered waste from which the gross receipts are derived, and (ii) estimated selling expenses. However, in determining expenses to be subtracted neither depreciation nor amortization of the facility is to be taken into account. Estimated profits shall not include any estimated savings to the taxpayer by reason of the taxpayer's reuse or recycling of wastes or other items recovered in connection with the operation of the plant or other property served by the treatment facility.

The following algorithm can be used to determine the Section 169 amortizable basis under Section 169(e):

$$A = UA - (\frac{P}{AB} \times UA)$$

where $A$ is the amortizable basis, $UA$ is the unadjusted amortizable basis, *i.e.*, amortizable basis determined without adjustment for either estimated profits or for a useful life greater than 15 years, $AB$ is the adjusted basis for purposes of determining a gain on disposition, and $P$ is the amount of estimated profits.[85]

What follows is an example of the determination of the Section 169(e) amortizable basis adapted from the Treasury regulations[86]:

Tejas Corporation installed a waste water treatment facility on June 30, 1981. This facility is used exclusively in connection with a petro-chemical plant that was placed in service in 1975. The useful life of the facility is 10 years. Estimated revenue from sale of wastes recovered in operating the treatment facility over its 10-year life will be $70,000. Annual operating and maintenance costs will be $1,500, while selling costs will be $500 annually. Tejas Corporation's amortizable basis for purposes of a Section 169 election is computed as follows:

| | |
|---|---:|
| Cost/adjusted basis before reduction for profits* | $400,000 |
| Estimated gross receipts from waste recovery | 70,000 |
| Less: Estimated deductible expenses** | 20,000 |
| | |
| Estimated profits | 50,000 |
| Less: Adjustment in basis for profits*** | 50,000 |
| | |
| Amortizable basis | $350,000 |

---

*Unadjusted amortizable basis
**The sum of annual operating expenses ($1,500) and the annual selling expenses ($500) for 10 years.
***Computation of adjustment:

$$\frac{P}{AB} \times UA = \frac{\$\ 50,000}{\$400,000} \times \$400,000 = \$50,000$$

## Leased Facilities

What if a treatment facility is owned by one taxpayer and the plant or other property is owned by another, for example, where the latter taxpayer is providing treatment facilities or pollution control services to the other taxpayer in a leasing arrangement or under a service contract? Section

169(e) provides that a facility does not fully qualify for rapid amortization where there are profits derived through the recovery of wastes or "otherwise." The term "otherwise" is not explained in the Treasury regulations. The EPA Guidelines, however, in defining estimated profits as "amounts derived through recovery of wastes or otherwise," specify that situations where the taxpayer is in the business of renting (leasing) a pollution control facility, s/he or it should theoretically qualify for Section 169 rapid amortization to the extent the cost of the facilities is greater than the profits derived therefrom. The Treasury Department makes the final decision, however.[87] The Treasury regulations are also silent on this point. Therefore, each case seems to be evaluated on its own merits. Why, however, would the lessor operate at a loss? With regard to the provision of treatment facilities or pollution control services for government or tax-exempt (not-for-profit) organizations, there is no requirement that the plant or other property be used in a trade or business or for production of income.[88] Finally, the definition of estimated profits for a treatment facility does not include fees (rent) received from polluters.[89] Thus, it would seem that such leased facilities qualify for Section 169 amortization to the extent that facility costs exceed recovery profits.

## Multiple Function Facilities

Only that portion of the multiple function facility's cost that is certified by the EPA Regional Administrator as attributable to controlling or abating atmospheric or water pollution will be considered an identifiable treatment facility. Neither the Treasury regulations[90] nor the EPA Guidelines[91] provide much guidance as to how the cost allocation should be made. Therefore, general cost accounting methods are presumably applicable to the allocation of costs between the identifiable treatment facility and the other part of the facility. In practice, the taxpayer must include a cost allocation method along with a justification thereof in his, her, or its certification application. However, if the EPA Regional Administrator does not approve the cost allocation method, then the Administrator must make the allocation.[92]

Similarly, neither the Treasury regulations nor the EPA Guidelines provide a definition of a "function other than pollution control." The Senate Finance Committee specifies that the cost of increasing the height of a smokestack to disperse the pollutants over a broader area does not qualify as an identifiable treatment facility because it serves only to disperse rather than collect or abate pollutants.[93] But what if the stack is equipped with a smog arrester that collects harmful pollutants? If the smog arrester is either tangible personal property or the structural component of a building (chimney) for which the exclusive function is air pollution control,[94] it would seem to qualify for Section 169 amortization.

## Certification of Pollution Control Facilities and Electing Section 169

As indicated above, a taxpayer must initially apply to both Federal and State agencies to certify that the facility meets the applicable Federal and State pollution control standards. Also as noted earlier, the Federal certifying authority is the EPA Regional Administrator. (See Appendix V.) In addition to certifying the pollution control facility, the EPA Regional Administrator monitors the restrictions included in Section 169.[95] A list of State certifying authorities is provided in Appendixes VI, VII, and VIII.

After filing a certification application with the Federal and State authorities, a taxpayer makes an election to amortize under Section 169 by attaching a statement of the election on the tax return for the first taxable year in which the deduction is taken.[96] The election is effective before certification has occurred as long as a certification application has been made.[97]

The EPA Regional Administrator will not approve a certification application until a taxpayer has obtained certification from the appropriate State certifying authority. Thus, the Regional Administrator will evaluate the facility's conformance with Federal pollution control regulations and determine if the facility is consistent with the objectives of the Water Pollution Control Act[98] and the Clean Air Act[99] after receiving notification of certification from the State authority that the facility is in conformance with the State's pollution control program. The same policy would apply in respect to other pollution control laws, for example, the Toxic Substances Control Act or State laws.

In the application to the State authority, the facility should be identified and sufficient evidence provided so that its conformity with State pollution control standards can be determined. Where the State has not established standards in a specific area, the certification application should include a request that the authority certify that the facility is not inconsistent with State requirements, although none has been specified for the area in question. In general, a State authority should verify a taxpayer's application by an on-site inspection. Insofar as there are often significant differences in State certification procedures, specific details on each State's application procedures should be obtained before applying for certification.[100]

After applying for State certification, a taxpayer may file a certification application with the EPA Regional Administrator. Appendix V presents a list of EPA Regional Administrators along with their jurisdictional responsibilities. The application to the EPA must include the following information:

(1)  Name, address, and IRS identification number of applicant.

(2)  Detailed description of the treatment facility. This description should identify the type of facility and explain its operation, with a copy of the schematic or engineering drawings attached.

(3)  Location of the facility.

(4)  A description (general and specific) of the process that is producing the pollutants being controlled by the facility.

(5)  Effect of the facility in removing, altering, storing, or disposing of pollutants in terms of the type and quantity of pollutants involved.

(6)  Cost of the facility, including an estimate of maintenance costs.

(7)  Date of acquisition or completion and date facility was placed in operation.

(8)  Identification of the applicable State and local water or air pollution control standards.

(9)  Estimated useful life.[101]

In the event that any of the restrictions of Section 169 are applicable, additional information may be required:

(1)  Where the facility serves a function in addition to pollution control, there must be a description of all functions provided by the facility along with both an allocation of total costs to the pollution control portion, and an explanation of the basis for making the allocation.

(2)  Where the facility is used in connection with both pre-1976 (or 1969) and post-1975 (or 1968) plants (or other property), there must be a description of the facility's operations with respect to each plant along with both an allocation of total costs to each plant, and a description of the basis for making the allocation.

(3)  Where it is anticipated that the facility will generate some profits from resource recovery, there must be an analysis of the estimated profits to be derived from the recovery of wastes or otherwise over the facility's useful life.

As long as a taxpayer specifies that a facility has not yet been completed or acquired and provides the expected date of completion or acquisition, the taxpayer may file a certification application with the EPA Regional Administrators, prior to the facility's completion or acquisition.[102] However, the certification application will be considered incomplete until certification has been received from the State.[103] An early application can be advantageous because the taxpayer may receive notification of intent to certify from the EPA Regional Administrator in anticipation of State certification assuming the EPA will eventually approve of the proposed facility.[104] Such early applications are coordinated with the State authority inasmuch as the EPA will inform the State authority of any action the EPA finally takes on a taxpayer's request for notice of intent to certify.[105]

Based on the EPA Guidelines, the Regional Administrator's decision to certify is generally going to be based on the application to the EPA and on State certification.[106] Although the Regional Administrator may make an on-site inspection, such field inspections are generally limited to situations involving very dangerous pollution sources or large investments. Finally, the EPA Administrator is not bound by State certification where there is evidence that the State authority has obviously ignored violations of air, water, or other pollution control standards.[107]

The election statement filed with the taxpayer's tax return for the first taxable year in which the deduction is to be taken will duplicate much of the information provided in the certification application. Consequently, such information need not be included in the election statement inasmuch as the certification application must be attached thereto. The following information must be included either in the election statement or in the certification application:

(1) A description clearly identifying each certified pollution control facility for which an amortization deduction is claimed.

(2) The date on which such facility was completed or acquired (see paragraph (b)(2)(iii) of Section 1.169-2).

(3) The period referred to in paragraph (a)(6) of Section 1.169-2 for the facility as of the date the property is placed in service.

(4) The date as of which the amortization period is to begin.

(5) The date the plant or other property to which the facility is connected began operating (see paragraph (a)(5) of Section 1.169-2).

(6) The total costs and expenditures paid or incurred in the acquisition, construction, and installation of such facility.

(7) A description of any wastes which the facility will recover during the course of its operation, and a reasonable estimate of the profits which will be realized by the sale of such wastes whether pollutants or otherwise, over the period referred to in paragraph (a)(6) of Section 1.169-2 as to the facility. Such estimate shall include a schedule setting forth a detailed computation illustrating how the estimate was arrived at, including every element prescribed in the definition of estimated profits in paragraph (d)(2) of Section 1.169-2;

(8) A computation showing the amortizable basis (as defined in Section 1.169-3) of the facility as of the first month for which the amortization deduction provided for by Section 169(a) is elected.

(9) A statement that the facility has been certified by the Federal certifying authority, together with a copy of such certification, and a copy of the application for certification that was filed with and approved by the Federal certifying authority. If the facility has not been certified by the Federal certifying

authority, a statement that application has been made to the proper State certifying authority (see paragraph (c)(2) of Section 1.169-2) together with a copy of such application and (except in the case of an election to which subparagraph (4) of this paragraph applies) a copy of the application filed or to be filed with the Federal certifying authority. Within ninety days after receipt by the taxpayer, the certification from the Federal certifying authority shall be filed by the taxpayer with the district director, or with the director of the internal revenue service center, with whom the return referred to in this subparagraph was filed.[108]

If a taxpayer makes a pre-certification application to the EPA, a copy of the certification must be filed with the Internal Revenue Service within ninety days of its receipt from the EPA Regional Administrator.[109] Similarly, the Regional Administrator must forward a copy of its certification to the IRS District Director and the State certification authority at the same time the taxpayer is notified.[110] Presumably, a taxpayer must file an amended return without a Section 169 amortization election if the certification application is not approved by the EPA Regional Administrator.

### Electing and Terminating the Amortization Period

The taxpayer who elects Section 169 amortization must also specify the month with which the rapid amortization period begins.[111] The taxpayer may elect to begin the sixty-month amortization period with either the first month after the month in which the facility is completed or acquired, or the first month of the following taxable year.[112] If the taxpayer elects to begin the amortization period in the following taxable year, then Section 167 depreciation may be available for the period between completion or acquisition and the beginning of the amortization period.[113] Since Section 179 additional first-year depreciation may no longer be available for the short period until the rapid amortization period begins,[114] it is unlikely that a taxpayer would elect to defer the beginning of the sixty-month period to the following taxable year because the depreciation deduction in the first year would seldom be more than the amortization deduction for the months the facility is used during the first year. Moreover, the wording of Section 168(e)(2)(B) may preclude depreciation deductions for a short year. Hopefully, the IRS will issue a ruling to clarify this conflict.

There are at least two nontax reasons for deferring the beginning of the amortization period. If a taxpayer anticipates that his (her or its) certification application will be either delayed or rejected, deferral would save on the administrative costs associated with the filing of an amended tax return. Furthermore, where a taxpayer anticipates improving or enlarging the facility prior to the inception of the next taxable year, postponing the certification application and the election would be cost-effective.[115]

At any time after electing the Section 169 amortization, a taxpayer may terminate that election by filing a statement indicating that intent with the IRS District Director. This statement must clearly identify the facility, be filed before the beginning of the month with which the termination takes effect, and specify the termination date. Where a taxpayer has made a pre-certification election, such taxpayer must file a copy of the certification within ninety days of receipt thereof for the termination election to be effective. Conversely, the Section 169 election is automatically terminated when the EPA notifies the taxpayer and the IRS that the pollution control facility is not in compliance with the Federal and State pollution control standards that were in effect at the time the certification was granted.[116] Unfortunately, a taxpayer may not elect Section 169 amortization again in respect to the property for which he or she has terminated a sixty-month amortization election. As a result of the enactment of Section 168 ACRS deductions, Section 179 expensing, and more liberal ITC provisions,[117] it may be advantageous not to make a Section 169 election for pollution control facilities placed in service after 1980.

### Capital Recovery After Termination of an Election

Upon deciding to terminate a Section 169 election, a taxpayer can take Section 168 depreciation deductions on the unamortized portion of the facility's amortizable basis.[118] Section 168 ACRS deductions seem to be available upon termination of a Section 169 election although such an election excludes property from ACRS recovery property status.[119] The Section 168 depreciation deductions begin with the first month for which Section 169 amortization deductions are not taken.

Finally, the depreciation deduction is based on the remaining portion of the period authorized under Section 168 for the facility as determined as of the first day of the first month that the amortization deduction is not applicable.[120] Insofar as the useful life is initially determined when the facility is placed in service, the period amortized is generally subtracted from the original useful life.

### Facilities With Useful Lives in Excess of Fifteen Years

If a pollution control facility has a useful life in excess of fifteen years, the amortizable basis for Section 169 is limited to that portion of the facility's cost attributable to the first fifteen years of its useful life. For purposes of this limitation, useful life is determined as of the first day of the first month for which a deduction is *allowable* under Section 169.[121] The amortizable basis is determined by multiplying the amount otherwise amortizable by a fraction. The numerator of the fraction is 15, and the denominator is the useful life of the facility.[122] The unamortizable basis is depreciated as provided in Section 168.[123]

The useful life for Section 169 is the shortest period allowable under Section 167 (and the regulations related thereto) if the rapid amortization election were not made.[124] As noted, the useful life is determined as of the first day of the first month for which a deduction is allowable under a Section 169 election.[125] Although both a Section 169 and a Section 167(m) Class Life-Asset Depreciation Range (CLADR) election could never be in effect for a pollution control facility,[126] the shortest useful life under Section 167 for such a facility is generally going to be determined by relying on Revised Procedure 77-10[127] (as modified by subsequent revenue procedures), that is, the useful lives for the CLADR election. The property's CLADR designation determines its recovery period, that is, an asset's (pollution control facility's) recovery property designation.

### Improvements and Additions to Pollution Control Facilities

Improvements or additions made after the amortization period has begun do not increase the amortizable basis of a pollution control facility.[128] If such improvements or additions are made before the amortization period begins, however, they do increase amortizable basis.[129] In addition, when the improvements or additions are so significant that they result in a new facility, then such costs should be included in the amortizable basis of a new facility. Consequently, the taxpayer must repeat the certification and election procedures for the new facility just as he (she or it) did for the old facility.[130]

### Determining the Amortization Deduction

In practice, a pollution control facility consists of several components or items of property rather than a single piece of equipment. Thus, each component of a facility may have a different useful life,[131] and a taxpayer may decide to establish a composite account for the facility to the extent that depreciation is taken on that portion of the property not amortizable under Section 169.[132] The determination of a single useful life for a pollution control facility with several components having different useful lives is not specified in either Treasury or EPA regulations. Presumably, the taxpayer should use an average useful life, which could be computed by determining a weighted average of the useful lives of the components of the facility where the weighting is based on the relative costs of the component properties.[133]

A comprehensive example of the determination is deferred until after the following discussion of the availability of the ITC for pollution control facilities. However, the example does not contrast ACRS treatment with Section 169 amortization where both would be eligible for ITC.

### The Investment Tax Credit and Section 169 Amortization

Originally in 1969, Section 169(h) provided that the Section 38 credit for investments in tangible personal property would not be allowed in the case of pollution control facilities for which rapid amortization was elected. Then in 1971 with the reinstatement of the ITC that had been terminated in 1969, Section 169(h) was repealed and Section 46(c)(5) added so that in the case of property with respect to which an election under Section 169 applies, and the useful life of which (determined without regard to Section 169) is not less than five years, 50 percent constituted the applicable percentage for purposes of determining the amount of qualified property eligible for the investment credit. In other words, an election to use Section 169 rapid amortization was accompanied by one-half of the regular ITC if the property's useful life was at least 5 years.

The Revenue Act of 1978 amended Section 46(c)(5) so that for pollution control facilities acquired or constructed after December 31, 1978, the ITC may be claimed on the entire amount of purchase. This assuming the useful life is not less than five years, with transition rules provided for facilities in the process of construction as of December 31, 1978, to permit the more liberal credit for the part of construction after that date. Pollution control facilities that had a useful life of three or four years continued to be subject to the pre-1981 law that limited the credit to one-third of the full credit, insofar as Section 169 would not be elected for them. In addition, where pollution abatement facilities subject to the Section 169 rapid amortization election are financed with Section 103(b) tax-exempt industrial development bonds (IDB), the 50 percent rule remains in effect. Finally, the temporary (1975-79) increase in the investment credit rate from 7 percent to 10 percent was made permanent by the Revenue Act of 1978. Thus, for qualified pollution abatement facilities, the maximum allowable credit will be 10 percent, and an effective rate of 5 percent will be available where the IDB exception applies.

The Economic Recovery Tax Act of 1981 amended the ITC provisions for most property by implicitly reducing the useful life required for a 10 percent credit from seven to five years, as well as providing for a 6 percent credit for property with a useful life of three (or four) years.[134] Thus, the one-third credit (3⅓ percent) for property with a useful life of three or four years and two-thirds credit (6⅔ percent) for property with a useful life of five or six years have been effectively eliminated where the property is placed in service after 1980 and is Section 168 ACRS recovery property. Since an election under Section 169 precludes the classification of a pollution control facility as Section 168 ACRS recovery property, the 1981 tax law does not affect the ITC available for pollution control facilities with a useful life of five years or more and for which Section 169 is elected. Alternatively, pollution control facilities for which rapid amortization is not elected would be Section 168 ACRS recovery property. Therefore, the post-1980 rules for ITC would be applicable.

### Determining Useful Life

For purposes of the ITC, useful life is determined at the time a property is placed in service. Although "useful life" is not defined for ITC purposes, it is the same as the useful life for computing depreciation.[135] Thus, the useful life may be determined either under the CLADR system or by the "facts and circumstances" method. However, for property that is amortized rather than depreciated, that is, ACRS deductions, the useful life is the amortization period.[136]

Initially, the useful life of a pollution control facility is determined by its CLADR asset classification, which is usually based on the type of industrial plant in which the facility is installed.[137] Alternatively, a shorter (or longer) useful life can be adopted where the taxpayer relies on "facts and circumstances" for the determination thereof. If a Section 169 election is effectively made, then the facility's useful life will be five years, that is, the amortization period, for that portion of a facility's basis that is eligible for rapid amortization. After 1980, the unamortizable portion of the facility's basis would have a recovery period based on its CLADR designation.

### Accelerated Cost Recovery System and Section 169

When the cost of a pollution control facility is recovered under the provisions of Section 168 rather than Sections 169 and 167, the CLADR designation of the facility determines its ACRS recovery property designation, that is, three-year, five-year, ten-year or fifteen-year public utility recovery property. Furthermore, the facility's useful life no longer determines the applicable percentage of ITC for which it qualifies because the Section 46(c)(2) percentages (10, 6⅔, and 3⅓) would not be applicable for recovery property insofar as Section 46(c)(7) provides for the following ITC percentages:

| Type of Recovery Property | ITC Percentage |
|---|---|
| 5-year, 10-year, and 15-year public utility | 10 |
| 3-year | 6 |

### Qualified Property

A property that qualifies for the Section 169 election should also be a Section 38 ITC property. More specifically, the ITC property includes principally tangible personal property (other than an air conditioning or heating unit); certain tangible real property (except a building and its structural components); qualified rehabilitated buildings; and storage facilities used in connection with the distribution of petroleum or any primary product of petroleum, such as gasoline, diesel fuel, or heating oil.

ITC property includes only recovery property (within the meaning of Section 168 without regard to useful life) and any other property in respect to which depreciation (or amortization in lieu of depreciation) is allowable and having a useful life (determined as of the time such property is placed in service) of three years or more. Property used by a tax-exempt organization or government does not generally qualify. However, there are exceptions for property used in the taxable trade or business operations of such organizations, as well as for leased, qualified, rehabilitated buildings.[138]

There is at least one major difference between what constitutes a pollution control facility and the definition of ITC qualified property. This difference is related to the exclusion of real property from eligibility for the Section 169 election and the Section 38 credit. Generally, the exclusion of such property from ITC qualified property is more restrictive than the same exclusion for pollution control facilities. More specifically, special purpose buildings or their structural components and land improvements used exclusively for the treatment of pollutants[139] are eligible for Section 169 amortization. In contrast, neither buildings nor their structural components (except for qualified rehabilitated buildings) qualify for the Section 38 credit, although tangible real property does qualify for the ITC where such property is used as an integral part (including the storage of bulk commodities) of manufacturing, production, or extraction, or of furnishing transportation, communications, electrical energy, gas, water, or sewage disposal services.[140] In addition, the ITC is generally not available for property leased to tax-exempt organizations and government; a notable exception is qualified rehabilitated buildings.[141] Certified pollution control facilities leased to the government may be eligible for Section 169 amortization at the discretion of the IRS.[142] There are, of course, several other more subtle differences between ITC property and qualified pollution control facilities, but they are too limited in scope to be discussed here.

### Investment Credit and Industrial Development Bonds

Although Section 46(c)(5) was amended by Section 313 of the Revenue Act of 1978 to allow a full ITC for Section 169 property acquired or constructed after 1978,[143] the Section 38 investment tax credit for certified pollution control facilities financed with proceeds from tax-exempt IDBs remains at 50 percent. That is, only 50 percent of the amortizable basis qualifies as Section 38 property. Thus, the Tax Reform Act of 1976 amendment of Section 48(a)(8) still applies to Section 169 property financed with IDBs.[144]

## Comprehensive Examples

The Tejas Oil and Chemicals Corporation (TOX), which is an accrual-basis, calendar-year taxpayer, began installation of a facility on December 1, 1980.[145] The facility, which cost $800,000, was completed and placed in service on June 30, 1981. It was installed at TOX's Port Tejas, Texas, petrochemical plant, which was completed and placed in service in 1975. The facility serves two functions—recirculating cooling water and treating that water by removing greases, oils, and chemicals from it. One-half of the cost of the facility is attributable to its water treatment function; $100,000 of the total cost of the facility is attributable to construction prior to January 1, 1981.

The water treatment portion of the facility has been certified by the Texas State Department of Water Resources and the Dallas Regional Administrator of the EPA. In filing its 1981 Federal income tax return on March 15, 1982, TOX elected Section 169 rapid amortization for the certified water pollution control facility.

The estimated useful life of the facility determined as provided in Regulation Section 1.169-2(a)(6) could be based on the following useful lives and costs for the components of the water treatment unit: a straining machine with a chemical filtering system that has a useful life of ten years and costs $200,000; separating tanks with a useful life of thirty years that costs $125,000; and an aerator with a useful life of thirty years that costs $75,000. Over the useful life of this water treatment facility, it is estimated that $50,000 in profits will be realized from the sale of wastes recovered in the operation thereof.

Because only half of the cost of the dual-function facility is a pollution control facility, only $400,000 of its cost qualifies in principle for the Section 169 rapid amortization election. The other portion of the facility, which recirculates the cooling water, is Section 168 ACRS property that falls under Class 28.0 of the CLADR system. Consequently, the water recirculation facility is classified as five-year property because its shortest useful life is eleven years, notwithstanding an apparent useful life of fifteen years based on the useful lives and relative costs of its components.[146] Thus, the adjusted basis of both the pollution control facility and the water recirculation facility is $400,000 each. Although part of the construction of the facility occurred in 1980, it was not completed and placed in service until mid-1981. Therefore, all of the adjusted basis of the facility is depreciable or amortizable in 1981.[147]

The new ACRS capital recovery provisions may preclude the election of Regulation Section 1.169-3(b)(2), which allows deferral of rapid amortization to the first taxable year following the year in which a certified pollution control facility is placed in service.[148] Thus, neither Section 167 depreciation (and Section 179 additional first year depreciation), nor Section 168 deductions would be available for the short year a pollution control facility is placed in service, in anticipation of electing Section 169 the next year.

Because the pollution control facility is expected to recover $50,000 of costs from the sale of wastes recovered in water treatment, an adjustment of its amortizable basis for Section 169 purposes of the amount of estimated profits from recovery is necessary. Inasmuch as the useful life of the facility is less than sixteen years, no Section 169(f)(2) adjustment is necessary. Nevertheless, Section 168 depreciation is available on that portion of the adjusted basis of the pollution control facility that does not qualify for Section 169 rapid amortization—$50,000. Thus, the amortizable basis is $350,000. If the plant had been a liquefied natural gas operation, then its shortest useful life under CLADR would have been 17.5 years, and the amortizable basis would have to be reduced by one-seventh of the basis remaining after the adjustment for estimated profits from waste recovery. Thus, amortizable basis would be $300,000 rather than $350,000.

Inasmuch as none of the pollution control facility was financed with the proceeds of the sale of industrial development bonds, the Section 38 investment tax credit would be available for the $350,000 of amortizable basis. In addition, the ITC would be available for the balance of the adjusted basis of the pollution control facility ($50,000) because its useful life is greater than six years. The amortizable basis, amortization, depreciation, and investment tax credit for the pollution control facility are shown in the accompanying calculations.

*Amortizable Basis*

| | |
|---|---|
| Cost of (adjusted basis) of water treatment unit | $400,000 |
| Less: Adjustment for profit-making estimate* | 50,000 |
| Amortizable basis | $350,000 |

*Amortizable Deduction*

$$\frac{\$350,000}{60 \text{ months}} \times 6 \text{ months in } 1981 =$$

$35,000 Amortization deduction** for 1981

The rapid amortization Section 169 deduction for 1981–85 would be $70,000, while the deduction in 1986 would be $35,000.

*Depreciation Deduction*

$$\$50,000 \times 15\% = \$7500 \text{ ACRS deduction for } 1981$$

The depreciation deduction as provided in Section 168 is taken for the year using the table for five-year recovery property. The depreciation deduction would increase in 1982 to 22 percent and then 21 percent per year for 1983-85.

*Investment Tax Credit*

$350,000 × 10% = $35,000 Investment tax credit
for amortized portion

50,000 × 10% = $ 5,000 Investment tax credit
for depreciated portion

_____

$40,000 Total investment tax credit
for water treatment unit

_____

*Multiply unadjusted amortizable basis ($400,000) by estimated profits from waste recovery ($50,000) and divide by the adjusted basis as provided in the formula:

$$\frac{UA \times P}{AB} = \frac{\$400,000 \times \$50,000}{\$400,000} = \$50,000.$$

To contrast capital recovery for the pollution control facility for which a Section 169 election is made with capital recovery without the amortization election, computation of the Section 168 deduction and ITC for the water recirculation facility is provided here.

*Depreciation Deduction (Section 168 Accelerated Cost Recovery System deduction for five-year recovery property)*

$400,000 × 15% = $60,000 ACRS deduction for 1981

The ACRS deduction for 1982 would be $88,000, while the deduction for 1983, 1984, and 1985 would be $84,000 annually.

*Investment Tax Credit*

$400,000 × 10% = $40,000 Investment tax credit
for water recirculation unit.

_____

If the hypothetical firm, Tejas Oil and Chemicals Corporation, is profitable and pays Federal income taxes at the highest marginal rate of 46 percent, it should not elect Section 169 rapid amortization because Section 168 ACRS deductions yield a more rapid and greater capital recovery. If, however, the firm has been operating at a loss for a few years but expects to be profitable in one or two years, then Section 169 rapid amortization may be advantageous because it defers more capital recovery deductions to future, profitable years when the deductions would have greater tax benefit. Of course, a more rigorous financial analysis of the relative value of the tax benefits of the two alternatives would require determination of the present value of the capital recovery deductions based on a reasonable assumption about the firm's discount rate.

## Notes

1. P.L. 97-34, August 13, 1981, is explained but the Tax Equity and Fiscal Responsibility Act, P.L. 97-248, September 3, 1982, is not included in the analysis.

2. Section 168(a).

3. Reg. Sec. 1.167(a)-11(b)(4)(iii)(b).

4. Section 168(b)(3)(B)(iii).

5. Section 168(b)(3)(A).

6. Section 168(e)(1).

7. Section 179(b)(1).

8. Section 168(d)(1)(A) and Section 179(c)(9).

9. Section 179(d)(1).

10. Section 179(d)(5).

11. Section 46(c)(7) and Section 46(a)(2)(A) and (B).

12. Section 46(a)(2)(A) and (C).

13. Section 46(a)(2)(A) and (E).

14. Reg. Sec. 1.48-1(c).

15. Section 48(c)(2).

16. Rev. Rul. 80-313, 1980-2 C.B. 25.

17. Rev. Rul. 79-40, 1979-1 C.B. 13.

18. Section 46(d).

19. *LTV Corporation*, 63 T.C. 39 (1974).

20. Rev. Rul. 79-40, *supra* note 17.

21. Section 263.

22. Rev. Proc. 77-10, 1977 C.B. 548.

23. See Table 8-1 for summary of state exemptions and credits.

24. Section 48(d).

25. Rev. Rul. 55-540, 1955-2 C.B. 39; and Rev. Proc. 75-21, 1975-1 C.B. 715, modified by Rev. Proc. 75-28, 1975-1 C.B. 752 and Rev. Proc. 79-48, 1979-2 C.B. 529.

26. Conference Rep. (House Rep.) No. 97-215, 97th Cong., 1st Sess. 217 (1981).

27. Section 168(f)(8)(A), (B)(ii), and (B)(iii).

28. Section 168(f)(8)(D).

29. Senate Rep. No. 97-144, 97th Cong., 1st Sess. 62 (1981).

30. Section 46(e)(3). Business expenses are those allowed solely under Section 162. That is, interest expenses (Section 163), taxes (Section 164), and capital cost recovery allowances (Section 168), among others, are not part of the 15 percent test for expenses.

31. See the temporary regulations applicable to Section 168(f)(8) issued in October and November 1981.

32. Reg. Sec. 1.48-4.

33. Section 1245(a).

34. Section 168(d)(2)(A) specifies an elective alternative rule for multiple asset accounts.

35. Section 1245(a)(3).

36. Section 1245(a)(1) and (2).

37. Section 168(d)(2)(C).

38. Section 47(a)(5).

39. EPA Reg. Sec. 20.8(a), (b), and (c). See Appendix III for EPA Regulations.
40. EPA Guidelines, Section 2(a). See Appendix IV for EPA Guidelines.
41. Reg. Sec. 1.169-2(a)(3). See Appendix II for Section 169 Regulations.
42. Section 169(d)(1)(C).
43. H. Rep. No. 1515, 94th Cong., 2d Sess., 1976.
44. EPA Guidelines, Sections 2 and 3.
45. EPA Guidelines, Section 2(b).
46. Ibid.
47. EPA Guidelines, Section 3(a).
48. Reg. Sec. 1.169-2(b)(1)(i).
49. EPA Guidelines, Section 2(b).
50. Section 169(d)(4)(A).
51. Reg. Sec. 1.169-2(b)(1)(i).
52. Reg. Sec. 1.169-2(b)(1).
53. Section 169(d)(4)(A).
54. Section 169(d)(1).
55. H. Rep. 91-321, 91st Cong., 1st Sess. (1969), 197.
56. Tax Reform Act of 1976, P.L. 94-455, Section 2112.
57. Reg. Sec. 1.169-2(a)(5)(i).
58. Ibid.
59. Reg. Sec. 1.169-2(a)(4).
60. Reg. Sec. 1.169-2(a)(5)(ii)(b).
61. Reg. Sec. 1.169-2(a)(5)(ii)(a).
62. Section 169(d)(4)(A).
63. Section 169(d)(4)(A)(ii).
64. Section 169(d)(4)(A)(i).
65. Reg. Sec. 1.169-2(b)(2)(iii)(g).
66. Reg. Sec. 1.167(c)-1(a)(2), and Reg. Sec. 1.48-2(b)(7).
67. Reg. Sec. 1.167(c)-1(a) and (b), and Reg. Sec. 1.48-2(b).
68. Reg. Sec. 1.169-2(b)(2)(iii).
69. Reg. Sec. 1.169-2(b)(2)(iii)(a).
70. Reg. Sec. 1.169-2(b)(2)(iii)(b), (c) and (d).
71. Reg. Sec. 1.169-2(b)(2)(iii)(f).
72. See Rolf Auster, *Depreciation Desk Book* (Englewood Cliffs, N.J.: Institute for Business Planning, Inc., 1980), Chapter 2.
73. Section 48(a)(2).
74. Reg. Sec. 1.169-2(b)(2)(i)(a).
75. Reg. Sec. 1.169-2(b)(2)(i)(b).
76. Section 169(d)(4) and (e), and Reg. Sec. 1.169-2(b)(2)(ii).
77. Reg. Sec. 1.169-2(b)(2)(i)(a).
78. Section 169(d).
79. EPA Reg. Sec. 20.8(e).
80. EPA Reg. Sec. 20.5(e) and EPA Guidelines, Section 5.
81. EPA Guidelines, Section 5.
82. EPA Reg. Sec. 20.9 and EPA Guidelines, Section 8.
83. Reg. Sec. 1.169-2(d)(3).
84. EPA Reg. Sec. 20.9.
85. Reg. Sec. 1.169-3(c).
86. Reg. Sec. 1.169-3(e).
87. EPA Guidelines, Section 8.
88. Reg. Sec. 1.169-2(a)(4).
89. Reg. Sec. 1.169-2(d) and 1.169-3(c).
90. Reg. Sec. 1.169-2(a)(3).

91. EPA Guidelines, Section 4.
92. EPA Reg. Sec. 20.8(e).
93. S. Rept. 91-552, 91st Cong., 1st Sess. (1969), 250.
94. Section 169(d)(4) and Reg. Sec. 1.169-2(b)(2)(ii).
95. Section 169(d)(1)(A) and (B).
96. Reg. Sec. 1.169-4(a)(1).
97. Reg. Sec. 1.169-4(a)(1)(ix) and Information Release 1177 (1971).
98. 33 U.S.C. 466, et. seq.
99. 42 U.S.C. 1857, et seq.
100. Michael J. McIntyre, 254 T.M., *Amortization of Pollution Control Facilities*—Section 169, p. A-39.
101. EPA Reg. Sec. 20.5.
102. EPA Reg. Sec. 20.3(d).
103. EPA Reg. Sec. 20.3(c).
104. EPA Reg. Sec. 20.4(a).
105. EPA Guidelines, Section 9.
106. EPA Guidelines, Section 6.
107. Ibid.
108. Reg. Sec. 1.169-4(a).
109. Reg. Sec. 1.169-4(a)(1)(ix).
110. EPA Reg. Sec. 602.3(f), 18 C.F.R. 602.
111. Reg. Sec. 1.169-4(a)(1).
112. Reg. Sec. 1.169-1(a)(1).
113. Reg. Sec. 1.169-3(b)(2).
114. Section 179.
115. McIntyre, 254 T.M., *Amortization of Pollution Control Facilities*, p. A-9.
116. Reg. Sec. 1.169-4(b)(1).
117. Economic Recovery Tax Act of 1981, P.L. 97-34, August 13, 1981.
118. Reg. Sec. 1.169-1(a)(6).
119. Section 168(e).
120. Reg. Sec. 1.169-1(a)(6).
121. Section 169(f)(2)(A).
122. Reg. Sec. 1.169-3(d).
123. Section 169(g).
124. Reg. Sec. 1.169-2(a).
125. Section 169(g).
126. Reg. Sec. 1.167(a)-11(b)(5).
127. 1977-1 C.B. 4. See Auster, *Depreciation Desk Book*, Chapters 16 and 17.
128. Section 169(f)(2)(B) and Reg. Sec. 1.169-3(f)(1).
129. Reg. Sec. 1.169-3(f)(1).
130. Reg. Sec. 1.169-3(f)(2).
131. EPA Guidelines, Section 2.
132. Reg. Sec. 1.167(a)-7.
133. McIntyre, 254 T.M., *Amortization of Pollution Control Facilities*, p. A-36.
134. P.L. 97-34, Sections 211 and 212.
135. Section 46(c)(2).
136. Reg. Sec. 1.46-3(e)(4).
137. See Rev. Proc. 77-10, 1977 C.B. 4.
138. Section 48(a)(1), (4) and (5).
139. Reg. Sec. 1.169-2(b)(1).
140. Section 48(a)(1).
141. Section 48(a)(4) and (5).
142. See the explanation of the tax implications of facilities to such groups provided above.

143. P.L. 95-600, November 6, 1978.
144. P.L. 94-455, October 4, 1976.
145. The example given in this section is adapted from the examples found in Reg. Sec. 1.169-3(e).
146. See Rev. Proc. 77-10, 1977-1 C.B. 548, for asset guideline classes and lives.
147. See Reg. Sec. 1.46-3(d)(1) and (2) for a definition of "placed in service."
148. Section 168(e).

# 10

# Environmental Excise Taxes on U.S. Crude Oil, Chemicals, and Hazardous Wastes

In an atmosphere of political compromise and concern for the environment, the "lame-duck" session of the 96th Congress enacted the Comprehensive Environmental Response Compensation and Liability Act of 1980 (CERCLA).[1] This law, more popularly known as "the Superfund," established a $1.6 billion fund to be used to finance the clean-up of both abandoned hazardous waste dumpsites and accidental spills of toxic materials. CERCLA does not, however, apply to crude oil spills. The fund is not to be used to compensate any individuals harmed by emissions from either a dumpsite or a spill. Therefore, such victims must sue the parties responsible for the hazardous or toxic waste in the State in which the personal injury occurred. The fund is to be reimbursed by the party responsible for the abandoned dumpsite or spill for either the clean-up cost incurred by the government or the damages caused to government-owned natural resources. Finally, $1.38 billion of the $1.6 billion Hazardous Substances Response Fund is to be derived from an environmental excise tax on petroleum, a variety of organic chemicals, and certain inorganic chemicals (but not on natural gas which is primarily methane). Consequently, the balance of the fund will be provided by Federal government appropriations.

## The Tax on Crude Oil

Since April 1, 1981, a new excise tax on crude oil has been imposed by the Federal government. The collection of this tax is authorized by Section 4611 of Chapter 38, a new chapter of the Internal Revenue Code of 1954 added thereto upon enactment of CERCLA. This tax is imposed at a rate of 79¢ per barrel (or fraction of a barrel), and it is imposed on both domestic crude oil production and imported petroleum products.

The tax is imposed either on the operator of any U.S. refinery receiving crude oil or on any importer of petroleum products that are brought into the United States for consumption, use, or warehousing. Where domestic crude oil is exported from or used in the United States before the 79¢ per barrel tax is imposed, the exporter or "consumer" of the crude oil is liable for the same tax that would have been imposed had a U.S. refiner received that oil. However, there is an exclusion from taxation for domestic crude oil production that is used for extraction of additional crude oil or natural gas on the premises where the oil was produced.[2] An example would be oil used for powerhouse fuel or for reinjection as part of a tertiary recovery process.

For purposes of this petroleum excise tax, the term "crude oil" includes crude oil condensate and natural gasoline, but not the other condensates from natural gas, that is, butane, ethane, or propane. However, butane is subject to the chemical excise tax explained below. Similarly, the term "petroleum products" includes crude oil, crude oil condensate, and natural gasoline. In addition, "petroleum products" includes refined gasoline, refined and residual oil, and any other hydrocarbon product derived from crude oil or natural gasoline that enters the United States in liquid form. Finally, the term "crude oil" does not include synthetic petroleum, that is, shale oil, liquids from coal, tar sands or biomass, or refined oil.

The definition of "refining" as any operation by which the physical or chemical characteristics of crude oil or its products are changed is not meant to encompass simple on-site operations such as passing crude oil through separators or placing crude oil in settling tanks. Thus, a U.S. refinery is any facility in the United States that under the existing Treasury regulations engages in refining (as explained above),[3] that is, crude oil processing.

As noted above, the environmental excise tax on domestic crude oil production is imposed on the operator of any U.S. refinery when that refinery receives the oil. However, where a U.S. refinery produces natural gasoline from natural gas, the tax moment is deemed to be the time at which natural gasoline actually is produced[4] rather than the time at which the refinery received the natural gas. Alternatively, where natural gasoline is recovered in a gas separation plant instead of at a refinery, the petroleum tax would be imposed when the condensate is received at a refinery or when it is used or exported without passing through a refinery. In summary, a refiner must pay the same tax whether it purchases natural gasoline or wet gas from which it extracts natural gasoline.[5]

The term "United States" (often abbreviated "U.S.") is very broadly defined for purposes of the petroleum excise tax. More specifically, "United States" encompasses the fifty States, the District of Columbia, Puerto Rico, U.S. Possessions, the Northern Mariana Islands, and the Trust Territory of the Pacific Islands. In addition, the term includes foreign trade zones of the United States, as well as Continental Shelf areas (as explained in Section 638).[6] With regard to areas that are not subject to general U.S. customs laws, for example, foreign trade zones, an "entry" into the United States occurs at any time that entry would have occurred *if customs laws were applicable thereto.* There is no exception for bonded petroleum products entering a foreign trade zone. However, an "entry" does not occur where fuel is stored on a ship entering U.S. waters or in the gas tank of an automobile entering the United States.[7]

Section 4612(b) should eliminate the double taxation problems inherent in taxing crude oil and petroleum products. If a person who would otherwise be liable for the excise tax on petroleum products establishes that the tax has already been imposed for such products, then that person is not subject to a second tax on the products.[8] If a second tax were paid on petroleum products, the taxpayer making the second payment would be entitled to a credit for the amount of second payment on subsequent payments of the petroleum excise tax.

## The Tax on Chemicals

Since April 1, 1981, a new Federal excise tax on chemicals has been in effect. The collection of this tax is authorized by Section 4661, which is also included in the new Chapter 38 addition to the 1954 Internal Revenue Code as provided in CERCLA. This tax is imposed on specified organic and inorganic chemicals on a per ton basis, and it is imposed on both domestically produced chemicals and imported chemicals.

The tax on most organic chemicals is $4.87 per ton, with methane being taxed at a lower rate of $3.44 per ton. Table 10-1 is a rate schedule for the tax on petrochemical feedstocks taken from Section 4661:

**Table 10-1**
**Tax on Petrochemicals**

| Chemical | Tax/Ton[a] | Chemical | Tax/Ton[a] |
|----------|-----------|----------|-----------|
| Acetylene | $4.87 | Methane | $3.44 |
| Benzene | 4.87 | Naphthalene | 4.87 |
| Butane | 4.87 | Propylene | 4.87 |
| Butylene | 4.87 | Toluene | 4.87 |
| Butadiene | 4.87 | Xylene | 4.87 |
| Ethylene | 4.87 | | |

[a]In respect to those taxable chemicals that are normally in a gaseous state, Section 4662(a)(4) defines the term "ton" as that amount of gas in cubic feet which is equivalent to 2,000 pounds on a molecular weight basis.

*SOURCE:* Section 4661(b).

Methane and butane are not taxable when they are used as a fuel.[9] In contrast, the tax on inorganic chemicals and heavy metals is imposed at a variety of rates, as summarized in Table 10-2.

## Table 10-2
## Tax on Inorganic Chemicals and Heavy Metals

| Chemical | Tax/Ton* | Chemical | Tax/Ton* |
|---|---|---|---|
| Ammonia | $2.64 | Hydrochloric acid | $0.29 |
| Antimony | 4.45 | Hydrogen fluoride | 4.23 |
| Antimony trioxide | 3.75 | Lead oxide | 4.14 |
| Arsenic | 4.45 | Mercury | 4.45 |
| Arsenic trioxide | 3.41 | Nickel | 4.45 |
| Barium sulfide | 2.30 | Phosphorous | 4.45 |
| Bromine | 4.45 | Stannous chloride | 2.85 |
| Cadmium | 4.45 | Stannic chloride | 2.12 |
| Chlorine | 2.70 | Zinc chloride | 2.22 |
| Chromium | 4.45 | Zinc sulfate | 1.90 |
| Chromite | 1.52 | Potassium hydroxide | 0.22 |
| Potassium dichromate | 1.69 | Sodium hydroxide | 0.28 |
| Sodium dichromate | 1.87 | Sulfuric acid | 0.26 |
| Cobalt | 4.45 | Nitric acid | 0.24 |
| Cupric sulfate | 1.87 | | |
| Cupric oxide | 3.59 | | |
| Cuprous oxide | 3.97 | | |

*In respect to those taxable chemicals that are normally in a gaseous state, Section 4662(a)(4) defines the term "ton" as that amount of gas in cubic feet which is equivalent to 2,000 pounds on a molecular weight basis.

*SOURCE:* Section 4661(b).

In the case of the sale of a fraction of a ton, a *pro rata* portion of the tax is exacted from the vendor.[10] In summary, the environmental excise tax on chemicals applies only to those chemicals listed above and at the rates specified in the two tables above.

The tax is imposed either because taxable chemicals have been manufactured or produced in the United States or where taxable chemicals have been "entered into" the United States for consumption, use, or warehousing.[11] The tax is actually imposed when a taxable chemical is sold by its producer, manufacturer, or importer.[12] In addition, use of a chemical by its producer, manufacturer, or importer is treated as a sale,[13] except where methane or butane is used as a fuel.

In addition to the exclusion of methane or butane used as a fuel, selected chemicals used for fertilizer production, sulfuric acid resulting from air pollution control, and chemicals derived from coal are tax exempt. When ammonia, methane, nitric acid, or sulfuric acid are used by their producer, manufacturer, or importer to produce or manufacture fertilizer, then these chemicals so used are exempt from the Section 4661 tax on chemicals. This exclusion from taxation is also applicable both where such chemicals are sold for use by the purchaser in the production of fertilizer and where such chemicals are sold for resale to a second purchaser who will use the substances in the production of fertilizer.

There is also an exclusion for sulfuric acid produced *solely* as a by-product of air pollution control. The production of such sulfuric acid must occur on the site upon which all the air pollution control equipment is located to qualify for tax-exempt status.[14]

For purposes of the chemical tax, the term "United States" (often abbreviated "U.S.") encompasses the areas as specified for purposes of the petroleum excise tax.[15] Thus, there would seem to be no opportunity for a tax avoidance scheme where otherwise taxable chemicals are imported for processing in a foreign trade zone before exporting the finished product to a foreign country for sale or consumption. Conversely, the tax would apply to chemicals processed in "border" plants on the Mexican-U.S. border and then returned to the United States.

Double taxation could be even more of a problem for the tax on chemicals as compared to the tax on petroleum products and crude oil insofar as several of the taxable chemicals are used in processes to produce other taxable chemicals. However, the statute is quite clear as to either the creditability or refund of taxes paid on chemicals used to produce either another taxable chemical or a fertilizer. The Treasury regulations prescribing the refund or crediting of such taxes when available in late 1982 should provide detailed procedures for ensuring proper refunding or creditability for tax previously paid on chemicals.

An amount equal to the tax previously paid is allowed as a credit or refund (without interest) to any person using a taxable chemical to produce either another taxable chemical or a fertilizer in the same manner as if the tax previously paid were an overpayment of tax imposed under Section 4661. In addition, the amount of any such credit or refund may not exceed the amount of tax imposed by Section 4661 on the other taxable chemical being produced.[16] Finally, subsequent purchasers of tax-paid chemicals for use in producing fertilizer should be able to obtain credits for or refunds of the tax paid on those chemicals that qualify for exemption when used to produce a fertilizer.[17]

## Sunset Provisions for Termination of Environmental Excise Taxes

Unless Congress decides to extend the taxes on crude oil and chemicals, both environmental excise taxes are due to terminate at the latest on September 30, 1985. Moreover, no Section 4611 tax will be imposed during the first calendar year beginning after 1983 or 1984 if the unobligated balance in the Hazardous Substance Response Trust Fund exceeds $900 million on either September 30, 1983, or September 30, 1984, and if the

Treasury Secretary after consulting with the EPA Administrator determines that the unobligated balance will be greater than $500 million on September 30 of the following year (1984 or 1985) if no Section 4611 taxes were imposed during the year following.[18] If the Section 4611 tax on crude oil and petroleum products has been terminated earlier than September 30, 1985 (on December 31 of either 1983 or 1984), then the Section 4661 tax on chemicals would also be terminated at that pre-September 30, 1985, date when the Section 4611 tax was terminated.[19] Alternatively, the Section 4611 tax on chemicals would terminate if and when the sum of the amounts received under Sections 4661 and 4611 total $1,380 million at any time before September 30, 1985. The Internal Revenue Service is responsible both for estimating when the $1,380 million level is reached and for promulgating regulations that provide procedures for the early termination of both taxes.[20]

### Administrative Requirements

No special administrative requirements are applicable to either the crude oil tax or the chemicals tax. Generally, the existing administrative provisions applicable to excise taxes are also pertinent to both environmental excise taxes. Therefore, the IRS has its full range of administrative and compliance powers available for administering and collecting the taxes, including the penalties for noncompliance with depositary and reporting requirements.[21]

Depositary requirements for one or both of the environmental excise taxes are determined under the general rules for determining if excise tax deposits must be made on a semimonthly, monthly, or quarterly basis. Quarterly reporting requirements include the filing of a Form 720 Quarterly Excise Tax Return along with a Form 6627 Environmental Tax Return and those supporting forms required by the IRS for reporting other excise taxes. The first quarterly filing due date was July 31, 1981, for the second quarter of the 1981 calendar year.[22] However, temporary regulations issued on July 22, 1981,[23] provide for a deferral of the deposit of taxes payable for the calendar quarter ending June 30, 1981.[24] In summary, the temporary regulations in T.D. 7782 establish collection procedures that are similar to those currently in force for the collection of the Manufacturers Excise Taxes.

### Tax on Hazardous Wastes

Although it is not effective until October 1, 1983, a tax of $2.13 per dry weight ton on "hazardous wastes" received by a "qualified hazardous waste disposal facility" must necessarily be described as a postscript to the taxes on crude oil and chemicals because all three exactions are authorized by

CERCLA as sources of revenue to fund $1,380 million of the $1.6 billion Hazardous Substance Response Fund.[25] The tax, which is imposed on the owner or operator of the hazardous waste-disposal site,[26] does not apply to those hazardous wastes that will not remain at the waste-disposal facility after it is closed.[27] Finally, the hazardous waste tax is to terminate either on September 30, 1985, or earlier if specified funding levels for the Hazardous Substance Response Fund are reached before September 30, 1985.[28] If funding levels were reached before October 1, 1983, the hazardous waste tax would not be imposed without congressional action.

This brief explanation of the hazardous waste tax is sufficient at this time for at least two reasons. First, the tax may never be imposed if the other two environmental taxes generate sufficient funds to reach specified maximum levels before October 1, 1983. Second, further explanation of the two terms most important to an understanding of this third environmental tax, that is, definitions for hazardous waste[29] and qualified hazardous waste-disposal facility,[30] require a rudimentary understanding of the Solid Waste Disposal Act wherein such terms are defined.[31]

## Some Issues to Be Resolved in the Regulations

The temporary regulations for the Superfund taxes on crude oil and chemicals were expected in late 1982. Consequently, the author did not have the benefit of such regulations in preparing the above analysis of the Superfund. What follows is a list of some of the issues that the regulations should resolve:

(1) Crediting procedures for refineries that receive crude oil or petroleum products consisting of a mixture of both domestic and imported oil or petroleum products.

(2) Documentation necessary to substantiate a credit or refund for Section 4661 taxes on chemicals used as a feedstock for a second taxable chemical.

(3) Procedure for determining the tax on the production or importation of a mixture of taxable and nontaxable chemicals, especially where the taxable portion is *de minimus*.

(4) Treatment of *de minimus* amounts of taxable chemicals, for example, benzene, butadiene, xylene, and butane, processed or blended with gasoline (or other fuels) produced in a refinery to enhance the performance of the gasoline (or other fuels).

The answer to the fourth issue may have been provided by Congressmen Al Ullman (Dem.-Ore.) and Barber Conable (Rep.-N.Y.) and Senator Russel B. Long (Dem.-La.) in *The Congressional Record* in mid-December 1980. A copy of their joint statement is provided below in lieu of future regulations resolving this issue.

3. A question has also arisen regarding the application of the excise tax on chemicals to gasoline and other refinery products. As I understand the petroleum industry, oil which is received at a refinery is first distilled to produce a light hydrocarbon stream which eventually is blended with other products and sold as gasoline. The products from the distillation process may then be further processed in a catalytic cracker or in a reformer. The cracking and reformation processes produce, in part, mixed streams of light hydrocarbons which are blended with distillates and sold as gasoline. The light hydrocarbon streams produced in the distillation, cracking, or reformation processes may contain various amounts of substances, such as benzene, butadiene and xylene, which are listed as taxable chemicals in title II of H.R. 7020. We further understand that before any of these substances can be used for chemical processes such as the production of pesticides, or other hazardous substances they must be isolated from the hydrocarbon stream. The question which has arisen is whether the Ways and Means Committee in drafting the original superfund tax intended to tax specified chemicals present in the gasoline refining process if those chemicals are not isolated for chemical use but remain instead in the gasoline.

The tax in H.R. 7020 is imposed upon the sale or use of any taxable chemical by the manufacturer, producer or importer of that taxable chemical; the relevant language is contained in both the House and Senate versions of the bill. If the process of producing gasoline and of isolating particular substances is as it has been described to us, then we would say that the Ways and Means Committee did not intend to tax named substances present in mixed hydrocarbon streams used to produce gasoline. When a refinery operates a distillation, cracking, or reformation process for the purpose of obtaining a light hydrocarbon stream that will be sold as gasoline it cannot be said to be engaged in the manufacture or production of any taxable chemical. In contrast, if that gasoline is not sold but is further processed to isolate one or more taxable chemicals then the refiner is the producer of those chemicals and the excise tax on chemicals would be imposed.

It should also be noted that although an importer of gasoline is not importing a named taxable chemical it is importing a petroleum product and would be subject to the excise tax on petroleum provided for in H.R. 7020.

4. The superfund legislation generally imposes a tax under section 4661 of the Code on the sale or use of certain chemicals. However, section 4662(b)(1) states that methane or butane are treated as a taxable chemical only if "used otherwise than as a fuel." These words appear in both the House and Senate versions of H.R. 7020.

It has been indicated that one important way in which butane is used is as a component of gasoline and other fuels produced in an oil refinery. As we understand it, butane is dissolved (without any change in its chemical composition) into the gasoline to enhance the overall performance of the gasoline, which is, of course, used as a fuel. The specific question which arises is whether, in this case, the butane has been used "otherwise than as a fuel" and thus, whether such use is subject to tax.

If the facts with respect to gasoline production are as represented above— that the chemical composition of the butane does not change and that it is merely dissolved in or mixed with other substances, all of which are used as a fuel, then the Congress did not intend that this butane should be subject to the tax. If, in some other situation, the butane is repeatedly mixed with other substances in an extended sequence of mixtures or its chemical composition is changed, then it is intended that it be "used otherwise than as a fuel," and thus, subject to the tax at that point.[32]

## Hazardous Substance Response Trust Fund

The purpose of the Hazardous Substance Response Trust Fund is to provide for the costs of remedial actions necessitated by hazardous waste releases. The fund is to be comprised of the tax revenues from Subtitle A, that is, general revenue from income taxes, the sums collected from violations of the Clean Water Act,[33] environmental excise taxes, penalties and punitive damages under the Superfund Act, and those amounts recovered on behalf of this fund.

In the event of a release or just the threat of a release of hazardous substances into the environment, the fund shall make expenditures to cover response costs, those claims not satisfied under the Clean Water Act, and injuries to natural resources, and to cover other remedial (clean-up) costs incurred. The Secretary of the Treasury is the trustee of the fund. As trustee, it is his duty to make a report to Congress on or after September 30 of each year beginning in 1981 on the financial condition of the fund and what occurred during the previous year. The Secretary is also under a duty to invest the excess funds in public debt securities so that the monies collected will not be idle until used for hazardous waste clean-up.

## Environmental Excise Taxes and Pollution Taxation

In theory, the Section 4661 tax should be imposed on all chemicals that are toxic when disposed of in the environment, but the Section 4661 definition of taxable chemicals does not include all potentially hazardous substances. Moreover, there is no apparent economic reason for taxing specific inorganic chemicals at different rates insofar as neither the relative cost of clean-up nor the location or amount of each type of toxic chemical is known.

Similarly, crude oil and petroleum products should be taxed when produced or imported. The potential for environmental damage exists at the time oil enters the biosphere. When imported oil enters U.S. waters, the possibility of environmental damage exists. Consequently, the Section 4611 tax on oil and petroleum products should be imposed upon entering U.S. waters or when oil is produced, so that the damage of oil spills will be correlated with domestic production and importation. If the Congress decides to extend the Section 4611, 4661, and 4681 taxes in 1985, imposition of such environmental excise taxes should more closely reflect the potential environmental cost of toxic substances.

## Notes

1. Pub.L. 96-510, December 11, 1980.
2. Section 4611(b) and 4611(c). "Premises" is defined in Section 4612(a)(7) to have the same meaning as when used for purposes of determining gross income from property under Section 613.
3. H. Rept. 96-1016, Part II, p. 6.
4. Section 4612(a)(6).
5. H. Rept. 96-1016, Part II, p. 7.
6. Section 4612(a)(4).
7. H. Rept. 96-1016, Part II, pp. 6-7.
8. Section 4612(b).
9. Section 4662(b)(1).
10. Section 4662(a)(5).
11. Section 4662(a)(1) and H. Rept. 96-1016, Part II, p. 8, which provides that "entered into" does not occur in case of fuel stores on a ship or the casual use of items such as fuel in an auto's gas tank during a visit to the United States.
12. Section 4-61(a).
13. Section 4662(c).
14. Section 4662(b). Section 301 of CERCLA, Pub.L. 96-510, December 11, 1980, requires the President to submit a report to Congress by December 11, 1984, wherein the following issues will be addressed: (1) What would be the economic impact of taxing coal-derived substances and recycled metals? (2) Should other chemicals used in making fertilizers be excluded from taxation just as ammonia, methane, nitric acid, and sulfuric acid are? and (3) What other changes in the coverage of the environmental excise tax on chemicals would benefit the environment?
15. Section 4662(a)(2).
16. Section 4662(d).
17. H. Rept. 96-1016, Part II, p. 8.
18. Section 4611(d) and Section 4661(c).
19. Section 4661(c).
20. Section 303, Pub.L. 96-510, December 11, 1980.
21. H. Rept. No. 96-1016, Part II, pp. 7 and 9.
22. Announcement 81-59, IRB 1980-12, March 23, 1980, at 55.
23. T.D. 7782, 46 *Fed. Reg.* 140 at 37,631, July 22, 1981.

24. Temp. Reg. §57.6302(c)-1(a)(4).

25. Sections 4681, 4661, and 4611.

26. Section 4682(b).

27. Section 4682(c).

28. Section 303, Pub.L. 96-510, December 11, 1980, and Section 4681(d).

29. See Sections 3001, 3002, and 3004 of the Solid Waste Disposal Act; 42 U.S.C. 6921, 6922, and 6924.

30. See Section 3005 of the Solid Waste Disposal Act; 42 U.S.C. 6925.

31. Section 4682(a).

32. 126 Cong. Rec H12,400 (daily edition, December 12, 1980) (remarks of Rep. Ullman and Rep. Conable); and 126 Cong. Rec. S16,681 (daily edition, December 16, 1980) (remarks of Sen. Long).

33. Federal Water Pollution Control Act, 33 U.S.C. §1321 (1980).

# 11

# A Comparative Analysis of Pollution Control Policy

As explained in Chapters 1 and 2, the six countries included in this study exhibit significant similarities as well as material differences. Therefore, a comparison of each country's response to the Environmental Crisis should prove helpful in the ultimate determination of an optimal mix for regulation, subsidization, and taxation in the pollution control policy of any developed country. In reviewing the vital statistics summarized in Table 11-1, one can see that there are material similarities and differences in the social, economic, and political systems of the major countries of North America and Western Europe.

Even though the population density varies significantly among the six countries, all are very urban, with probably no more than 30 percent of the population of any one of the nations residing in nonurban areas. The per capita income in five countries is comparable, but with the British experiencing a level of income that is significantly lower than that generated in Sweden and West Germany. In addition, the overwhelming majority of the population of each country speaks the same language and shares the same cultural values, although a significant number of Americans (Spanish) and Canadians (French) primarily speak a Romance language rather than English.

There are two types of political and legal traditions in the six countries included in this study. One-half of the nations are Federal States (Canada, West Germany, and the United States), while the other three have a strong central government. In addition, one-half rely heavily on the common law, whereas the other three are civil law countries. The common law is an Anglo-American concept of jurisprudence wherein the courts have traditionally had a great deal of authority to review legislative and administrative law. Therefore, judicial or case law is an important component of the body of law in Canada, the United Kingdom, and the United States.

## Table 11-1
## Comparison of Vital Statistics, 1977–1978

| Statistic | Canada | France | West Germany | Sweden | United Kingdom | United States |
|---|---|---|---|---|---|---|
| Urban population | 75.5% | 73.0% | a | 82.7% | 75.2% | 73.5% |
| Population[b] (Millions) | 24[c] | 53 | 62 | 8.3[d] | 56 | 220 |
| Area/KM$^2$ (Thousands) | 9,972 | 551.7 | 248.6 | 448.1 | 244 | 9,363[e] |
| GNP/Capita* (1978/US$) | 8,574 (10,580) | 7,150 (12,140) | 10,416 (13,310) | 10,528[f] (14,760) | 4,360 (9,340) | 7,860 (11,360) |
| Political system | Federal-Provincial/ Parliamentary | Central/ National Assembly[g] | Federal-State/ Parliamentary[g] | Constitutional Monarchy/ Parliamentary | Constitutional Monarchy/ Parliamentary[g] | Federal-State/ Congress |
| Language | English[h] | French | German | Swedish[i] | English | English[j] |
| Legal tradition | Common[h] | Civil | Civil | Civil | Common[k] | Common |

[a]Not available, but West Germany is more urbanized than France, with one-fourth of the population living in the Ruhr and another one-eighth in Berlin, Hamburg, Munich, Cologne, Frankfort, Dortmund, and Stuttgart.

[b]Less than 15 percent of each country's population is considered non-European with the possible exception of the United States where Mexican-Americans may be considered both European and Indian (North America).

**Table 11–1 (continued)**
**Comparison of Vital Statistics, 1977–1978**

cThe overwhelming majority of the population resides within 100 to 200 miles of the Canadian-U.S. border in urban areas such as Toronto, Montreal, and Vancouver.

dMost Swedes live in or near the three major metropolitan areas of Stockholm, Gotenberg, and Malmo.

eIncluding the population and area of Alaska and Hawaii.

fFor Sweden, the calculation is based on gross domestic product (GDP) rather than gross national product (GNP) for 1977.

gFrance, West Germany, and the United Kingdom are also members of the European Economic Community. Therefore, they are subject to decisions of the European Commission, Council of Ministers, and Parliament, which was popularly elected for the first time in the summer of 1979.

hBoth English and French are official, although the majority speak only English; both French language and French civil law prevail in Quebec.

iThe Swedish educational system requires proficiency in a second language. Hence, most Swedes can speak English, German, or French in addition to their native language.

jA significant and rapidly growing segment of the population speaks Spanish because of their Hispanic (Mexico, Puerto Rico, and Cuba) cultural heritage.

kThere can be no judicial review of Acts of Parliament.

SOURCES: *Demographic Yearbook 1978* (New York: United Nationa, 1979); and U.S. Central Intelligence Agency, *National Basic Intelligence Factbook* (Washington, D.C.: U.S. Government Printing Office, 1979).

*The amounts in parentheses are per capita GDP amounts for 1980 as compiled by the OECD and reported in the *OECD Observer*, March, 1982, pp. 26–27. The material changes reflect exchange rate fluctuations more than they reflect variations in economic growth in 1979–80, except that United Kingdom's North Sea oil and gas income has temporarily stimulated economic growth. Thus, the recent strength of the American dollar would deflate all amounts except that for the United States.

Inasmuch as this study is ultimately a comparative analysis of the use of tax policy to control pollution, knowledge of the structure of the six national fiscal systems is a prerequisite for any evaluation of the efficacy of tax incentives for pollution control and pollution taxation. Upon analyzing the summary in Table 11-2, one can see that taxes amount to 31 to 41 percent of the gross domestic product in each of the countries except Sweden. Over half of the tax revenue of each country is raised by the central or Federal government. In the Anglo-American countries, the income tax is the major source of revenue for the national government, with the employment tax, the Canadian Federal sales tax, and the British value added tax (VAT) also generating considerable revenue. In France, West Germany, and Sweden, the employment tax is the major revenue source, while the VAT, income tax, and excise taxes are also productive sources of revenue. In all six countries, the major type of national government expenditure, which encompasses environmental regulation to the extent that the public sector spending thereon relates to public health, is categorized as health, education, and welfare. The relative importance of other types of expenditures, such as general public services, economic services, and defense, varies significantly among countries.

## Pollution Control Policy

Having delineated the social, economic, political, and fiscal systems of the six nations under study, our attention now returns to their pollution control policy. In Chapters 3 through 8, the objective has been to provide a perspective on the public's specific response to the Environmental Crisis through legislation and government programs. Each summary of a country's pollution control policy was begun with an explanation of the significant pollution problems experienced by that country, with special emphasis on the unique problems that have materially affected the country's design of its pollution control apparatus. Then for each of the six countries there has been an analysis of government legislation and policies to determine the extent to which concern for the environment has been translated into specific regulatory and taxation programs. Furthermore, there has been an examination of the subsidization or government assistance—direct grants and tax expenditures—made available to polluters with Chapter 9 focusing on the U.S. taxation of pollution control investments. Finally, Chapter 10 provides an explanation of the new U.S. tax on hazardous substances.

Subsidies are allegedly provided only to alleviate inequities in the incidence of pollution control costs so that each nation complies with the spirit of the "polluter pays principle." This principle provides that, while some public funds may be used to subsidize the private sector's pollution control costs, the major portion of such costs should be paid for through increases in the prices of goods and services in accordance with a producer's use of environmental assimilative capacity (EAC) and/or through lower profits or wages in highly polluting industries.[1] Consequently, although all six nations have adopted the notion that the polluter should pay for his, her, or its own use of EAC, each country has implemented that principle in its own way for legitimate social, economic, and political reasons.

The response to the Environmental Crisis has been the same in all six nations in that all have relied heavily on regulatory programs to correct the market failure that results from the nonexistence of private property rights in air and water. Although the inefficient and inequitable use of common property resources, such as air and water, can also be controlled through reliance on markets, legislative bodies seem predisposed to enact laws that regulate the behavior of natural and legal persons rather than providing for public utilities that sell air and water, subsidies to reward persons for not polluting, or taxes on the pollution of air and water. In addition to this natural legislative bias for regulation, concern in the international economic community for the trade and economic growth implications of subsidization of polluters prompted the OECD, the major international economic policy organization in the noncommunist world, to establish the "polluter pays principle" for its members in 1972. Thus, the OECD sanctions both regulation and taxation, while it severely limits the subsidization of polluters.

Although all six countries rely heavily on the regulatory approach in managing environmental quality, their manner of implementation differs significantly. The most obvious source of such differences ensues from each country's enabling legislation, which reflects not only the structure and dynamics of their domestic political systems, but also their legal tradition and history. Thus, many differences in administrative procedures are predictable because each country's enabling legislation is unique. However, other more subtle variations in public administration may probably be attributed more to national style. In other words, one would expect the Americans, Canadians, French, Germans, Swedes, and British to operationalize the same set of regulatory rules of law somewhat differently, even if there were no differences in their respective political and economic systems because of each society's unique combination of cultural values. The following comparison of the countries' pollution control policies begins with a commentary on the use of environmental planning; includes an evaluation of both the water and air pollution abatement programs of each country; and concludes with an analysis of each nation's program for disposal of solid, hazardous, and/or toxic wastes.

## Environmental Planning

Land use planning has traditionally been the responsibility of local government. Nonetheless, the realization that selected location decisions by industrial, commercial, and public institutions can have a significant

**Table 11-2**

**Comparison of Fiscal Systems**

| | Canada | France | West Germany | Sweden | United Kingdom | United States |
|---|---|---|---|---|---|---|
| Taxes/GDP[a] | 31 | 41 | 37 | 50 | 34 | 31 |
| Central government taxes[b] | 50–59.9 | over 95.0 | 60–69.9 | 60–69.9 | 80–89.9 | 60–69.9 |
| *Revenue Source*[c] | | | | | | |
| Income tax | 54.5 | 18.0 | 19.4 | 21.7 | 42.7 | 55.6 |
| Employment tax | 10.9 | 41.4 | 51.4 | 34.7 | 18.4 | 28.3 |
| Value added tax | 10.4[d] | 24.9 | 13.2 | 15.5 | 8.6 | 0.0 |
| Excise tax | 5.4 | 8.3 | 11.2 | 13.7 | 6.1 | 5.4 |
| Customs duties | 7.3 | 0.01 | 0.03 | 1.3 | 9.6 | 2.0 |
| Other taxes | 0.2 | 1.8 | 0.6 | 1.3 | 1.8 | 1.7 |
| Nontax revenue | 11.3 | 5.5 | 4.1 | 11.8 | 12.8 | 7.0 |
| Corporate | | | | | | |
| Income tax | 14.2 | 6.2 | 3.5 | 1.0 | 4.7 | 13.3 |
| Individual | | | | | | |
| Income tax | 39.1 | 11.8 | 15.7 | 20.6 | 38.0 | 42.3 |
| *Expenditure Function*[e] | | | | | | |
| General public services | 9.7 | 4.0[f] | 4.8 | 10.0 | 7.8 | 3.9 |
| HEW[g] | 47.2 | 65.7[f] | 69.5 | 64.6 | 38.7 | 49.2 |

**Table 11-2** (continued)
**Comparison of Fiscal Systems**

| | Canada | France | West Germany | Sweden | United Kingdom | United States |
|---|---|---|---|---|---|---|
| Economic services[h] | 16.8 | 8.2[f] | 8.1 | 11.6 | 9.9 | 9.4 |
| Defense | 8.0 | 8.0[f] | 10.1 | 9.9 | 14.1 | 22.2 |
| Other functions | 18.3 | 14.1[f] | 7.5 | 3.9 | 29.5 | 15.3 |

[a]Taxes as a percentage of gross domestic product for 1979.

[b]Central government tax revenue as a percentage of total revenue for 1979, except in the case of the United States for which the percentage is based on 1977 data.

[c]Revenue by source as a percentage of total revenues for 1976.

[d]Canada imposes a Federal sales tax rather than value added tax.

[e]Expenditure by function as a percentage of total expenditures for 1976.

[f]France's expenditure analysis is for 1975 rather than 1976.

[g]HEW is an acronym for health, education, and welfare.

[h]Economic services includes road and other transportation; communications; agriculture, forestry, fishing, and hunting; and general administration and regulation.

SOURCES: *Revenue Statistics of OECD Member Countries, 1965–80: A Standardized Classification* (Paris: Organization for Economic Cooperation and Development, 1981); and *Government Finance Statistics Yearbook: Volume II/1979* (Washington, D.C.: International Monetary Fund, 1980).

impact on environmental quality beyond the jurisdiction of any local government wherein a facility is located has precipitated the enactment of laws that provide for an assessment of the environmental impact of major projects (EIS). The assessment is made by national, regional, State, or provincial authorities in an effort to mitigate the potential for environmental damage during the planning stages of an industrial, commercial, or public project. There may also be a public hearing where individuals and groups voice their support of or objection to a project before construction begins. Therefore, the environmental impact assessment survey or statement procedure can be more democratic than a regular bureaucratic review is.

Ideally, a national or regional environmental (or physical) planning system would delineate the long-term, preferred use of a country's land and water resources through a political process wherein both social and economic values are considered and conflicts are resolved through explicit tradeoffs or optimization. Such an ideal planning system should only have to be compromised or amended on those rare occasions in the future when either the overriding national interest or substantial local objection would be better served. In other words, environmental (or comprehensive land use) planning is always limited by both national or regional criteria and local interests, neither of which remains immutable in the long run. Sweden's system of national physical planning comes very close to satisfying the requirements of the ideal environmental planning system outlined above.

In reviewing the summary presented in Table 11-3, one can also see a material difference between the structure of environmental planning in Federal States and the non-Federal States. Canada, West Germany, and the United States have no authority for comprehensive physical planning at the Federal level, although there is an EIS for those projects in which the national government has a valid legal or fiscal interest. In all six nations, local government has traditionally been given the responsibility for land use planning and public health within its jurisdiction. In Canada, West Germany, and the United States, it has been the States or provinces that have relinquished their responsibilities to local government. In France, Sweden, and the United Kingdom, the central government has deferred to local authorities where the central government's legislative body has determined that the local authorities can be most effective.

Table 11-3 demonstrates that the concept of environmental impact assessment is well established in Canada, France, West Germany, Sweden, and the United States, although there is considerable variation in the implementation thereof. In the three Federal States (Canada, West Germany, and the United States), the scope of the EIS process is limited to projects in which the Federal government has a legal or fiscal interest. Sweden and France have no such limits inasmuch as environmental impact review applies to any public or private project that would materially affect the management

**Table 11-3**
**Comparison of Environmental Planning Systems**

| Country | Environmental Impact | National Central or Federal Planning |
|---------|---------------------|--------------------------------------|
| Canada | Yes, Federal Environmental Assessment and Review Process (EARP) since 1972, with less than 30 of over 4,000 projects reviewed requiring an environmental impact statement (EIS) before approval. | No, because the provinces have original jurisdiction over property (and civil) rights, but Federal fiscal power, EARP, and residual environmental protection authority for air, water, and waste (solid/hazardous). |
| West Germany | Yes, since 1974 but only for Federal planning and projects, especially highway construction and nuclear power plants. | No, although Federal licensing enables it to control the air, water, and waste pollution of any new (or altered) facility, the States continue to have original jurisdiction for land-use and for nature and landscape protection, but there is formal Federal-State cooperation on matters of national environmental policy and development planning. |
| United States | Yes, since the 1969 enactment of NEPA, an EIS has been required for any Federal program that materially affects the environment, including the allocation of Federal funds and the issuance of permits and licenses. | No, but the development of statewide planning for air quality, water resources, and waste management mandated by Federal law along with the EIS process, effluent standards for both sources and selected bodies of water, and ambient air quality standards could have the effect of a national environmental planning system. |

**Table 11-3** (continued)
**Comparison of Environmental Planning Systems**

| Country | State or Provincial Planning | Local Planning |
|---|---|---|
| Canada | Yes, to an extent but the provinces have delegated much of the responsibility to local and/or regional government while providing laws and bureaucracies that regulate and coordinate implementation of both land use planning and the management of air and water resources at the local and regional level. | Yes, to the extent that provincial governments have relinquished their responsibility for land use planning, water supply, wastewater treatment, solid waste disposal, and public health. |
| West Germany | Yes, concurrent licensing authority with the Federal government and mandatory water use and land use planning necessitated by Federal "framework" legislation should eventually result in comprehensive planning on a statewide and regional basis. | Yes, inasmuch as the State development programs are supposed to encompass both regional and local development planning, including zoning plans. |
| United States | No, except in a few States (Oregon, Hawaii, and Florida); however, planning requirements of Federal environmental law (air and water quality, and waste management) could effect an implicit, piecemeal environmental planning system in each State. | Yes, through their land use planning authority, municipal (and to some extent county) governments control the physical development within their jurisdiction by relying on zoning and in some cases master planning. |

**Table 11-3**  (continued)
**Comparison of Environmental Planning Systems**

| Country | Environmental Impact | National Central or Federal Planning |
| --- | --- | --- |
| France | Yes, since January 1, 1978, an EIS has been required for all public and private projects that may have a detrimental effect on the inhabitants, resources, or natural surroundings of a proposed location. | Yes, piecemeal through national administration of the regional (Prefect) licensing of "classified installations" in concert with the environmental quality goals established in the five-year national economic plans, through the CIANE, and recently through the environmental impact review process initiated in 1978. |
| Sweden | Yes, since January 1, 1973, the central government has reviewed all proposals for new or altered facilities that would materially affect the management of land, water resources, forests, or energy. | Yes, comprehensive physical planning based on national guidelines for land and water use that are based on *geographical* location and type of *activity* in addition to the more specific limitations inherent in the NEPB/NFB licensing of new (or altered) facilities generating any pollution. |
| United Kingdom | No, although an environmental impact review for any "registrable works" has been proposed by the Royal Commission on Environmental Pollution as a method of monitoring the decisions of local planners. | Yes, not only regional economic development planning (new towns), but also inasmuch as local governments with land use planning responsibilities are required to submit a "structure plan" for "areawide" development to the central government on a continuous basis, so that each change therein can be reviewed. |

**Table 11-3** (continued)
**Comparison of Environmental Planning Systems**

| Country | State or Provincial Planning | Local Planning |
|---|---|---|
| France | N.A. | Yes, to the extent both that Prefects rely on the recommendations of Municipal and Health Councils in issuing permits (or licenses) to "classified installations," and that local government influences the national economic planning process. |
| Sweden | N.A. | Yes, national *activity* and *geographical* guidelines have been designed as an overall limit on the existing system of County (regional administration of cenral government) and local planning inasmuch as the primary responsibility and authority for physical planning has remained at the local level where there is both comprehensive or general planning (master plans) for municipalities and specific planning (district or subdivision plans) within municipalities. |
| United Kingdom | N.A. | Yes, once a county, regional (Scotland), or district government's "structure plan" has been approved, more detailed "action area" (or local) planning for the use and development of land is prepared. Consequently, local planning authorities must approve most forms of construction, resource extraction, land use changes, and regional development sponsored by the central government as well as locally initiated development. |

N.A. = Not applicable.

of natural resources (land, water, forests, or energy) in a public project or that may have a detrimental effect on the inhabitants, resources, or natural surroundings near a proposed private project. Although the United Kingdom is the only one of the six subjects of this study that has not established some type of environmental impact review, the European Economic Commission's proposal to include environmental impact assessment in the Common Market's 1977-81 environmental program should eventually result in adoption of an EIS process in the United Kingdom. Therefore, there should be a convergence of environmental impact assessment policy in France, West Germany, and the United Kingdom—all EEC members. The current EIS proposal was not passed at the June 1982 meeting of EEC environment ministers.[2]

As noted above, Sweden's national physical planning system provides for comprehensive environmental planning by the central government while maintaining the traditional role of the local governments in general and specific planning within their jurisdiction. Although France has implemented no explicit system of national environmental planning except the regional development of provincial cities, environmental quality goals are explicitly delineated in its five-year national economic plans. Moreover, the central government not only controls the location of polluting (classified) installations through the regional (Prefect) licensing thereof, but also reviews those major private and public projects with a detrimental environmental impact.

Like the French national government, the central government in the United Kingdom has relied on regional planning for urban areas both near and far from the capital city of London. However, there is neither national economic planning nor environmental impact review to complement regional development planning. In the Federal States, superior fiscal resources and national environmental policy should enable the Federal governments to effect piecemeal environmental planning at the State or provincial level, while at the same time preserving local government's traditional role in the land use planning.

## Water Pollution Control

Insofar as water resources generally exhibit a finite, though somewhat arbitrary, physical geography, water use (abstraction and discharge) should be viewed as a water resource management problem. Therefore, regulation thereof should be a logical extension of environmental planning. Again, such an ideal may only exist in Sweden because none of the other five countries being studied has implemented comprehensive physical planning at the national level in concert with the traditional land use planning responsibilities of local government.

In reviewing Table 11-4, there are apparently few systematic relation-

## Table 11-4
### Comparison of Water Pollution Control Policy

| Country | Quality Standards | | Licensing or Permit Procedure | Effluent Taxes/Charges | Subsidization (Grants) |
| | Source/Content of Effluents | Ambient or Water Body | | | |
|---|---|---|---|---|---|
| Canada | Yes, for most major industries (private and public) with others in process. | No, except for the Great Lakes and in selected provinces. | Yes, standards or most practicable technology required by the provinces. | No, although authorized by the 1970 Water Act and used by some cities. | No, except for grants for municipal wastewater treatment. |
| West Germany* | Yes, national minimum effluent standards for licensing, which vary for municipalities and industries. | No, except for the *Genossenschaften*. | Yes, with "pollution protection officers" in most facilities. | Yes, with *Genossenschaften* in operation and wastewater charges beginning in 1981. | No, except for grants to water management associations and municipalities. |
| United States | Yes, for over 50 major industries with others in process and for public facilities. | Yes, for selected bodies of water, especially the Great Lakes. | Yes, the NPDES permit program is being implemented by the States with Federal authority in reserve. | No, although the penalty for non-compliance with future effluent standards is marginal cost. | Yes, loan guarantees for small businesses borrowing from States or local authorities and through grants for municipal wastewater treatment. |

*France, West Germany, and the United Kingdom are also subject to the directives on water pollution control and water quality promulgated by the European Economic Community.

**Quality Standards**

| Country | Source/Content of Effluents | Ambient or Water Body | Licensing or Permit Procedure | Effluent Taxes/Charges | Subsidization (Grants) |
|---|---|---|---|---|---|
| France* | No, each case dealt with separately through licensing, but effluent standards are being developed. | No, except for those implicit in the national economic plan. | Yes, for any facility that generates pollution. | Yes, imposed by 6 water basin finance agencies. | Yes, from both the 6 basin finance agencies and the national government. |
| Sweden | No, each case dealt with separately through licensing. | Yes, but only implicitly in the national physical planning system. | Yes, for about 40 types of facilities with notification required of 30 other types. | No. | Yes, but only for a transitional period (mid-1969 through 1975) and for municipal wastewater treatment. |
| United Kingdom* | No, each case dealt with separately through licensing with ambient standards being developed. | No, but water quality objectives should be delineated when the economy improves. | Yes, with issuance by local (or regional) authorities reflecting their attitudes and conditions. | No, but the "trades" have traditionally been encouraged to use (for a charge) public sewage systems. | No, except for nationalized industries that operate at a deficit and/or that receive long-term, low-interest loans. |

*France, West Germany, and the United Kingdom are also subject to the directives on water pollution control and water quality promulgated by the European Economic Community.

ships between the regulatory approaches used by the six subjects of this investigation. Negative generalizations are more obvious than positive. None of our subjects relies explicitly on ambient water quality standards for all bodies of water. Nevertheless, such standards are implicit in the Swedish physical planning and French economic planning systems, and they exist in some of the regions of Canada, West Germany, the United Kingdom, and the United States.

The Federal states have established source effluent standards for most major industries and public facilities. The French, Swedes, and British rely on a case-by-case evaluation of each facility's potential for water pollution before issuance at the regional level of an operating permit that specifies the technology to be used and effluent standards to be maintained. In the Federal states, the permit process is based on established source effluent standards or on best reasonable-cost technology available. Enforcement is generally carried out by the States or provinces. In West Germany, the polluter may be required by law to hire and financially support a "certified pollution protection officer" who monitors and evaluates the firm's water pollution control activities. Similarly, in the United Kingdom, a polluting installation must include facilities for use by a government pollution inspector. In all countries except the United Kingdom, there is a subsidy program for construction of municipal wastewater treatment facilities. Finally, a form of effluent tax or charge is being used in several countries. However, a comparative analysis of both this use of tax policy and the use of tax incentives (or implicit subsidization) is deferred to the final chapter.

### Air Pollution Control

Upon an initial review of Table 11-5, one is likely to arrive at the same conclusion about air pollution control policy that one did above for water pollution control. In general, there would appear to be very little similarity between the regulatory approaches utilized by the six nations to effect improvements in air quality, except that all have adopted emission standards for a major mobile source of air pollution—new motor vehicles. Otherwise, air pollution control policy in a specific country is similar to water pollution control policy in that country. In France, Sweden, and the United Kingdom, there is heavy reliance again on the case-by-case licensing process for polluting installations. However, the Swedish system includes regulations that specify the amount, control, and reporting of emissions.

Predictably, Federal states have established regulatory systems that are remarkably similar in principle, although they differ somewhat in practice. In Canada, West Germany, and the United States, there are both ambient air quality standards and stationary-source emission standards, with the States and provinces primarily responsible for enforcement of the stationary-

**Table 11-5**
**Comparison of Air Pollution Control Policy**

| Country | Emission Standards | | Limits on Fuel |
| | *Stationary Sources* | *Mobile Sources* | Composition |
|---|---|---|---|
| Canada | Yes, both guidelines for most industries drafted by the Federal government to aid provinces and Federal regulations for Pb, Hg, asbestos, and so on. | Yes, since 1970 for new motor vehicles in respect to CO, hydrocarbons, and particulates. | Yes, Pb in gasoline. |
| France[a] | No, except in "special protection zones," for each case is dealt with separately through licensing. | Yes, for new and used motor vehicles EEC limits on hydrocarbons, particulates, and CO. | Yes, sulfur content of gas oil and home heating oil and Pb in gasoline. |
| West Germany[a] | Yes, for those facilities that must obtain permits and best technology for others. | Yes, 1980 models must emit 10 percent (or less) of 1969 levels of hydrocarbons, CO, and particulates. | Yes, sulfur content of fuel oil and diesel oil and Pb in gasoline. |
| Sweden | Yes, since 1970 regulations have specified amount, control, and reporting of emissions. | Yes, motor vehicles manufactured after 1968 and after 1974 adopted U.S. standards. | Yes, sulfur content of fuel oil. |
| United Kingdom[a] | No, for over 60 major sources of emissions either limits or "best practicable means." | Yes, for motor vehicles manufactured for model years 1973 and later. | Yes, sulfur content of fuel oil and Pb in gasoline. |
| United States | Yes, for over 25 major sources of emissions and 5 types of hazardous pollutants. | Yes, for new motor vehicles manufactured for post-1975 standards established and periodically revised. | Yes, Pb in gasoline. |

[a]France, West Germany, and the United Kingdom are also subject to the directives on air pollution control and air quality promulgated by the EEC.

**Table 11-5**   (continued)
**Comparison of Air Pollution Control Policy**

| Country | Ambient Air Quality Standards | Licensing or Permit Procedures | Subsidization Grants |
|---------|-------------------------------|--------------------------------|----------------------|
| Canada | Yes, for 5 pollutants desirable, acceptable, and tolerable levels. | No, except in provinces where licensing has been established. | No, except in some provinces, for example, Ontario's loans for small business. |
| France* | No, except for "special protection zones" at Paris, Lyon, and Lille and for "alert zones" in other areas. | Yes, for any facility that generates pollution. | Yes, low-interest loans and in the future proposed air agencies that make loans and grants. |
| West Germany* | Yes, with each State being responsible for measuring levels and planning improvements for ten pollutants. | Yes, standards, "emission control officer," and technical requirements for each facility. | No, except for loans in some States. |
| Sweden | No, except for the ambient level of sulfur dioxide. | Yes, for about 40 types of facilities, with notification required of 30 other types. | Yes, but only for a transitional period between mid-1969 and mid-1975. |
| United Kingdom* | No, except for "smoke control areas" wherein smoke from combustion is regulated. | Yes, at either the national or local level, depending on the type of facility. | No, except for nationalized firms that have a deficit or that receive low-interest loans. |
| United States | Yes, primary and secondary for 6 pollutants with non-degradation for permits and State plans. | Yes, from either a State with an approved implementation plan or the EPA. | No, except loan guarantees for small business borrowing from State or local authorities. |

*France, West Germany, and the United Kingdom are also subject to the directives on air pollution control and air quality promulgated by the EEC.

source standards and planning for meeting or maintaining the former. In addition, both the Germans and the Americans have included a State licensing process with their system of standards. Finally, all six countries are regulating the composition of household and/or motor vehicle fuel in respect to sulfur and/or lead content.

## Solid Waste Pollution Control

Traditionally, the responsibility and strategy for handling solid and hazardous waste has been left to the local governments concerned about public health. Consequently, wastes have been disposed of without much consideration of recycling. In reviewing Table 11-6, the only other generalization that can be made about waste disposal in the six countries concerns the problem of toxic substances. All six countries have a national law that provides for some degree regulation of the production, use, and disposal of hazardous or toxic substances. However, harmonization of such regulation must still be accomplished. Over the past few years, the OECD has been the forum for effecting harmonization of toxic substances regulation.[3]

One pattern that continues is the comparability in the response of the Federal states to a pollution problem — the waste-disposal problem. Canada, West Germany, and the United States have established a sharing arrangement for the regulation of the disposal of solid and hazardous waste wherein the Federal government specifies guidelines or standards and the State or province enforces them. In addition, there is a State licensing procedure in West Germany with permits issued by districts. Finally, several provinces and American states have implemented licensing procedures.

In the non-Federal states, a permit or license is required for disposal of all wastes regardless of who generated them. In France and Sweden, the license is issued by a regional branch of the central government, whereas the British local government is authorized to issue waste-disposal permits. As noted above, local governments have traditionally been preoccupied with the public health implications of their waste-disposal responsibilities. Today the strategy in France, West Germany, Sweden, and the United States is similar in its emphasis on waste reduction and materials recovery (or recycling). Therefore, the national governments provide financial and technical assistance to their local governments for programs to effect waste reduction and/or materials recovery.

### Table 11-6
### Comparison of Solid and Hazardous Waste Control Policy

| Country | Solid Waste | | Hazardous Waste |
| | Responsibility | Strategy | |
| --- | --- | --- | --- |
| Canada | Local and industrial with provincial regulation that varies somewhat and with regional management in Ontario, Quebec, and British Columbia. | Primarily treatment and disposal to protect public health and avoid water pollution. | Federal-provincial sharing of regulation with the Federal controlling toxics. |
| France* | Local with central government regulation through the licensing of waste producers and of waste disposers by the Prefect. | Traditionally, treatment and disposal to protect health and control air and water pollution, but future policy will stress waste recovery and reduction, as well as conservation. | Central government regulation and regional licensing and control of the disposal of hazardous and toxic wastes through the Prefect. |
| West Germany* | Local, district (or intrastate regional), and industrial, with State licensing, enforcement, and planning based on Federal regulatory legislation. | Traditionally, disposal by local government for a service charge and private recycling if supported by the market, but now waste reduction and recycling are stressed. | Federal-State sharing of regulation with the State responsible for disposal and the Federal government enacting toxic waste legislation. |

**Table 11-6** (continued)
**Comparison of Solid and Hazardous Waste Control Policy**

| Country | Solid Waste | | Hazardous Waste |
| | *Responsibility* | *Strategy* | |
| --- | --- | --- | --- |
| Sweden | Local and industrial, with central government licensing or review of the latter entities, waste production and disposal | Traditionally, disposal necessary to protect public health and avoid water pollution, but both waste reduction and materials recovery are emphasized now. | Local regulation with residual powers to the central government through its licensing authority, toxics control law, and joint ventures with private sector through SAKAB and INTERKAB. |
| United Kingdom* | Local with central government goals and guidelines, along with national standards where there are extralocal effects. | Traditional treatment and disposal to protect public health and to control land use within a local jurisdiction. | Local regulation of collection and disposal as well as licensing and planning with central government regulation, especially for extralocal effects. |
| United States | Local with State (or Federal if necessary) regulation, site inspection, and enforcement and Federal guidelines, technical assistance, and financial aid for State planning. | Primarily disposal to protect public health, but State must establish comprehensive solid waste plans wherein resource recovery and conservation are integrated with sanitary disposal. | State regulation (or Federal where necessary) of generation, use, and disposal, along with Federal regulation of use, production, development, testing, and disposal of toxic substances. |

Table 11-6 (continued)
**Comparison of Solid and Hazardous Waste Control Policy**

| Country | Permit or Licensing Procedures | Subsidization | Recycling |
|---------|-------------------------------|---------------|-----------|
| Canada | No, except in some provinces. | Yes, loans and grants by some provinces. | Mandatory for most beverage containers in Alberta, BC, Ontario, and other provinces, but otherwise subject to market conditions. |
| France* | Yes, for facilities that generate waste and for waste-disposal and recovery facilities, with permits issued by the Prefect. | Yes, grants and loans for public and private investments in and the operating expenses of waste-disposal facilities. | Central government authority to ensure that there are no fiscal or legal impediments to the use of recycled materials as an alternative to new materials. |
| West Germany* | Yes, for facilities that generate waste and for waste-disposal and recovery facilities, with permits issued for the State by the district. | Yes, State grants and low-interest loans for regional facilities. | Market conditions support the extensive recycling of autos, tires, and glass beverage containers and waste oil recycled through taxation. |
| Sweden | Yes, for about 40 types of facilities, with notification required of 30 other types. | Yes, for municipalities, while direct aid for industry available only for a transitional period from mid-1969 to mid-1975 and indirect aid through SAKAB/INTERKAB. | Local government must recycle wastepaper by 1980, and the central government effects recovery of junk autos and beverage containers through taxation. |

## Table 11-6   (continued)
## Comparison of Solid and Hazardous Waste Control Policy

| Country | Permit or Licensing Procedures | Subsidization | Recycling |
|---------|-------------------------------|---------------|-----------|
| United Kingdom* | Yes, local government specifies operating conditions and types of waste for each site. | No, except for nationalized industries that operate at a deficit and/or that receive low-interest loans. | Market conditions support extensive recycling of paper and scrap metal, but central government only provides exchange information for other industrial wastes. |
| United States | No, but *de facto* permit requirement is inevitable upon full implementation of the Resource Conservation and Recovery Act (RCRA). Moreover, some States have imposed licensing requirements for hazardous waste dumpsites. | No, but Federal grants to States for costs of compliance with RCRA. | Market conditions support extensive recycling of steel and aluminum, and the RCRA requires State plans wherein materials recovery is integrated with conservation and disposal. |

*France, Germany, and the United Kingdom are also subject to the directives on solid and hazardous waste, as well as toxic substances, promulgated by the European Economic Community.

## Note

1. OECD Environment Committee, *The Polluter Pays Principle*, (Paris: Organization for Economic Cooperation and Development, 1974), pp. 5-7.
2. *International Environment Reporter—Current Report* July 14, 1982, pp. 277-78.
3. Ibid., May, 12, 1982, pp. 167, 172-73, and 183-85.

# 12

# Summary and Conclusions

The natural response of government to public pressure for a solution to the Environmental Crisis has been to regulate polluters, but the regulatory approach to controlling pollution has not developed systemically or in any pattern. Nevertheless, one can make a strong case for convergence inasmuch as the general problem of pollution control is not unique to any one of the six countries investigated here and there is a free flow of information among them through the OECD. The piecemeal, incremental development of regulation has been in response to increasing public pressure. However, the evolution has been constrained by legal tradition—civil or common with judicial review—and the national political system—Federal-State or central government. Traditionally, pollution problems have been handled at the local level when public health effects necessitate their resolution. When pollution problems become too complex, serious, or pervasive to be handled by local government, the regional (State, provincial, or subdivision of the central government) or national (Federal or central) government is forced to take direct control through its own regulatory system and/or indirect control through guidelines and financial assistance for local or regional government. Federal governments have had to rely more on indirect control because of State or provincial resistance. Transitional financial assistance for firms and households is also commonplace, usually in the form of tax incentives.

It may be that the body politic in Canada, France, West Germany, Sweden, the United Kingdom, and the United States has reached the point of diminishing returns in respect to their benefit-cost analysis of increased reliance on regulation. Regulation and prohibition may remain the most cost-effective method in the short term for controlling the production, use, and disposal of toxic substances; the generation and disposal of hazardous wastes; health and safety in the working environment; and the safety of selected consumer products such as food and drugs. Support for this notion is implicit in the development of water pollution taxes or effluent charges in France and West Germany, the excise taxes on beverage containers in various jurisdictions, and the German waste oil tax. Other proposals for use of a market mechanism—pollution taxes or user charges—to effect marginal improvements in air and water quality and reduced production of solid waste are gaining in popularity.

## Use of Direct Grant Subsidization

Direct financial aid or subsidization has not been an important element of the pollution control policy of any of the six nations. Sweden probably established the most unique financial aid program by relying on direct grants to existing industry for a five-year transitional period between 1969 and 1974. These grants assisted the private sector in financing the pollution control investments necessitated by the enactment of the Swedish Environmental Protection Act in 1969. While this financial aid program was extended for one more year for fiscal policy reasons, Swedish financial aid for pollution control is now limited to the funding of industrial research and development of technology. As one can see from reviewing Tables 11-4, 11-5, and 11-6, the other five countries have established a variety of subsidy programs from which the private sector may benefit. Most of the benefits to the private sector are indirect. For example, the subsidization of wastewater treatment facilities is by far the most common form of grant program, with only the United Kingdom not providing some type of financial assistance for construction of municipal and water basin wastewater treatment facilities. As explained above, such subsidy programs are not in violation of the OECD's "polluter pays principle" where they are either temporary or reflect a country's unique social and economic circumstances and where the financial aid does not cause distortions in international trade.

## Use of Tax Incentive Subsidization

The use of tax incentives (implicit subsidization) is also limited by the "polluter pays principle." Along with taxation itself, such tax incentives are considered an element of tax policy for purposes of this study. These two uses of tax policy are diametric in that tax incentive is a "carrot" or positive reinforcement, whereas the taxation is a "stick" or negative reinforcement. One can see, upon even the most cursory review of Table 12-1, that tax incentives, like direct financial aid, have been used sparingly. The tax incentives have been temporary, transitional provisions designed to encourage industry and commerce to voluntarily comply with pollution control regulations in respect to facilities already in existence when emission or

**Table 12-1**

**Comparison of the Use of Tax Incentives for Investment in Pollution Control**

| Country | Special Income Tax Deduction | Investment Credit | Other Tax Incentive |
|---|---|---|---|
| Canada | Yes, rapid amortization for new facilities at establishments in operation before 1974. | Yes | Yes, exemption from the Federal sales tax. |
| France | Yes, rapid amortization for facilities constructed before 1977. | N/A | No |
| West Germany | Yes, rapid amortization for facilities purchased or manufactured before 1979. | N/A | No |
| Sweden | No | Yes | No |
| United Kingdom | Not applicable inasmuch as most equipment qualifies for 100% deduction in year of acquisition. | N/A | No |
| United States | Yes, rapid amortization for facilities added to establishments in operation before 1976. | Yes | Yes, tax-exempt bonds,[a] and many States have provided special deductions, credits, and exemptions in respect to their income, sales and property taxes. (See Table 8–10) |

[a]To the extent that private firms can lower their borrowing costs by financing pollution control facilities from the proceeds of tax-exempt industrial development revenue bonds.

N/A = Not applicable.

411

effluent standards are either enacted or made more stringent. The major exceptions to the preceding generalization are the exemptions from the Canadian Federal sales tax and the various deductions, credits, and exemptions provided by American States.

### Emergence of Pollution Taxation

The summary in Table 12-2 supports the notion that, although taxation of pollution has not been implemented in a "pure" form anywhere in the six countries investigated, "second-best" systems of effluent taxation are operational in West Germany and France. Neither the *Genossenschaften* of the Ruhr Basin nor the Water Basin Finance Agencies of France impose an effluent tax directly based on the damage or social cost of the pollution. The rate structures for both charge systems are determined from the level of expenditures necessary to carry out the water quality management programs of the *Genossenschaften* and the financial aid program for water resource management in the French five-year national economic plan in which Water Basin Finance Agencies fulfill the function of a financial intermediary. However, the new German system of wastewater charges implemented in 1981 will be much more like the ideal pollution tax system wherein exactions are based on the damage of an amount of a pollutant to society, that is, marginal social cost of a specific amount and type of pollution.

Based on both their experience with the *Genossenschaften* and their measurements of the amount of, the composition of, and the cost to clean up wastewater discharges, the Germans have established a proportional rate structure. The rate is scheduled to increase from 1981 (1976 DM 12) to 1986 (1976 DM 40) in anticipation of effecting a significant decrease in water pollution if their calculations have been accurate. The rates were determined by reference to the approximate cost of the average size wastewater treatment plant designed for full biological purification with at least 90 percent efficiency where average cost includes both operating and capital costs. The base for effluent tax is also consistent with the purist's theoretical notion of imposing an exaction according to the damage caused by a specific amount of pollution. As explained in Chapter 5, the base includes suspended solids, oxidizable substances, and toxicity. However, the anticipated use of much of the effluent tax revenue to finance the wastewater treatment facilities of municipalities and water management associations detracts somewhat from the "purity" of the pollution tax system, especially to the extent that polluters can reduce their pollution control costs and where pollution damage is not reimbursed.

The Germans have developed another novel idea for using pollution taxation to improve environmental quality. Since January 1, 1969, there has been a Federal excise tax (DM 7.50 per 100 kg) on lubricants sold in

## Table 12-2
## Comparison of Effluent Charge/Tax Systems

|  | Present Status | Damage Based Taxes | Commentary |
|---|---|---|---|
| Canada | Water Act of 1970 authorizes the establishment of effluent charge or tax systems by water quality management associations to be established for water basins. | Not Known | Charges could be based on the damage of discharges inasmuch as the law does not specify base computation. |
| France | Water Basin Finance Agencies have been operational since January 1969. The 1969 Water Management Act, which established these agencies, is based on the *Genossenschaften*. | No | In practice, charges are based on revenue needs of the Water Basin Finance Agencies as part of the national economic planning goals. |
| West Germany | Waste Water Charges Act of 1976 provides for an effluent charge or tax system that will be operational in 1981 with law taking effect on January 1, 1978. | Yes | *Genossenschaften* in the Ruhr Basin have been operating an effluent charge system based on revenue needs of the association. |
| Sweden | Although as there is a well-developed national land use planning system and a case-by-case licensing process for polluters, the economic penalty has been increased to effect compliance with specifications in discharge permits. | N/A | |
| United Kingdom | The Water Act of 1973 effected a reorganization of water supply and wastewater treatment institutions into a few water authorities with regional jurisdictions and discharge permits related to water basins, but private sector pays cost of its discharges into public facilities. | No | Regional Water Authorities could impose effluent taxes or charges, but at present they charge only for wastewater treatment. |

Table 12-2   (continued)
Comparison of Effluent Charge/Tax Systems

|  | Present Status | Damage Based Taxes | Commentary |
|---|---|---|---|
| United States | Although stationary source effluent standards and ambient water quality guidelines are being used, effluent charges or taxes are still possible in the long run. | N/A | |

N/A = Not applicable.

West Germany for use therein. Inasmuch as the revenues are used to finance the mandatory collection and disposal (or recycling) of waste oil, the purist would argue that the exaction cannot be a pollution tax because it is not determined by reference to the damage or social cost caused by the pollutant. The EEC has promulgated a directive that provides for implementation of a waste oil tax throughout the Common Market. However, other EEC countries have been slow to adopt the German-inspired excise tax. France is imposing a temporary tax on waste oil.

Another excise tax designed to minimize solid waste is only imposed nationally in Sweden. Taxes or deposits on both returnable and nonreturnable beverage containers may be immaterial or low as they are in Sweden. However, a material exaction is more common in the American States and Canadian provinces that impose such excise taxes. In all cases, the tax on returnable containers is refundable when the consumer returns it for reuse.

The Swedes also impose an excise tax or deposit on every new motor vehicle sold in the country for the purpose of encouraging the recycling thereof when the vehicle is inevitably junked. If a Swede finally disposes of his vehicle at a licensed salvage yard, his tax or deposit is refunded and the salvage yard also receives a subsidy to supplement its revenue from the sale of parts and recoverable materials.

There is also an innovative solid waste pollution tax in the United States where the Surface Mining Control and Reclamation Act of 1977 provides for an excise tax on strip-mined coal that is accumulated in a trust fund. Accordingly, the future reclamation of surface mines will be carried out using the trust fund, even if the mining firm is financially unable to do so. Finally, there is a new environmental excise tax on petroleum, chemicals, heavy metals, and hazardous wastes that is explained in Chapter 10.

In conclusion, it seems apparent that, in spite of the alleged superior economic efficiency of pollution taxes, there have been no comprehensive uses of them. Of course, the German effluents charges due to became operational in 1981, and the French system can be adapted to taxation. The EEC Commission has proposed a directive to implement an effluent tax system throughout the Common Market.[1] If the experimental French emission charge system, which is based on the current Dutch system, is successful, the Germans and the rest of the EEC may adopt such a system. The time may come, however, when there will have to be serious consideration in Canada, Sweden, the United Kingdom, and the United States of water and/or air pollution taxation. Certainly, the taxpayer's revolt in the United States and fiscal conservatism in four of the other five industrialized countries (Socialist France excepted) are in large part a reflection of the public's frustration with the recent limited success of regulation and its accompanying bureaucracy in alleviating our societal ills, including pollution. So maybe "the pendulum will swing" in favor of pollution taxation as a *complement* to regulation now that its alleged superior economic efficiency is being put into practice.

## Conclusions

The third research question posed in Chapter 1 relates to how the United States can benefit from the experiences of the other five countries. One method of addressing this issue of transferability involves an extension of, or variation on, the analysis of the political economy of pollution control that is provided at the end of Chapter 2. Inasmuch as the development of pollution control policy is a result of political and economic considerations, the underlying issue is concerned with the preference of public policymakers for the use of regulation when the other two alternatives—taxation and subsidization—are in theory equally efficient in effecting any specified level of environmental quality. In practice, the three strategies for pollution control do differ somewhat in terms of their economic efficiency, equity, and political and administrative feasibility. Therefore, the reluctance of public policymakers to rely more on tax policy—indirect subsidization and taxation—and direct subsidization can be explained by analyzing the political economy of pollution control policy.

When a social problem becomes critical, the legislative branch of government naturally responds with regulation and/or prohibition to control and subsidization to compensate those harmed by the problem and/or the resolution thereof. Taxation has traditionally provided the revenue necessary for financing public goods and services, such as national security (defense), domestic security (fire, police, and judiciary), education, transportation, housing, other social welfare programs, and pollution control. Consequently, the use of taxation to simulate a market for common property resources would be a significant departure from the traditional revenue-raising function of taxation. In other words, the proposal to use taxation to control pollution was not politically feasible when the Environmental Crisis surfaced in the late 1960s.

The other approach to simulating the market system through economic incentives not to pollute may never be politically feasible. The alternative of paying polluters not to pollute—subsidizing polluters—is not acceptable to the body politic on the principle that there is no right to pollute, whereas there is a "birthright" to a clean, quality environment. Bribing individuals and institutions not to pollute is somehow inherently wrong, even though in practice it may be the most cost-effective method of effecting the preferred level of environmental quality when compared to regulation and taxation. Thus, the use of subsidization has been limited to that agreed upon by the OECD ministers in 1972—transitional financial assistance both for equity reasons and for social and/or economic (nontrade) objectives where unique circumstances deem it appropriate.

Administrative feasibility is a problem or issue that was originally considered to be unique to the use of taxation to control pollution. The establishment of uniform standards or best practicable and economically achievable technology requirements for polluters is not only administratively feasible, but is also guaranteed to bring results—a reduction of pollution by those affected. The same would be true of subsidies for polluting less than one did during a prior period. However, the information allegedly necessary to establish the pollution tax base and rates is adjudged to be an insurmountable hurdle. In theory, the incomplete information problem is equally shared by regulation, taxation, and subsidization. However, during a crisis the body politic is more concerned with immediate results than with the most cost-effective method for attaining a given level of pollution control. Inasmuch as subsidies or "bribes" are not politically feasible, the public policymakers are forced to enact regulatory programs with subsidization justified for equity reasons where regulation has the effect of increasing costs (lowering income) retroactively.

The "polluter pays principle" is equitable inasmuch as it internalizes the cost of pollution through regulation or pollution taxation. Thus, the price of goods and services reflects the cost of pollution control. Without pollution control, the costs of pollution are passed on to the society, and everyone pays for pollution rather than the consumers of goods and services the production and use of which generates the pollution. Although limited subsidization to correct the inequitable incidence of pollution control costs is acceptable, the use of subsidization to simulate a market system for environmental quality would allow the polluter to pass on its costs to the taxpayer. Because the incidence of taxes and the consumption of goods and services the production and use of which generates pollution will seldom result in an equitable incidence of pollution control costs, taxpayers and their elected representatives have another basis upon which to prefer regulation or pollution taxation.

Having relied heavily on regulatory programs to produce substantial early improvements in pollution control, the public becomes disillusioned and frustrated when the rate of improvement decreases, while the costs of regulation are being paid for in both higher taxes used to pay for public administration of regulatory programs and inflation caused by the increased cost of pollution control. Ironically, the taxpayer's revolt in the United States and fiscal conservatism in Canada, West Germany, and the United Kingdom are responses to the cost of Big Government's regulatory and welfare (subsidy) program. In contrast, the public's attitude to pollution taxation has grown more favorable, with the French and Germans relying on effluent tax or charge systems, proposals or provisions for similar systems in Canada and the United Kingdom, and favorable public opinion in the United States.[2] Even in Sweden, which has the highest taxes and is probably the most regulated country among the industrialized nations of the noncommunist world, there has been a "change of heart." More specifically, in June 1978 the Swedish Governmental Commission on Environmental Costs, after over six years of investigation, made the following proposals for introducing economic considerations into public policy decisions on environmental protection:

> A special charge should be levied on emissions over and above those licensed by administrative permits and regulations, regardless of whether such excess emissions are intentional, accidental, or the result of negligence. There should be strict application of the "Polluter Pays Principle" by way of a one-time fee to cover the costs of administrative licensing, and a yearly fee to cover the costs of administrative supervision and control. Such licensing and control fees should be introduced beginning in 1980, and should give the state an income of Skr 100 Million yearly.
>
> Intensified research is needed on methods for evaluating the total socio-economic costs and benefits of environmental policy programs and activities.
>
> Introduction of a more elaborate system of long-term environmental policy planning is recommended. This should be based on explicitly stated political goals and values, but guided as far as possible by economic evaluations of the socio-economic costs and benefits of different policy alternatives.[3]

In summary, there is overwhelming evidence that the resistance to pollution taxation is decreasing. Objections because it is neither administratively feasible nor politically feasible are no longer a serious hurdle in view of the cost-effectiveness, administrative, and political problems inherent in more regulation. As a result of over a decade of regulation, we have learned much about pollution control. Therefore, the incomplete information problem is not the hurdle it once was, especially if we are willing to implement a "second best" system of effluent, emission, and excise taxes where the objective is marginal improvements and management of environmental quality. While complementing the present regulatory system with a system of pollution taxes will result in inequities, the limited use of subsidies is a well-accepted method of alleviating any short-term problems of equity.

## Postscript

After the Arab oil embargo of 1973-74, the economies of Western Europe and North America began suffering from stagnant growth and inflation. This phenomenon, which has been characterized as stagflation, persisted through the middle 1970s. Subsequently, the trebling of world crude oil prices between 1978 and 1981 triggered an extended recessionary period that spread throughout the world. Restrictive monetary policies that began as an attempt to control inflation became especially troublesome during the recent recessionary period because it made the financing of large government budget deficits expensive. The deficits were precipitated by the increased cost of social programs devised to alleviate the social problems related to high unemployment and increases in the cost of imported oil, as well as by revenue losses resulting from the recession and tax incentives to stimulate economic activity. Although unemployment continued to rise through 1981, significant reductions in inflation and oil imports by 1982 provided a basis for economic recovery in 1983 inasmuch as less restrictive monetary policies and stable prices for world crude oil were anticipated.[4]

The economic problems of the past decade would have provided the governments of Western Europe and North America with the perfect opportunity to relax or delay environmental protection measures that increased production costs and consumed scarce capital resources. However, the governments of all six countries were steadfast in their efforts to maintain and improve environmental quality through the 1970s, but budget cuts in the early 1980s began to affect the enforcement and research activities of many government agencies with environmental protection responsibilities. Similarly, the private sector invested heavily in pollution control technology and equipment during the 1970s, but such investments were adversely affected by the worldwide recession of the 1980s.

Fortunately, opinion polls in all six countries indicate widespread support for environmental protection, especially with regard to toxic substances and hazardous wastes where the dangers to human and environmental health are self-evident. Notwithstanding strong public support for environmental protection measures, many of the low-cost measures necessary to mitigate the most obvious pollution have been implemented in full or at least partially. Therefore, future improvements in environmental quality will be more expensive and inflationary.[5] Thus, public support may wane in the face of the increasing private and public costs of more stringent pollution control standards. This erosion of public support is more likely if economic recovery remains elusive, while industry and conservative political groups lobby for deregulation as a means of "freeing up" scarce capital for investment in productive capacity that creates jobs and economic progress.

Consensus on more stringent requirements for the production, use, and disposal of toxic or hazardous substances seems assured, although the determination of what substances are toxic or hazardous will continue to be controversial. Conversely, air quality and emissions standards will be subject to considerable controversy because of their tenuous relationship to public and environmental health and because the public remains enamored with automobiles that consume expensive oil-refined fuels. In addition, the conversion from oil to coal as a source of energy for power generation aggravates air pollution problems and, therefore, requires costly investment in pollution control equipment that removes $SO_2$ and $NO_x$ from emissions.

Even with the consensus on toxic or hazardous substances and the uncertain future of more stringent air pollution control standards, there remains the issue of water quality. Temporary or persistent shortages of quality water for a variety of uses will soon become a major recurring problem in many regions of all the countries studied herein except perhaps Sweden. North America and the densely populated, heavily industrialized regions of Western Europe already suffer from periodic shortages of quality surface water. Moreover, groundwater resources are being rapidly depleted or contaminated. Water pollution aggravates the problem of water shortages. Therefore, the traditional treatment of water as a free or inexpensive resource must inevitably be reformed.

Most certainly, the water pollution problem would be worse if considerable monies had not been expended both on subsidies for investment in public wastewater treatment facilities and on investments by the private sector in the facilities necessary for compliance with water pollution control standards. However, public subsidies and private investment for wastewater treatment decreased in either real or actual terms during the extended recessionary period of the early 1980s. Insofar as modern economic activity seems to be subject to a cyclical pattern of intermittent periods of slow growth and moderate-to-serious recessions where inflation persists, it would seem advantageous to insulate environmental protection measures, and especially water resource management—quantity and quality—from the vagaries of the economic "rollercoaster," as long as the long-term benefits of environmental quality improvements outweigh the short-term costs. Notwithstanding the overwhelming preference of public policymakers for regulation with a modicum of subsidization as a means of solving social problems, including pollution, obvious cost savings would seem probable if the "market mechanisms" were used more often to complement the other methods of social control, that is, regulation or prohibition and subsidization.

In the case of water resource management, some success in using the market to allocate surface water supply and to regulate its quality has occurred in France and the Ruhr Valley of Germany. The experience in the Ruhr was the basis upon which the French Water Basin Finance Agencies were established.[6] Furthermore, the Germans have relied on the Ruhr experience as the basis for establishing a national wastewater charge system as of January 1, 1981. The reorganization of water authorities to basin-wide institutions in the United Kingdom has stimulated the development of cost-based charges for wastewater treatment therein. Canadian Federal water law authorizes the establishment of basin-wide wastewater charge systems, although only a few municipalities have adopted wastewater charges based on estimated treatment costs. Basin-wide wastewater charge systems have been recommended in the United States by academicians[7] and by at least one distinguished national association of industrialists and economists.[8] However, current U.S. Federal water pollution control legislation does not authorize the establishment of such systems. In spite of the absence or lack of a national mandate, many States would seem to have the legal authority to establish wastewater charge systems. Only Sweden would seem to be resistant to the establishment of wastewater charge systems, even though the recommendations of the Governmental Commission on Environmental Costs included proposals for charging polluters for licensing and permit violations.[9] Finally, the Economic Commission of Europe has recommended that wastewater charge systems be established.[10]

Uniform effluent standards, whether national, State, provincial, or local in scope, are inefficient for at least two reasons. First, such standards are seldom devised to allow periodic adjustments to reflect variations in the environmental assimilative capacity (EAC) of a body of water. In addition, they are inconsistent with the fact that polluters incur different costs in attaining a specified level of discharge for various contaminants. Although consensus on the prohibition of toxic or hazardous substances in wastewater can be based on public or environmental health criteria, reasonable persons can seldom agree on a set of uniform effluent standards that reflect both variations in the EAC of a body of water and the optimal marginal cost of pollution control for all polluters. However, agreement on minimum uniform effluent standards should be easier to consummate where there is both a private cost—a proportional or graduated wastewater charge or effluent tax—for discharges above that minimum level and a water basin agency (utility) authorized to manage water quality and quantity for a variety of uses by individuals and institutions, both public and private.

Currently, the OECD Environment Committee is evaluating the U.S. "bubble concept" and other economic approaches to environmental regulation. In 1983, the OECD will sponsor an international conference on environmental economics as a forum for research on other economic approaches to pollution control, cost-benefit analysis of environmental regulation, the interrelationship of economic and environmental policies, and responses to emerging environmental problems.[11]

In summary, the optimal strategy for managing water resources should integrate all three methods of social control available to public policymakers. So far a fixation on regulation, which includes prohibition, along with subsidization for equity and public wastewater treatment facilities has been moderately successful in controlling the grossest water pollution problems. However, cost efficiency and the continuity of efforts to manage water resources can best be satisfied by complementing regulation and subsidization with effluent taxation. Thus, polluters can decide on their lowest cost alternative for handling residuals, and the affected public can expend relatively stable tax revenues to ensure water resource quality and quantity in concert with variations in the EAC of bodies of water and regardless of the state of the national economy. Eventually, emission taxation and excise taxes on solid wastes may also be developed, but there is no reason to defer adoption of effluent taxation any longer.

## Notes

1. *International Environment Reporter—Current Report*, March 10, 1978, p. 59.

2. "Perspectives on Current Developments: Should Polluters Pay," *Regulation*, Vol. 2, No. 4 (July-August 1978), p. 14.

3. Lennart J. Lundqvist, "New Proposals for Sweden's Environmental Policy," *Ambio*, Vol. 7, No. 4, (1978), p. 186. As noted in Chapter 6, most of the recommendations of the Commission were adopted in amendments to the Swedish Environmental Protection Act enacted in early 1981. However, pollution taxes were not included in the Commission's recommendations.

4. Lawrence R. Klein, et al., "Recovery: A Slow Start Now, But Big Gains in '83," *Business Week*, February 15, 1982, pp. 97-116; and David Brand, "Western Europe, Beset by Economic Maladies, Faces Painful Recovery," *The Wall Street Journal*, February 18, 1982, pp. 1 and 15; Lawrence R. Klein, et. al., "An Eye on Recovery," *Business Week*, August 9, 1982, pp. 84-94.

5. *International Environment Reporter—Current Report*, January 14, 1981, pp. 591-92. See also Environment Directorate, *The State of the Environment* (Paris: Organization for Economic Cooperation and Development, 1979).

6. See Blair T. Bower, et al., *Incentives in Water Quality Management* (Washington, D.C.: Resources for the Future, 1981) for a detailed description and analysis of the wastewater charge systems in France and the Ruhr area.

7. See Allen V. Kneese and Blair T. Bower, *Environmental Quality and Residuals Management* (Baltimore: Johns Hopkins University Press, 1979).

8. CED Research and Policy Committee, *More Effective Programs for a Cleaner Environment* (New York: Committee for Economic Development, 1974).

9. Lundqvist, "New Proposals for Sweden's Environmental Policy," p. 186.

10. *International Environment Reporter—Current Report*, January 14, 1981, p. 592.

11. Ibid., March 10, 1982, pp. 122-23; and July 14, 1982, pp. 293-94.

# Waste Water Charges Act September 13 1976 Act Pertaining To Charges Levied For Discharging Waste Water Into Waters

The following Act has been passed by the Federal Parliament (Bundestag):

## Section 1 General Regulations

### Article 1 Principle

A charge shall be paid for discharging waste water into waters within the meaning of Article 1, para 1 of the Federal Water Act passed on July 27, 1957 (Federal Law Gazette I p. 1110) last amended by the Fourth Amendment to the Federal Water Act, passed on April 26, 1976 (Federal Law Gazette I p. 1109), such charge to be referred to as the Waste Water Charge. The charge shall be levied by the federal states (Länder).

### Article 2 Definitions

(1) Within the meaning of this Act, waste water shall be deemed to be water changed in its properties by domestic, commercial, agricultural or other use and the water running off in conjunction therewith in dry weather (polluted water), as well as water running off from builtup or paved surfaces following precipitation (hereinafter referred to as rain water).

(2) Within the meaning of this Act discharging shall be deemed to be the immediate and direct conveyance of waste water into a water body; conveyance into the subsoil shall be regarded as discharging into a water, with the exception of

conveyance into the ground within the framework of agricultural soil treatment.

(3) Within the meaning of this Act a waste water treatment plant shall be deemed to be a facility used to reduce or eliminate the noxiousness of waste water. Facilities serving to prevent the generation of waste water either in full or in part shall also be regarded as waste water treatment plant.

## Article 3 System of Assessment

(1) The amount of the waste water charge shall depend upon the noxiousness of the waste water, which shall be determined on the basis of the volume of waste water, the settleable solids contained therein, the oxidizable substances and the toxicity of the waste water, expressed in units of noxiousness in accordance with the Annex to this Act.

(2) In such cases as described under Article 9, para 3 (river sewage plant), the charge levied shall be calculated according to the number of units of noxiousness in the water body downstream from the river sewage plant.

(3) The Länder shall be free to determine that the noxiousness of waste water shall be left out of account to the extent that it is eliminated in secondary settling ponds directly connected with a waste water treatment plant.

(4) The Länder shall be free to have the noxiousness of settleable solids determined by the weight of such solids on request of the parties liable to pay the waste water charge, provided the number of cubic metres of such solids generated annually is more than five times larger than the number of tons of dry substance generated annually.

(5) The Federal Government shall be authorized to adapt to the respective state of science and technology, by statutory ordinance with the consent of the Bundesrat, the regulations pertaining to procedures for the determination of noxiousness, as set forth in Part B of the Annex, in order to refine such procedures or to reduce the personal or material effort and expenditure required for the determination of noxiousness, provided that this does not cause any substantial change in the assessment of noxiousness.

## Section 2 Determination of Noxiousness

## Article 4 Determination on the Basis of an Official Notice

(1) With the exception of rain water (Article 7) and small waste water discharges (Article 8), the values to be applied for determining the number of units of noxiousness shall be taken from the official notice licensing the waste water discharge. Such official notice shall at least provide data on the maximum amount of polluted water permitted annually, the settleable solids, the oxidizable substances and the degree of toxicity as set forth in Article 3, para 1, such data to be distinguished according to the mean values to be maintained (standard values) and the values which may not be exceeded under any circumstances whatsoever (maximum values). The standard values or, respectively, at least 50 per cent of the maximum values, shall be applied as a basis for determining the number of units of noxiousness (reference values). Should there be no reason to expect the presence of settleable solids, oxidizable substances or a degree of toxicity as defined under Article 3, para 1, in the waste water, or should the amount of mercury to be

expected in the waste water be less than one kilogram and the amount of cadmium less than 10 kilograms annually, the requirement to fix definite values in the official notice may be waived. Should such official notice nevertheless specify mercury or cadmium values, such values shall be left out of account in determining the degree of noxiousness.

(2) In cases such as those specified under Article 9, para 3, (river sewage plant) para 1 shall apply accordingly.

(3) Should any water extracted directly from a water resource already possess noxious properties as set forth under Article 3, para 1 prior to use (pre-pollution), such pre-pollution shall be estimated at the request of the party liable to pay waste water charges the pre-pollution so estimated not being attributable to such party. The Länder shall be free to determine the pre-pollution for waters or parts thereof on a uniform level.

(4) Within the framework of water supervision, state or state-acknowledged agencies shall supervise compliance with the official notice issued; in so doing they shall proceed according to the pertinent provisions of water legislation. Should such supervision reveal that one of the maximum values determined in an official notice is exceeded more than one a year, a higher reference value shall be applied in determining the number of units of noxiousness. Such higher reference value shall be calculated as the aggregate of the reference value indicated in the official notice and the arithmetical mean of the differences by which the values measured exceed the maximum value set forth in the official notice.

(5) Should a party discharging waste water inform the authority that, for a certain period of not less than three months, he will be discharging a reduced volume of waste water or will maintain reduced standard values while not exceeding correspondingly lower maximum values, the number of units of noxiousness for such period shall be determined in accordance with the reduced volume of waste water or the reference values indicated. To be considered in this context, the deviation obtained must be at least 25 per cent of the waste water volume or of the standard values applicable. Paras 3 and 4 shall apply accordingly.

### Article 5 Determination on the Basis of Measurements.

(1) Should a party liable to pay waste water charges show by submitting values obtained on the basis of a measuring programme approved by the responsible authority, that the weighted mean of measuring results obtained in the preceding period of assessment deviates from the standard value specified under Article 4, para 1 by more than 25 per cent, such weighted mean of the measuring values, but at least 50 per cent of the highest value measured, shall be applied in determining the number of units of noxiousness. The measuring programme must comprise at least one daily sample taken at different times of the day in addition to the continual volume measurement.

(2) Article 4, paras 3 and 4 shall apply accordingly. Should the supervision reveal that the maximum measured value submitted in accordance with para 1, sentence 1 was exceeded more than once a year, Article 4, para 4, sentence 2 shall apply accordingly in determining the number of units of noxiousness. In this process at least 50 per cent of the maximum value specified in the official notice shall be applied as a basis for calculation.

## Article 6 Determination on Other Cases

(1) In cases where a value of substantial importance for determining the number of units of noxiousness in accordance with Article 3, para 1 is not specified in an official notice referred to in Article 4, para 1, although such value is not dispensable as provided in Article 4, para 1, sentence 4, it shall be determined on the basis of the results obtained by way of official supervision. Should no result of official supervision be available, the responsible authorities shall provide an estimate of such value.

(2) Article 4, para 3 shall apply accordingly.

## Article 7 Comprehensive Assessment for Discharge of Polluted Rain Water

(1) The number of units of noxiousness of rain water discharged through a public sewerage system shall be deemed to be 12 per cent of the number of inhabitants connected to such system. The number of these inhabitants may be estimated.

(2) The Länder shall determine in how far the number of units of noxiousness is reduced either by the retention of rain water or by treatment of such water in a sewage treatment plant. In such cases the Länder may determine that the discharge shall be exempt from waste water charges.

## Article 8 Comprehensive Assessment in the Case of Small Discharges of Domestic and Other Types of Sewage

Unless prescribed otherwise by the Länder, the number of units of noxiousness of domestic sewage and similar types of sewage for which a public corporation shall be liable to pay duties in accordance with Article 9, para 2, shall be deemed to be half the number of inhabitants not connected to the public sewerage system. Where it is impossible to determine the number of inhabitants, or where this would involve unreasonable expenditure, the number may be estimated.

## Section 3 Liability to Pay Charges

## Article 9 Liability to Pay Charges Rates

(1) Whoever shall discharge waste water (discharger) shall be liable to pay waste water charges.

(2) The Länder may determine that public corporations shall be liable to pay waste water charges in lieu of dischargers. Public corporations to be designated by the Länder shall be liable to pay waste water charges in lieu of dischargers who discharge, on an annual average, less than 8 cubic metres of sewage per day from domestic households or similar sources. The Länder shall determine how waste water charges levied in this way may be passed on to the originators.

(3) Where the water of a water body is purified by means of a river sewage plant, the Länder may determine that the operator of the river sewage plant shall be liable to pay waste charges in lieu of dischargers in a catchment area to be defined in advance. Para 2, sentence 3 shall apply accordingly.

(4) Liability to pay waste water charges shall not become effective prior to December 31, 1980. The annual rate levied per unit of noxiousness shall be

| DM 12.– | as of January 1, 1981 |
|---------|------------------------|
| DM 18.– | as of January 1, 1982 |
| DM 24.– | as of January 1, 1983 |
| DM 30.– | as of January 1, 1984 |
| DM 36.– | as of January 1, 1985 |
| DM 40.– | as of January 1, 1986 |

(5) Except in the case of rain water (Article 7) or small discharges (Article 8), the rate specified in para 4, sentence 2 shall be reduced by 50 per cent for those units of noxiousness which cannot be avoided despite compliance with the minimum requirements set forth under Article 7 a, para 1, sentence 3 of the Federal Water Act. Should the official notice specify more stringent requirements in respect of values within the meaning of Article 4, para 1, the above mentioned reduction shall only be applicable if these requirements are observed.

(6) In order to avert any significantly detrimental economic developments, the Federal Government shall be authorized to exempt, by statutory ordinance with the consent of the Bundesrat, the parties liable to pay waste water charges or regional or sectoral groups of such parties carrying out, or causing to be carried out, measures for reducing the noxiousness of waste water, from such liability, the exemption not to remain in force, either in full or in part, until later than December 31, 1989.

### Article 10 Exemption from Liability to Pay Charges

(1) The following exemptions from liability to pay waste water charges shall be granted:

1. for the discharge of water already polluted when extracted from a water resource prior to use and not showing any further noxiousness within the meaning of this after use,

2. for the discharge of polluted water into a surface water created during the extraction of mineral raw materials, provided that such water is only used for washing the product obtained at such location and does not contain any noxious substances other than those extracted, and as far as it is guaranteed that noxious substances will not reach other waters,

3. for the discharge of sewage from watercraft where it is generated,

4. for the discharge of rain water if a public sewerage system is not used for this purpose.

(2) The Länder may determine that the discharge of waste water into underground layers in which the groundwater, on account of its natural properties, is not suitable for the extraction of drinking water by way of conventional methods of treatment, shall not entail liability to pay waste water charges.

(3) A waste water treatment plant shall be exempt from liability for the period of three years prior to the planned date of commissioning, in an amount corresponding to the anticipated reduction in the number of units of noxiousness upon discharge into the water concerned, provided such reduction is at least 20 per cent. Should the plant not become operative, liability to pay charges in their full amount shall arise with retroactive effect. Should the actual degree of purification fall short of the anticipated reduction in the number of units of noxiousness, liability to pay charges shall to that extent arise with retroactive effect.

## Section 4 Establishment, Levy and Use of Waste Water Charges

### Article 11 Period of Assessment, Obligation of Disclosure

(1) The calendar year shall be deemed to be the period of assessment.

(2) In the cases referred to under Articles 7 and 8 the party liable to pay waste water charges shall calculate the units of noxiousness of the waste water and submit the associated documents to the responsible authority. Should such party not simultaneously be the discharger (Article 9, paras 2 and 3), the discharger shall supply the liable party with the requisite data and documents.

(3) The Länder may determine that the liable party shall calculate the number of units of noxiousness of the waste water also in other cases, that such party shall provide the data required for an estimate, and submit the associated documents to the responsible authority.

Para 2, sentence 2 shall apply accordingly.

### Article 12 Infringement of the Obligation of Disclosure

(1) Should the liable party fail to comply with its obligations as set forth under Article 11, para 2, sentence 1 and the supplementary regulations issued by the Länder, the responsible authority shall be free to estimate the number of units of noxiousness.

(2) A discharger who is not liable to pay waste water charges in accordance with Article 9, paras 2 or 3 may nevertheless be rendered liable by way of an estimate to pay such charges if he fails to comply with his obligations as set forth under Article 11, para 2, sentence 2 and the supplementary regulations issued by the Länder. Under such circumstances the liable party and the discharger shall be liable jointly and severally.

### Article 13 Use of Charges Levied

(1) The revenue accruing from waste water charges shall only be used for specific purposes connected with measures for maintaining or improving water quality. The Länder may determine that the administrative expenditure associated with the enforcement of this Act and of their own supplementary regulations shall be paid for out of the revenue accruing from waste water charges.

(2) Particularly, the following shall be deemed to be measures as provided in para 1 above:

1. The construction of waste water treatment plant.

2. The construction of rain retention basins and facilities for the purification of rain water.

3. The construction of ring-shaped and holding canals at and along dams, lake and sea shores and of main connecting sewers permitting the erection of jointly-operated sewage treatment facilities.

4. The construction of plant for the disposal of sewage sludge.

5. Measures taken in and at water bodies for observing and improving water quality (such as raising the level of low-water flow or providing for oxygen enrichment) and for maintaining such water bodies.

6. Research on and development of suitable plant or techniques for improving water quality.

7. Basic and further training of operating staff for water treatment plant and other facilities designed to maintain and improve water quality.

## Section 5 Common and Final Regulations

### Article 14 Application of Regulations on Fines and Penalties under the Fiscal Code

The penal clauses set forth in Article 370, paras 1, 2 and 4 and in Article 371 of the Fiscal Code (Abgabenordnung – AO 1977) shall apply accordingly to any act of evasion involving waste water charges; the penalty provision set forth in Article 378 of the Fiscal Code (AO 1977) shall apply accordingly to any unlawful reduction of the waste water charges payable.

### Article 15 Breaches of Regulations

(1) Whoever intentionally or negligently
1. contrary to Article 5, para 1, sentence 1, submits measuring values not conforming to the measuring programme,
2. contrary to Article 11, para 2, sentence 1, fails to submit the calculations or documents or does not submit them in an accurate or complete condition,
3. contrary to Article 11, para 2, sentence 2, fails to provide or does not provide the liable party with the requisite data or documents in an accurate or complete condition, shall be deemed to commit a breach of regulations.

(2) A penalty not exceeding five thousand Deutschmarks may be imposed for any such breach of regulations.

### Article 16 Clause Pertaining to Cities with the Status of a Land (City States)

Article 1 shall also apply if the Länder of Berlin and Hamburg are themselves liable to pay waste water charges. Article 9, para 2, sentence 1 and 2 shall apply to the Länder of Berlin and Hamburg with the proviso that they may also determine themselves to be liable to pay waste water charges.

### Article 17 Berlin Clause

Under the terms of Article 13, para 1 of the Third Transitional Act (Drittes Überleitujgsgesetz) passed on January 4, 1952 (Federal Law Gazette 1 p. 1), the present Act shall also be applicable in the Land of Berlin. Statutory ordinances issued by virtue of the present Act shall apply in the Land of Berlin in accordance with Article 14 of the Third Transitional Act.

### Article 18 Coming into Force

The present Act shall come into force on January 1, 1978.

### Annex to Article 3

A.

(1) In determining the noxiousness of waste water, 0.1 millilitres per litre of waste water shall be deducted in advance from the settleable solids, and 15 milligrams per litre of waste water from the oxidizable substances. Should the

figures obtained in this way be below zero, they shall remain unconsidered. The number of units of noxiousness may be inferred from the following table:

| Pollutants and groups of pollutants assessed | Number of units of noxiousness for each full measuring unit | |
|---|---|---|
| | Unit of noxiousness | Measuring unit (s) |
| 1. Settleable solids containing at least 10 per cent organic matter | 1 | Annual volume in cubic metres or, respectively, in tons should Article 3, para 4 be applicable |
| 2. Settleable solids containing less than 10 per cent organic matter | 0.1 | Annual volume in cubic metres or, respectively in tons should Article 3, para 4 be applicable |
| 3. Oxidizable substances expressed as Chemical Oxygen Demand (COD) | 2.2 | Annual volume of 100 kilograms |
| 4. Mercury and its compounds | 5 | Annual volume of 100 grams of mercury |
| 5. Cadmium and its compounds | 1 | Annual volume of 100 grams of cadmium |
| 6. Toxicity for fish | 0.3 $G_F$+ | Annual waste water volume of 1,000 cubic metres |

+)$G_F$ represents the dilution factor at which waste water loses its toxic effect on fish. When $G_F = 2$, the figure applied shall be 0.

(2) Should waste water be discharged into coastal waters, the toxic effect of such waste water on fish shall remain unconsidered to the extent that it is due to the content of such salts as are similar to the principal components of sea water. The same shall apply to the discharge of waste water into the estuaries of surface waters leading into the sea, provided such estuaries have a natural salt content similar to that of coastal waters.

B.

(1) The volume of settleable solids shall be determined after a two-hour settling period.

(2) The chemical oxygen demand shall be determined in accordance with the dichromate procedure, silver sulphate being applied as a catalyst.

(3) Mercury and cadmium shall be determined by way of atomic absorption spectrometry.

(4) The toxic effect in fish tests is determined by using the species orfe (Leuciscus idus melanotus) as a test fish and applying various degrees of waste water dilution.

# Sec. 169 Regulations.

§ 1.169-1. **Amortization of pollution control facilities.** —(a) *Allowance of deduction* —(1) *In general.* Under section 169(a), every person, at his election, shall be entitled to a deduction with respect to the amortization of the amortizable basis (as defined in § 1.169-3) of any certified pollution control facility (as defined in § 1.169-2), based on a period of 60 months. Under section 169(b) and paragraph (a) of § 1.169-4, the taxpayer may further elect to begin such 60-month period either with the month following the month in which the facility is completed or acquired or with the first month of the taxable year succeeding the taxable year in which such facility is completed or acquired. Under section 169(c), a taxpayer who has elected under section 169(b) to take the amortization deduction provided by section 169(a) may, at any time after making such election and prior to the expiration of the 60-month amortization period, elect to discontinue the amortization deduction for the remainder of the 60-month period in the manner prescribed in paragraph (b)(1) of § 1.169-4. In addition, if on or before May 18, 1971, an election under section 169(a) has been made, consent is hereby given to revoke such election without the consent of the Commissioner in the manner prescribed in paragraph (b)(2) of § 1.169-4.

(2) *Amount of deduction.* With respect to each month of such 60-month period which falls within the taxable year, the amortization deduction shall be an amount equal to the amortizable basis of the certified pollution control facility at the end of such month divided by the number of months (including the month for which the deduction is computed) remaining in such 60-month period. The amortizable basis at the end of any month shall be computed without regard to the amortization deduction for such month. The total amortization deduction with respect to a certified pollution control facility for a taxable year is the sum of the amortization deductions allowable for each month of the 60-month period which falls within such taxable year. If a certified pollution control facility is sold or exchanged or otherwise disposed of during one month, the amortization deduction (if any) allowable to the original holder in respect of such month shall be that portion of the amount to which such person would be entitled for a full month which the number of days in such month during which the facility was held by such person bears to the total number of days in such month.

(3) *Effect on other deductions.* (i) The amortization deduction provided by section 169 with respect to any month shall be in lieu of the depreciation deduction which would otherwise be allowable under section 167 or a deduction in lieu of depreciation which would otherwise be allowable under paragraph (b) of § 1.162-11 for such month.

(ii) If the adjusted basis of such facility as computed under section 1011 for purposes other than the amortization deduction provided by section 169 is in excess of the amortizable basis, as computed under § 1.169-3, such excess shall be recovered through depreciation deductions under the rules of section 167. See section 169(g).

(iii) See section 179 and paragraph (e)(1)(ii) of § 1.179-1 and paragraph (b)(2) of § 1.169-3 for additional first-year depreciation in respect of a certified pollution control facility.

(4) [**Deleted.**]

(5) *Special rules.* (i) In the case of a certified pollution control facility held by one person for life with the remainder to another person, the amortization deduction under section 169(a) shall be computed as if the life tenant were the absolute owner of the property and shall be allowable to the life tenant during his life.

(ii) If the assets of a corporation which has elected to take the amortization deduction under section 169(a) are acquired by another corporation in a transaction to which section 381 (relating to carryovers in certain corporate acquisitions) applies, the acquiring corporation is to be treated as if it were the distributor or transferor corporation for purposes of this section.

(iii) For the right of estates and trusts to amortize pollution control facilities see section 642(f) and § 1.642(f)-1. For the allowance of the amortization deduction in the case of pollution control facilities of partnerships, see section 703 and § 1.703-1.

(6) *Depreciation subsequent to discontinuance or in the case of revocation of amortization.* A taxpayer which elects in the manner prescribed under paragraph (b)(1) of § 1.169-4 to discontinue amortization deductions or under paragraph (b)(2) of § 1.169-4 to revoke an election under section 169(a) with respect to a certified pollution control facility is entitled, if such facility is of a character subject to the allowance for depreciation provided in section 167, to a deduction for depreciation (to the extent allowable) with respect to such facility. In the case of an election to discontinue an amortization deduction, the deduction for depreciation shall begin with the first month as to which such amortization deduction is not applicable and shall be computed on the adjusted basis of the property as of the beginning of such month (see section 1011 and the regulations thereunder). Such depreciation deduction shall be based upon the remaining portion of the period authorized under section 167 for the facility, as determined, as of the first day of the first month as of which the amortization deduction is not applicable. If the taxpayer so elects to discontinue the amortization deduction under section 169(a), such taxpayer shall not be entitled to any further amortization deduction under this section and section 169(a) with respect to such pollution control facility. In the case of a revocation of an election under section 169(a), the deduction for depreciation shall begin as of the time such depreciation deduction would have been taken but for the election under section 169(a). See paragraph (b)(2) of § 1.169-4 for rules as to filing amended returns for years for which amortization deductions have been taken.

(7) *Definitions.* Except as otherwise provided in § 1.169-2, all terms used in section 169 and the regulations thereunder shall have the meaning provided by this section and §§ 1.169-2 through 1.169-4.

(b) *Examples.* This section may be illustrated by the following examples:

*Example (1).* On September 30, 1970, the X Corporation, which uses the calendar year as its taxable year, completes the installation of a facility all of which qualifies as a certified pollution control facility within the meaning of paragraph (a) of § 1.169-2. The cost of the facility is $120,000 and the period referred to in paragraph (a)(6) of § 1.169-2 is 10 years. In accordance with the rules set forth in paragraph (a) of § 1.169-4, on its income tax return filed for 1970, X elects to take amortization deductions under section 169(a) with respect to the facility and to begin the 60-month amortization period with October 1970, the month following the month in which it was completed. The amortizable basis at the end of October 1970 (determined without regard to the amortization deduction under section 169(a) for that month) is $120,000. The allowable amortization deduction with respect to such facility for the taxable year 1970 is $6,000, computed as follows:

Monthly amortization deductions:

| | |
|---|---:|
| October: $120,000 divided by 60 ......................... | $ 2,000 |
| November: $118,000 (that is, $120,000 minus $2,000) divided by 59. ...................................... | 2,000 |
| December: $116,000 (that is, $118,000 minus $2,000) divided by 58. ...................................... | $2,000 |
| Total amortization deduction for 1970 ......................... | $6,000 |

*Example (2).* Assume the same facts as in example (1). Assume further than on May 20, 1972, X properly files notice of its election to discontinue the amortization deductions with the month of June 1972. The adjusted basis of the facility as of June 1, 1972, is $80,000, computed as follows:

Yearly amortization deductions:

| | |
|---|---:|
| 1970 (as computed in example (1)). ........................ | $ 6,000 |
| 1971 (computed in accordance with example (1)). ............. | 24,000 |
| 1972 (for the first five months of 1972 computed in accordance with example (1)) .......................... | 10,000 |
| Total amortization deductions for 20 months. ............ | 40,000 |
| Adjusted basis at beginning of amortization period ............. | 120,000 |
| Less: Amortization deductions ........................... | 40,000 |
| Adjusted basis as of June 1, 1972. ........................ | $ 80,000 |

Beginning as of June 1, 1972, the deduction for depreciation under section 167 is allowable with respect to the property on its adjusted basis of $80,000. [Reg. § 1.169-1.]

§ 1.169-2. **Definitions.**—(a) *Certified pollution control facility*—(1) *In general.* Under section 169(d), the term "certified pollution control facility" means a facility which—

(i) The Federal certifying authority certifies, in accordance with the rules prescribed in paragraph (c) of this section, is a "treatment facility" described in subparagraph (2) of this paragraph, and

    (ii) Is "a new identifiable facility" (as defined in paragraph (b) of this section).

For profitmaking abatement works limitation, see paragraph (d) of this section.

    (2) *Treatment facility.* For purposes of subparagraph (1)(i) of this paragraph, a "treatment facility" is a facility which (i) is used to abate or control water or atmospheric pollution or contamination by removing, altering, disposing, or storing of pollutants, contaminants, wastes, or heat and (ii) is used in connection with a plant or other property in operation before January 1, 1969. Determinations under subdivision (i) of this subparagraph shall be made solely by the Federal certifying authority. See subparagraph (3) of this paragraph. For meaning of the phrases "plant or other property" and "in operation before January 1, 1969," see subparagraphs (4) and (5), respectively, of this paragraph.

    (3) *Facilities performing multiple functions or used in connection with several plants, etc.* (i) If a facility is designed to perform or does perform a function in addition to abating or controlling water or atmospheric pollution or contamination by removing, altering, disposing or storing pollutants, contaminants, wastes, or heat, such facility shall be a treatment facility only with respect to that part of the cost thereof which is certified by the Federal certifying authority as attributable to abating or controlling water or atmospheric pollution or contamination. For example, if a machine which performs a function in addition to abating water pollution is installed at a cost of $100,000 in, and is used only in connection with, a plant which was in operation before January 1, 1969, and if the Federal certifying authority certifies that $30,000 of the cost of such machine is allocable to its function of abating water pollution, such $30,000 will be deemed to be the adjusted basis for purposes of determining gain for purposes of paragraph (a) of § 1.169-3.

    (ii) If a facility is used in connection with more than one plant or other property, and at least one such plant or other property was not in operation before January 1, 1969, such facility shall be a treatment facility only to the extent of that part of the cost thereof certified by the Federal certifying authority as attributable to abating or controlling water or atmospheric pollution in connection with plants or other property in operation before January 1, 1969. For example, if a machine is constructed after December 31, 1968, at a cost of $100,000 and is used in connection with a number of plants only some of which were in operation before January 1, 1969, and if the Federal certifying authority certifies that $20,000 of the cost of such machine is allocable to its function of abating or controlling water pollution in connection with the plants or other property in operation before January 1, 1969, such $20,000 will be deemed to be the adjusted basis for purposes of determining gain for purposes of paragraph (a) of § 1.169-3. In a case in which the Federal certifying authority certifies the percentage of a facility which is used in connection with plants or other property in operation before January 1, 1969, the adjusted basis for the purposes of determining gain for purposes of paragraph (a) of § 1.169-3 of the portion of the facility so used shall be the adjusted basis for determining gain of the entire facility multiplied by such percentage.

    (4) *Plant or other property.* As used in subparagraph (2) of this paragraph, the phrase "plant or other property" means any tangible property whether or not such property is used in the trade or business or held for the production of income. Such term includes, for example, a papermill, a motor vehicle, or a furnace in an apartment house.

(5) *In operation before January 1, 1969.* (i) For purposes of subparagraph (2) of this paragraph and section 169(d), a plant or other property will be considered to be in operation before January 1, 1969, if prior to that date such plant or other property was actually performing the function for which it was constructed or acquired. For example, a papermill which is completed in July 1968, but which is not actually used to produce paper until 1969 would not be considered to be in operation before January 1, 1969. The fact that such plant or other property was only operating at partial capacity prior to January 1, 1969, or was being used as a standby facility prior to such date, shall not prevent its being considered to be in operation before such date.

(ii) (*a*) A piece of machinery which replaces one which was in operation prior to January 1, 1969, and which was a part of the manufacturing operation carried on by the plant but which does not substantially increase the capacity of the plant will be considered to be in operation prior to January 1, 1969. However, an additional machine that is added to a plant which was in operation before January 1, 1969, and which represents a substantial increase in the plant's capacity will not be considered to have been in operation before such date. There shall be deemed to be a substantial increase in the capacity of a plant or other property as of the time its capacity exceeds by more than 20 percent its capacity on December 31, 1968.

(*b*) In addition, if the total replacements of equipment in any single taxable year beginning after December 31, 1968, represent the replacement of a substantial portion of a manufacturing plant which had been in operation before such date, such replacement shall be considered to result in a new plant which was not in operation before such date. Thus, if a substantial portion of a plant which was in existence before January 1, 1969, is subsequently destroyed by fire and such substantial portion is replaced in a taxable year beginning after that date, such replacement property shall not be considered to have been in operation before January 1, 1969. The replacement of a substantial portion of a plant or other property shall be deemed to have occurred if, during a single taxable year, the taxpayer replaces manufacturing or production facilities or equipment, which comprises such plant or other property, and which has an adjusted basis (determined without regard to the adjustments provided in section 1016(a)(2) and (3)) in excess of 20 percent of the adjusted basis (so determined) of such plant or other property determined as of the first day of such taxable year.

(6) *Useful life.* For purposes of section 169 and the regulations thereunder, the terms "useful life" and "actual useful life" shall mean the shortest period authorized under section 167 and the regulations thereunder if an election were not made under section 169.

(b) *New identifiable facility*—(1) *In general.* For purposes of paragraph (a)(1)(ii) of this section, the term "new identifiable facility" includes only tangible property (not including a building and its structural components referred to in subparagraph (2)(i) of this paragraph, other than a building and its structural components which under subparagraph (2)(ii) of this paragraph is exclusively a treatment facility) which—

(i) Is of a character subject to the allowance for depreciation provided in section 167,

(ii) (*a*) Is property the construction, reconstruction, or erection (as defined in subparagraph (2)(iii) of this paragraph) of which is completed by the taxpayer after December 31, 1968, or (*b*) Is property acquired by the taxpayer after December 31, 1968, if the original use of the property commences with the taxpayer and commences after such date (see subparagraph (2)(iii) of this paragraph), and

(iii) Is placed in service (as defined in subparagraph (2)(iv) of this paragraph) prior to January 1, 1975.

(2) *Meaning of terms.* (i) For purposes of subparagraph (1) of this paragraph, the terms "building" and "structural component" shall be construed in a manner consistent with the principles set forth in paragraph (e) of § 1.48-1. Thus, for example, the following rules are applicable:

(*a*) The term "building" generally means any structure or edifice enclosing a space within its walls, and usually covered by a roof, the purpose of which is, for example, to provide shelter or housing, or to provide working, office, parking, display, or sales space. The term includes, for example, structures such as apartment houses, factory and office buildings, warehouses, barns, garages, railway or bus stations, and stores. Such term includes any such structure constructed by, or for, a lessee even if such structure must be removed, or ownership of such structure reverts to the lessor, at the termination of the lease. Such term does not include (*1*) a structure which is essentially an item of machinery or equipment, or (*2*) an enclosure which is so closely combined with the machinery or equipment which it supports, houses, or serves that it must be replaced, retired, or abandoned contemporaneously with such machinery or equipment, and which is depreciated over the life of such machinery or equipment. Thus, the term "building" does not include such structures as oil and gas storage tanks, grain storage bins, silos, fractioning towers, blast furnaces, coke ovens, brick kilns, and coal tipples.

(*b*) The term "structural components" includes, for example, chimneys, and other components relating to the operating or maintenance of a building. However, the term "structural components" does not include machinery or a device which serves no function other than the abatement or control of water or atmospheric pollution.

(ii) For purposes of subparagraph (1) of this paragraph, a building and its structural components is exclusively a treatment facility if the Federal certifying authority certifies that its only function is the abatement or control of air or water pollution. However, the incidental recovery of profits from wastes or otherwise shall not be deemed to be a function other than the abatement or control of air or water pollution. A building and its structural components which serve no function other than the treatment of wastes will be considered to be exclusively a treatment facility even if its contains areas for employees to operate the treatment facility, rest rooms for such workers, and an office for the management of such treatment facility. However, for example, if a portion of a building is used for the treatment of sewage and another portion of the building is used for the manufacture of machinery, the building is not exclusively a treatment facility. The Federal certifying authority will not certify as to what is a building and its structural components within the meaning of subdivision (i) of this subparagraph.

(iii) For purposes of subparagraph (1)(ii) (*a*) and (*b*) of this paragraph (relating to construction, reconstruction, or erection after December 31, 1968, and original use after December 31, 1968) and paragraph (b)(1) of § 1.169-3 (relating to definition of amortizable basis), the principles set forth in paragraph (a)(1) and (2) of § 1.167(c)-1 and in paragraphs (b) and (c) of § 1.48-2 shall be applied. Thus, for example, the following rules are applicable:

(*a*) Property is considered as constructed, reconstructed, or erected by the taxpayer if the work is done for him in accordance with his specifications.

(*b*) The portion of the basis of property attributable to construction, reconstruction, or erection after December 31, 1968, consists of all costs of construction, reconstruction, or erection allocable to the period after December 31, 1968, including the cost or other basis of materials entering into such work (but not including, in the case of reconstruction of property, the adjusted basis of the property as of the time such reconstruction is commenced).

(*c*) It is not necessary that materials entering into construction, reconstruction or erection be acquired after December 31, 1968, or that they be new in use.

(*d*) If construction or erection by the taxpayer began after December 31, 1968, the entire cost or other basis of such construction or erection may be taken into account for purposes of determining the amortizable basis under section 169.

(*e*) Construction, reconstruction, or erection by the taxpayer begins when physical work is started on such construction, reconstruction, or erection.

(*f*) Property shall be deemed to be acquired when reduced to physical possession or control.

(*g*) The term "original use" means the first use to which the property is put, whether or not such use corresponds to the use of such property by the taxpayer. For example, a reconditioned or rebuilt machine acquired by the taxpayer after December 31, 1968, for pollution control purposes will not be treated as being put to original use by the taxpayer regardless of whether it was used for purposes other than pollution control by its previous owner. Whether property is reconditioned or rebuilt property is a question of fact. Property will not be treated as reconditioned or rebuilt merely because it contains some used parts.

(iv) For purposes of subparagraph (1) (iii) of this paragraph (relating to property placed in service prior to January 1, 1975), the principles set forth in paragraph (d) of § 1.46-3 are applicable. Thus, property shall be considered placed in service in the earlier of the following taxable years:

(*a*) The taxable year in which, under the taxpayer's depreciation practice, the period for depreciation with respect to such property begins or would have begun; or

(*b*) The taxable year in which the property is placed in a condition or state of readiness and availability for the abatement or control of water or atmospheric pollution.

Thus, if property meets the conditions of (*b*) of this subdivision in a taxable year, it shall be considered placed in service in such year notwithstanding that the period for depreciation with respect to such property begins or would have begun in a succeeding taxable year because, for example, under the taxpayer's depreciation practice such property is or would have been accounted for in a multiple asset account and depreciation is or would have been computed under an "averaging convention" (see § 1.167(a)-10), or depreciation with respect to such property would have been computed under the completed contract method, the unit of production method, or the retirement method. In the case of property acquired by a taxpayer for use in his trade or business (or in the production of income), property shall be considered in a condition or state of readiness and availability for the abatement or control of water or atmospheric pollution if, for example, equipment is acquired for the abatement or control of water or atmospheric pollution and is operational but is undergoing testing to eliminate any defects. However, materials and parts acquired to be used in the construction of an item of equipment shall not be considered in a condition or state of readiness and availability for the abatement or control of water or atmospheric pollution.

(c) *Certification*—(1) *In general.* For purposes of paragraph (a)(1) of this section, a facility is certified in accordance with the rules prescribed in this paragraph if—

(i) The State certifying authority (as defined in subparagraph (2) of this paragraph) having jurisdiction with respect to such facility has certified to the Federal certifying authority (as defined in subparagraph (3) of this paragraph) that the facility was constructed, reconstructed, erected, or acquired in conformity with the State program or requirements for the abatement or control of water or atmospheric pollution or contamination applicable at the time of such certification, and

(ii) The Federal certifying authority has certified such facility to the Secretary or his delegate as (*a*) being in compliance with the applicable regulations of Federal agencies (such as, for example, the Atomic Energy Commission's regulations pertaining to radiological discharge (10 CFR Part 20)) and (*b*) being in furtherance of the general policy of the United States for cooperation with the States in the prevention and abatement of water pollution under the Federal Water Pollution Control Act, as amended (33 U.S.C. 1151-1175) or in the prevention and abatement of atmospheric pollution and contamination under the Clean Air Act, as amended (42 U.S.C. 1857 et seq.).

(2) *State certifying authority.* The term "state certifying authority" means—

(i) In the case of water pollution, the State water pollution control agency as defined in section 23(a) of the Federal Water Pollution Control Act, as amended (33 U.S.C. 1173(a)),

(ii) In the case of air pollution, the air pollution control agency designated pursuant to section 302(b)(1) of the Clean Air Act, as amended (42 U.S.C. 1857h(b)), and

(iii) Any interstate agency authorized to act in place of a certifying authority of a State. See section 23(a) of the Federal Water Pollution Control Act, as amended (33 U.S.C. 1173(b)) and section 302(c) of the Clean Air Act, as amended (42 U.S.C. 1857h(c)).

(3) *Federal certifying authority.* The term "Federal certifying authority" means the Administrator of the Environmental Protection Agency (see Reorganization Plan No. 3 of 1970, 35 F. R. 15623).

(d) *Profitmaking abatement works, etc.*—(1) *In general.* Section 169(e) provides that the Federal certifying authority shall not certify any property to the extent it appears that by reason of estimated profits to be derived through the recovery of wastes or otherwise in the operation of such property its costs will be recovered over the period referred to in paragraph (a)(6) of this section for such property. The Federal certifying authority need not certify the amount of estimated profits to be derived from such recovery of wastes or otherwise with respect to such facility. Such estimated profits shall be determined pursuant to subparagraph (2) of this paragraph. However, the Federal certifying authority shall certify—

(i) Whether, in connection with any treatment facility so certified, there is potential cost recovery through the recovery of wastes or otherwise, and

(ii) A specific description of the wastes which will be recovered, or the nature of such cost recovery if otherwise than through the recovery of wastes.

For effect on computation of amortizable basis, see paragraph (c) of § 1.169-3.

(2) *Estimated profits.* For purpose of this paragraph, the term "estimated profits" means the estimated gross receipts from the sale of recovered wastes reduced by the sum of the (i) estimated average annual maintenance and operating expenses, including utilities and labor, allocable to that portion of the facility which is certified as a treatment facility pursuant to paragraph (a)(1)(i) of this section which produces the recovered waste from which the gross receipts are derived, and (ii) estimated selling expenses. However, in determining expenses to be subtracted neither depreciation nor amortization of the facility is to be taken into account. Estimated profits shall not include any estimated savings to the taxpayer by reason of the taxpayer's reuse or recycling of wastes or other items recovered in connection with the operation of the plant or other property served by the treatment facility.

(3) *Special rules.* The estimates of cost recovery required by subparagraph (2) of this paragraph shall be based on the period referred to in paragraph (a)(6) of this section. Such estimates shall be made at the time the election provided for by section 169 is made and shall also be set out in the application for certification made to the Federal certifying authority. There shall be no redetermination of estimated profits due to unanticipated fluctuations in the market price for wastes or other items, to an unanticipated increase or decrease in the costs of extracting them from the gas or liquid released, or to other unanticipated factors or events occurring after certification.

§ 1.169-3. **Amortizable basis.**—(a) *In general.* The amortizable basis of a certified pollution control facility for the purpose of computing the amortization deduction under section 169 is the adjusted basis of such facility for purposes of determining gain (see part II (section 1011 and following) subchapter O, chapter 1 of the Code), as modified by paragraphs (b), (c), and (d) of this section. For the adjusted basis for purposes of determining gain (computed without regard to such modifications) of a facility which performs a function in addition to pollution control, or which is used in connection with more than one plant or other property, or both, see paragraph (a)(3) of § 1.169-2. For rules as to additions and improvements to such a facility, see paragraph (f) of this section.

(b) *Limitation to post-1968 construction, reconstruction, or erection.* (1) If the construction, reconstruction, or erection was begun before January 1, 1969, there shall be included in the amortizable basis only so much of the adjusted basis of such facility for purposes of determining gain (referred to in paragraph (a) of this section) as is properly attributable under the rules set forth in paragraph (b)(2)(iii) of § 1.169-2 to construction, reconstruction, or erection after December 31, 1968. See section 169(d)(4). For example, assume a certified pollution control facility for which the shortest period authorized under section 167 is 10 years has a cost of $500,000, of which $450,000 is attributable to construction after December 31, 1968. Further, assume such facility does not perform a function in addition to pollution control and is used only in connection with a plant in operation before January 1, 1969. The facility would have an amortizable basis of $450,000 (computed without regard to paragraphs (c) and (d) of this section). For depreciation of the remaining portion ($50,000) of the cost, see section 169(g) and paragraph (a)(3)(ii) of § 1.169-1. For the definition of the term "certified pollution control facility," see paragraph (a) of § 1.169-2.

(2) If the taxpayer elects to begin the 60-month amortization period with the first month of the taxable year succeeding the taxable year in which such facility is completed or acquired and a depreciation deduction is allowable under section 167 (including an additional first-year depreciation allowance under section 179) with respect to the facility for the taxable year in which it is completed or acquired, the amount determined under subparagraph (1) of this paragraph shall be reduced by an amount equal to (i) the amount of such allowable depreciation multiplied by (ii) a fraction the numerator of which is the amount determined under subparagraph (1) of this paragraph, and the denominator of which is its total cost. The additional first-year allowance for depreciation under section 179 will be allowable only for the year in which the facility is completed or acquired and only if the taxpayer elects to begin the amortization deduction under section 169 with the taxable year succeeding the taxable year in which such facility is completed or acquired. See paragraph (e)(1)(ii) of § 1.179-1.

(c) *Modification for profitmaking abatement works, etc.* If it appears that by reason of estimated profits to be derived through the recovery of wastes or otherwise (as determined by applying the rules prescribed in paragraph (d) of § 1.169-2) a portion or all of the total costs of the certified pollution control facility will be recovered over the period referred to in paragraph (a)(6) of § 1.169-2, its amortizable basis (computed without regard to this paragraph and paragraph (d) of this section) shall be reduced by an amount equal to (1) its amortizable basis (so computed) multiplied by (2) a fraction the numerator of which is such estimated profits and the denominator of which is its adjusted basis for purposes of determining gain. See section 169(e).

(d) *Cases in which the period referred to in paragraph (a)(6) of § 1.169-2 exceeds 15 years.* If as to a certified pollution control facility the period referred to in paragraph (a)(6) of § 1.169-2 exceeds 15 years (determined as of the first day of the first month for which a deduction is allowable under the election made under section 169(b) and paragraph (a) of § 1.169-4), the amortizable basis of such facility shall be an amount equal to (1) its amortizable basis (computed without regard to this paragraph) multiplied by (2) a fraction the numerator of which is 15 years and the denominator of which is the number of years of such period. See section 169(f)(2)(A).

(e) *Examples.* This section may be illustrated by the following examples:

*Example (1).* The X Corporation, which uses the calendar year as its taxable year, began the installation of a facility on November 1, 1968, and completed the installation on June 30, 1970, at a cost of $400,000. All of the facility qualifies as a certified pollution control facility within the meaning of paragraph (a) of § 1.169-2. $40,000 of such cost is attributable to construction prior to January 1, 1969. The X Corporation elects to take amortization deductions under section 169(a) with respect to the facility and to begin the 60-month amortization period with January 1, 1971. The corporation takes a depreciation deduction under sections 167 and 179 of $10,000 (the amount allowable, of which $2,000 is for additional first year depreciation under section 179) for the last 6 months of 1970. It is estimated that over the period referred to in paragraph (a)(6) of § 1.169-2 (20 years) as to such facility, $80,000 in profits will be realized from the sale of wastes recovered in its operation. The amortizable basis of the facility for purposes of computing the amortization deduction as of January 1, 1971, is $210,600, computed as follows:

| | | |
|---|---|---|
| (1) Portion of $400,000 cost attributable to post-1968 construction, reconstruction, or erection.............. | | $360,000 |
| (2) Reduction for portion of depreciation deduction taken for the taxable year in which the facility was completed: | | |
| (a) $10,000 depreciation deduction taken for last 6 months of 1970 including $2,000 for additional first year depreciation under section 179.......................... | $10,000 | |
| (b) Multiplied by the amount in line (1) and divided by the total cost of the facility ($360,000/$400,000)................... | 0.9 | $9,000 |
| (3) Subtotal....................................... | | $351,000 |
| (4) Modification for profitmaking abatement works: Multiply line (3) by estimated profits through waste recovery ($80,000) and divide by the adjusted basis for determining gain of the facility ($400,000) | | |
| (5) Reduction..................................... | | 70,200 |
| (6) Subtotal....................................... | | $280,800 |
| (7) Modification for period referred to in paragraph (a)(6) of § 1.169-2 exceeding 15 years: Multiply by 15 years and divide by such period (determined in accordance with paragraph (d) of this section) (20 years)............... | | $    0.75 |
| (8) Amortizable basis................................ | | $210,600 |

*Example (2).* Assume the same facts as in example (1), except that the facility is used in connection with a number of separate plants some of which were in operation before January 1, 1969, that the Federal certifying authority certifies that 80 percent of the capacity of the facility is allocable to the plants which were in operation before such date, and that all of the waste recovery is allocable to the portion of the facility used in connection with the plants in operation before January 1, 1969. The amortizable basis of such facility, for purposes of computing the amortization deduction as of January 1, 1971, is $157,950 computed as follows:

| | |
|---|---|
| (1) Adjusted basis for purposes of determining gain: Multiply percent certified as allocable to plants in operations before January 1, 1969 (80 percent) by cost of entire facility ($400,000)........................ | $320,000 |
| (2) Portion of adjusted basis for determining gain attributable to post-1968 construction, reconstruction, or erection: Multiply line (1) by portion of total cost of facility attributable to post-1968 construction, reconstruction, or erection ($360,000) and divide by the total cost of the facility ($400,000)................................ | $288,000 |

(3) Reduction for portion of depreciation deduction taken for the
taxable year in which the facility was completed:

 (a) $10,000 depreciation deduction taken for last
  6 months of 1970 including $2,000 for
  additional first year depreciation under
  section 170 . . . . . . . . . . . . . . . . . . . . . . . . . . $10,000

 (b) Multiplied by the amount in line (2) and
  divided by the total cost of the facility
  ($288,000/$400,000) . . . . . . . . . . . . . . . . . . . 0.72   7,200

(4) Subtotal . . . . . . . . . . . . . . . . . . . . . . . . . . . . . . . . . . . $280,800

(5) Modification for profit making abatement works; Multiply
line (4) by estimated profits through waste recovery
($80,000) and divide by the amount in line (1) ($320,000)

(6) Reduction . . . . . . . . . . . . . . . . . . . . . . . . . . . . . . . . . $70,200

(7) Subtotal . . . . . . . . . . . . . . . . . . . . . . . . . . . . . . . . . . $210,600

(8) Modification for period referred to in paragraph (a)(6)
of § 1.169-2 exceeding 15 years: Multiply by 15 years
and divide by such period (determined in accordance with
paragraph (d) of this section) (20 years). . . . . . . . . . . . . . . 0.75

(9) Amortizable basis. . . . . . . . . . . . . . . . . . . . . . . . . . . . . . $157,950

(f) *Additions or improvements.* (1) If after the completion or acquisition of a certified pollution control facility further expenditures are made for additional construction, reconstruction, or improvements, the cost of such additions or improvements made prior to the beginning of the amortization period shall increase the amortizable basis of such facility, but the cost of additions or improvements made after the amortization period has begun shall not increase the amortizable basis. See section 169(f)(2)(B).

(2) If expenditures for such additional construction, reconstruction, or improvements result in a facility which is new and is separately certified as a certified pollution control facility as defined in section 169(d)(1) and paragraph (a) of § 1.169-2, and, if proper election is made, such expenditures shall be taken into account in computing under paragraph (a) of this section the amortizable basis of such new and separately certified pollution control facility. [Reg. § 1.169-3.]

§ 1.169-4. **Time and manner of making elections.** — (a) *Election of amortization* — (1) *In general.* Under section 169(b), an election by the taxpayer to take an amortization deduction with respect to a certified pollution control facility and to begin the 60-month amortization period (either with the month following the month in which the facility is completed or acquired, or with the first month of the taxable year succeeding the taxable year in which such facility is completed or acquired) shall be made by a statement to that effect attached to its return for the taxable year in which falls the first month of the 60-month amortization period so elected. Such statement shall include the following information (if not otherwise included in the documents referred to in subdivision (ix) of this subparagraph):

(i) A description clearly identifying each certified pollution control facility for which an amortization deduction is claimed;

(ii) The date on which such facility was completed or acquired (see paragraph (b)(2)(iii) of § 1.169-2);

(iii) The period referred to in paragraph (a)(6) of § 1.169-2 for the facility as of the date the property is placed in service;

(iv) The date as of which the amortization period is to begin;

(v) The date the plant or other property to which the facility is connected began operating (see paragraph (a)(5) of § 1.169-2);

(vi) The total costs and expenditures paid or incurred in the acquisition, construction, and installation of such facility;

(vii) A description of any wastes which the facility will recover during the course of its operation, and a reasonable estimate of the profits which will be realized by the sale of such wastes whether pollutants or otherwise, over the period referred to in paragraph (a)(6) of § 1.169-2 as to the facility. Such estimate shall include a schedule setting forth a detailed computation illustrating how the estimate was computed including every element prescribed in the definition of estimated profits in paragraph (d)(2) of § 1.169-2;

(viii) A computation showing the amortizable basis (as defined in § 1.169-3) of the facility as of the first month for which the amortization deduction provided for by section 169(a) is elected; and

(ix) (a) A statement that the facility has been certified by the Federal certifying authority, together with a copy of such certification, and a copy of the application for certification which was filed with and approved by the Federal certifying authority or (b), if the facility has not been certified by the Federal certifying authority, a statement that application has been made to the proper State certifying authority (see paragraph (c)(2) of § 1.169-2) together with a copy of such application and (except in the case of an election to which subparagraph (4) of this paragraph applies) a copy of the application filed or to be filed with the Federal certifying authority.

If subdivision (ix)(b) of this subparagraph applies, within 90 days after receipt by the taxpayer, the certification from the Federal certifying authority shall be filed by the taxpayer with the district director, or with the director of the internal revenue service center, with whom the return referred to in this subparagraph was filed.

(2) *Special rule.* If the return for the taxable year in which falls the first month of the 60-month amortization period to be elected is filed before November 16, 1971, without making the election for such year, then on or before December 31, 1971 (or if there is no State certifying authority in existence on November 16, 1971, on or before the 90th day after such authority is established), the election may be made by a statement attached to an amended income tax return for the taxable year in which falls the first month of the 60-month amortization period so elected. Amended income tax returns or claims for credit or refund must also be filed at this time for other taxable years which are within the amortization period and which are subsequent to the taxable year for which the election is made. Nothing in this paragraph should be construed as extending the time specified in section 6511 within which a claim for credit or refund may be filed.

(3) *Other requirements and considerations.* No method of making the election provided for in section 169(a) other than that prescribed in this section shall be permitted on or after May 18, 1971. A taxpayer which does not elect in the manner prescribed in this section to take amortization deductions with respect to a certified pollution control facility shall not be entitled to such deductions. In the case of a

taxpayer which elects prior to May 18, 1971, the statement required by subparagraph (1) of this paragraph shall be attached to its income tax return for either its taxable year in which December 31, 1971 occurs or its taxable year preceding such year.

(4) *Elections filed before February 29, 1972.* If a statement of election required by subparagraph (1) of this paragraph is attached to a return (including an amended return referred to in subparagraph (2) of this paragraph) filed before February 29, 1972, such statement of election need not include a copy of the Federal application to be filed with the Federal certifying authority but a copy of such application must be filed no later than February 29, 1972, by the taxpayer with the district director, or with the director of the internal revenue service center, with whom the return or amended return referred to in this subparagraph was filed.

(b) *Election to discontinue or revoke amortization*—(1) *Election to discontinue.* An election to discontinue the amortization deduction provided by section 169(c) and paragraph (a)(1) of § 1.169-1 shall be made by a statement in writing filed with the district director, or with the director of the internal revenue service center, with whom the return of the taxpayer is required to be filed for its taxable year in which falls the first month for which the election terminates. Such statement shall specify the month as of the beginning of which the taxpayer elects to discontinue such deductions. Unless the election to discontinue amortization is one to which subparagraph (2) of this paragraph applies, such statement shall be filed before the beginning of the month specified therein. In addition, such statement shall contain a description clearly identifying the certified pollution control facility with respect to which the taxpayer elects to discontinue the amortization deduction, and, if a certification has previously been issued, a copy of the certification by the Federal certifying authority. If at the time of such election a certification has not been issued (or if one has been issued but has not been filed as provided in paragraph (a)(1) of this section), the taxpayer shall file, with respect to any taxable year or years for which a deduction under section 169 has been taken, a copy of such certification within 90 days after receipt thereof. For purposes of this paragraph, notification to the Secretary or his delegate from the Federal certifying authority that the facility no longer meets the requirements under which certification was originally granted by the State or Federal certifying authority shall have the same effect as a notice from the taxpayer electing to terminate amortization as of the month following the month such facility ceased functioning in accordance with such requirements.

(2) *Revocation of elections made prior to May 18, 1971.* If on or before May 18, 1971, an election under section 169(a) has been made, such election may be revoked (see paragraph (a)(1) of § 1.169-1) by filing on or before August 16, 1971, a statement of revocation of an election under section 169(a) in accordance with the requirements in subparagraph (1) of this paragraph for filing a notice to discontinue an election. If such election to revoke is for a period which falls within one or more taxable years for which an income tax return has been filed, amended income tax returns shall be filed for any such taxable years in which deductions were taken under section 169 on or before August 16, 1971. [Reg. § 1.169-4.]

# EPA Regulations

§ 20.1. **Applicability.** — The regulations of this part apply to certifications by the Administrator of water or air pollution control facilities for purposes of section 169 of the Internal Revenue Code of 1954, as amended, 26 U.S.C. 169. Applicable regulations of the Department of the Treasury are set forth at 26 CFR 1.169 et seq. [Reg. § 20.1.]

§ 20.2. **Definitions.** — As used in this part, the following terms shall have the meaning indicated below:

(a) "Act" means, when used in connection with water pollution control facilities, the Federal Water Pollution Control Act, as amended (33 U.S.C. 1151 et seq.) or, when used in connection with air pollution control facilities, the Clean Air Act, as amended (42 U.S.C. 1857 et seq.).

(b) "State certifying authority" means:

(1) For water pollution control facilities, the State health authority, except that, in the case of any State in which there is a single State agency, other than the State health authority, charged with responsibility for enforcing State laws relating to the abatement of water pollution, it means such other State agency; or

(2) For air pollution control facilities, the air pollution control agency designated pursuant to section 302(b)(1) of the Act; or

(3) For both air and water pollution control facilities, any interstate agency authorized to act in place of the certifying agency of a State.

(c) "Applicant" means any person who files an application with the Administrator for certification that a facility is in compliance with the applicable regulations of Federal agencies and in furtherance of the general policies of the United States for cooperation with the States in the prevention and abatement of water or air pollution under the Act.

(d) "Administrator" means the Administrator, Environmental Protection Agency.

(e) "Regional Administrator" means the Regional designee appointed by the Administrator to certify facilities under this part.

(f) "Facility" means property comprising any new identifiable treatment facility which removes, alters, disposes of or stores pollutants, contaminants, wastes, or heat.

(g) "State" means the States, the District of Columbia, the Commonwealth of Puerto Rico, the Canal Zone, Guam, American Samoa, the Virgin Islands, and the Trust Territory of the Pacific Islands. [Reg. § 20.2.]

§ 20.3. **General Provisions.** —(a) An applicant shall file an application in accordance with this part for each separate facility for which certification is sought: *Provided,* That one application shall suffice in the case of substantially identical facilities which the applicant has installed or plans to install in connection with substantially identical properties: *Provided further,* That an application may incorporate by reference material contained in an application previously submitted by the applicant under this part and pertaining to substantially identical facilities.

(b) The applicant shall, at the time of application to the State certifying authority, submit an application in the form prescribed by the Administrator to the Regional Administrator for the region in which the facility is located.

(c) Applications will be considered complete and will be processed when the Regional Administrator receives the completed State certification.

(d) Applications may be filed prior or subsequent to the commencement of construction, acquisition, installation, or operation of the facility.

(e) An amendment to an application shall be submitted in the same manner as the original application and shall be considered a part of the original application.

(f) If the facility is certified by the Regional Administrator, notice of certification will be issued to the Secretary of the Treasury or his delegate, and a copy of the notice shall be forwarded to the applicant and to the State certifying authority. If the facility is denied certification, the Regional Administrator will advise the applicant and State certifying authority in writing of the reasons therefor.

(g) No certification will be made by the Regional Administrator for any facility prior to the time it is placed in operation and the application, or amended application, in connection with such facility so states.

(h) An applicant may appeal any decision of the Regional Administrator which:

(1) Denies certification;

(2) Disapproves the applicant's suggested method of allocating costs pursuant to § 602.8(e) [20.8(e)]; or

(3) Revokes a certification pursuant to § 602.10 [20.10].

Any such appeal may be taken by filing with the Administrator within 30 days from the date of the decision of the Regional Administrator a written statement of objections to the decision appealed from. Within 60 days, the Administrator shall affirm, modify, or revoke the decision of the Regional Administrator, stating in writing his reasons therefor. [Reg. § 20.3.]

§ 20.4. **Notice of intent to certify.** —(a) On the basis of applications submitted prior to the construction, reconstruction, erection, acquisition, or operation of a facility, the Regional Administrator may notify applicants that such facility will be certified if:

(1) The Regional Administrator determines that such facility, if constructed, reconstructed, erected, acquired, installed, and operated in accordance with such application, will be in compliance with requirements identified in § 602.8 [20.8]; and if

(2) The application is accompanied by a statement from the State certifying authority that such facility, if constructed, reconstructed, acquired, erected, installed,

and operated in accordance with such application, will be in conformity with the State program or requirements for abatement or control of water or air pollution.

(b) Notice of actions taken under this section will be given to the appropriate State certifying authority. [Reg. § 20.4.]

§ 20.5. **Applications.**—Applications for certification under this part shall be submitted in such manner as the Administrator may prescribe, shall be signed by the applicant or agent thereof, and shall include the following information:

(a) Name, address, and Internal Revenue Service identifying number of the applicant;

(b) Type and narrative description of the new identifiable facility for which certification is (or will be) sought, including a copy of schematic or engineering drawings, and a description of the function and operation of such facility;

(c) Address (or proposed address) of facility location;

(d) A general description of the operation in connection with which such facility is (or will be) used and a description of the specific process or processes resulting in discharges or emissions which are (or will be) controlled by the facility;

(e) If the facility is (or will be) used in connection with more than one plant or other property, one or more of which were not in operation prior to January 1, 1969, a description of the operations of the facility in respect to each plant or other property, including a reasonable allocation of the costs of the facility among the plants being serviced, and a description of the reasoning and accounting method or methods used to arrive at such allocation;

(f) Description of the effect of such facility in terms of type and quantity of pollutants, contaminants, wastes or heat, removed, altered, stored, or disposed of by such facility;

(g) If the facility performs a function other than removal, alteration, storage, or disposal of pollutants, contaminants, wastes or heat, a description of all functions performed by the facility, including a reasonable identification of the costs of the facility allocable to removal, alteration, storage, or disposal of pollutants, contaminants, wastes or heat, and a description of the reasoning and the accounting method or methods used to arrive at such allocation;

(h) Date when such construction, reconstruction, or erection will be completed or when such facility was (or will be) acquired;

(i) Date when such facility is placed (or is intended to be placed) in operation;

(j) Identification of the applicable State and local water or air pollution control requirements and standards, if any;

(k) Expected useful life of facility;

(l) Cost of construction, acquisition, installation, operation, and maintenance of the facility;

(m) Estimated profits reasonably expected to be derived through the recovery of wastes or otherwise in the operation of the facility over the period referred to in paragraph (a)(6) of 26 CFR 1.169-2;

(n) Such other information as the Administrator deems necessary for certification. [Reg. § 20.5.]

§ 20.6. **State certification.**—The State certification shall be by the State certifying authority having jurisdiction with respect to the facility in accordance with 26 U.S.C. 169(d)(1)(A) and (d)(2). The certification shall state that the facility described in the application has been constructed, reconstructed, erected, or acquired in conformity with the State program or requirements for abatement or control of water or air pollution. It shall be executed by an agent or officer authorized to act on behalf of the State certifying authority. [Reg. § 20.6.]

§ 20.7. **General policies.**—(a) The general policies of the United States for cooperation with the States in the prevention and abatement of water pollution are: To enhance the quality and value of our water resources; to eliminate or reduce the pollution of the nation's waters and tributaries thereof; to improve the sanitary condition of surface and underground waters; and to conserve such waters for public water supplies, propagation of fish and aquatic life and wildlife, recreational purposes, and agricultural, industrial, and other legitimate uses.

(b) The general policy of the United States for cooperation with the States in the prevention and abatement of air pollution is to cooperate with and to assist the States and local governments in protecting and enhancing the quality of the Nation's air resources by the prevention and abatement of conditions which cause or contribute to air pollution which endangers the public health or welfare. [Reg. § 20.7.]

§ 20.8. **Requirements for certification.**—(a) Subject to § 602.9 [20.9], the Regional Administrator will certify a facility if he makes the following determinations:

(1) It has been certified by the State certifying authority.

(2) It removes, alters, disposes of, or stores pollutants, contaminants, wastes or heat, which, but for the facility, would be released into the environment.

(3) The applicant is in compliance with all regulations of Federal agencies applicable to use of the facility, including conditions specified in any permit issued to the applicant under section 13 of the Rivers and Harbors Act of 1899, as amended.

(4) The facility furthers the general policies of the United States and the States in the prevention and abatement of pollution.

(5) The applicant has complied with all the other requirements of this part and has submitted all requested information.

(b) In determining whether use of a facility furthers the general policies of the United States and the States in the prevention and abatement of water pollution, the Regional Administrator shall consider whether such facility is consistent with the following, insofar as they are applicable to the waters which will be affected by the facility:

(1) All applicable water quality standards, including water quality criteria and plans of implementation and enforcement established pursuant to section 10(c) of the Act or State laws or regulations;

(2) Recommendations issued pursuant to section 10(e) and (f) of the Act;

(3) Water pollution control programs established pursuant to section 3 or 7 of the Act.

(c) In determining whether use of a facility furthers the general policies of the United States and the States in the prevention and abatement of air pollution, the Regional Administrator shall consider whether such facility is consistent with and meets the following requirements, insofar as they are applicable to the air which will be affected by the facility;

(1) Plans for the implementation, maintenance, and enforcement of ambient air quality standards adopted or promulgated pursuant to section 110 of the Act;

(2) Recommendations issued pursuant to sections 103(e) and 115 of the Act

which are applicable to facilities of the same type and located in the area to which the recommendations are directed;

(3) Local government requirements for control of air pollution, including emission standards;

(4) Standards promulgated by the Administrator pursuant to the Act.

(d) A facility which removes elements or compounds from fuels which would be released as pollutants when such fuels are burned may not be certified whether or not such facility is used in connection with the applicant's plant or property where such fuels are burned.

(e) Where a facility is used in connection with more than one plant or other property, one or more of which were not in operation prior to January 1, 1969, or where a facility will perform a function other than the removal, alteration, storage or disposal of pollutants, contaminants, wastes, or heat, the Regional Administrator will so indicate on the notice of certification and will approve or disapprove the applicant's suggested method of allocation of costs. If the Regional Administrator disapproves the applicant's suggested method, he shall identify the proportion of costs allocable to each such plant, or to the removal, alteration, storage or disposal of pollutants, contaminants, wastes, or heat. [Reg. § 20.8.]

§ 20.9. **Cost recovery.** — Where it appears that, by reason of estimated profits to be derived through the recovery of wastes, through separate charges for use of the facility in question, or otherwise in the operation of such facility, all or a portion of its costs may be recovered over the period referred to in paragraph (a)(6) of 26 CFR 1.169-2, the Regional Administrator shall so signify in the notice of certification. Determinations as to the meaning of the term "estimated profits" and as to the percentage of the cost of a certified facility which will be recovered over such period shall be made by the Secretary of the Treasury, or his delegate: *Provided,* That in no event shall estimated profits be deemed to arise from the use or reuse by the applicant of recovered waste. [Reg. § 20.9.]

§ 20.10. **Revocation.** — Certification hereunder may be revoked by the Regional Administrator on 30 days' written notice to the applicant, served by certified mail, whenever the Regional Administrator shall determine that the facility in question is no longer being operated consistent with the § 602.8(b) and (c) [20.8(b) and (c)] criteria in effect at the time the facility was placed in service. Within such 30-day period, the applicant may submit to the Regional Administrator such evidence, data or other written materials as the applicant may deem appropriate to show why the certification hereunder should not be revoked. Notification of a revocation under this section shall be given to the Secretary of the Treasury or his delegate. See 26 CFR 1.169-4(b)(1). [Reg. § 20.10.]

# EPA Guidelines
# 36 FR 189

## Pollution Control Facilities

Guidelines for Certification

On May 26, 1971, the Environmental Protection Agency published final regulations in the *FEDERAL REGISTER* (36 F.R. 9509) pursuant to section 169 of the Internal Revenue Code of 1954, added by section 704 of the Tax Reform Act of 1969, Public Law 91-172. The EPA regulations are complementary to final regulations issued under section 169 by the Treasury Department, published on May 18, 1971 (36 F.R. 9010).

The 10 Regional Offices of EPA will be primarily responsible for administration of the certification procedures. In order to insure that similar applications for certification receive similar treatment in different regional offices, the interpretative guidelines printed below are being issued to the regional offices and are published in the *FEDERAL REGISTER* for the information of affected members of the public.

Dated: September 23, 1971.

WILLIAM D. RUCKELSHAUS,
*Administrator, Environmental
Protection Agency.*

GUIDELINES

1. General.
2. Air pollution control facilities.
a. Pollution control or treatment facilities normally eligible for certification.
b. Air pollution control facility boundaries.

c. Examples of eligibility limits.
d. Replacement of manufacturing process by another nonpolluting process.
3. Water pollution control facilities.
a. Pollution control or treatment facilities normally eligible for certification.
b. Examples of eligibility limits.
4. Multiple-purpose facilities.
5. Facilities serving both old and new plants.
6. State certification.
7. Dispersal of pollutants.
8. Profit-making facilities.
9. Multiple applications.


GUIDELINES

1. *General.* Section 704 of the Tax Reform Act of 1969 (Public Law 91-172, December 30, 1969), added a new section 169, "Amortization of Pollution Control Facilities", to the Internal Revenue Code. The new section provides for the amortization of the cost of certified pollution control facilities over a sixty-month period, if certain conditions are met.

The Act defines a "certified pollution control facility" as a "new identifiable treatment facility" which is:

(a) Used in connection with a plant or other property in operation before January 1, 1969, to abate or control pollution by removing, altering, disposing of, or storing pollutants, contaminants, wastes or heat;

(b) Constructed, reconstructed, erected or (if purchased) first placed in service by the taxpayer after December 31, 1968;

(c) Placed in service before January 1, 1975; and

(d) Certified by both State and Federal authorities, as provided in section 169(d)(1)(A) and (B).

If the facility is a building, moreover, the statute requires that it be exclusively devoted to pollution control. Most questions as to whether a facility is a "building", and, if so, whether it is "exclusively" devoted to pollution control are resolved by §1.169-2(b)(2) of the Treasury Department regulations.

2. *Air pollution control facilities*—a. *Pollution control or treatment facilities normally eligible for certification.* The following devices may constitute such facilities for the removal, alteration, disposal or storage of air pollution:

(1) Inertial separators (cyclones, etc.).
(2) Wet collection devices (scrubbers).
(3) Electrostatic precipitators.
(4) Cloth filter collectors (baghouses).
(5) Direct fired afterburners.
(6) Catalytic afterburners.
(7) Gas absorption equipment.
(8) Gas adsorption equipment.
(9) Vapor condensers.
(10) Vapor recovery ss tems.
(11) Floating roofs for storage tanks.
(12) Combinations of the above.

(b) *Air pollution control facility boundaries.* Most facilities are systems consisting of several parts. The facility need not start at the point where the gaseous effluent leaves the last unit of processing equipment, nor will it in all cases extend to the point where the effluent is emitted to the atmosphere or existing stack, breeching, ductwork or vent. It includes all the auxiliary equipment used to operate the control system, such as fans, blowers, ductwork, valves, dampers, electrical equipment, etc. It also includes all equipment used to handle, store, transport, or dispose of the collected pollutant material.

c. *Examples of eligibility limits.* The amortization deduction is limited to any new identifiable treatment facility which removes, alters, disposes of, or stores contaminants or wastes. It is not available for all expenditures for air pollution control and is limited to devices which actually remove, alter, destroy, dispose of or store air pollutants.

(1) *Boiler modifications or replacements.* Modifications of boilers to accommodate "cleaner" fuels are not eligible for amortization: e.g., removal of stokers from a coal-fired boiler and the addition of gas or oil burners. The purpose of the burners is to produce heat, and they do not qualify as air pollution control facilities. A new gas or oil fired boiler that replaced a coal-fired boiler would also be ineligible for certification.

(2) *Fuel processing.* Eligible air pollution control facilities do not include preprocessing equipment which removes potential air pollutants from fuels prior to their combustion. Thus, a desulfurization facility would not be eligible, irrespective of whether it is installed at the plant where the desulfurized coal is burned.

(3) *Incinerators.* The addition of an afterburner, secondary combustion chamber or particulate collector would be eligible.

(4) *Collection devices used to collect product or process material.* In some manufacturing operations, collection devices are used to collect product or process material, as in the case of the manufacture of carbon black. The baghouse would be eligible for certification, but the certification would alert the Treasury Department of the profitable waste recovery involved. See paragraph 8, below.

d. *Replacement of manufacturing process by another nonpolluting process.* An installation will not qualify for certification where it utilizes a process known to be "cleaner" than an alternative, but where it does not actually remove, alter or dispose of pollution; as, for example, a minimally polluting electric induction furnace to melt cast iron which replaces, or is installed instead of, a heavily polluting iron cupola furnace. However, if the replacement equipment has an air pollution control device added to it, the control device would be eligible while the process device would not. For example, in the case where a primary copper smelting reverberatory furnace is replaced by a flash smelting furnace, followed by the installation of a contact sulfuric acid plant, the contact sulfuric acid plant would qualify (since it is a control device not necessary to the process), while the flash smelting furnace would not qualify, as its purpose is to produce copper matte.

3. *Water Pollution Control Facilities.—*a. *Pollution control or treatment facilities normally eligible for certification.* The following types of equipment may constitute such facilities for the removal, alteration, disposal or storage of water pollution:

(1) Pretreatment facilities such as those which neutralize or stabilize industrial and/or sanitary waste, from a point immediately preceding the point of such treatment to a point of disposal to, and acceptance by, a municipal waste treatment

facility for final treatment, including the necessary pumping and transmitting facilities.

(2) Treatment facilities such as those which neutralize or stabilize industrial and or sanitary waste, in compliance with established Federal, State, and local effluent or water quality standards, from a point immediately preceding the point of such treatment to a point of disposal, including the necessary pumping and transmitting facilities.

(3) Ancillary devices and facilities such as lagoons, ponds, and structures for the storage and/or treatment of wastewaters or waste from a plant or other property.

(4) Devices, equipment, or facilities constructed or installed for the primary purpose of recovering a byproduct of the operation (saleable or otherwise), previously lost either to the atmosphere or to the waste effluent. Examples are:

(a) A facility to concentrate and recover vaporous byproducts from a process stream for reuse as raw feedstock or for resale, unless the estimated profits from resale exceed the cost of the facility. See paragraph 8, below.

(b) A facility to concentrate and/or remove "gunk" or similar type "tars" or polymerized tar-like materials from the process waste effluent previously discharged in the plant effluents.

(c) A device used to extract or remove a soluble constituent from a solid or liquid by use of a selective solvent; an open or closed tank or vessel in which such extraction or removal occurs; a diffusion battery of tanks or vessels for countercurrent decantation, extraction, or leaching, etc.

(d) A skimmer or similar device for the removal of greases, oils and fat-like materials from an effluent stream.

b. *Examples of eligibility limits.* (1) Inplant process changes which may prevent the production of pollutants, contaminants, wastes, or heat, but which by themselves cannot be said to remove, alter, dispose of, or store pollutants, contaminants, wastes, or heat, will not be considered eligible for certification as a water pollution control facility.

(2) Any device, equipment and/or facility which is associated with or included in a disposal system for subsurface injection of untreated or inadequately treated industrial or sanitary waste waters or effluent containing pollutants, contaminants or wastes will not be eligible.

4. *Multiple-purpose facilities.* As the regulations make clear, a facility can qualify for favorable tax treatment if it serves a function other than the abatement of pollution (unless it is a building). Otherwise, the effect might have been to discourage installation of sensible pollution abatement facilities in favor of less efficient single-function facilities which qualified for the deduction.

Accordingly, EPA must decide what percentage of a given facility's cost is properly allocable to its abatement function. The regulations require the applying taxpayer to make such an allocation in his application, and to justify his grounds therefor. The regional offices will review those allocations. Although not generally necessary or desirable for the purpose of such review, on-site inspections may be appropriate in cases involving large sums of money and or unusual types of equipment.

5. *Facilities serving both old and new plants.* As noted above, the statute requires that a pollution control facility must be used in connection with a plant or other property that was in operation prior to January 1, 1969. When a facility is used in connection with pre-1969 properties as well as in connection with newer ones, it

may qualify for the rapid amortization to the extent it is used in connection with pre-1969 facilities.

Again, the taxpayer will submit his theory of the allocation of the cost of the facility as between old and new plants or properties, and the regional offices will review that allocation. Such an allocation will result in a percentage. The most appropriate method of making such an allocation is to compare the effluent capacity of the pre-1969 plant to the treatment capacity of the control facility. For example, if the old plant has a capacity of 80 units of effluent (but an average output of 60 units), the new plant has a capacity of 40 units (but an average output of 20 units), and the control facility a capacity of 150 units, then 80/150 of the cost of the control facility would be eligible for rapid amortization.

Should a taxpayer present a seemingly reasonable method of allocation different from the foregoing, Regional Office personnel are invited to consult with the Office of Water Programs or the Air Programs Office, whichever is appropriate, and with the Office of the General Counsel.

6. *State certification.* In order to qualify for rapid amortization under section 169, a facility must first be certified by the State as being installed "in conformity with the State program or requirements for abatement or control of water or atmospheric pollution or contamination." Significantly, the statute does not say that installation of a facility must be required by a State. Accordingly, assuming that use of a facility will not contravene any applicable State requirements, it will be eligible for accelerated depreciation. One example would be a facility installed in order to comply with regulations of the Atomic Energy Commission on emissions of radioactive particulates. The same result would obtain in cases where the certifying State had not yet adopted an implementation plan under the Clean Air Act to meet national ambient air quality standards.

It is contemplated that the facts contained in the taxpayer's application, plus the certification from the State agency, will form the basis for EPA certification. By heavily relying on the State's certification, the administrative task of the regional offices can and should be minimized. It is not contemplated that on-site inspection will be necessary or desirable in the majority of cases. Exceptions to the foregoing must of course depend on the exercise of sound judgment by Regional Office personnel.

Of obvious relevance to the exercise of such judgment would be: The volume and toxicity of the discharge sought to be controlled by the facility in question; the amount of money at stake; and experience on the basis of which it may be said that the certifying State agency is in fact ignoring obvious violations of applicable water or air quality standards.

It should be noted that certification of a facility does not constitute the personal warranty of the certifying official that the conditions of the statute have been met; as is the case with a ruling from the Internal Revenue Service itself, EPA certification is only binding on the Government to the extent the submitted facts are accurate and complete.

7. *Dispersal of pollutants.* Section 169 applies to facilities which remove, alter, dispose of, or store pollutants—including heat. As the legislative history of the section makes clear, a facility which merely disperses pollutants cannot qualify. It is not felt, however, that the legislative intent to disqualify facilities which only disperse pollution is applicable to facilities which dissipate heat; there is no way to "dispose of" heat other than by transferring B.t.u.'s to the environment. A cooling

tower will be eligible for certification, therefore, provided it is used in connection with a pre-1969 plant, as will a cooling pond, or an addition to an outfall structure, the result of which is to lessen the amount by which the temperature of the receiving stream is elevated, and which meets applicable State standards.

8. *Profit-making facilities.* The statute denies favorable tax treatment to facilities the cost of which will be recovered from profits derived through the recovery of waste, "or otherwise."

If an abatement facility recovers marketable wastes, estimated profits on which are not sufficient to recover the entire cost of the facility, the amortizable basis of the facility will be reduced in accordance with Treasury Department regulations. The responsibility of the regional offices will be to identify for the Treasury Department's benefit those cases in which estimated profits will in fact arise; their amount, and the extent to which they can be expected to result in cost recovery, will be determined by the Treasury Department. Accordingly, the responsibility of the regional offices is, for all practical purposes, to notify the Treasury Department when marketable byproducts are recovered by the facility. Such notification will be included in EPA's form of certification.

The phrase "or otherwise" also encompasses situations where the taxpayer is in the business of renting his facility for a fee. In such a case, the facility may theoretically qualify for certification by EPA, the decision as to the extent of its profitability being left to the Treasury Department. Situations may also arise where use of a facility is furnished at no additional charge to a number of users, or to the public, as part of a package of other services. In such a case, no profits will be deemed to arise from operation of the facility unless the other services included in the package are merely ancillary to use of the facility.

It should be noted that §602.9 of the EPA regulations is not meant to affect general principles of Federal income tax law. An individual other than the title holder of a piece of property may be entitled to take depreciation deductions on it, if the arrangements by which he has use of the property may, for all practical purposes, be viewed as a purchase. In any such case, the facility could qualify for full rapid amortization, notwithstanding that the title holder charges a separate fee for the use of the facility, as long as the taxpayer—in such a case, the user—does not himself charge a separate fee for the use of the facility.

9. *Multiple applications.* Under EPA regulations, a multiple application may be submitted by a taxpayer who applies for certification of substantially identical pollution abatement facilities used in connection with substantially identical properties. It is not contemplated that use of the multiple application option will be used with respect to facilities in different States, since each such facility would require separate applications for certification from the different States involved. EPA regulations also permit an applicant to incorporate by reference in an application material contained in an application previously filed. The purpose of the provision for incorporation by reference is to avoid the burden of furnishing detailed information (which may in some instances include portions of catalogs or process flow diagrams) which the certifying official has previously received. Accordingly, material filed with a Regional Office of EPA may be incorporated by reference only in an application subsequently filed with the same regional office.

[FR Dec. 71-14267 Filed 8-28-71; 8:45 am]

# EPA Regional Offices

Region I
Room 2203, John F. Kennedy Building
Boston, Massachusetts 02203
(617) 223-7210

States: Connecticut, Maine, Massachusetts, New Hampshire, Rhode Island, and Vermont

Region II
Room 1009, 26 Federal Plaza
New York, New York 10007
(212) 264-2525

States: New Jersey, New York, Puerto Rico, and Virgin Islands

Region III
Curtis Building
6th and Walnut Streets
Philadelphia, Pennsylvania 19106

States: Delaware, Maryland, Pennsylvania, Virginia, West Virginia, and Washington, D.C.

Region IV
345 Courtland Street, N.E.
Atlanta, Georgia 30308
(404) 881-4727

States: Alabama, Florida, Georgia, Kentucky, Mississippi, North Carolina, South Carolina, and Tennessee

Region V
Federal Building
230 South Dearborn
Chicago, Illinois 60604
(312) 353-2000

States: Illinois, Indiana, Minnesota, Michigan, Ohio, and Wisconsin

Region VI
First International Building
Suite 1110
1201 Elm Street
Dallas, Texas 75201
(214) 767-2600

States: Arkansas, Louisiana, New Mexico, Texas, and Oklahoma

Region VII
1735 Baltimore Avenue
Kansas City, Missouri 64108
(816) 374-5493

States: Iowa, Kansas, Missouri, and Nebraska

Region VIII
Suite 900, 1860 Lincoln Street
Denver, Colorado 80203
(303) 837-3895

States: Colorado, Montana, North Dakota, South Dakota, Utah, and Wyoming

Region IX
  215 Fremont Street
  San Francisco, California 94111
  (415) 556-2320

  States: Arizona, California, Nevada,
    Hawaii, Guam, American Samoa,
    Trust Territories, and Wake Island

Region X
  1200 Sixth Avenue
  Seattle, Washington 98101
  (206) 442-1220

  States: Idaho, Oregon, Washington,
    and Alaska

# State Water Quality Agencies

*Alabama*

Water Improvement Commission
State Office Building
Montgomery, Alabama 36130

*Alaska*

Department of Environment Conservation
Division of Environmental Quality
  Management
Pouch "O"
Juneau, Alaska 99801

*Arizona*

Department of Health
Division of Environmental Health
  Services
Bureau of Water Quality Control
1740 West Adams Street
Phoenix, Arizona 85007

*Arkansas*

Department of Pollution Control and
  Ecology
Water Division, 208 Planning Section
8001 National Drive
Little Rock, Arkansas 72209

*California*

Water Resources Control Board
P.O. Box 100
Sacramento, California 95801

*Colorado*

Department of Health
Water Quality Control Division
4210 East 11th Avenue
Denver, Colorado 80220

*Connecticut*

Department of Environmental Protection
Water Compliance and Hazardous Substances
122 Washington Street
Hartford, Connecticut 06115

*Delaware*

Department of Natural Resources and
  Environmental Control
Division of Environmental Control, Water
  Resources Management
Tatnall Building
Dover, Delaware 19901

*District of Columbia*

Department of Environmental Services
Bureau of Air and Water Quality
5010 Overlook Avenue S.W.
Washington, D.C. 20032

*Florida*

Department of Environmental Regulation
Division of Environmental Programs
Water Quality Planning Section
2500 Blair Stone Road
Twin Towers Office Building
Tallahassee, Florida 32301

*Georgia*

Department of Natural Resources
Environmental Protection Division,
   Water Protection Branch
270 Washington Street, S.W., Room 822
Atlanta, Georgia 30334

*Hawaii*

Department of Health
Environmental Health Division
P.O. Box 3378
Honolulu, Hawaii 96801

*Idaho*

Department of Health and Welfare
Bureau of Water Quality
State House
Boise, Idaho 83720

*Illinois*

Environmental Protection Agency
2200 Churchill Road
Springfield, Illinois 62706

*Indiana*

Stream Pollution Control Board/Water
   Pollution Control Division
1330 West Michigan Street
Indianapolis, Indiana 46206

*Iowa*

Department of Environmental Quality
Chemicals and Water Quality Division
3920 Delaware Avenue, Henry A.
   Wallace Building
900 E. Grand Avenue
Des Moines, Iowa 50319

*Kansas*

Department of Health and Environment
Division of Environment
Bureau of Water Quality
Forbes AFB Bldg. No. 740
Topeka, Kansas 66620

*Kentucky*

Department of Natural Resources and
   Environmental Protection
Bureau of Environmental Protection
Division of Water Quality
Century Plaza—U.S. 127 South
Frankfort, Kentucky 40601

*Louisiana*

Division of Environmental Services
   Water Quality Section
P.O. Box 60630
Baton Rouge, Louisiana 70160

*Maine*

Department of Environmental Protection
Bureau of Water Quality Control
State House
Augusta, Maine 04333

*Maryland*

Department of Natural Resources
Water Resources Administration
Tawes State Office Building
Annapolis, Maryland 21401

*Massachusetts*

Department of Environmental Quality
   Engineering
Division of Water Pollution Control
110 Tremont Street
Boston, Massachusetts 02108

*Michigan*

Department of Natural Resources
Water Resources Commission
Bureau of Water Quality Division
P.O. Box 30028
Lansing, Michigan 48909

*Minnesota*

Minnesota Pollution Control Agency
Division of Water Pollution Control
1935 West County Road B2
Roseville, Minnesota 55113

*Mississippi*

Department of Natural Resources
Water Division
P.O. Box 827, Robert E. Lee Building
Jackson, Mississippi 39205

*Missouri*

Department of Natural Resources
Water Quality Program
Division of Environmental Quality
P.O. Box 1368
Jefferson City, Missouri 65101

*Montana*

Department of Health and Environmen-
tal Sciences
Division of Environmental Sciences
Water Quality Bureau
Cogswell Building
Helena, Montana 59601

*Nebraska*

Department of Environmental Control
Water Pollution Control Division
P.O. Box 94877—301 Centennial Mall
Lincoln, Nebraska 68509

*Nevada*

Department of Conservation and
Natural Resources
Water Resources Division
201 South Fall Street
Carson City, Nevada 89710

*New Hampshire*

Water Supply and Pollution Control
Commission
Hazen Drive
P.O. Box 95
Concord, New Hampshire 03301

*New Jersey*

Department of Environmental Protection
Division of Water Resources
P.O. Box CN029
Trenton, New Jersey 08625

*New Mexico*

Environmental Improvement Agency
Water Quality Division
P.O. Box 968, Crown Building
Santa Fe, New Mexico 87501

*New York*

Department of Environmental
Conservation
Division of Waters
50 Wolf Road
Albany, New York 12233

*North Carolina*

Department of Natural Resources and
Community Development
Division of Environmental Management
Water Quality Section
P.O. Box 27687
Raleigh, North Carolina 27611

*North Dakota*

Department of Health
Division of Water Supply and Pollution
Control
State Capitol
Bismarck, North Dakota 58505

*Ohio*

Environmental Protection Agency
Waste Water Pollution Control
361 E. Broad Street
P.O. Box 1049
Columbus, Ohio 43216

*Oklahoma*

Department of Pollution Control
  (planning and coordination)
P.O. Box 53504
NE 10th and Stonewall
Oklahoma City, Oklahoma 73152

Water Resources Board
Water Quality Division
1000 NE 10th, 12th Floor
Oklahoma City, Oklahoma 73105

Department of Health
Environmental Health Services
Water Quality Services
NE 10th and Stonewall
Oklahoma City, Oklahoma 73152

*Oregon*

Department of Environmental Quality
Water Quality Division
P.O. Box 1760
Portland, Oregon 97205

*Pennsylvania*

Department of Environmental Resources
Bureau of Water Quality Management
P.O. Box 2063
Harrisburg, Pennsylvania 17120

*Puerto Rico*

Environmental Quality Board
Water Quality Bureau
P.O. Box 11488
Santurce, Puerto Rico 00910

*Rhode Island*

Department of Environmental
  Management
Division of Water Resources
75 Davis Street, Room 209
Providence, Rhode Island 02908

*South Carolina*

Department of Health and Environmen-
  tal Control
Environmental Quality Control
Bureau of Wastewater and Stream
  Quality Control
2600 Bull Street
J. Marion Sims Building
Columbia, South Carolina 29201

South Dakota

Department of Water and Natural
  Resources
Office of Drinking Water
Joe Foss Building, Room 411
Pierre, South Dakota 57501

*Tennessee*

Department of Public Health
Bureau of Environmental Health
Water Quality Control Division
621 Cordell Hull Building
Nashville, Tennessee 37219

*Texas*

Department of Water Resources
P.O. Box 13087
Capitol Station
Austin, Texas 78711

*Utah*

Health Services Branch
Bureau of Water Quality
150 W.N. Temple, Room 426
Salt Lake City, Utah 84110

*Vermont*

Agency of Environmental Conservation
Department of Water Resources
Montpelier, Vermont 05602

Virginia

State Water Control Board
211 N. Hamilton Street
Richmond, Virginia 23220

*Washington*

Department of Ecology
Office of Water Programs
Olympia, Washington 98504

*West Virginia*

Department of Natural Resources
Division of Water Resources
1800 Washington Street, Room 669
Charleston, West Virginia 25311

*Wisconsin*

Department of Natural Resources
Division of Environmental Standards
Bureau of Water Quality
P.O. Box 7921
Madison, Wisconsin 53707

*Wyoming*

Department of Environmental Quality
Water Quality Division
Hathaway Building
Cheyenne, Wyoming 82001

# State Air Quality Agencies

*Alabama*

Air Pollution Control Commission
645 S. McDonough Street
Montgomery, Alabama 36130

*Alaska*

Department of Environmental Conservation
Air Lands Section
Pouch "O"
Juneau, Alaska 99811

Arizona

Department of Health Services
Bureau of Air Quality Control
1740 West Adams Street
Phoenix, Arizona 85007

*Arkansas*

Department of Pollution Control and
    Ecology
Air Division
8001 National Drive
Little Rock, Arkansas 72209

*California*

Air Resources Board
P.O. Box 2815
1102 Q Street
Sacramento, California 95812

*Colorado*

Department of Health
Air Pollution Control Division/
    Commission
4210 East 11th Avenue
Denver, Colorado 80220

*Connecticut*

Department of Environmental Protection
Air Compliance Unit
State Office Building
165 Capitol Avenue
Hartford, Connecticut 06115

*Delaware*

Department of Natural Resources and
    Environmental Control
Division of Environmental Control, Air
    Resources Section
Tatnall Building, P.O. Box 1401
Dover, Delaware 19901

*District of Columbia*

Department of Environmental Services
Bureau of Air and Water Quality
5010 Overlook Avenue, S.W.
Washington, D.C. 20032

*Florida*

Department of Environmental Regulation
Division of Environmental Programs
Air Quality Management Bureau
Montgomery Building
2500 Blair Stone Road
Twin Towers Office Building
Tallahassee, Florida 32301

*Georgia*

Department of Natural Resources
Environmental Protection Division, Air
    Protection Branch
270 Washington Street, S.W., Room 822
Atlanta, Georgia 30334

*Hawaii*

Department of Health
Environmental Protection and Health
    Services Division
1250 Punch Bowl Street
Honolulu, Hawaii 96813

*Idaho*

Department of Health and Welfare
Division of Environment
Bureau of Air Quality
State House
Boise, Idaho 83720

*Illinois*

Environmental Protection Agency
Division of Air Pollution Control
2200 Churchill Road
Springfield, Illinois 62706

*Indiana*

Board of Health
Air Pollution Control Board/Division
1330 West Michigan Street
Indianapolis, Indiana 46206

*Iowa*

Department of Environmental Quality
Air and Land Quality Division
900 E. Grand Avenue, Henry A. Wallace
    Building
Des Moines, Iowa 50319

*Kansas*

Department of Health and Environment
Division of Environment
Air Quality and Occupational Health
Forbes AFB Building No. 740
Topeka, Kansas 66620

*Kentucky*

Department for Natural Resources and
    Environmental Protection
Division of Air Pollution
U.S. 127 South
Frankfort, Kentucky 40601

*Louisiana*

Department of Health and Human
    Resources
Air Quality Section
325 Loyola Avenue, Room 409
New Orleans, Louisiana 70160

*Maine*

Department of Environmental Protection
Bureau of Air Quality Control
State House
Augusta, Maine 04333

*Maryland*

Department of Health and Mental
    Hygiene
Bureau of Air Quality Control
201 West Preston Street
Baltimore, Maryland 21201

*Massachusetts*

Department of Environmental Quality
   Engineering
Division of Air & Hazardous Materials
600 Washington Street
Boston, Massachusetts 02111

*Michigan*

Department of Natural Resources
Air Quality Division
P.O. Box 30028
Lansing, Michigan 48909

*Minnesota*

Minnesota Pollution Control Agency
Division of Air Quality
1935 West County Road, B2
Roseville, Minnesota 55113

*Mississippi*

Department of Natural Resources
Bureau of Pollution Control
Robert E. Lee Building P.O. Box 827
Jackson, Mississippi 39205

*Missouri*

Department of Natural Resources
Division of Environmental Quality
Air Quality Program
P.O. Box 1368
Jefferson City, Missouri 65101

*Montana*

Department of Health and Environmen-
   tal Sciences
Division of Environmental Sciences
Air Quality Bureau
Cogswell Building
Helena, Montana 59601

*Nebraska*

Department of Environmental Control
Air Pollution Division
P.O. Box 94877
301 Centennial Mall
Lincoln, Nebraska 68509

*Nevada*

Department of Conservation and
   Natural Resources
Air Quality Office
201 South Fall Street
Carson City, Nevada 89710

*New Hampshire*

Air Resources Agency
Health and Welfare Building
Hazen Drive
Concord, New Hampshire 03301

*New Jersey*

Department of Environmental Protection
Division of Environmental Quality,
   Bureau of Air Pollution Control
John Fitch Plaza
Labor and Industry Building, Room 1110
Trenton, New Jersey 08625

*New Mexico*

Environmental Improvement Agency
Air Quality Division
P.O. Box 968, Crown Building
Santa Fe, New Mexico 87503

*New York*

Department of Environmental Conservation
Division of Air Resources
50 Wolf Road
Albany, New York 12233

*North Carolina*

Department of Natural Resources and
   Community Development
Division of Environmental Management
Air Quality Section
P.O. Box 27687
Raleigh, North Carolina 27611

*North Dakota*

Department of Health
Division of Environmental Engineering
1200 Missouri Avenue, Room 304
Bismarck, North Dakota 58505

*Ohio*

Environmental Protection Agency
Office of Air Pollution Control
361 E. Broad Street
P.O. Box 1049
Columbus, Ohio 43216

*Oklahoma*

Department of Health
Environmental Health Services
Air Quality Service
NE 10th and Stonewall, P.O. Box 53551
Oklahoma City, Oklahoma 73152

*Oregon*

Department of Environmental Quality
Air Quality Division
P.O. Box 1760
Portland, Oregon 97207

*Pennsylvania*

Department of Environmental Resources
Bureau of Air Quality
P.O. Box 2063
Harrisburg, Pennsylvania 17120

*Puerto Rico*

Environmental Quality Board
Air Quality Bureau
P.O. Box 11488
Santurce, Puerto Rico 00910

*Rhode Island*

Department of Environmental
   Management
Division of Air Resources
204 Cannon Building
Providence, Rhode Island 02908

*South Carolina*

Department of Health and Environmental Control
Bureau of Air Quality Control
2600 Bull Street
Columbia, South Carolina 29201

*South Dakota*

Department of Health
Division of Air and Solid Waste
Joe Foss Bldg., 120 E. Capitol
Pierre, South Dakota 57501

*Tennessee*

Department of Public Health
Bureau of Environmental Health Services
Air Pollution Control Division
256 Capitol Hill Building
Nashville, Tennessee 37219

*Texas*

Air Control Board
6330 Highway 290 East
Austin, Texas 78723

*Utah*

State Department of Health
Bureau of Air Quality
150 W.N. Temple
P.O. Box 2500
Salt Lake City, Utah 84110

*Vermont*

Agency of Environmental Conservation
Air Pollution Control
State Office Building
Montpelier, Vermont 05602

*Virginia*

Air Pollution Control Board
Ninth Street Office Building
Richmond, Virginia 23219

*Washington*

Department of Ecology
Office of Air Programs
Olympia, Washington 98504

*West Virginia*

Air Pollution Control Commission
1558 Washington Street, East
Charleston, West Virginia 25311

*Wisconsin*

Department of Natural Resources
Division of Environmental Standards
Air Pollution Control Section
P.O. Box 7921
Madison, Wisconsin 53707

*Wyoming*

Department of Environmental Quality
Air Quality Division
Hathaway Building
Cheyenne, Wyoming 82001

# State Solid Waste Agencies

*Alabama*

Department of Public Health
Environmental Health Admini-
  stration
Division of Solid Waste
State Office Building
Montgomery, Alabama 36130

*Alaska*

Department of Environmental Con-
  servation
Solid Waste Program
Pouch "O"
Juneau, Alaska 99811

*Arizona*

Bureau of Sanitation
411 North 24th Street
Phoenix, Arizona 85008

*Arkansas*

Department of Pollution Control and
  Ecology
8001 National Drive
Little Rock, Arkansas 72209

*California*

Solid Waste Management Board
1020 9th Street, Suite 300
Sacramento, California 95814

*Colorado*

Department of Health
Solid Waste Management Project
4210 East 11th Avenue
Denver, Colorado 80220

*Connecticut*

Department of Environmental Protection
Solid Waste Management Unit
122 Washington Street
Hartford, Connecticut 06115

*Delaware*

Department of Natural Resources and
  Environmental Control
Division of Environmental Control, Solid
  Waste Section
P.O. Box 1401
Dover, Delaware 19901

*District of Columbia*

Solid Waste Administration
415 12th Street, N.W., Room 303
Washington, D.C. 20004

*Florida*

Department of Environmental Regu-
  lation
Solid Waste Management Program
2500 Blair Stone Road
Twin Towers Office Building
Tallahassee, Florida 32301

*Georgia*

Department of Natural Resources
Environmental Protection Division, Solid
  Waste Management Section
270 Washington Street, S.W., Room 822
Atlanta, Georgia 30334

*Hawaii*

Department of Health
Environmental Health Division
P.O. Box 3378
Honolulu, Hawaii 96801

Department of Planning and Economic
  Development
Land Use Commission
250 South King Street
Honolulu, Hawaii 96813

*Idaho*

Department of Health and Welfare
Division of Environment
Solid Waste Management Section
Statehouse
Boise, Idaho 83720

*Illinois*

Environmental Protection Agency
Division of Land-Noise Pollution Control
2200 Churchill Road
Springfield, Illinois 62706

*Indiana*

Board of Health
Solid Waste Section
1330 West Michigan Street
Indianapolis, Indiana 46206

*Iowa*

Department of Environmental Quality
Air and Land Quality Management
  Division
3920 Delaware Avenue
P.O. Box 3326
Des Moines, Iowa 50316

*Kansas*

Department of Health and Environment
Division of Environment
Bureau of Environmental Sanitation
Forbes AFB Building, No. 321
Topeka, Kansas 66620

*Kentucky*

Department for Natural Resources and
  Environmental Protection
Division of Solid Waste
Pine Hill Plaza, U.S. 60
Frankfort, Kentucky 40601

*Louisiana*

Department of Health and Human
  Resources
Solid Waste and Vector Control Unit
P.O. Box 60630
New Orleans, Louisiana 70160

*Maine*

Department of Environmental Protection
Division of Solid Waste Management
State House
Augusta, Maine 04333

*Maryland*

Department of Health
Division of Solid Waste
P.O. Box 13387
Baltimore, Maryland 21201

*Massachusetts*

Department of Environmental Management
Division of Solid Waste Disposal
Room 1905, Leverett Saltonstall Building
100 Cambridge Street
Boston, Massachusetts 02202

*Michigan*

Department of Natural Resources
Resource Recovery Division
P.O. Box 30028
Lansing, Michigan 48909

*Minnesota*

Pollution Control Agency
Solid Waste Division
1935 West County Road B2
Roseville, Minnesota 55113

*Mississippi*

Bureau of Environmental Health
State Board of Health
Solid Waste Management
880 Lakeland Drive
Jackson, Mississippi 39205

*Missouri*

Department of Natural Resources
Division of Environmental Quality
Solid Waste Program
P.O. Box 1368
Jefferson City, Missouri 65102

*Montana*

Department of Health and Environmental Sciences
Solid Waste Management Bureau
Cogswell Building
Helena, Montana 59601

*Nebraska*

Department of Environmental Control
Solid Waste Management Division
P.O. Box 94877
301 Centennial Mall
Lincoln, Nebraska 68509

*Nevada*

Department of Conservation and
  Natural Resources
Solid Waste Management Program
201 South Fall Street
Carson City, Nevada 89701

*New Hampshire*

Department of Health and Welfare
Division of Public Health Services
Bureau of Solid Waste Management
State Laboratory Building
Hazen Drive
Concord, New Hampshire 03301

*New Jersey*

Department of Environmental Protection
Solid Waste Administration
32 E. Hanover Street
Trenton, New Jersey 08625

*New Mexico*

Environmental and Improvement
  Division
P.O. Box 968, Crown Building
Santa Fe, New Mexico 87503

*New York*

Department of Environmental
  Conservation
Division of Solid Waste Management
50 Wolf Road
Albany, New York 12233

*North Carolina*

Department of Natural Resources and
   Community Development
Division of Environmental Management
Land Quality Section
P.O. Box 27687
Raleigh, North Carolina 27611

Department of Human Resources
Division of Health Services
Solid Waste and Vector Control Division
P.O. Box 2091
Raleigh, North Carolina 27602

*North Dakota*

Department of Health
Division of Waste Management and
   Research
1200 Missouri Avenue
Bismarck, North Dakota 58505

*Ohio*

Environmental Protection Agency
Land Pollution Control
361 E. Broad Street
P.O. Box 1049
Columbus, Ohio 43216

*Oklahoma*

Department of Health
Industrial and Solid Waste Division
P.O. Box 53551
Oklahoma City, Oklahoma 73152

*Oregon*

Department of Environmental Quality
Solid Waste Management Division
P.O. Box 1760
Portland, Oregon 97207

Land Conservation and Development
   Commission
1175 Court Street, N.E.
Salem, Oregon 97310

*Pennsylvania*

Department of Environmental Resources
Bureau of Solid Waste Management
P.O. Box 2063
Harrisburg, Pennsylvania 17120

*Rhode Island*

Department of Environmental
   Management
Division of Solid Waste Management
Health Building
75 Davis Street
Providence, Rhode Island 02908

*South Carolina*

Department of Health and Environmental Control
Solid Waste Management Division
2600 Bull Street
Columbia, South Carolina 29201

*South Dakota*

Department of Health
Solid Waste Program
120 E. Capitol
Pierre, South Dakota 57501

*Tennessee*

Department of Public Health
Division of Sanitation and Solid Waste
   Management
Solid Waste Management Section
320 Capitol Hill Building
Nashville, Tennessee 37219

*Texas*

Department of Health (municipal solid
   waste)
Division of Solid Waste Management
1100 West 49th Street
Austin, Texas 78756

*Utah*

State Department of Health
Bureau of Solid Waste Management
150 W.N. Temple
Salt Lake City, Utah 84110

*Vermont*

Agency of Environmental Conservation
Air and Solid Waste Program
State Office Building
Montpelier, Vermont 05602

*Virginia*

Health Department
Division of Solid and Hazardous Waste
109 Governor Street
Richmond, Virginia 23219

*Washington*

Department of Ecology

Solid Waste Management Division
Olympia, Washington 98504

*West Virginia*

Department of Health
Bureau of Solid Waste Disposal
1800 Washington Street, Room 520
Charleston, West Virginia 25305

*Wisconsin*

Department of Natural Resources
Bureau of Waste Management
P.O. Box 7921
Madison, Wisconsin 53707

*Wyoming*

Department of Environmental Quality
Solid Waste Management Program
Hathaway Building
Cheyenne, Wyoming 82002

# Environmental Excise Taxes

TAX ON PETROLEUM

Code Sec. 4611. IMPOSITION OF TAX (a) GENERAL RULE.—There is hereby imposed a tax of 0.79 cent a barrel on—

(1) crude oil received at a United States refinery, and

(2) petroleum products entered into the United States for consumption, use, or warehousing.

(b) TAX ON CERTAIN USES AND EXPORTATION.—

(1) IN GENERAL.—If—

(A) any domestic crude oil is used in or exported from the United States, and

(B) before such use or exportation, no tax was imposed on such crude oil under subsection (a),

then a tax of 0.79 cent a barrel is hereby imposed on such crude oil.

(2) EXCEPTION FOR USE ON PREMISES WHERE PRODUCED.—Paragraph (1) shall not apply to any use of crude oil for extracting oil or natural gas on the premises where such crude oil was produced.

(c) PERSONS LIABLE FOR TAX.—

(1) CRUDE OIL RECEIVED AT REFINERY.—The tax imposed by subsection (a)(1) shall be paid by the operator of the United States refinery.

(2) IMPORTED PETROLEUM PRODUCT.—The tax imposed by subsection (a)(2) shall be paid by the person entering the product for consumption, use, or warehousing.

(3) TAX ON CERTAIN USES OR EXPORTS.—The tax imposed by subsection (b) shall be paid by the person using or exporting the crude oil, as the case may be.

(d) TERMINATION.—The taxes imposed by this section shall not apply after September 30, 1985, except that if on September 30, 1983, or September 30, 1984—

(1) the unobligated balance in the Hazardous Substance Response Trust Fund as of such date exceeds $900,000,000, and

(2) the Secretary, after consultation with the Administrator of the Environmental Protection Agency, determines that such unobligated balance will exceed $500,000,000 on September 30 of the following year if no tax is imposed under section 4611 or 4661 during the calendar year following the date referred to above,

then no tax shall be imposed by this section during the first calendar year beginning after the date referred to in paragraph (1).

.05 Added by P.L. 96-510.

Code Sec. 4612 [1954 Code]. DEFINITIONS AND SPECIAL RULES (a) DEFINITIONS. — For purposes of this subchapter —

(1) CRUDE OIL. — The term "crude oil" includes crude oil condensates and natural gasoline.

(2) DOMESTIC CRUDE OIL. — The term "domestic crude oil" means any crude oil produced from a well located in the United States.

(3) PETROLEUM PRODUCT. — The term "petroleum product" includes crude oil.

(4) UNITED STATES. —

(A) IN GENERAL. — The term "United States" means the 50 states, the District of Columbia, the Commonwealth of Puerto Rico, any possession of the United States, the Commonwealth of the Northern Mariana Islands, and the Trust Territory of the Pacific Islands.

(B) UNITED STATES INCLUDES CONTINENTAL SHELF AREAS. — The principles of section 638 shall apply for purposes of the term "United States".

(C) UNITED STATES INCLUDES FOREIGN TRADE ZONES. — The term "United States" includes any foreign trade zone of the United States.

(5) UNITED STATES REFINERY. — The term "United States refinery" means any facility in the United States at which crude oil is refined.

(6) REFINERIES WHICH PRODUCE NATURAL GASOLINE. — In the case of any United States refinery which produces natural gasoline from natural gas, the gasoline so produced shall be treated as received at such refinery at the time so produced.

(7) PREMISES. — The term "premises" has the same meaning as when used for purposes of determining gross income from the property under section 613.

(8) BARREL. — The term "barrel" means 42 United States gallons.

(9) FRACTIONAL PART OF BARREL. — In the case of a fraction of a barrel, the tax imposed by section 4611 shall be the same fraction of the amount of such tax imposed on a whole barrel.

(b) ONLY 1 TAX IMPOSED WITH RESPECT TO ANY PRODUCT. — No tax shall be imposed by section 4611 with respect to any petroleum product if the person who would be liable for such tax establishes that a prior tax imposed by such section has been imposed with respect to such product.

(c) DISPOSITION OF REVENUES FROM PUERTO RICO AND THE VIRGIN ISLANDS. — The provisions of subsections (a)(3) and (b)(3) of section 7652 shall not apply to any tax imposed by section 4611.

.05 Added by P.L. 96-510.

TAX ON CHEMICALS

Code Sec. 4661. IMPOSITION OF TAX (a) GENERAL RULE.—There is hereby imposed a tax on any taxable chemical sold by the manufacturer, producer, or importer thereof.

(b) AMOUNT OF TAX.—The amount of the tax imposed by subsection (a) shall be determined in accordance with the following table:

| In the case of: | The tax is the following amount per ton |
|---|---|
| Acetylene | $4.87 |
| Benzene | 4.87 |
| Butane | 4.87 |
| Butylene | 4.87 |
| Butadiene | 4.87 |
| Ethylene | 4.87 |
| Methane | 3.44 |
| Naphthalene | 4.87 |
| Propylene | 4.87 |
| Toluene | 4.87 |
| Xylene | 4.87 |
| Ammonia | 2.64 |
| Antimony | 4.45 |
| Antimony trioxide | 3.75 |
| Arsenic | 4.45 |
| Arsenic trioxide | 3.41 |
| Barium sulfide | 2.30 |
| Bromide | 4.45 |
| Cadmium | 4.45 |
| Chlorine | 2.70 |
| Chromium | 4.45 |
| Chromite | 1.52 |
| Potassium dichromate | 1.69 |
| Sodium dichromate | 1.87 |
| Cobalt | 4.45 |
| Cupric sulfate | 1.87 |
| Cupric oxide | 3.59 |
| Cuprous oxide | 3.97 |
| Hydrochloric acid | 0.29 |
| Hydrogen fluoride | 4.23 |
| Lead oxide | 4.14 |
| Mercury | 4.45 |
| Nickel | 4.45 |
| Phosphorus | 4.45 |
| Stannous chloride | 2.85 |
| Stannic chloride | 2.12 |
| Zinc chloride | 2.22 |
| Zinc sulfate | 1.90 |
| Potassium hydroxide | 0.22 |
| Sodium hydroxide | 0.28 |
| Sulfuric acid | 0.26 |
| Nitric acid | 0.24 |

(c) TERMINATION.—No tax shall be imposed under this section during any period during which no tax is imposed under section 4611(a).

Sec. 4662 [1954 Code]. (a) DEFINITIONS.—For purposes of this subchapter—

(1) TAXABLE CHEMICAL.—Except as provided in subsection (b), the term "taxable chemical" means any substance—

(A) which is listed in the table under section 4661(b), and

(B) which is manufactured or produced in the United States or entered into the United States for consumption, use, or warehousing.

(2) UNITED STATES.—The term "United States" has the meaning given such term by section 4612(a)(4).

(3) IMPORTER.—The term "importer" means the person entering the taxable chemical for consumption, use, or warehousing.

(4) TON.—The term "ton" means 2,000 pounds. In the case of any taxable chemical which is a gas, the term "ton" means the amount of such gas in cubic feet which is the equivalent of 2,000 pounds on a molecular weight basis.

(5) FRACTIONAL PART OF TON.—In the case of a fraction of a ton, the tax imposed by section 4661 shall be the same fraction of the amount of such tax imposed on a whole ton.

(b) EXCEPTIONS; OTHER SPECIAL RULES.—For purposes of this subchapter—

(1) METHANE OR BUTANE USED AS A FUEL.—Under regulations prescribed by the Secretary, methane or butane shall be treated as a taxable chemical only if it is used otherwise than as a fuel (and, for purposes of section 4661(a), the person so using it shall be treated as the manufacturer thereof).

(2) SUBSTANCES USED IN THE PRODUCTION OF FERTILIZER.—

(A) IN GENERAL.—In the case of nitric acid, sulfuric acid, ammonia, or methane used to produce ammonia which is a qualified substance, no tax shall be imposed under section 4661(a).

(B) QUALIFIED SUBSTANCE.—For purposes of this section, the term "qualified substance" means any substance—

(i) used in a qualified use by the manufacturer, producer, or importer,

(ii) sold for use by the purchaser in a qualified use, or

(iii) sold for resale by the purchaser to a second purchaser for use by such second purchaser in a qualified use.

(C) QUALIFIED USE.—For purposes of this subsection, the term "qualified use" means any use in the manufacture or production of a fertilizer.

(3) SULFURIC ACID PRODUCED AS A BYPRODUCT OF AIR POLLUTION CONTROL.—

In the case of sulfuric acid produced solely as a byproduct of and on the same site as air pollution control equipment, no tax shall be imposed under section 4661.

(4) SUBSTANCES DERIVED FROM COAL.—For purposes of this subchapter, the term "taxable chemical" shall not include any substance to the extent derived from coal.

(c) USE BY MANUFACTURER, ETC., CONSIDERED SALE.—If any person manufactures, produces, or imports a taxable chemical and uses such chemical, then such person shall be liable for tax under section 4661 in the same manner as if such chemical were sold by such person.

(d) REFUND OR CREDIT FOR CERTAIN USES.—

(1) IN GENERAL.—Under regulations prescribed by the Secretary, if—

(A) a tax under section 4661 was paid with respect to any taxable chemical, and

(B) such chemical was used by any person in the manufacture or production of any other substance the sale of which by such person would be taxable under such section,

then an amount equal to the tax so paid shall be allowed as a credit or refund (without interest) to such person in the same manner as if it were an overpayment of tax imposed by such section. In any case to which this paragraph applies, the amount of any such credit or refund shall not exceed the amount of tax imposed by such section on the other substance manufactured or produced.

(2) USE AS FERTILIZER.—Under regulations prescribed by the Secretary, if—

(A) a tax under section 4661 was paid with respect to nitric acid, sulfuric acid, ammonia, or methane used to make ammonia without regard to subsection (b)(2), and

(B) any person uses such substance, or sells such substance for use, as a qualified substance,

then an amount equal to the excess of the tax so paid over the tax determined with regard to subsection (b)(2) shall be allowed as a credit or refund (without interest) to such person in the same manner as if it were an overpayment of tax imposed by this section.

(e) DISPOSITION OF REVENUES FROM PUERTO RICO AND THE VIRGIN ISLANDS.
—The provisions of subsections (a)(3) and (b)(3) of section 7652 shall not apply to any tax imposed by section 4661.

## TAX ON HAZARDOUS WASTE

Code Sec. 4681. IMPOSITION OF TAX (a) GENERAL RULE.—There is hereby imposed a tax on the receipt of hazardous waste at a qualified hazardous waste disposal facility.

(b) AMOUNT OF TAX.—The amount of the tax imposed by subsection (a) shall be equal to $2.13 per dry weight ton of hazardous waste.

Code Sec. 4682. DEFINITIONS AND SPECIAL RULES (a) DEFINITIONS.—For purposes of this subchapter—

(1) HAZARDOUS WASTE.—The term "hazardous waste" means any waste—

(A) having the characteristics identified under section 3001 of the Solid Waste Disposal Act, as in effect on the date of the enactment of this Act (other than waste the regulation of which under such Act has been suspended by Act of Congress on that date), or

(B) subject to the reporting or recordkeeping requirements of sections 3002 and 3004 of such Act, as so in effect.

(2) QUALIFIED HAZARDOUS WASTE DISPOSAL FACILITY.—The term "qualified hazardous waste disposal facility" means any facility which has received a permit or is accorded interim status under section 3005 of the Solid Waste Disposal Act.

(b) TAX IMPOSED ON OWNER OR OPERATOR.—The tax imposed by section 4681 shall be imposed on the owner or operator of the qualified hazardous waste disposal facility.

(c) TAX NOT TO APPLY TO CERTAIN WASTES. — The tax imposed by section 4681 shall not apply to any hazardous waste which will not remain at the qualified hazardous waste disposal facility after the facility is closed.

(d) APPLICABILITY OF SECTION. — The tax imposed by section 4681 shall apply to the receipt of hazardous waste after September 30, 1983, except that if, as of September 30 of any subsequent calendar year the unobligated balance of the Post-closure Liability Trust Fund exceeds $200,000,000, no tax shall be imposed under such section during the following calendar year.

# SELECTED BIBLIOGRAPHY

Ambassade de France. *Protection of the Environment of France*. New York: Service de Presse et d'Information, 1974.

Anderson, Frederick R., et al. *Environmental Improvement Through Economic Incentives*. Baltimore: Johns Hopkins University for Resources for the Future, 1979.

Anderson, Robert J., and Wilen, James E. "The Proposed Pure Air Tax of 1972," *National Tax Journal*, Vol. 27, No. 1 (1974), pp. 151-62.

Avelar, Linda S. (ed.). *Business Study: United Kingdom*. New York: Touche Ross International, 1978.

Baram, Michael S. *Alternatives to Regulation: Managing Risks to Health, Safety and the Environment*. Lexington, Mass.: D.C. Heath and Company, 1982.

Baumol, William J., and Oates, Wallace E. *The Theory of Environmental Policy*. Englewood Cliffs, N.J.: Prentice-Hall, Inc., 1975.

Bengtsson, Ingemund. *The Act of Products Hazardous to Health and to the Environment*. Stockholm: Royal Ministry of Agriculture, 1973.

Boddewyn, Jean. *Comparative Management and Marketing*. Glenview, Ill.: Scott, Foresman, & Co., 1969.

Bower, Blair T. et al. *Incentives in Water Quality Management: France and Ruhr Area*. Washington, D.C.: Resources for the Future, Inc., 1981.

*Britain, 1974: An Official Handbook*. London: Her Majesty's Stationery Office, 1974.

British Central Office of Information. *The New British System of Taxation*. London: Her Majesty's Stationery Office, 1973.

British Department of Environment. *The Human Environment: The British View*. London: Her Majesty's Stationery Office, 1972.

Bureau d'Etudes Fiscales et Juridiques Francis Lefebre. *Business Operations in France*, Tax Management Portfolio No. 39-4th. Washington, D.C.: Bureau of National Affairs, Inc., 1972.

*Business Study — Canada*. New York: Touche Ross International, 1975.

Canadian Department of Industry, Trade, and Commerce. *Doing Business in Canada: Federal Incentives to Industry*. Ottawa: Queen's Printer for Canada, 1972.

_____. *Doing Business in Canada: Taxation — Income, Business, Property*. Ottawa: Queen's Printer for Canada, 1972.

Canadian Department of Secretary of State. *Canadian System of Government*. Ottawa: Queen's Printer for Canada, 1970.

479

Carlson, David. *Revitalizing North American Neighborhoods: Comparison of Canadian and U.S. Programs*. Washington, D.C.: U.S. Government Printing Office, 1978.

*CCH State Tax Guide — All States 1982*. Chicago: Commerce Clearing House, Inc., 1982.

Clawson, Marion, and Hall, Peter. *Planning and Urban Growth*. Baltimore: Johns Hopkins University Press, 1973.

Committee on Chemistry and Public Affairs. *Cleaning Our Environment: The Chemical Basis for Action*. Washington, D.C.: American Chemical Society, 1969.

Commoner, Barry. *The Closing Circle*. New York: Alfred A. Knopf, 1971.

Crandall, Robert W., and Lave, Lester B., editors. *The Scientific Basis of Health and Safety Regulation*. Washington, D.C.: Brookings Institution, 1981.

D'Arge, Ralph C.; Ayres, Robert U.; and Kneese, Allen V. *Economics and the Environment*. Baltimore: Johns Hopkins University for Resources for the Future, 1970.

Davies, J.C., III. *The Politics of Pollution*. New York: Pegasus, 1970.

*Demographic Yearbook 1976*. New York: United Nations, 1977.

*Doing Business in Germany*. New York: Price Waterhouse & Co., 1975.

Dorfman, Robert, and Dorfman, Nancy S., editors. *Economics of the Environment: Selected Readings*, Second Edition. New York: W.W. Norton & Company, Inc., 1979.

Dykes, G.K., and Tomsett, E. *France: Business Study*. New York: Touche Ross International, 1979.

Edel, Matthew. *Economies and the Environment*. Englewood Cliffs, N.J.: Prentice-Hall, Inc., 1973.

Elkington, John. *The Ecology of Tomorrow's World*. London: Associated Business Press, 1980.

Environment Canada. *Inland Waters Directorate*. Ottawa: Queen's Printer for Canada, 1973.

Erlander, Tage. *Sweden's National Report to the United Nations on the Human Environment*. Stockholm: Royal Ministry of Foreign Affairs, 1971.

Flacker, Ake, and Holm, Lennart. *Urbanization and Planning in Sweden*. Stockholm: Royal Ministry of Agriculture, 1972.

*France: A National Profile*. New York: Ernst & Ernst International, 1975.

Freeman, A. Myrick, III; Haveman, Robert H.; and Kneese, Allen V. *The Economics of Environmental Policy*. New York: John Wiley & Sons, Inc., 1973.

German Press and Information Office. *Protection of the Environment and Conservation of Nature*. Bonn: Federal Republic of Germany, 1972.

Goldsmith, Jean-Claude. *Business Operations in France*, Tax Management Portfolio No. 39-5th. Washington, D.C.: Bureau of National Affairs, Inc., 1981.

Gomeche, Eugene L. *Business Operations in the United Kingdom*, Tax Management Portfolio No. 68-5th. Washington, D.C.: Bureau of National Affairs, Inc., 1973.

*Government Finance Statistics Yearbook: Volume II/1978*. Washington, D.C.: International Monetary Fund, 1979.

Grace, Richard, and Fisher, Jonathan. *Beverage Containers: Re-Use or Recycling*. Paris: Organization for Economic Cooperation and Development, 1978.

Gumpel, Henry J. *World Tax Series: Taxation in the Federal Republic of Germany*. 2d ed. Chicago: Commerce Clearing House, Inc., 1969.

Hancock, M. Donald. *Sweden: The Politics of Post-Industrial Change*. Hinsdale, Ill.: Dryden Press, 1972.

Henning, Daniel H. *Environmental Policy and Administration*. New York: American Elsevier Publishing Co., Inc., 1973.

Hite, James C., et al. *The Economics of Environmental Quality*. Washington, D.C.: American Enterprise Institute for Public Policy Research, 1972.

*International Environmental Guide — 1975*. Washington, D.C.: Bureau of National Affairs, Inc., 1975.

*International Environment Reporter*. Washington, D.C.: Bureau of National Affairs, Inc., 1978.

Irwin, William A. *Changes on Effluents in the United States and Europe*. Washington, D.C.: Environmental Law Institute, 1974.

Jarrett, Henry (ed.). *Environmental Quality in a Growing Economy*. Baltimore: Johns Hopkins University for Resources for the Future, 1966.

Kelman, Steven. *What Price Incentives? Economists and the Environment*. Boston: Auburn House Publishing Company, 1981.

Kilius, Juergen. *Business Operations in West Germany*, Tax Management Portfolio No. 174-4th. Washington, D.C.: Bureau of National Affairs, Inc., 1978.

_____, and Stiefel, Ernest C. *Business Operations in West Germany*, Tax Management Portfolio No. 174-2nd. Washington, D.C.: Bureau of National Affairs, Inc., 1971.

Kneese, Allen V., and Bower, Blair T. *Environmental Quality and Residuals Management*. Baltimore: Resources for the Future by Johns Hopkins University Press, 1979.

_____. *Managing Water Quality: Economics, Technology, Institutions*. Baltimore: Johns Hopkins University for Resources for the Future, 1968.

_____, and Schultze, Charles L. *Pollution, Prices and Public Policy*. Washington, D.C.: Brookings Institution, 1975.

Landau, Norman J., and Rheingold, Paul D. *The Environmental Law Handbook*. New York: Ballantine Books for Friends of the Earth, 1971.

Lave, Lester B. *The Strategy of Social Regulation: Decision Frameworks for Policy*. Washington, D.C.: Brookings Institution, 1981.

_____, and Seskin, Eugene P. *Air Pollution and Human Health*. Baltimore: Johns Hopkins University for Resources for the Future, 1977.

MacAvoy, Paul (ed.). *The Crisis of the Regulatory Commissions*. New York: Grossman Publishers, 1970.

McHale, John. *The Ecological Context*. New York: George Braziller, Inc., 1970.

Mansfield, Edwin. *Microeconomics: Theory and Applications*. New York: W. W. Norton & Co., Inc., 1976.

*A Market Approach to Air Pollution Control Could Reduce Compliance Costs Without Jeopardizing Clean Air Goals*, Washington, D.C.: Government Accounting Office, 1982.

Marx, Wesley. *Man and His Environment: Waste*. New York: Harper & Row, Publishers, Inc., 1971.

Nadel, S. F. *The Foundations of Social Anthropology*. London: Cohen & West, Ltd., 1951.

Norr, Martin, et al. *The Tax System in Sweden*. Stockholm: Skandinaviska Enskilda Banken, 1972.

_____. *World Tax Series: Taxation in Sweden*. Boston: Little, Brown & Co., 1959.

OECD Economic Policy Committee. *Economic Implication of Pollution Control*. Paris: Organization for Economic Cooperation and Development, 1974.

OECD Environment Committee. *The Polluter Pays Principle*. Paris: Organization for Economic Cooperation and Development, 1975.

OECD Environment Directorate, *Economic and Ecological Interdependence*. Paris: Organization for Economic Cooperation and Development, 1982.

_____. *Economic and Policy Instruments for Water Management in Canada*. Paris: Organization for Economic Cooperation and Development, 1976.

_____. *Economic and Policy Instruments for Water Management in France*. Paris: Organization for Economic Cooperation and Development, 1976.

_____. *Economic and Policy Instruments for Water Management in Germany*. Paris: Organization for Economic Cooperation and Development, 1976.

_____. *Economic and Policy Instruments for Water Management in the United Kingdom*. Paris: Organization for Economic Cooperation and Development, 1976.

_____. *Economic Instruments in Solid Waste Management*. Paris: Organization for Economic Cooperation and Development, 1981.

_____. *Economic Measurement of Environmental Damage*. Paris: Organization for Economic Cooperation and Development, 1976.

_____. *Environmental Impact Assessment*. Paris: Organization for Economic Cooperation and Development, 1979.

_____. *Glossaire de L'Environment*. Paris: Organization for Economic Cooperation and Development, 1981.

_____. *Macro-economic Evaluation of Environmental Programmes*. Paris: Organization for Economic Cooperation and Development, 1978.

_____. *Pollution Charges: An Assessment*. Paris: Organization for Economic Cooperation and Development, 1976.

_____. *Pollution Charges in Practice*. Paris: Organization for Economic Cooperation and Development, 1980.

_____. *The Costs and Benefits of Sulphur Oxide Control*. Paris: Organization for Economic Cooperation and Development, 1981.

_____. *The Environment: Challenges for the 1980s*. Paris: Organization for Economic Cooperation and Development, 1981.

_____. *The State of the Environment*. Paris: Organization for Economic Cooperation and Development, 1979.

_____. *Transfrontier Pollution and the Role of States*. Paris: Organization for Economic Cooperation and Development, 1981.

_____. *Waste Management in OECD Member Countries*. Paris: Organization for Economic Cooperation and Development, 1976.

_____. *Water Management Policies and Instruments*. Paris: Organization for Economic Cooperation and Development, 1977.

_____. *Water Management in Industrialized River Basins*. Paris: Organization for Economic Cooperation and Development, 1980.

OECD Environment Secretariat, *Environment Policies for the 1980s*. Paris: Organization for Economic Cooperation and Development, 1980.

Pearson, Charles. "Environmental Control Costs and Border Adjustments," *National Tax Journal*, Vol. 27, No. 4 (1974), pp. 599-608.

Persson, Lennart. *Environmental Protection Act and Marine Dumping Prohibition Act*. Stockholm: National Environment Protection Board, 1972.

*Revenue Statistics of OECD Member Countries, 1965-76: A Standardized Classification*. Paris: Organization for Economic Cooperation and Development, 1978.

Russel, Clifford S. "What Can We Get from Effluent Charges?" *Policy Analysis*, Vol. 5, No. 2 (1979), pp. 155-80.

Secretariat General du Hait Comite de l'Environnement *L'Etat de l'Environnement: Rapport Annual 1976-77*, Tome 1. Paris: La Documentation Francaise, 1978.

Seneca, Joseph J., and Taussig, Michael K. *Environmental Economics*. Englewood Cliffs, N.J.: Prentice-Hall, Inc., 1974.

Service de l'Environnement Industriel. *Industrialization and Environmental Protection in France*. Paris: Ministere de l'Environnement et du Cadre de Vie, 1979.

_____. *Installations Registered for Purposes of Environmental Protection*. Paris: Ministere de l'Environnement et du Cadre de Vie, 1978.

Service d'Information et de Diffusion. *Des Actions Pour la Qualite de la Vie*. Paris: Premier Ministre, 1976.

Smith, Dan Throop, et al. *Report of the Tax Policy Advisory Committee to the Council on Environmental Quality*. Washington, D.C.: Council on Environmental Quality, February, 1973.

Soell, Hermann. *Beitrage Zur Umweltgestaltung: Depreciation Allowances or Subsidies*. Berlin: Verlag, 1975.

*Steel—The Recyclable Material*. New York: Council on Economic Priorities, 1973.

Surrey, Stanley S. *Pathways to Tax Reform*. Cambridge, Mass.: Harvard University Press, 1973.

Swartzman, Daniel, et al., editors, *Cost-Benefit Analysis and Environmental Regulations*. Washington, D.C.: Conservation Foundation, 1982.

*Sweden's Reply to the United Nations Inquiry in Connection with the U.N. Conference on the Human Environment.* Stockholm: Royal Ministry for Foreign Affairs, 1970.

Swedish Delegation, OECD Environment Committee. *Environmental Policy in Sweden.* Paris: Organization for Economic Cooperation and Development, 1977.

Swedish Ministry of Physical Planning and Local Government. *Management of Land and Water Resources.* Stockholm: Royal Ministry of Foreign Affairs, 1971.

*Tax and Trade Guide—Canada.* Chicago: Arthur Anderson & Co., 1973.

Thompson, Donald N. *The Economics of Environmental Protection.* Cambridge, Mass.: Winthrop Publishers, Inc., 1973.

U.S. Central Intelligence Agency. *National Basic Intelligence Factbook.* Washington, D.C.: U.S. Government Printing Office, 1979.

U.S. Congress, Joint Economic Committee. *Achieving Price Stability Through Economic Growth.* Washington, D.C.: U.S. Government Printing Office, 1974.

U.S. Council on Environmental Quality. *Environmental Quality: Annual Reports*, First-Ninth. Washington, D.C.: U.S. Government Printing Office, 1970-78.

U.S. Department of Commerce. *The Effects of Pollution Abatement on International Trade.* Washington, D.C.: U.S. Government Printing Office, 1973.

Vinde, Pierre. *Swedish Government Administration.* Stockholm: Swedish Institute, 1971.

Weissert, Carol S. *State Mandating of Local Expenditures.* Washington, D.C.: Advisory Commission on Intergovernmental Relations, 1978.

Zwick, David, and Benstock, Mary (eds.). *Water Wasteland.* New York: Grossman Publishers, 1971.


Aharoni, Yair. *The No-Risk Society.* Chatham, N.J.: Chatham House, 1981.

Anthony, Richard P. "How Effective was the Green Vote?" *Environment*, Vol. 24 No. 10 (December 1982), pp. 2-4

———. "Polls, Pollution and Politics", *Environment*, Vol. 24 No. 4 (May 1982), pp. 14-19.

Bardach, Eugene and Robert A. Kagan. *Going by Book: The Problem of Regulatory Unreasonableness.* Philadelphia: Temple University Press, 1982.

Basta, Daniel J., and Bower, Blair T., editors. *Analyzing Natural Systems: Analysis for Regional Residuals—Environmental Quality Management.* Washington, D.C.: Resources for the Future, Inc., 1982.

Bohm, Peter. *Deposit-Refund Systems: Theory and Application to Environmental Conservation and Consumer Policy.* Baltimore: Johns Hopkins University Press, 1981.

Boris, Constance M., and Krutilla, John V. *Water Rights and Energy Development in the Yellowstone River Basin.* Baltimore: Johns Hopkins University Press for Resources for the Future, Inc., 1980.

Boulding, Kenneth. "The Role of Government in a Free Society", *Technology Review*, Vol. 85 No. 6 (August/September 1982), pp. 6-7.

Burmaster, David R. "The New Pollution: Groundwater Contamination", *Environment*, Vol. 24 No. 2 (March 1982), pp. 6-13.

Clark II, Edwin H., editor. *State of the Environment 1982.* Washington, D.C.: The Conservation Foundation, 1982.

Crandal, Robert W. "Reagulation—The First Year: The Environment," *Regulation* (January/February 1982), pp. 29-31.

Downing, Paul B., and Hanf, Kenneth, editors. *Implementing Pollution Laws: International Comparisons.* Tallahassee, Fla.: Florida State University Policy Sciences Program, 1981.

Durso-Hughes, Katherine, and Lewis, James. "Recycling Hazardous Waste," *Environment*, Vol. 24 No. 2 (March 1982), pp. 14-19.

Ferguson, Allen R., and LeVeen, E. Phillip. *The Benefits of Health and Safety Regulation.* Cambridge, Mass.: Ballinger Publishing Company, 1981.

Fisher, Anthony C. *Resource and Environmental Economics.* Cambridge. Cambridge University Press, 1981.

Frederick, Kenneth D., and Harrison, James C. *Water for Western Agriculture*. Washington, D.C.: Resources for the Future, Inc., 1982.

Freeman, III, Myrick A. *The Benefits of Environmental Improvement: Theory and Practice*. Baltimore: Johns Hopkins University Press for Resources for the Future, Inc., 1979.

Fromm, Gary. *Studies in Public Regulation*. Cambridge, Mass.: MIT Press, 1981.

Gianessi, Leonard P., et al. "Analysis of National Water Pollution Control Policies—A National Network Model," *Water Resources Research*, Vol. 17 No. 4 (August 1981), pp. 796-801.

Harris, Louis. "Campaign '82: Public Opinion Revisited," *The Amicus Journal*, Vol. 4 No. 1 (Summer 1982), pp. 8-9.

Kazis, Richard, and Grossman, Richard L. "Environmental Protection: Job-Taker or Job-Maker?" *Environment*, Vol. 24 No. 9 (November 1982), pp. 12-20, 43 and 44.

Kelman, Steven. *Regulating America, Regulating Sweden*. Cambridge, Mass.: MIT Press, 1981.

Kneese, Allen V., and Brown, F. Lee. *The Southwest Under Stress*. Baltimore: Johns Hopkins University Press for Resources for the Future, Inc., 1981.

Kurlansky, Mark J. "Who is Killing the Rhine?" *Environment*, Vol. 24 No. 7 (September 1982), pp. 41-42.

K. Leman, Christopher, and Nelson, Robert H. "Ten Commandments for Policy Economists," *Journal of Policy Analysis and Management*, Vol. 1 No. 1 (1981), pp. 97-117.

Nager, Glen D. "Buraucrats and the Cost-Benefit Chamelon," *Regulation* (September/October 1982), pp. 37-42.

Peskin, Henry M., et. al., editors. *Environmental Regulation and U.S. Economy*. Baltimore: Johns Hopkins University Press, 1981.

Portney, Paul R., editor. *Current Issues in Natural Resource Policy*. Baltimore: Johns Hoplins University Press for Resources for the Future, Inc., 1982.

Portney, Paul R. "How Not to Create a Job," *Regulation* (November/December 1982), pp. 35-38

Price, Kent A. *Regional Conflict and National Policy*. Baltimore: Johns Hopkins University for Press Resources for the Future, Inc., 1982.

*Protecting the Environment; Politics, Pollution and Federal Policy*. Washington, D.C.: Advisory Commission on Intergovernmental Relations, 1981.

Russell, Clifford S. "Controlled Trading of Pollution Permits," *Environmental Science and Technology*, Vol. 15 no. 1 (January 1981), pp. 24-28.

Seldman, Neil. "Resource Recovery: The 1982 Synthetic Fuels Corporation Amendment," *Environment*, Vol. 24 No. 9 (November 1982), pp. 2-4.

Simon, Julian L. *The Ultimate Resource*. Princeton: Princeton University Press, 1981.

Strom. David. "Improving Environmental Programs through Public Policy Models," *Environment*, Vol. 24, No. 10 (December, 1982) pp. 26-29.

Swanson, R. L. and M. Devine. "Ocean Dumping Policy," *Environment*, Vol. 24, No. 5 (June 1982), pp. 14-21.

U.S. Council on Environmental Quality. *Environmental Quality: Twelfth Annual Report*. Washington, DC: U.S. Government Printing Office, 1982.

U.S. Government Accounting Office. *Cleaning Up the Environment: Progress Achieved But Major Unresolved Issues Remain*, CED-82-72, 2 volumes. Washington, D.C.: U.S. Government Printing Office, 1982.

White, Lawrence J. *The Regulation of Air Pollutant Emissions from Motor Vehicles*. Washington, D.C.: American Enterprise Institute, 1982.

Zamora, Jennifer, et al. "Pricing Urban Water: Theory and Practice in Three Southwestern Cities," *The Southwestern Review of Management and Economics*, Vol. 1 No. 1 (1981), pp. 89-113.

# INDEX

Accelerated Cost Recovery System. *See* Section 168

Acid rain: Canada, 49, 74-75, 78, 81-82; Germany, 179, 185; Sweden, 193, 207, 210, 224-25; United States, 306-7, 312, 314

Administrative feasibility. *See* Imperfect information problem

Ad valorem taxes. *See* Property taxes

*Agence Financière de Bassin. See* Water Basin Finance Agency

Agriculture, Ministry of. *See* Ministry of Agriculture

Agricultural run-off, 21, 180, 265, 289

Air Agency (France), 107-8

Air pollutants, 16, 17

Air pollution: definition, 15-19; effects, 18; measurement, 17; strategies for controlling, 18

Air pollution control legislation: Air Pollution Control Act (France), 101-3, 116; Alkali Act (United Kingdom), 248-50; Clean Air Act (Canada), 62-63; Clean Air Act (United Kingdom), 248, 250-53; Clean Air Act (United States), 4, 75, 286-88, 298, 302, 304-8, 313-16, 345, 356; Emission Control Act (Germany), 146, 157-60, 176, 179, 184-85; Environmental Protection Act (Sweden), 195-96; Pollution Control Act (United Kingdom), 248, 250-52, 254

Air pollution control programs: Canada, 62-64; comparative analysis, 400-403; France, 101-4, 107-8, 115-16, 128-29;

Germany, 157-60; Italy, 101; Sweden, 194-95, 197, 199, 201, 206, 207, 210; United Kingdom, 248-54, 261-62, 268-70, 272-74; United States, 286-88, 291, 305-8, 313-16, 345, 356

Air pollution inspectors (United Kingdom), 249, 269

Air pollution monitoring: France, 101-2, 115; Germany, 158; United Kingdom, 249-54; United States, 272-74. *See* Pollutants Standard Index

Air pollution standards. *See* Emission standards

Air pollution taxes. *See* Air Agency

Air Quality Control Regions (U.S.), 286-87

Air quality objectives. *See* Ambient air quality standards

Air resources management (Sweden), 206

Airshed concept, 19

Alert areas (France), 102, 121

Ambient air quality, 17, 19

Ambient air quality standards: Canada, 62-64; France, 101-2, 116; Germany, 157, 178, 184-85; United Kingdom, 250-52, 270; United States, 286-87, 305-8

*Amoco Cadiz*, 88, 126-27

Amortization of pollution control facilities. *See* Section 169

Anglo-American: legal tradition, 385; tax system, 39

Antipollution contracts (France), 110

485

Approach, comparative. *See* Comparative approach
Arab oil embargo, 3, 226, 418
Assimilative capacity. *See* Environmental assimilative capacity
Automobile emissions. *See* Lead pollution; Motor vehicle emission control
Automobiles, junked. *See* Junked automobiles
Automotive scrap. *See* Wrecked auto tax
Ayres-Kneese model, 29-30

Baltic Sea, 139
Barre, Raymond, 116
Basin Committee (France), 95-100
Basin-wide firm. *See Genossenschaften*
Baum, Gerhardt Rudolf, 184
Best available technology standard: Sweden, 209, 225; United States, 286-87, 308
Best practicable means standard: Sweden, 199, 206; United Kingdom, 238, 249-53, 261, 269, 273
Best practicable technology standard: Canada, 54-56; Germany, 148; Sweden, 199, 206; United Kingdom, 238, 249-53, 256; United States, 284
Best technology economically achievable: France, 92, 107, 110, 123-24
Beverage containers, pollution control and taxation: Canada, 65-67; Germany, 163-64; Oregon, 291; Sweden, 211-12, 229, 414; United States, 291
Biochemical oxygen demand (BOD), 21, 59, 96-97, 153, 156, 216, 246
Bower, Blair T., 151
Brandt, Willy, 140-41
Broyhill, James T., 307
British North America Act, 71
BUA. *See* Federal Environment Office
Bubble concept (U.S.), 306, 315-16, 420
Budget for environmental protection: comparative analysis, 388, 390-91; France, 112-13, 128-29; Germany, 182-83; United Kingdom, 259, 264-65, 269-70; United States, 82, 302-4, 312-13, 318
*Bund* (German: Federal government), 141-48
Bush, George, 297, 301-2, 307

Cadmium poisoning, 172, 177-78
California pollution control regulation, 291
Canada, 49-86
Carbon monoxides, 16
Carcinogens, 184, 281, 300

Carter Administration, 300-303, 311
CEEQ. *See* Council of Experts for Environmental Questions
Center-Liberal coalition, 224
Central government: comparative analysis, 385-87; France, 88-91; Sweden, 196-99, 202, 204-8, 223-24; United Kingdom, 237-40
CEQ. *See* Council on Environmental Quality
Charge system. *See* Effluent charges
Chemical conversion process, 23
Chemical oxygen demand (*COD*), 96-97, 153, 156
Chemicals Control Act. *See* Toxic substances control legislation
Chemicals Law. *See* Toxic substances control legislation
Christian Democrat-Free Democrat coalition (Germany), 183-84
Clark, Joe, 71
Classified installations (France), 91-93
Class Life Asset Depreciation Range (CLADR). *See* Section 167
Clean Air Act (U.S.). *See* Air pollution control legislation
Coal. *See* Energy; Surface Mining Conservation and Reclamation Act
Coase, Ronald H., 35
Commoner, Barry, 3
Common property resources, 31, 389, 415
Common Market. *See* European Economic Community
Comparative approach, 8-9
Compliance cost, 39
Composting, 23
Conable, Barber, 379
Conceptual framework, 9; for analyzing environmental policy, 13-45
"Concerted Action" (Germany), 141
Congress (U.S.), 298, 305, 307, 347-48, 373; House Energy and Commerce Committee, 307, 323; House Subcommittee on Health and Environment, 305; Senate Environment and Public Works Committee, 305, 313
Conservation movement, 13
Conservation of mass, 25
Conservative party, 258-59, 264
Constitution: Canada, 71; Germany, 146-47
Cost. *See* Compliance cost; Marginal cost; Noncompliance cost; Social costs and values

Council of Experts for Environmental Questions (CEEQ), Germany, 144, 180

Council on Environmental Quality (CEQ), United States, 5, 144, 282, 303, 308, 310-11

County administration (Sweden), 197-206, 219-23

"Cowboy economy", 13

Crepeau, Michel, 125-29, 131

Customs duties, 147, 258

Davis, Jack, 51-52

de Gaulle, Charles, 88

Demographic factors, comparative analysis of, 385-87

*Department* (France), 91-93, 98, 100, 129-32

Department of Environment (United Kingdom), 238-40, 259-60, 262-63

Department of the Interior (U.S.), 283, 297-99, 303, 311, 323

Depreciation. *See* Section 167

Deregulation, 264, 297, 301-2, 310, 312, 323-24

Deterioration, prevention of. *See* Prevention of significant deterioration

Development, regional. *See* Regional development (United Kingdom)

Dingell, John D., 307

Direct grants: to households, 108, 256-57; to industry, 105-10, 133, 167, 214-16, 218, 256, 292; to municipalities. *See* Municipal wastewater treatment subsidies; to Regional Water Authorities (United Kingdom), 257; for research and development, 69, 108-9, 167, 170, 215, 219, 257, 283; to State governments (U.S.), 284, 292-93, 303-4, 313; to water management associations (*Genossenschaften*), 150

Discharge permits: Canada, 54-55; France, 91-95, 117-18; Germany, 148, 155-56; Sweden, 198-200, 219-20; United Kingdom, 245-48; United States, 283-85, 309-10

d'Ornano, Michel, 111-12, 125

EAC. *See* Environmental assimilative capacity

Earth Week, 3

Economic Commission of Europe (United Nations) (ECE), 179, 210, 224-25, 269, 420

Economic planning (France), 97-100, 105-6, 111-13, 121-24, 130-31

Economic policy, 6

Economic Recovery Tax Act of 1981. *See* Tax legislation (U.S.)

Economic theory, 34-44

EEC. *See* European Economic Community

Effluent. *See* Pollution problems; Water pollution

Effluent charges or taxes: Canada 56-60; France, 95-100, 105-6; Germany, 149-56, 174; United Kingdom, 246; United States, 283-86, 290-91; 293-95

Effluent guidelines and standards, 420; Canada, 54-56; France, 95, 117; Germany, 148, 174; Sweden, 199, 206, 207, 225; United Kingdom, 245, 247-48, 256; United States, 283-86, 317

EIS. *See* Environmental impact statement

Emission charges. *See* Air Agency

Emission Control Act. *See* Water pollution control legislation

Emission control officer (Germany), 158

Emission guidelines and standards, 419; Canada, 62-64; France, 102, 115; Germany, 158-60; Sweden, 206-10; United States, 286-88, 305-8

Emissions. *See* Air pollution; Pollution problems

Energy: coal, 178, 226-28; district heating (Sweden), 226-28; "gas guzzler" tax, 69-70, 290; National Energy Program (Canada), 71, 78-79, 81; nuclear, 125, 127, 146, 171, 175, 178, 182, 194, 205, 225. *See also* Low-sulfur fuel

Endangered Species Act (U.S.), 308, 310

Energy Crisis, 4, 23, 89, 97, 141, 226

Enforcement problems (Sweden), 200-1, 217-23

Environment, state of. *See* State of the environment

Environmental Assessment and Review Process (Canada), 52

Environmental assimilative capacity (EAC), 19, 22, 25, 27-29, 32-37, 389, 420-21

Environmental Consciousness, 3

Environmental Contaminants Act. *See* Toxic substances control legislation

Environmental Crisis, 3, 140-41, 152, 389-90, 409

Environmental excise taxes: coal, 291; chemicals, 375-77, 379-81; France, 104-5; petroleum, 373-75, 379-81

Environmental groups: France, 131; Sweden, 194, 223; United States, 298-99, 308, 310-12

Environmental impact assessment: Canada, 52; comparative analysis, 392-97; France, 89, 91-93, 107; Germany, 145-47; Sweden, 223; United Kingdom, 270-71

Environmental impact statement (EIS): United States, 282, 303

Environmental information system: Sweden, 213; United Kingdom, 249-52

Environmental law, New York, 291

Environmental Lobby, 299, 303, 312

Environmental party. *See* Green parties

Environmental planning, comparative analysis of, 389, 391-97

Environmental protection administration: Canada, 51-54, 64-65, 67, 71-73, 76-77, 81; France, 89-90, 93-100, 111-13, 120-21; Germany, 143-45, 172-73, 182-84; national style, 389; Sweden, 195-201, 204-11, 219-23; United Kingdom, 73, 238-42, 244-51, 259, 264, 269-70

Environmental Protection Administration (EPA), United States, 282-89, 300-303, 308-10, 312-13, 315-22, 344-47, 352-60, 365

Environmental resources, 25-29, 33-34

Environmental Statistics Law (Germany), 165

Environment Canada, 49-52, 72, 74, 77-80

Environment Ministry. *See* Department of the Environment (United Kingdom)

EPA. *See* Environmental Protection Administration

Equity, 40-41

European Economic Community (EEC), 87, 139, 165, 170, 223, 228, 254

Environmental directives, 89, 93, 100-101, 103-5, 117, 120, 146-48, 156, 165, 171, 174, 177-79, 185, 187, 254, 261, 263, 266, 268-70, 272

Excise taxes: Canada, 50; United Kingdom, 258

Externalities, 25, 32-33

Federal Environment Office (BUA), Germany, 143-44, 173, 182-84

Federal Environment Ministry (FEM). *See* Environment Canada

Federal government: Canada, 50-51; comparative analysis, 385-407; Germany,
141-48; United States, 282-83, 285-91, 303-4, 308-10, 312

*Federal Register*, 299, 301

Federal Water Pollution Control Act (U.S.). *See* Water pollution control legislation

FEM. *See* Environment Canada

Fiscal policy, 7, 264, 302-4, 417

Fiscal systems: Canada, 50, 68-69; comparative analysis, 388, 390-91; France, 109; Germany, 147, 156; United Kingdom, 257-58; United States, 293, 295

Fisheries Act. *See* Water pollution control legislation

Fish kills (Rhine River), 140

Florida, land-use planning, 291

Five Year Plan. *See* Economic planning

France, 87-138, 420

Free Democratic party, 141, 175, 183-84

Free goods, 25

Fuels. *See* Energy

"Gas guzzler" tax, 69-70, 290

GAO. *See* General Accounting Office

Gasoline taxes, 69-70, 290

Gaullist party, 88, 130

General Accounting Office (GAO): Germany, 172; United States, 311-12, 317

General equillibrium economic models, 29-30

*Genossenschaften* (Water management associations), 60, 149-53, 167, 412, 413-15

Genscher, Hans Dietrich, 140

German chemical industry, 171, 178, 185

Germany. *See* West Germany

Giscard d'Estaing, Valery, 88-89, 111-12, 118, 123, 125-29

Government assistance: Canada, 67-70; comparative analysis, 397-407, 410; France, 105-10; Germany, 165-70; Sweden, 214-18; United Kingdom, 256-58; United States, 292-95

Gorsuch, Anne M., 298-300, 303, 306, 313-14, 318, 321-22, 324

Gothenberg (Gotenberg), 193, 210, 227-28

Grants. *See* Direct grants

Great Lakes (Water Quality Agreement): Canada, 53, 55-56, 60-62, 75-76, 82; United States, 317

Green parties, 175, 183-84, 224

Greenhouse effect, 312

Gross domestic/national product, 388, 390-91

Hartkopf, Gunter, 176
Hazardous Products Act. *See* Toxic substances control legislation
Hazardous substances pollution control: Canada, 67, 72, 76-77; comparative analysis, 403-7, 418-19; France, 96, 105, 110-11, 113-15, 121-22; Germany, 160-61, 163-65, 170-74, 176-77, 185-87; Sweden, 212-14, 228-29; United Kingdom, 255-56, 260, 262-63, 271, 274-76; United States, 284-86, 288-90, 298, 309, 318-21
Hazardous Substances Response Trust Fund, 373, 377-79, 381
Health and safety inspectors (United Kingdom), 249-53, 269, 272-73
Hubbert, M. King, 25
Hydrocarbons, 16

Imperfect information problem, 42-43, 416-18
Incentives. *See* Research Incentives; Subsidies; Tax incentives for pollution control
Incineration, 23
Income taxes: Canada, 50, 68-69; France, 109; Germany, 147, 166; United Kingdom, 257-58; United States, 293, 295-97
Incomplete information. *See* Imperfect information
Industrial Development/Revenue Bonds. *See* Internal Revenue Code: Section 103
Industrialization: France, 87-89; Germany, 139; Sweden, 193; United Kingdom, 235-37
Industrial solid waste, 24
Industry sectoral contracts and programs (*contracts de branche* and *programmes de branche*), 107, 109-10, 113
Information problem. *See* Imperfect information problem
Interest groups. *See* Environmental groups
Interior, Ministry of. *See* Ministry of Interior
Interior, U.S. Department of the. *See* Department of the Interior
Internal Revenue Code (U.S.): Section 38, Investment Tax Credit, 293-94, 334-36, 338-40, 342, 350-51, 361-68; Section 44F, Tax Credit for Research and Experimentation, 295; Section 103, Tax Exempt Industrial Development/Revenue Bonds, 293-95, 329, 364; Section 167, Depreciation and Class Life Asset Depreciation Range, 330, 334-37, 343, 350; Section 169,

Amortization of Pollution Control Facilities, 293-94, 330, 343-68; Section 168, Accelerated Cost Recovery System and Leasing, 330-34, 337-44, 360-68; Section 179, Election to Expense Depreciable Investments, 334, 359-61, 365
International Joint Commission, 61, 75-76, 317
Investment reserve system (Sweden), 217
Investment tax credit. *See* Section 38

Junked automobiles, 24

Keynes, Lord, 258
King, Tom, 265, 270
Kneese, Allen V., 29-30, 151
Kneese-Ayres model, 29-30

Labor party, 258, 274
Lake Constance (Germany), 139, 157, 168-69
*Länder* (Germany). *See* State government
Land-use planning: comparative analysis, 389; Florida, 291; France, 91-93; Germany, 140, 145-48; Sweden, 201-6; United Kingdom, 240-43, 270-71
Law of the Sea Treaty. *See* United Nations
Lead pollution: Canada, 64; France, 103-4, 119, 121-22; Germany, 157, 160, 172, 184-85; United Kingdom, 254, 259, 269, 273-74; United States, 286-88, 314-15
Leasing. *See* Section 168
Legislative veto (U.S.), 302
Liberal party: Canada, 50-51, 71, 77-79; United Kingdom, 274
Licensing of polluters/polluting facilities: France, 91-95, 102, 104, 107, 111, 113, 117-18, 129; Germany, 145-46, 148, 151, 157-58, 162, 164-65, 186; Sweden, 195-200, 204-6, 207, 210, 213-14, 219-20, 222-23; United Kingdom, 242, 245, 247-50, 252-55, 270-72; United States, 283-85, 287-89, 305-6, 309-10, 319-20
Limitations of the study, 10
Loan guarantees and low-interest loans: Canada, 70; France, 105-8; Germany, 167-69; United Kingdom, 256; United States, 292-93
Local government: Canada, 50-51, 58-59, 64-66; France, 102, 104, 130-31; Germany, 146, 175; Sweden, 197, 201-3; United Kingdom, 238-43, 270-71; United

Local government (*cont'd*)
   States, 283, 289, 291, 293-95. *See also*
   Municipal responsibility for waste disposal
London, 237, 243, 250, 255
Long, Russell B., 379
Love Canal, 76, 82, 114, 122, 229, 271, 276,
   310, 318, 321
Low-flow augmentation, 150
Low-sulfur fuels, 36, 74, 103, 115-16, 175,
   178, 207, 210, 224-27, 235, 261, 268,
   299, 306
Luken, Thomas A., 307

Macroeconomic theory, 25-30
Marginal cost, 33, 35-41
Marine pollution: Canada, 53, 55, 60; France,
   88; Germany, 139; Sweden, 206; United
   Kingdom, 236-37, 263-68; United States,
   286, 317. *See also Amoco Cadiz*; Baltic Sea;
   Great Lakes; Mediterranean Sea; North
   Sea; *Torrey Canyon*
Market failure, 31-33
Market forces, 31-42, 389
Materials balance approach, 25-30
Materials recovery. *See* Resource recovery
Materials recovery processes, 23
Mauroy, Pierre, 125, 132
Mediterranean Action Plan (United Nations),
   119-20, 133
Mediterranean Sea, 88, 100-101, 118-21,
   133
Methodology of the study, 8-9
Michigan: beverage containers, 291; conserva-
   tion law, 291
Microeconomic theory, 31-34
Minimum discharge control standards, 98,
   149-56, 247-48
Ministry of Agriculture (Sweden), 196, 208-9
Ministry of Environment (France), 89-90,
   112-13, 125-29
Ministry of Interior (Germany), 140-41,
   143-45, 176, 180, 184
Mitterrand, Francois, 124-32
Monitoring pollution abatement or control:
   France, *See* Air pollution monitoring;
   Germany, *See* Air pollution monitoring;
   Water protection officers; United Kingdom,
   *See* Air pollution monitoring; Health and
   Safety Inspectors; United States, *See*
   Pollutants Standards Index
Most efficient technology (Sweden), 199,
   206

Motor vehicle emission control: Canada, 64;
   comparative analysis, 400-403; France,
   103-5, 119, 121-22; Germany, 157, 159,
   179; Sweden, 210; United Kingdom, 254,
   259-60, 269; United States, 286, 288, 300,
   305-8
Municipal effluent charges: Canada, 58;
   United Kingdom, 245
Municipalities. *See* Local government
Municipal responsibility for waste disposal:
   Canada, 64-66; France, 104; Germany,
   162-63; Sweden, 210-12; United
   Kingdom, 254-55, 262, 275-76; United
   States, 289
Municipal Sanitation Act (Sweden). *See* Solid
   waste pollution control legislation
Municipal wastewater treatment subsidies:
   Canada, 68, 70; France, 106-8, 118, 120;
   Germany, 167-69; Sweden, 214-16, 218;
   United Kingdom, 259; United States,
   283-84, 292-93, 313-14, 316

National Board of Urban Planning (Sweden),
   197, 201
National economic planning: France, 96-101;
   Sweden, 202
National Environmental Policy Act (NEPA),
   U.S., 5, 282, 303
National Environmental Protection Board
   (NEPB), Sweden, 195-201, 206-9, 213-16,
   219-22
National government. *See* Central govern-
   ment; Federal government
National Pollution Discharge Elimination Sys-
   tem (NPDES), U.S., 285, 309-10
National Water Council (United Kingdom),
   244, 247-48
Natural resource production wastes, 25
Nature Conservancy Act (Sweden). *See* Solid
   waste pollution control legislation
*Naturfreunde* (friend of nature), 140
NEPA. *See* National Environment Protection
   Act
NEPB. *See* National Environment Protection
   Board
New Federalism, 299-300, 303-4, 310
New York: environmental law, 291
Nitrogen oxides, 16
Noise charges or taxes, 129
Noise pollution, 122-23, 129, 132, 159, 179,
   185, 195, 197, 199, 260
Noncompliance cost, 39

Nondegradation policy. *See* Significant deteriorization prevention

Northern Ireland. *See* United Kingdom

North Sea, 139, 179-82, 185, 187, 236, 267-68

NPDES. *See* National Pollution Discharge Elimination System

Nuclear power. *See* Energy

Objectives of the study, 6

OECD. *See* Organization for Economic Cooperation and Development

Office of Management and Budget (U.S.), 299, 301-2, 322

Office of Surface Mining, 299, 301, 303, 308-9

Offset policy (U.S.), 306, 315-16

Oil discharges or pollution, 88, 101, 286

OPEC. *See* Organization of Petroleum Exporting Countries

Open dumps, 22-23

Optimal prices, 31-42

Oregon, beverage container charges on taxes, 291

Organization for Economic Cooperation and Development (OECD), 15, 67, 81, 105, 165, 168, 170, 178, 256, 292, 389, 409, 416, 420

Organization of Petroleum Exporting Countries (OPEC), 4, 226

Organization of the study, 8

"Pareto optimality," 30

Paris, 87-89, 101-2, 115, 121, 129-30

Particulates, 16

Penalties: Canada, 55-56, 62; France, 92, 115; Germany, 148-49, 156-57, 163; Sweden, 200-201, 222; United Kingdom, 245, 269; United States, 284-86, 290-91

Permits. *See* Discharge permits; Licensing

Planning. *See* Economic planning; Land-use planning; National economic planning

Policy. *See* Economic policy; Fiscal policy; Public policy; Tax policy

Political economy of pollution control, 37-40, 107-8, 115, 133, 385, 389, 415-21

Political feasibility. *See* Political economy of pollution control

Pollutants Standards Index (PSI), U.S., 287

"Polluter pays principle," 67-68, 105, 113, 141, 165, 168, 170, 256, 292, 388-89, 416-17

Pollution. *See* Air pollution; Solid waste pollution; Water pollution

Pollution abatement programs: Canada, 49-53; evolution, 409; France, 89-93; Germany, 141-165; Sweden, 194-214; United Kingdom, 237-56; United States, 281-291

Pollution Control Act (United Kingdom). *See* Solid waste pollution control legislation; Toxic substances control legislation; Water pollution control legislation

Pollution control incentives. *See* Effluent charges or taxes; Subsidies; Tax incentives for pollution control

Pollution control regulation. *See* California

Pollution control strategy, 35-43, 385, 389

Pollution problems: Canada, 49, 53, 60-67, 72; France, 87-88, 101-3, 107, 110; Germany, 139-40, 152, 179; Sweden, 193-94, 207-13, 228-29; United Kingdom, 235-38, 259-62, 265, 267-76; United States, 281, 284-90, 309-12, 319

Pollution tax agency, 39-40, 42

Pollution tax system, 39-40, 290, 293

Pollution taxes. *See* Air Agency; Beverage containers; Effluent charges or taxes; Environmental excise taxes; Waste oil tax; Wrecked auto tax.

Pompidou, Georges, 88-89, 126

Population: Canada, 49; comparative analysis, 385-87; France, 87; Germany, 139; Sweden, 193; United Kingdom, 237; United States, 385

Prefect (France), 91-94, 129-32

Prevention of pollution acts. *See* Water pollution control legislation

Prevention of significant deterioration, 287, 305-8

Private property, 31-35

Progressive Conservative party, 71

Property, common, resources. *See* Common property resources

Property rights. *See* Private property

Property taxes: Canada, 50; United Kingdom, 245-46, 256; United States, 295, 337

Proposition 13, 5

Provincial effluent charges, 57

Provincial government, 50-51

PSI. *See* Pollutants Standard Index.

Public expenditures. *See* Budget for environmental protection: comparative analysis, 388, 390-91

Public goods, 33-34
Public Health Act (United Kingdom). *See* Solid waste pollution control legislation
Public health laws: Canada, 64-65; Germany, 160-63; Sweden, 194, 210-11, 213; United Kingdom, 243, 245, 248, 250-51; United States, 283
Public opinion, 3, 5, 194-95, 224, 418-19
Public policy, 7
*Pyrolysis process*, 23

RCEP. *See* Royal Commission on Environmental Pollution
Reagan Administration, 298-313, 316, 318-19, 323-24
Reagan, Ronald, 295, 297-98, 303-4, 310, 312-13
Recession, 4, 89, 97, 141, 169
Recycling. *See* Resource recovery
Refuse Act of 1899. *See* Water pollution control legislation
Region (French economic planning), 95, 98, 100, 131
Regional development (United Kingdom), 241
Regional residuals management, 66
Regional Water Authority (United Kingdom), 245-47, 257
Regulatory approach to pollution control, 4-5, 37-40, 389-407, 409-10, 415-21
Research incentives: Canada, 69; France, 108-9; Germany, 167, 170; Sweden, 215, 219; United Kingdom, 257; United States, 283, 295
Residuals, 25-29
Resource Conservation Act (U.S.). *See* Solid waste pollution control legislation
Resource recovery, 23-24; Canada, 65; France, 104, 108, 114-15, 121, 123; Germany, 164; Sweden, 211-13, 217, 226-29; United Kingdom, 255-56; United States, 291
Resources, common property. *See* Common property resources
Resources, environmental. *See* Environmental resources
Revenue Act of 1978. *See* Tax legislation (U.S.)
Rhine River, 88, 100-101, 118-19, 126-27, 140, 156-57, 168-69, 180
River authorities (United Kingdom), 244-47
River basin management. *See* Basin Committee; *Genossenschaften*; River

authorities; Water authorities; Water Basin Finance Agency
Roberts, John, 81-82
Royal Commission on Environmental Pollution (RCEP), United Kingdom, 240, 265, 267-68, 274
Ruhr Valley, 149-152

Safe Drinking Water Act. *See* Water pollution control legislation
"Sagebrush Rebellion," 298
Sales taxes; Canada, 50, 69; United States, 293, 295-97, 337
Sanitary landfills, 22-24
Schmidt, Helmut, 140-41, 175, 182-83
Scotland. *See* United Kingdom
Section 38, Investment Tax Credit, 293-94, 334-36, 338-40, 342, 350-51, 361-68
Section 44F, Tax Credit for Research and Development, 295
Section 103, Tax-exempt Industrial Development/Revenue Bonds, 293-95, 329, 364
Section 167, Depreciation and Class Life Asset Depreciation Range, 330, 334-37, 343, 350
Section 168, Accelerated Cost Recovery System and Leasing, 330-34, 337-44, 360-68
Section 169, Amortization of Pollution Control Equipment, 293-94, 330, 343-68
Section 179, Election to Expense Depreciable Investments, 334, 359-361, 365
Self-reporting. *See* Voluntary compliance
"Separation of powers" doctrine (U.S.), 302
Setting charges (tax rates), 42-43; France, 96-100; United Kingdom, 245-46
Seveso (Italy), 126, 177, 186-87
Sewer charges. *See* Municipal effluent charges
Significant deteriorization prevention. *See* Prevention of significant deteriorization
SM. *See* Suspended solids.
Smog, 16, 18
Smoke control areas (United Kingdom), 250-51
Social costs and values, 31-33, 35-37
Social damages, charge estimates, 381-82
Social Democratic party: Germany, 141, 175, 180; Sweden, 221, 223-24; United Kingdom, 274
Social Democrat-Free Democrat coalition (Germany), 141, 175, 186-87
Socialists (France), 124, 130-32

Solid waste pollutants, 22-24

Solid waste pollution: control strategy, 22-23; definition, 22; effects, 23-24; measurement, 22; sources, 22-23

Solid waste pollution control: Canada, 64-67, 76-77; comparative analysis, 403-7; France, 104-5, 111, 113-15, 121; Germany, 160-65, 176-77; Sweden, 194-95, 197, 210-14, 217; United Kingdom, 254-56, 260, 262-63, 275-76; United States, 288-91, 318-321

Solid waste pollution control legislation: Municipal Sanitation Act (Sweden), 210-13; Nature Conservancy Act (Sweden), 211-13; Pollution Control Act (United Kingdom), 255-56, 262-63, 275-76; Public Health Act (United Kingdom), 255; Resource Conservation and Recovery Act (U.S.), 288-91, 304, 318-21, 346; Surface Mining Control and Reclamation Act (U.S.), 289, 291, 299, 304, 323, 414; Waste Disposal and Materials Recovery Act (France), 89, 104, 115; Waste Disposal/Management Act (Germany), 148, 160-65, 176, 185-86; Wrecked Automobile Act (Sweden), 212

Special protection zones (France), 101-2, 116

Stagflation, 4, 89, 97, 141, 169, 256, 264-65, 418

Standards. See Effluent and emission standards

Standards-permit-enforcement strategy, 141, 149, 397-400

State government: Canada. See Provincial government; Germany, 141-47, 157, 160, 175-76; United States, 283-91, 295-97, 303-5, 308-10, 313-14, 322, 344-46, 356-60, 365

State of the environment: France, 120-21; United Kingdom, 259-61, 264-69, 273-74; United States, 28, 303, 310-12

State implementation plans (U.S.), 286-97, 305-8, 314

State plans (Germany), 148

Stockholm, 193-94, 210, 225, 28

Stockman, David A., 299-300, 302

Stream-flow augmentation. See *Genossenschaften*

Subsidies. See Direct grants; Loan guarantees and low-interest loans; Municipal water treatment subsidies

Subsidization approach to pollution control, 4-5, 40-42, 409-11, 415-21

Sulfur-oxide (SO$_2$) emissions, 16; Canada, 74; France, 102, 115, 121, 127; Germany, 160, 178-79, 184-85; Sweden, 193, 207, 210, 224-26; United Kingdom, 235, 249, 259-60, 268-69; United States, 286, 306, 314

Superfund, 298, 310, 312, 321-22, 373-83

Surface Mining Conservation and Reclamation Act. See Solid waste pollution control legislation

Suspended solids (SM): Canada, 59-60; France, 96-97; Germany, 153, 156

Sweden, 193-233, 414, 417

Task Force on Regulatory Relief (U.S.), 301-2

Tax Equity and Fiscal Responsibility Act of 1982. See Tax legislation (U.S.)

Taxes. See Effluent charges or taxes; Environmental excise taxes; Excise taxes; "Gas guzzler" tax; Gasoline taxes; Income taxes; Noise charges or taxes; Property taxes; Provincial effluent charges; Sales taxes; Value added taxes; Waste oil tax; Wrecked auto tax

Tax expenditures, 44, 293

Tax incentives for pollution control: Canada, 68-69; France, 109; Germany, 166-67; United Kingdom, 257-58; United States, 293-97, 329-71

Tax legislation (U.S.): Economic Recovery Tax Act of 1981, 329, 338, 362, 368; Revenue Act of 1978, 362, 364; Tax Equity and Fiscal Responsibility Act of 1982, 330, 368; Tax Reform Act of 1969, 343; Tax Reform Act of 1976, 348, 364

Tax policy, defined, 6-8

Tax Reform Act of 1969. See Tax legislation (U.S.)

Tax Reform Act of 1976. See Tax legislation (U.S.)

Tax revolt, 5

Tax systems. See Fiscal systems

Taxation approach to pollution control, 4-5, 37-43, 389, 410-21

Technology standard. See Best available technology standard; Best practicable technology standard; Best technology economically achievable; Most efficient technology

Thames River, 237, 272

Thatcher, Margaret, 258-59, 264-65, 268-69

Theory, 9

*Torrey Canyon*, 88

Toulemon, Robert, 111-12
Toxic Substances Control Act. *See* Toxic substances control legislation
Toxic substances control legislation: Chemicals Control Act (France), 89, 105, 111-12, 114; Chemicals Law (Germany), 170-74, 177-78; Environmental Contaminants Act (Canada), 67, 72, 77; Hazardous Products Act (Sweden), 212-14; Pollution Control Act (United Kingdom), 262; Toxic Substances Control Act (United States), 289-90, 300-302, 304, 318-321, 346, 356
Toxic substances regulation. *See* Hazardous waste pollution control
Toxic substances taxation. *See* Environmental excise tax
Transaction costs, 35
Transnational pollution. *See* Acid rain; Great Lakes Water Quality Agreement; Mediterranean Sea; North Sea; Rhine River
Treaty, Law of the Sea. *See* United Nations
Trudeau, Pierre, 51, 71

Ullman, Al, 379
U.K. Department of Environment. *See* Department of Environment (United Kingdom)
United Kingdom, 235-80, 420
United Nations: Conference on the Human Environment, 3, 194, 225; Law of Sea Treaty, 268
United States, 28, 82, 281-327, 329-71
Urbanization: Canada, 49; France, 87-89; Germany, 139; Sweden, 193; United Kingdom, 235-37
Urban refuse, 23
U.S. Congress. *See* Congress, (U.S.)
U.S. Department of the Interior. *See* Department of the Interior (U.S.)
U.S. Environmental Protection Agency. *See* Environmental Protection Agency (U.S.)
U.S. Office of Management and Budget. *See* Office of Management and Budget (U.S.)

Value added (or national sales) taxes: Canada, 69; France, 109; Germany, 147; United Kingdom, 257-58
Voluntary compliance, 38-39, 269-70

Wales. *See* United Kingdom
Walrus-Cassel economic model, 29-30

Waste assimilative. *See* Environmental assimilative capacity
Waste Disposal and Materials Recovery Act (U.S.) *See* Solid waste pollution control legislation
Waste Disposal/Management Act (Germany). *See* Solid waste pollution control legislation
Waste oil tax: France, 105, 111, 114-15, 133; Germany, 160-61, 167, 412, 414
Waste Water Charges Act. *See* Water pollution control legislation
Wastewater treatment. *See* Municipal wastewater treatment subsidies
Water Act. *See* Water pollution control legislation
Water Act/Code. *See* Water pollution control legislation
Water Acts. *See* Water pollution control legislation
Water authorities or associations (Germany), 149
Water basin authority, 56-57
Water Basin Finance Agency, 93-100, 105-6, 121, 412, 413-15
Water management associations. See *Genossenschaften*
Water Management/Development Act. *See* Water pollution control legislation
Water pollutants, 20-21
Water pollution: control strategy, 22; definition, 19-20; effects, 21-22; measurement, 19-20; sources, 20-21
Water pollution charges or taxes. *See* Effluent charges or taxes
Water pollution control: Canada, 53-62, 72, 74-76, 80-82; comparative analysis, 397-400; France, 93-101, 117-20, 125-26, 128, 133; Germany, 147-157, 170, 174; Sweden, 194-95, 197, 199-201, 205-6, 223; United Kingdom, 243-48, 259-61, 265-68, 271-72; United States, 283-86, 290-91, 316-18
Water Pollution Control Act. *See* Water pollution control legislation
Water pollution control legislation: Emission Control Act (Germany), 146, 148; Environmental Protection Act (Sweden), 195; Fisheries Act (Canada), 54-56; Pollution Control Act (United Kingdom), 244-48, 266, 271-72; Prevention of Pollution Acts (United Kingdom), 244; Refuse Act of 1899 (U.S.), 283-84, 345; Safe Drinking

Water pollution control legislation (*cont'd*)
    Water Act (U.S.), 310, 317-18; Waste
    Water Charges Act, (Germany), 149-50,
    152-56, 174, 176; Water Act (Canada),
    56-58; Water Act/Code (Sweden), 194,
    200, 205, 211; Water Acts (United King-
    dom), 244-47; Water Management/
    Development Act (France), 93-95, 105;
    Water Pollution Control Act (U.S.), 4,
    283-85, 290-93, 298, 304, 316-17, 322,
    345, 356; Water Resources Acts (United
    Kingdom), 246; Water Resources Manage-
    ment Act (Germany), 148
Water protection officers (Germany), 148
Water quality management authority or
    agency, 56-57
Water quality objectives or standards:
    Canada, 55-56; United Kingdom,
    243-44, 247-48, 272; United States, 283,
    317

Water Resources Act. *See* Water pollution
    control legislation
Water Resources Management Act. *See* Water
    pollution control legislation
Water resources management: Canada, 79-80;
    France, 93-100, 128, 420; Germany,
    148-49, 179-80, 420; Sweden, 201-5;
    United Kingdom, 243-48, 272, 420; United
    States, 317-321
Watt, James G., 298-99, 301, 323-24
West Germany, 139-192, 420
White Papers on the Environment (France),
    131-32
*Wirtschaftswunder* (economic miracle), 139,
    182
Wrecked Automobile Act (Sweden). *See* Solid
    waste pollution control legislation
Wrecked auto tax (Sweden), 212, 414

Zero-discharge goal, 21, 283-84

# ABOUT THE AUTHOR

CRAIG E. REESE recently joined the faculty of Southwest Texas State University, San Marcos, as an Associate Professor of Tax Accounting. He is a graduate of the University of Texas doctoral program in taxation. He has written and spoken extensively on taxation, and his articles have appeared in publications such as *Tax Notes, Proceedings of the NTA/TIA Annual Conference on Taxation, Taxation for Accountants, Oil and Gas Tax Quarterly, Taxes,* and *Tax Adviser.*